ADVANCES IN CHEMICAL PHYSICS

VOLUME LVII

Advances in
CHEMICAL PHYSICS

EDITED BY

I. PRIGOGINE

University of Brussels
Brussels, Belgium
and
University of Texas
Austin, Texas

AND

STUART A. RICE

Department of Chemistry
and
The James Franck Institute
The University of Chicago
Chicago, Illinois

VOLUME LVII

AN INTERSCIENCE® PUBLICATION
JOHN WILEY & SONS
NEW YORK · CHICHESTER · BRISBANE · TORONTO · SINGAPORE

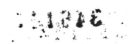

An Interscience® Publication

Copyright© 1984 by John Wiley & Sons, Inc.

Library of Congress Catalog Number: 58-9935

ISBN 0-471-87830-8

Printed in the United States of America

10 9 8 7 6 5 4 3 2 1

CONTRIBUTORS TO VOLUME LVII

L. S. CEDERBAUM, Theoretische Chemie, Institut für Physikalische Chemie, Universität Heidelberg, Heidelberg, West Germany

C. A. CHATZIDIMITRIOU-DREISMANN, I. N. Stranski Institute for Physical and Theoretical Chemistry, Technical University Berlin, West Germany

M. S. CHILD, Theoretical Chemistry Department, Oxford University, Oxford, United Kingdom

W. DOMCKE, Theoretische Chemie, Institut für Physikalische Chemie, Universität Heidelberg, Heidelberg, West Germany

L. HALONEN, Theoretical Chemistry Department, Oxford University, Oxford, United Kingdom

B. R. JUDD, Physics Department, The Johns Hopkins University, Baltimore, Maryland, U.S.A.

H. KÖPPEL, Theoretische Chemie, Institut für Physikalische Chemie, Universität Heidelberg, Heidelberg, West Germany

E. LIPPERT, I. N. Stranski Institute for Physical and Theoretical Chemistry, Technical University Berlin, West Germany

K.-H. NAUMANN, I. N. Stranski Institute for Physical and Theoretical Chemistry, Technical University Berlin, West Germany

INTRODUCTION

Few of us can any longer keep up with the flood of scientific literature, even in specialized subfields. Any attempt to do more and be broadly educated with respect to a large domain of science has the appearance of tilting at windmills. Yet the synthesis of ideas drawn from different subjects into new, powerful, general concepts is as valuable as ever, and the desire to remain educated persists in all scientists. This series, *Advances in Chemical Physics*, is devoted to helping the reader obtain general information about a wide variety of topics in chemical physics, which field we interpret very broadly. Our intent is to have experts present comprehensive analyses of subjects of interest and to encourage the expression of individual points of view. We hope that this approach to the presentation of an overview of a subject will both stimulate new research and serve as a personalized learning text for beginners in a field.

ILYA PRIGOGINE

STUART A. RICE

CONTENTS

ADVANCES IN CHEMICAL PHYSICS

VOLUME LVII

OVERTONE FREQUENCIES AND INTENSITIES IN THE LOCAL MODE PICTURE

M. S. CHILD AND L. HALONEN

Theoretical Chemistry Department
University of Oxford
Oxford, United Kingdom

CONTENTS

I. INTRODUCTION

The concept of the normal mode is so firmly engrained in the theory of molecular vibrations that it is only in recent years that analysis of the overtone and combination band spectra has demonstrated the advantages of a local mode description, particularly for X—H vibrations. This new insight is primarily stimulated by developments in conventional absorption,[1-15] thermal lensing,[16-23] and photoacoustic[24-32] spectroscopy, which have extended the observation range to progressively higher overtone levels. The

Present address of Dr. Halonen: Department of Chemistry, Helsinki University of Technology, SF-02150 Espoo 15, Finland.

1

greatest surprise is that localization develops even in symmetrical molecules and the following is devoted to such situations.

This is not the first review of the subject,[33-36] but it is convenient to begin with a brief historical introduction. We start from the observation that the normal mode picture stems from a harmonic approximation to the potential energy surface around the point of minimum energy. In the case of two coordinates labeled (Q_1, Q_3) for ease of identification with symmetric and antisymmetric X—H stretching modes, this leads after the introduction of anharmonic corrections to the familiar term value expansion

$$G(v_1, v_3) = \left(v_1 + \tfrac{1}{2}\right)\omega_1 + \left(v_3 + \tfrac{1}{2}\right)\omega_3 + \left(v_1 + \tfrac{1}{2}\right)^2 x_{11} + \left(v_3 + \tfrac{1}{2}\right)^2 x_{33}$$
$$+ \left(v_1 + \tfrac{1}{2}\right)\left(v_3 + \tfrac{1}{2}\right)x_{13} \tag{1.1}$$

To the extent that the x_{ij} can be neglected, this predicts for every value of $v_1 + v_3$ a pattern of overtone levels that are equally spaced by $|\omega_1 - \omega_3|$. It is known, however, that the spacings in many overtone manifolds are far from uniform and that Eq. (1.1) gives a progressively poor approximation to the spectrum as $v_1 + v_3$ increases. On the other hand, several studies by Henry, Siebrand, and their colleagues[2,37-41] have shown that a two-parameter local mode expansion

$$G(v_a, v_b) = \sum_{\nu = a, b} \left[\left(v_\nu + \tfrac{1}{2}\right)\omega - \left(v_\nu + \tfrac{1}{2}\right)^2 \omega x \right] \tag{1.2}$$

gives a better overall fit than Eq. (1.1) to the observed eigenvalues, particularly for the lowest frequency bands of the most highly excited overtone manifolds. This aspect of the theory is well documented by Henry,[35] who gives a table of $(\omega, \omega x)$ values (in the notation $\omega_i = \omega - \omega x$; $x_{ii} = -\omega x$) for the CH vibrations of different molecules.

Two deductions from Eq. (1.2) are (a) that a properly anharmonic description of the diagonal bond potential functions is essential for an adequate treatment of moderately to highly excited X—H vibrations and (b) that the Morse approximation is at least a useful starting point. The first of these conclusions has led to bond separable potential models for different molecules, notably by Wallace[42] and Elert et al.,[43] and the second to investigations of the analytical properties of the Morse oscillator.[44-53]

More recent studies have shown that it is not the bond separability of the potential function that gives rise to local model behavior, because models based on this approximation include interbond coupling via the kinetic energy operator. Thus the presence or absence of potential coupling can make a quantitative rather than a qualitative change in the nature of the spectrum. A study of SO_2 shows in fact that the dominant interbond coupling comes from the kinetic energy and that the spectrum is characteristically normal in its eigenvalue separations.[54,55] Thus we have an example of a completely separable potential model leading to normal mode behavior. In the case of H_2O the strength of interbond coupling, as measured by the splitting between the stretching frequency fundamental, is roughly equally attributable to potential and kinetic energy terms.[55] Finally it appears that the bond vibrations of SiH_4 and GeH_4 are coupled by both kinetic and potential energy terms[56] but that the two types of contribution have opposite signs, leading to almost exact cancellation. These molecules are therefore prototype X—H local mode oscillators, but their spectra would be poorly described in terms of a bond separable potential.

Further insight into the local mode phenomena comes from related classical[57-60] and quantum mechanical[61-63] studies on realistic potential energy surfaces. The key points from the classical investigations are, first, the coexistence of two types of trajectory (termed local and normal, as shown in Fig. 1 in Section II, below) at a given energy and, secondly, the overriding importance of the relative bond vibrational frequencies in determining the nature of the overall motion. The latter point relates to the anharmonicity of the bond potential, because the frequency determined by a realistically anharmonic potential decreases with increasing energy.

Strong near-resonant coupling arises if the frequencies of two modes are roughly equal, and this will occur if the available energy is roughly equally shared between them. The coupling becomes much weaker however if the same total energy is shared more unequally. Its strength is reflected in classical mechanics in the frequency of interbond energy transfer which translates in quantum mechanics to a frequency splitting of the degenerate eigenvalues given by Eq. (1.2). Semiclassical developments of the theory[57,58,60,63] therefore lead to the prediction of very small local mode doublet splittings for states which are classically local and much larger splittings for states at similar energies that are classically normal.

The pattern described above for water, for example, is fully confirmed by quantum mechanical calculations[61-63] and by experiment.[64-76] From the quantum mechanical viewpoint the effect of the anharmonicity parameter ωx in Eq. (1.2) is to lift the $[2(v_1 + v_3)+1]$-fold degeneracy of the overtone manifold. Thus the effect of the interbond coupling terms, measured say by a parameter λ may be quenched by a sufficiently large anharmonicity, ωx. The character of the eigenvalue spectrum therefore depends on the ratio $\lambda/\omega x$. It turns out that the highest terms in a given manifold with an odd value of $v_1 + v_3$ are split to order λ and progressively lower terms to order $\lambda(\lambda/\omega x)^2, \lambda(\lambda/\omega x)^4, \ldots$. In the case that $v_1 + v_3$ is even, the uppermost level is nondegenerate and successive splittings are to order $\lambda(\lambda/\omega x)$, $\lambda(\lambda/\omega x)^3, \ldots$.

This underlines the point that it is not the origin of λ, whether from potential or kinetic energy coupling, that gives rise to local mode behavior. Furthermore molecular vibrations do not fall into two exclusive classes, local or normal oscillations. The character of the eigenvalue spectrum depends on the magnitude of the continuous variable ($\lambda/\omega x$), with local and normal limits when $|\lambda| \ll \omega x$ and $|\lambda| \gg \omega x$, respectively. One can therefore locate individual molecules on a correlation diagram showing the change in the nature of the spectrum between these limits. Such diagrams are given below for different symmetry situations.

There has also been recent progress on the development of simple potential models for the interpretation and prediction of the overtone and combination spectra of a variety of molecules. Like the conventional anharmonic expansion (1.1), the familiar multiparameter force field expansion in powers of displacement coordinates is limited to relatively low vibrational states, although its usefulness can be extended by adding terms to ensure the correct dissociation behavior. Calculations for $H_2O^{61,62}$ on the Sorbie–Murrell potential,[77] which is of this type, have been quite successful, for example. Much simpler descriptions have been obtained by including optimized Morse oscillator terms for the individual bond potentials together with bilinear potential coupling terms. A three-parameter potential of this type has been used to fit 12 observed CH stretching vibrational eigenvalues of up to 13790 cm^{-1} of CH_4,[56] with a root-mean-square (rms) deviation of 8.8 cm^{-1}, for example. Similar calculations have been performed for other tetrahedral hydrides and deuterides[56] and mixed H and D substituents,[78] for benzene,[79] for various isotopic species of acetylene,[80,81] and even for UF_6, WF_6, and SF_6.[82]

Attention has also recently been given to calculation of the absorption intensities of $X—H$ stretching overtone and combination bands.[21,52,53,56,61,79,83-86] Here one must consider the effects of the bond potential anharmonicities, vibrational coupling terms, and electrical anharmonicity.[86] The first two may be reliably assessed from the potential used to fit the vibrational spectrum, but the latter relates to the form of the dipole function, which is less well known. A favorite model has been the bond dipole approximation, and this, together with the localization of vibrational motion, accounts well for the observation that the intensity is often predominantly concentrated into the lowest energy components of any stretching overtone manifold.[21,79,87]

Finally, many papers have addressed the kinetic and dynamic consequences of local mode behavior, with particular reference to the widths of the overtone band profiles for aromatic molecules. We treat this topic as outside the scope of this review, but a list of references[8,24,25,28,88-99] to relevant recent papers may be of interest.

This review is divided into nine sections. After the present introduction, Section II covers classical and semiclassical considerations and section III deals with the structure of the quantum mechanical theory. Section IV describes the calculation of intensities. These are described for simplicity in terms of the behavior of two coupled equivalent anharmonic oscillators. The remaining sections deal separately with different specific molecules according to the number of coupled oscillators. Thus, Section V on XY_2 molecules includes H_2O, C_2H_2, and CH_2Z_2 ($Z = Cl, Br, I$), and Sections VII, VIII, and IX deal respectively with AH_3, AH_4, XY_6, and C_6H_6 systems.

II. CLASSICAL PERSPECTIVES

The classical mechanics of anharmonic bond stretching vibrations brings out a sharp distinction between local mode and normal mode types of motion against which the subsequent discussion of quantum mechanical and experimental eigenvalues may be set. Attention is restricted to motion in two symmetry-related degrees of freedom, as exemplified by the stretching modes of H_2O.

Figure 1 shows the paths traced out by the two types of trajectory on the Sorbie–Murrell potential surface for H_2O. This behavior has been observed by a variety of authors for several model potentials.[57-60,100] the important distinction is that the normal mode trajectory retains the symmetry of the potential surface while the local mode one does not. This means that each local mode trajectory is strictly degenerate with its mirror image in the Q_1 axis, a classical degeneracy that shades to a near degeneracy, or local mode

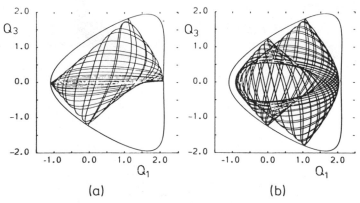

(a)   (b)

Fig. 1. (a) A local trajectory and (b) a normal trajectory at 26367 cm^{-1} on the Sorbie–Murrell potential for water. (Taken from Lawton and Child,[58] with permission.)

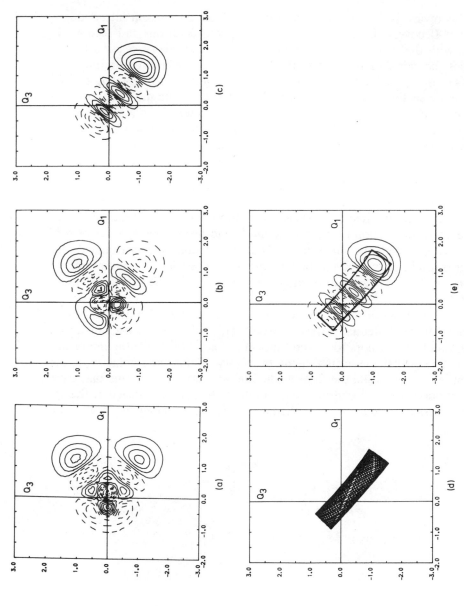

Fig. 2. (a) The $|v_1, v_3\rangle = |5, 0\rangle$ wave function, (b) the $|4, 1\rangle$ wave function, (c) the linear combination $|5, 0\rangle - |4, 1\rangle$, (d) the $|n, m\rangle = |5, 0\rangle$ local mode trajectory, and (e) a superposition of (c) and the caustics of (d). Dashed contours indicate a negative phase of the wave function. (Taken from Lawton and Child,[58] with permission.)

6

doubling, which is one of the most striking features of the quantum mechanical spectrum. The corresponding quantum mechanical wave functions may be regarded as symmetrized and antisymmetrized combinations of nonstationary local mode states, which are seen in Fig. 2 to be closely related to the appropriately quantized local mode classical trajectories.[58]

Lawton and Child[58] and Muckermann et al.[101] have made a detailed numerical study of the relative volumes of phase space occupied by the two types of motion at different energies. Jaffé and Brumer[59] have examined the same question from an analytical viewpoint, while Sibert et al.[60,63] have made an elegant extension of the classical analysis which ultimately yields a simple picture of the quantum mechanical eigenvalue spectrum. Details of this analysis are given below.

As a prelude we consider a perturbative solution to the exactly soluble problem of two harmonically coupled harmonic oscillators as described by

$$H = H_a(p_a, q_a) + H_b(p_b, q_b) + K_{ab}q_a q_b \qquad (2.1)$$

where

$$H_\nu(p_\nu, q_\nu) = \frac{p_\nu^2}{2\mu} + \frac{\mu \Omega_\nu^2 q_\nu^2}{2} \qquad (2.2)$$

This might model the coupling between the stretching vibrations of mutually perpendicular bonds because g matrix terms in $p_a p_b$ have been omitted for simplicity. Note that subscript letters are used to denote individual bonds to avoid confusion with conventional normal coordinate labels. Also capital letters K_{ab} and Ω_ν are employed to avoid confusion with similar quantities in Section III that are defined in terms of wave number rather than energy units. The conversion is $k_{ab} = K_{ab}/hc$ and $\omega = \Omega/hc$, etc. The frequencies Ω_a and Ω_b are taken in general to be different in order to bring out as simply as possible the importance of the resonance condition $\Omega_a = \Omega_b$.

The term $K_{ab}q_a q_b$ will be treated as a perturbation in the absence of which the unperturbed motion is most elegantly described in terms of angle–action variables most commonly denoted (J_ν, w_ν),[102] with the angles w_ν periodic over the interval $[0,1]$. It is however more convenient for the sake of contact with Sibert et al.[60,63] to employ (I_ν, θ_ν), with θ_ν periodic over 2π and $I_\nu = J_\nu/2\pi$. The transformation from (I_ν, θ_ν) to (p_ν, q_ν) in the case of a harmonic oscillator may be shown to be[102,103]

$$q_\nu = \left(\frac{2I_\nu}{\mu \Omega_\nu}\right)^{1/2} \cos \theta_\nu$$
$$p_\nu = -(2I_\nu \mu \Omega_\nu)^{1/2} \sin \theta_\nu \qquad (2.3)$$

so that

$$H^0 = H_a(p_a, q_a) + H_b(p_b, q_b) = I_a \Omega_a + I_b \Omega_b \qquad (2.4)$$

This means, according to Hamilton's equations,[102]

$$\dot{I}_\nu = -\frac{\partial H^0}{\partial \theta_\nu} = 0, \qquad \dot{\theta}_\nu = \frac{\partial H^0}{\partial I_\nu} = \Omega_\nu \qquad (2.5)$$

that

$$I_\nu = I_\nu^0 = \text{const}, \qquad \theta_\nu = \Omega_\nu t + \alpha_\nu \qquad (2.6)$$

In other words the two actions are constants of the motion, with corresponding energies $I_\nu^0 \Omega_\nu$, and the two angle variables increase linearly with time.

On introduction of the perturbation, the angle–action representation of the Hamiltonian becomes

$$H = I_a\Omega_a + I_b\Omega_b + K_{ab}\left(\frac{I_a I_b}{\mu^2 \Omega_a \Omega_b}\right)^{1/2}\left[\cos(\theta_a - \theta_b) - \cos(\theta_a + \theta_b)\right] \quad (2.7)$$

The procedure in nondegenerate classical perturbation theory[103] is to examine the effect of the perturbation on the previous constant I_ν, assuming θ_ν to be adequately approximated by Eq. (2.6). Thus,

$$\dot{I}_a = -\frac{\partial H}{\partial \theta_a} = K_{ab}\left(\frac{I_a I_b}{\mu^2 \Omega_a \Omega_b}\right)^{1/2}\left[\sin(\theta_a - \theta_b) - \sin(\theta_a + \theta_b)\right]$$

$$\dot{I}_b = -\frac{\partial H}{\partial \theta_b} = -K_{ab}\left(\frac{I_a I_b}{\mu^2 \Omega_a \Omega_b}\right)^{1/2}\left[\sin(\theta_a - \theta_b) + \sin(\theta_a + \theta_b)\right]$$

$$(2.8)$$

On introducing the time dependence of the θ_ν given by Eq. (2.6), ignoring that of the I_ν on the right-hand side of Eq. (2.8) and integrating with respect to time, it follows that

$$I_a = I_a^0 - K_{ab}\left(\frac{I_a^0 I_b^0}{\mu^2 \Omega_a \Omega_b}\right)^{1/2}\left(\frac{\cos\left[(\Omega_a - \Omega_b)t + \alpha_a - \alpha_b\right]}{\Omega_a - \Omega_b}\right.$$

$$\left. - \frac{\cos\left[(\Omega_a + \Omega_b)t + \alpha_a + \alpha_b\right]}{\Omega_b + \Omega_b}\right)$$

$$I_b = I_b^0 + K_{ab}\left(\frac{I_a^0 I_b^0}{\mu^2 \Omega_a \Omega_b}\right)^{1/2}\left(\frac{\cos\left[(\Omega_a - \Omega_b)t + \alpha_a - \alpha_b\right]}{\Omega_a - \Omega_b}\right.$$

$$\left. + \frac{\cos\left[(\Omega_a + \Omega_b)t + \alpha_a + \alpha_b\right]}{\Omega_a + \Omega_b}\right)$$

$$(2.9)$$

This shows that the perturbed actions and related energies I_ν, Ω_ν will follow small fluctuations about their unperturbed values I_ν^0 within an envelope determined by the ratio $K_{ab}/(\Omega_a - \Omega_b)$. The conditions for a valid perturbative description are therefore that $(K_{ab}^2/\mu^2\Omega_a\Omega_b)^{1/2} \ll |\Omega_a - \Omega_b|$. This condition clearly breaks down for $\Omega_a \simeq \Omega_b$, because physically two systems with the same resonant frequency are strongly coupled even by a small interaction between them.

A convenient procedure in this so-called Chirikov resonance[104] situation, when $\Omega_a = \Omega_b = \Omega$, is to transform to new angle–action variables[60]

$$I = I_a + I_b, \qquad \theta_I = \frac{\theta_a + \theta_b}{2}$$

$$J = I_a - I_b, \qquad \theta_J = \frac{\theta_a - \theta_b}{2} \tag{2.10}$$

where the factors 2^{-1} arise from the requirement that the Jacobian from $(I_a, I_b, \theta_a, \theta_b)$ to $(I, J, \theta_I, \theta_J)$ should be unity for a canonical transformation. The unperturbed Hamiltonian now depends only on the total action I

$$H^0 = (I_a + I_b)\Omega = I\Omega \tag{2.11}$$

Hence by solution of Hamilton's equations

$$I = I^0 = \text{const}, \qquad \theta_I = \Omega t + \text{const}$$

$$J = J^0 = \text{const}, \qquad \theta_J = \theta_J^0 = \text{const} \tag{2.12}$$

The energy is therefore determined by the total action I and the only motion resides in its conjugate angle θ_I, this motion being in fact around an ellipse in (q_a, q_b) space, with an orientation determined by θ_J^0 and ellipticity given by J^0/I^0.

The effect of the perturbation is to cause a precession of the axes of the ellipse, coupled with a change in its ellipticity. The dominant behavior is conveniently described by substitution for $(I_a, I_b, \theta_a, \theta_b)$ in Eq. (2.7), followed by averaging the resulting Hamiltonian over a cycle of the fast motion. The averaged result in the present case is

$$\overline{H} = I^0\Omega + \left(\frac{K_{ab}I^0}{2\mu\Omega}\right)\left[1 - \left(\frac{J}{I^0}\right)^2\right]^{1/2}\cos 2\theta_J \tag{2.13}$$

Since Hamilton's equations ensure that $\overline{H}$ is a constant of the motion, it fol-

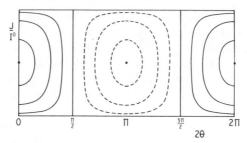

Fig. 3.   Phase space orbits given by Eq. (2.14). Continuous and dashed contours correspond to positive and negative values of C respectively. The central points indicate $C = \pm 1$.

lows that the motion generated in J and θ_J must lie on curves

$$\left[1 - \left(\frac{J}{I^0}\right)^2\right]^{1/2} \cos 2\theta_J = C \tag{2.14}$$

where

$$C = \frac{2\mu\Omega E_J}{K_{ab}I^0} \tag{2.15}$$

E_J being the energy associated with the (J, θ_J) motion. The physically allowed C values are clearly $-1 < C < 1$, and the forms of typical orbits in the (J, θ_J) phase space are shown in Fig. 3. It is evident that J oscillates symmetrically between positive and negative values with the implication, according to Eq. (2.10), that energy and action are symmetrically exchanged between the two degrees of freedom.

The significance of this discussion in relation to Fig. 1 is that substantial anharmonicities in even equivalent bond potentials will lead to markedly different vibrational frequencies Ω_v and hence to the almost independent

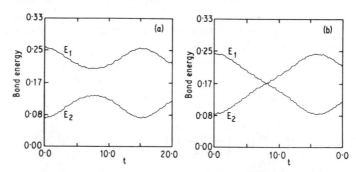

Fig. 4.   (a) Local mode and (b) normal mode time evolutions of the individual bond energies E_1 and E_2. (Taken from Silbert et al.,[60] with permission.)

oscillator picture described by Eq. (2.9) whenever the available energy is very unequally distributed between the oscillators. If, on the other hand, the two oscillators receive roughly equal shares of the same total energy, they will oscillate with roughly equal frequencies and transfer energy freely from one to the other. These two behaviors are nicely illustrated by the diagram from Sibert et al.[60] shown in Fig. 4.

The above brief account of classical perturbation theory may also serve as an introduction to the elegant analysis suggested by Jaffé and Brumer[59] and later pursued by Sibert et al.,[60,63] which is applied to the more interesting model of a pair of harmonically coupled identical Morse oscillators represented by the Hamiltonian

$$H = H^{(0)} + H^{(1)}$$

$$H^{(0)} = \sum_{\nu = a, b} \left[\tfrac{1}{2} G_{\nu\nu} p_\nu^2 + V(q_\nu) \right]; \qquad G_{aa} = G_{bb} \qquad (2.16)$$

$$H^{(1)} = G_{ab} p_a p_b + K_{ab} q_a q_b$$

$$V(q_\nu) = D \left[1 - \exp(-a q_\nu) \right]^2 \qquad (2.17)$$

The transformation to Morse angle–action variables is more complicated than in the harmonic case[59]

$$q_\nu = a^{-1} \ln \left[\frac{1 - \left(1 - \lambda_\nu^2\right)^{1/2} \cos\theta_\nu}{\lambda_\nu^2} \right]$$

$$p_\nu = \left(\frac{\lambda_\nu \Omega}{G_{\nu\nu} a} \right) \left[\frac{\left(1 - \lambda_\nu^2\right)^{1/2} \sin\theta_\nu}{1 - \left(1 - \lambda_\nu^2\right)^{1/2} \cos\theta_\nu} \right] \qquad (2.18)$$

with

$$\lambda_\nu = 1 - \frac{I_\nu}{I_D}$$

$$I_D = \left(\frac{2D}{G_{\nu\nu} a^2} \right)^{1/2} \qquad (2.19)$$

$$\Omega = \frac{2D}{I_D} = \left(2 D G_{\nu\nu} a^2 \right)^{1/2}$$

Here I_D is the action at the dissociation energy, which may be related to a Morse anharmonicity parameter ΩX by

$$\Omega X = \frac{\Omega}{2 I_D} = \frac{\Omega^2}{4D}. \qquad (2.20)$$

It is readily shown by substitution for (p_ν, q_ν) in Eq. (2.16) that

$$H^{(0)} = \sum_{\nu = a,b} \left[I_\nu \Omega - I_\nu^2 \Omega X \right] \qquad (2.21)$$

The absence of θ_ν from this equation implies that the unperturbed actions are constants of the motion (with quantized values $I_\nu = (v_\nu + \frac{1}{2})\hbar$ on passing to quantum mechanics) and the unperturbed classical frequencies become

$$\Omega_\nu = \dot{\theta}_\nu = \frac{\partial H^{(0)}}{\partial I_\nu} = \Omega - 2I_\nu \Omega X \qquad (2.22)$$

Explicit expressions for the interaction term $H^{(1)}$ as a function of the (I_ν, θ_ν) are given by Jaffé and Brumer[59] and in special cases $K_{ab} = 0$ by Sibert et al.[60] It is, however, unnecessary for present purposes to quote these in full because the previous discussion shows that this interaction will be of dominant importance only in situations when the two unperturbed frequencies Ω_a and Ω_b have similar values—in other words, according to Eq. (2.22) when $I_a \simeq I_b$. The major term is then the $1:1$ Fourier component of $H^{(1)}$, i.e., the term in $\cos(\theta_a - \theta_b)$. As in the previous discussion, the motion induced by this term is most conveniently handled by a further canonical transformation to the sum and difference variables

$$\begin{aligned} I_a + I_b &= I, & \theta_a + \theta_b &= 2\theta_I \\ I_a - I_b &= J, & \theta_a - \theta_b &= 2\theta_J \end{aligned} \qquad (2.23)$$

In view of the dominant importance of the $1:1$ resonance, it is sufficient for illustrative purposes to truncate the Hamiltonian after this term. Following Sibert et al.,[60] the resulting truncated Hamiltonian with $K_{ab} = 0$ may be expressed in the present notation as

$$H_T = H_I - H_J \qquad (2.24)$$

$$H_I = I\Omega - \frac{I^2 \Omega X}{2} \qquad (2.25)$$

$$H_J = \frac{J^2 \Omega X}{2} + V_0 \cos 2\theta_J \qquad (2.26)$$

$$V_0 = \left(\frac{4DG_{ab}}{G_{aa}} \right) \left(1 - \frac{I\Omega}{4D} \right)^2 \left(\frac{I\Omega}{8D - I\Omega} \right) \qquad (2.27)$$

This form for V_0 involves a further approximation in that it is evaluated at the resonance point $I_a = I_b$ or $J = 0$. Since neither H_I nor H_J depends on θ_I, it is evident that the total action I is a constant of the motion. Hence the effect of the transformation is to reduce the problem from motion in two degrees of freedom to motion in one degree of freedom. Note for comparison with the work of Sibert et al. that the quantity E used by these authors is the total action term H_I, not the total energy.

Equation (2.26) shows that the residual (J, θ_J) motion is that of a pendulum, which takes one of two distinct forms as shown by the phase space orbits in Fig. 5. Those orbits lying inside the heavy dashed separatrix give rise to periodic sign changes in the action difference $J = I_a - I_b$ and hence correspond to normal mode motions in which energy passes freely from one oscillator to the other. The orbits outside the separatrix, on the other hand, never cross the line $J = 0$; hence either $I_a > I_b$ for all t or vice versa: this is the characteristic of local mode behavior. The pendulum analog is that normal modes correspond to pendulum oscillations and the local modes to end over end rotations.

One immediate consequence suggested by the work of Jaffé and Brumer[59] is that the range of J values for any given I consistent with normal mode behavior is immediately determinable from the separatrix equation $H_J = 0$, the maximum and minimum values of J being given by

$$J_\pm = \pm \left(\frac{2V_0}{\Omega X} \right)^{1/2}$$

$$= \pm \left(\frac{G_{ab} I \Omega}{G_{aa} \Omega X} \right)^{1/2} = \pm \left(\frac{2 II_D G_{ab}}{G_{aa}} \right)^{1/2} \qquad (2.28)$$

The approximate forms here are obtained by neglecting terms of order $I\Omega/D$ in Eq. (2.27). The form containing I_D, which is obtained by substitution from Eq. (2.20), is given for comparison with Eq. (27) of Jaffé and Brumer,[59] who consider the case of $K_{ab} \neq 0$ and obtain limits equivalent in the present no-

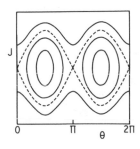

Fig. 5.   Phase space orbits derived from Eq. (2.26). The inner vibrational orbits correspond to normal modes and the outer rotational ones to local modes.

tation to

$$J_{\pm} = \pm \left[2\Pi_D \left\{ \frac{G_{ab}}{G_{aa}} + \frac{K_{ab}}{2Da^2} \right\} \right]^{1/2}$$

$$= \pm \left(\frac{2I\Lambda}{\Omega X} \right)^{1/2} \tag{2.29}$$

where

$$\Lambda = 0.5\Omega \left[\frac{G_{ab}}{G_{aa}} + \frac{K_{ab}}{2Da^2} \right] \tag{2.30}$$

the second form of Eq. (2.29) being convenient for comparison with the first two terms in Eq. (3.23), which is the quantum mechanical equivalent of the present result.

As a test of the validity of Eq. (2.29), one may compare the implied limits on the normal mode values of I_a and I_b—namely, for $I_a > I_b$

$$I_a = \frac{I + J_+}{2}, \qquad I_b = \frac{I + J_-}{2} \tag{2.31}$$

—with those obtained by a direct numerical investigation of the trajectories studied by Lawton and Child.[57,58] The test is not completely rigorous because the trajectory study employed a more complicated potential function than that in Eq. (2.16), but both this potential and the present model with the parameter fit $|\Lambda/X\Omega| = (49.5/84.4) = 0.587$ are known to give a good representation of the spectrum[55] (see Section III, below). Figure 6 based on

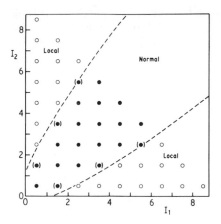

Fig. 6. Local and normal regions of action space. The dashed boundaries are given by Eq. (2.29). The open and closed circles correspond to the local mode and normal mode states determined by Lawton and Child.[58] Points in parentheses (●), were found[58] to be transitional.

this ratio compares the division of the numerically determined quantized trajectories between normal and local types with the perturbation model. The level of agreement is seen to be excellent.

A second test is to examine the onset of local mode behavior, which occurs for $I_b = 0$ in Eq. (2.31). It follows on substituting $J_- = -I$ in Eq. (2.29) that

$$I_{min} = \frac{2\Lambda}{\Omega X} \tag{2.32}$$

The analytical prediction $I_{min} = 1.2\hbar$ in this case is somewhat larger than the numerically determined value $I_{min} \simeq 0.36\hbar$.[59]

Sibert et al.[60,63] have taken the discussion further by comparing the eigenvalues implied by the quantum mechanical equivalent of Eq. (2.24) with those determined by diagonalization of the quantal analog of Eq. (2.16). To do this, these authors adopted the prescription

$$I = \left(v + \tfrac{1}{2}\right)\hbar \tag{2.33}$$

in order to diagonalize the Hamiltonian H_I; they also cast motion under the Hamiltonian H_J into the Mathieu equation

$$\frac{d^2\psi}{dz^2} + (\alpha - 2q\cos 2z)\psi = 0 \tag{2.34}$$

by means of the substitutions

$$z = \theta_J$$

$$\alpha = \left(\frac{8D}{\hbar^2\Omega^2}\right)E_J, \qquad q = \left(\frac{4D}{\hbar^2\Omega^2}\right)V_0. \tag{2.35}$$

It is then a simple matter to obtain the eigenvalues of H_J by scaling the published[105] eigenvalues of Eq. (2.34). Care is however required in the choice of physically relevant eigenvalues, because these must go over when $q = V_0 = 0$ to the free rotor levels

$$E_J = \frac{J^2\Omega X}{2} \tag{2.36}$$

with J values consistent, according to Eq. (2.32), with

$$I = (v_a + v_b + 1)\hbar$$
$$J = (v_a - v_b)\hbar \tag{2.37}$$

because in this uncoupled limit $I_\nu = (v_\nu + \tfrac{1}{2})\hbar$. It follows, for a given value, say, $I = (v + 1)\hbar$, that

$$j = \frac{J}{\hbar} = v - 2v_b \tag{2.38}$$

The pattern of resulting energy levels for $v = 5$ is shown in Fig. 7. These are labeled by the symbol $|j^{\pm}\rangle$, where the superscript $\pm$ indicates the character under sign reversal of z. The parameter values employed were the same as those used to calculate Fig. 6, in which it can be seen that there are four local mode states and two normal mode states along the line $I_a + I_b = 6\hbar$. The former correspond to the four levels above the barrier in Fig. 7, and the latter to the two levels below.

It should be noted that H_J appears in Eq. (2.24) with a negative sign, with the implication that the near-degenerate almost-free rotor states should be the lowest in any manifold. This pattern is clearly demonstrated in Table I, which shows the comparison made by Sibert et al.[63] between the present pendulum model eigenvalues and those obtained by diagonalization of the full Hamiltonian H in Eq. (2.16). The parameter values are appropriate to the stretching modes of H_2O, but with $K_{ab} = 0$; hence the local mode splittings are smaller than those of the experimental spectrum (see Table IV in Section V, below). It is clear, however, that the present pendulum model gives an excellent approximation to the exact eigenvalue spectrum. The semi-classical eigenvalues in the final column, which were derived from a JWKB type of solution to the twofold hindered rotor problem, are somewhat less satisfactory.

Finally, it is of some interest that other authors[57,58] have predicted the general form of the eigenvalue spectrum from the viewpoint that the small local mode splittings could be attributed to a classically forbidden dynamical channeling motion from one local mode to another. Lawton and Child[58]

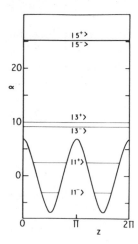

Fig. 7.   Restricted rotor states for $q = 3.226$ in Eq. (2.34), corresponding roughly to $v = 5$ levels for H_2O.

TABLE I

Quantum (E^Q, Δ^Q), Pendulum Model (E^{Pen}, Δ^{Pen}), and Semiclassical (E^{sc}, Δ^{sc}) Eigenvalues and Local Mode Splittings

v_a	$v_b^{\pm}$	E^Q (cm^{-1})	Δ^Q (cm^{-1})	E^{Pen} (cm^{-1})	Δ^{Pen} (cm^{-1})	E^{sc} (cm^{-1})	Δ^{sc} (cm^{-1})
1	0^+	3675.55	53.7	3675.80	53.1	3675.4	43.9
1	0^-	3729.27		3728.91		3719.3	
2	0^+	7220.74	16.9	7221.11	16.2	7224.2	13.5
2	0^-	7237.64		7237.29		7237.7	
1	1	7419.40		7418.93		7412.0	
3	0^+	10595.25	2.7	10595.58	2.1	10597.0	0.2
3	0^-	10597.97		10597.65		10597.2	
2	1^+	10895.73	97.6	10895.51	97.7	10888.2	97.3
2	1^-	10993.28		10993.19		10985.5	
4	0^+	13794.94	0.3	13794.93	0.1	13795.1	0.0
4	0^-	13795.26		13795.07		13795.1	
3	1^+	14271.82	40.5	14272.37	37.4	14276.8	38.9
3	1^-	14312.30		14309.74		14315.7	
2	2	14507.14		14509.16		14500.4	
5	0^+	16824.49	0.4	16824.64	0.0	16824.7	0.0
5	0^-	16824.88		16824.64		16824.7	
4	1^+	17488.37	8.8	17490.36	5.4	17490.3	2.1
4	1^-	17497.19		17495.78		17492.4	
3	2^+	17786.17	131.1	17784.62	133.6	17774.8	136.2
3	2^-	17917.30		17918.18		17911.0	

have obtained a qualitatively satisfactory account of the spectrum on this basis. The more elegant analysis of Sibert et al.,[60,63] on the other hand, attributes the near degeneracy to the approach to a free rotor limit. The canonical systems of variables employed from the two viewpoints are of course quite different, and it is not easy to see how they might be reconciled.

III. COUPLED QUANTUM MECHANICAL ANHARMONIC OSCILLATORS

The eigenvalue pattern suggested above on classical grounds, and explicitly demonstrated in Table I, is fully confirmed by quantum mechanical calculations. The main features are brought out by a model suggested by Watson et al.[54] and developed in more detail by Child and Lawton[55] and

Mortensen et al.[106] The formulation below follows the notation of Child and Lawton,[55] who develop the argument in greatest detail.

Consider a system of two identical Morse oscillators, with displacement coordinates r_a and r_b coupled both to each other and to a set of other oscillators $\{r_\mu\}$ which are treated as harmonic. The dominant terms are assumed to be represented by the Hamiltonian

$$H = H_{ab}^0 + H_\zeta^0 + V_{\text{dir}} + V_{\text{ind}} \tag{3.1}$$

where

$$\frac{H_{ab}^0}{hc} = \tfrac{1}{2} g_{aa}\left(p_a^2 + p_b^2 \right) + D\left[1 - \exp\left(- a r_a \right) \right]^2 + D\left[1 - \exp\left(- a r_b \right) \right]^2 \tag{3.2}$$

$$\frac{H_\zeta^0}{hc} = \tfrac{1}{2} \sum_\mu \left(g_{\mu\mu} p_\mu^2 + k_{\mu\mu} r_\mu^2 \right) \tag{3.3}$$

$$\frac{V_{\text{dir}}}{hc} = g_{ab} p_a p_b + k_{ab} r_a r_b \tag{3.4}$$

$$\frac{V_{\text{ind}}}{hc} = \sum_{\nu,\mu} \left(g_{\nu\mu} p_\nu p_\mu + k_{\nu\mu} r_\nu r_\mu \right), \qquad \nu = a, b \tag{3.5}$$

The eigenvalues of H_{ab}^0 and H_ζ^0 (in wavenumber units) are respectively

$$G_{ab}^0 = \sum_{\nu = a, b} \left[\omega\left(v_\nu + \tfrac{1}{2} \right) - \omega x \left(v_\nu + \tfrac{1}{2} \right)^2 \right] \tag{3.6}$$

$$G_\zeta^0 = \sum_\mu \omega_\mu \left(v_\mu + \tfrac{1}{2} \right) \tag{3.7}$$

where

$$\omega = \left(2 D g_{aa} \right)^{1/2} \left(\frac{a\hbar}{hc} \right) \tag{3.8}$$

$$\omega x = \frac{g_{aa} a^2 \hbar^2}{2hc} \tag{3.9}$$

$$\omega_\mu = \frac{\hbar \left(k_{\mu\mu} g_{\mu\mu} \right)^{1/2}}{hc} \tag{3.10}$$

The matrix elements of p_ν and r_ν between appropriate zeroth-order eigenstates are required to assess the effects of the coupling terms V_{dir} and V_{ind}. Those for the Morse oscillator have been given by Wallace,[50] Sage,[51] and

Carney,[48,49] and a factorization method for developing properties of the Morse oscillator has been described by Mohammadi.[107] The following expressions are taken from Sage[51] and Carney et al.[48,49] (N.B. Equation (3.12) differs from Sage's equation by factor j^{-1}.)

$$\langle v+j|p|v\rangle = i\hbar(-1)^{j+1}N_vN_{v+j}\left[\frac{\Gamma(k-v-j)}{2v!}\right](1-\delta_{j0}); \qquad j\geq 0$$

(3.11)

$$\langle v+j|r|v\rangle = (-1)^{j+1}a^{-2}N_vN_{v+j}\left[\frac{\Gamma(k-v-j)}{j(k-2v-j-1)v!}\right]; \qquad j\geq 1$$

(3.12)

$$\langle v|r|v\rangle = a^{-1}\left\{\ln k - \Phi(k-1-2v) + \sum_{j=1}^{v}\frac{1}{k-v-j}(1-\delta_{v0})\right\}$$

(3.13)

where

$$k = \frac{2(2D/g_{aa})^{1/2}}{a\hbar} = \frac{\omega}{\omega x}$$

(3.14)

$$N_v = \left[\frac{av!(k-2v-1)}{\Gamma(k-v)}\right]^{1/2}$$

(3.15)

Here, $\Phi(z)$ is the digamma function.[108]

The corresponding forms for the more familiar nonzero harmonic oscillator matrix elements are[109]

$$\langle v+1|p_\mu|v\rangle = i\left(\frac{k_{\mu\mu}}{g_{\mu\mu}}\right)^{1/4}\left[\frac{(v+1)\hbar}{2}\right]^{1/2} = \langle v|p_\mu|v+1\rangle^*$$

$$\langle v+1|r_\mu|v\rangle = \left(\frac{g_{\mu\mu}}{k_{\mu\mu}}\right)^{1/4}\left[\frac{(v+1)\hbar}{2}\right]^{1/2} = \langle v|r_\mu|v+1\rangle$$

(3.16)

The above Morse expressions are required for any numerically exact diagonalization of the full Hamiltonian. It is, however, convenient for present purposes to recognize that the parameter k is normally large compared with v for the states of interest; $k \simeq 46$ and 65 for H_2O and D_2O, for example. Hence it is permissible to neglect terms of order v/k appearing in Eqs. (3.11)–(3.13). This leads to the following approximations to Eqs.

(3.11)–(3.13):

$$\langle v+j|p|v\rangle \simeq i(-1)^{j+1}\left(\frac{k_{aa}}{g_{aa}}\right)^{1/4}\left[\frac{k^{1-j}(v+j)!\hbar}{2v!}\right]^{1/2}; \qquad j\geq 0$$

$$\langle v+j|r|v\rangle \simeq (-1)^{j+1}\left(\frac{g_{aa}}{k_{aa}}\right)^{1/4}\left[\frac{k^{1-j}(v+j)!\hbar}{2j^2v!}\right]^{1/2}; \qquad j\geq 1$$
(3.17)

$$\langle v|r|v\rangle = O(k^{-1})$$
(3.18)

where

$$k_{aa}=2a^2D$$
(3.19)

It is readily verified that Eqs. (3.17) go over to (3.16) for $j=1$ and that terms for $j>1$ fall off in magnitude as $k^{-(j-1)/2}$. This behavior is confirmed by numerical examples given by Mortensen et al.[106] for $k\simeq 50$, as appropriate to the dihalomethanes; the error in using the harmonic approximation varied from 3% at $v=0$ to 15% at $v=4$. The advantages of this harmonic approximation in the present context are that the coupling terms are limited to $\Delta v_\nu = \Delta v_\mu = \pm 1$ for all degrees of freedom and that the number of parameters required to account for the appearance of the spectrum is reduced (see below).

The significance of the selection rule $\Delta v_\nu = \pm 1$ is that the direct interaction term gives rise to coupling between states with $\Delta v = 0, \pm 2$, where v is the total vibrational quantum number

$$v=v_a+v_b$$
(3.20)

Given the large energy separation between different manifolds of given v, the dominant effect will come from the coupling within the $\Delta v = 0$. These states would be degenerate in a harmonic zeroth-order bond oscillator model, but they are split by the anharmonic terms involving ωx in Eq. (3.6).

The resulting coupling scheme is shown in Fig. 8 for manifolds with $v = 1, 2, 3$. It is clear that $V_{\rm dir}$ gives rise to a first-order splitting between the two components of $v=1$, a second-order splitting of the doublet for $v=2$, and third-order and first-order splittings of the lower and upper doublets for $v = 3$. Thus, if the anharmonic splittings are large compared with the coupling terms, the spectrum will develop the pattern of increasingly close local mode doublets at higher energies shown in Table I. The corresponding eigenstates are closely approximated by the symmetry-adapted local mode contributions:

$$|n \quad m^\pm\rangle = 2^{-1/2}(|n \quad m\rangle \pm |m \quad n\rangle), \qquad n<m$$
$$|n \quad m^+\rangle = |n \quad m\rangle, \qquad n=m$$
(3.21)

in which the first and second quantum numbers on the right-hand side refer

Model   Eigen values

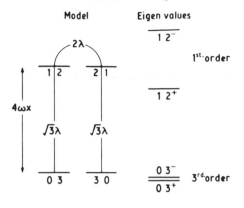

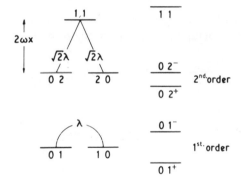

Fig. 8. Anharmonic oscillator coupling schemes for $v = 1$, 2, and 3 zeroth-order levels are labeled by $|v_a, v_b\rangle$; those arising from the coupling are approximately $|v_a, v_b \pm\rangle = 2^{-1/2}(|v_a, v_b\rangle \pm |v_b, v_a\rangle)$, if $v_a \neq v_b$ and $|v_a, v_b\rangle = |v_a, v_a\rangle$ if $v_a = v_b$.

to excitation of r_a and r_b, respectively. The reader will recall that Fig. 2 shows a coordinate plot of such wave functions.

The model outlined above has been cast into quantitative form by Child and Lawton[55] and Mortensen et al.,[106] of which the former also includes the second-order effect on the intermanifold coupling of the indirect coupling term V_{ind} in Eq. (3.5). It turns out that the intermanifold coupling can be expressed in terms of three parameters, namely, an effective frequency ω', the anharmonic parameter ωx, and a coupling parameter λ, the equations for ω' and λ being given by[55]

$$\omega' = \omega - \frac{(\alpha - \beta)^2}{2\omega} - \sum_\mu \left\{ \frac{(\alpha_\mu - \beta_\mu)^2}{\omega + \omega_\mu} - \frac{(\alpha_\mu + \beta_\mu)^2}{\omega - \omega_\mu} \right\} \qquad (3.22)$$

$$\lambda = \alpha + \beta + \sum_\mu \left\{ \frac{(\alpha_\mu + \beta_\mu)^2}{\omega - \omega_\mu} - \frac{(\alpha_\mu - \beta_\mu)^2}{\omega + \omega_\mu} \right\} \qquad (3.23)$$

where

$$\alpha = \omega \left(\frac{g_{ab}}{2g_{aa}} \right)$$

$$\beta = \omega \left(\frac{k_{ab}}{2k_{aa}} \right)$$

$$\alpha_\mu = \omega g_{a\mu} \left[\frac{\omega_\mu}{4 g_{aa} g_{\mu\mu} \omega} \right]^{1/2}$$

$$\beta_\mu = \omega k_{a\mu} \left[\frac{\omega_\mu}{4 k_{aa} k_{\mu\mu} \omega} \right]^{1/2}$$

(3.24)

Notice the close connection between the direct contribution $\beta + \alpha$ to λ and the form of Λ in Eq. (2.30).

The forms of the symmetry-factorized coupling matrices, $H^\pm(v)$ derived from the model are given below, energies being measured from the zero point level

$$G_{00}^0 = \omega' - \tfrac{1}{2}\omega x$$

(3.25)

$$H^{(\pm)}(1) = \omega' - 2\omega x \pm \lambda$$

$$H^{(+)}(2) = \begin{bmatrix} 2\omega' - 6\omega x & 2\lambda \\ 2\lambda & 2\omega' - 4\omega x \end{bmatrix}; \quad H^{(-)}(2) = 2\omega' - 6\omega x$$

$$H^{(\pm)}(3) = \begin{bmatrix} 3\omega' - 12\omega x & \sqrt{3}\,\lambda \\ \sqrt{3}\,\lambda & 3\omega' - 8\omega x \pm 2\lambda \end{bmatrix}$$

$$H^{(+)}(4) = \begin{bmatrix} 4\omega' - 20\omega x & 2\lambda & 0 \\ 2\lambda & 4\omega' - 14\omega x & \sqrt{12}\,\lambda \\ 0 & \sqrt{12}\,\lambda & 4\omega' - 12\omega x \end{bmatrix}$$

(3.26)

$$H^{(-)}(4) = \begin{bmatrix} 4\omega' - 20\omega x & 2\lambda \\ 2\lambda & 4\omega' - 14\omega x \end{bmatrix}$$

$$H^{(\pm)}(5) = \begin{bmatrix} 5\omega' - 30\omega x & \sqrt{5}\,\lambda & 0 \\ \sqrt{5}\,\lambda & 5\omega' - 22\omega x & \sqrt{8}\,\lambda \\ 0 & \sqrt{8}\,\lambda & 5\omega' - 18\omega x \pm 3\lambda \end{bmatrix}$$

The matrices for $v = 6$ are given by Mortensen et al.[106] in a slightly different notation.

One consequence of the simple structure of these matrices is that the splitting pattern for different manifolds may be expressed in the form of a correlation diagram by plotting the reduced eigenvalue

$$\varepsilon = \frac{E - \overline{E}(v)}{\left[\lambda^2 + (\omega x)^2\right]^{1/2}} \tag{3.27}$$

where $\overline{E}$ is the mean energy of a particular overtone manifold, as a function of what may be termed the local mode parameter

$$\xi = \frac{2}{\pi} \arctan\left(\frac{\lambda}{\omega x}\right) \tag{3.28}$$

Small values of $|\xi|$ correspond to local mode behavior, and values of ξ close to ± 1 to the normal mode picture, as illustrated in Fig. 9. Notice that ξ gives a more precise indication of local or normal mode character than the list of attributes suggested by Stannard et al.[86]

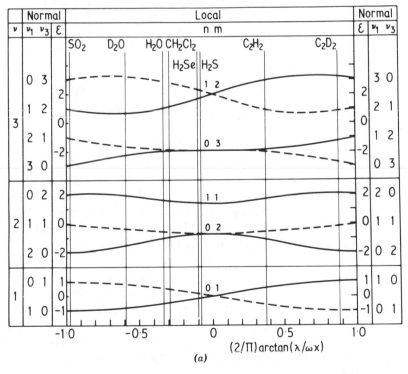

Fig. 9. A local to normal mode correlation diagram for two bilinearly coupled anharmonic oscillators: (a) $v = 1$–3 and (b) $v = 4, 5$. The positions of individual molecules are derived from the parameters in Table II.

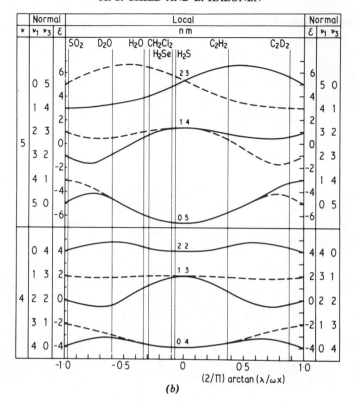

Fig. 9. (*Continued*).

The diagram also shows the positions of the different molecules along the ξ axis based on the parameter values ω', ωx, and λ given in Table II, which were obtained by fits to the experimental spectra.[55,106] Detailed investigation shows that the negative λ value for H_2O is roughly equally attributable to the direct momentum (α) and potential (β) contribution to λ in Eq. (3.23), both α and β being negative. The same must also be true for CH_2I_2; in the case of SO_2, on the other hand, λ arises almost entirely from momentum coupling. The cases of C_2H_2 and C_2D_2 are interesting in that the dominant contribution to λ is indirect momentum coupling via the C—C stretching coordinate.[55] Equation (3.23) shows that a term in α_μ alone necessarily gives rise to $\lambda > 0$ because $\omega > \omega_\mu$. The fact that λ for C_2D_2 is much larger than that for C_2H_2 is attributable to a smaller energy denominator $\omega_{CD} - \omega_{CC}$ compared with $\omega_{CH} - \omega_{CC}$.[55]

Another outcome of Eqs. (3.26) is that one can compare the present parameters ω', ωx, and λ with those used in the conventional anharmonic

TABLE II
Model Parameters for Selected Molecules

	ω' (cm^{-1})	ωx (cm^{-1})	λ (cm^{-1})
H_2O^a	3876.2	84.4	-49.5
$C_2H_2^a$	3450.3	58.4	39.0
$C_2D_2^a$	2619.4	23.6	132.9
SO_2^a	1271.6	7.5	-105.2
$CH_2Cl_2^b$	3143.3	63.2	-31.0
$CH_2Br_2^b$	3149.5	62.4	-36.0
$CH_2I_2^b$	3120.7	60.5	-37.5

[a] From Ref. 55.
[b] From Ref. 106.

expansion[64,110]

$$G(v_1, v_3) = \omega_1\left(v_1 + \tfrac{1}{2}\right) + \omega_3\left(v_3 + \tfrac{1}{2}\right) + x_{11}\left(v_1 + \tfrac{1}{2}\right)^2 + x_{33}\left(v_3 + \tfrac{1}{2}\right)^2$$
$$+ x_{13}\left(v_1 + \tfrac{1}{2}\right)\left(v_3 + \tfrac{1}{2}\right) \tag{3.29}$$

where Q_1 and Q_3 are the symmetric and antisymmetric stretching normal coordinates. It is readily found by comparison with the eigenvalues $H^{\pm}(1)$ and $H^{\pm}(2)$ for $|\lambda| \gg \omega x$ that in the near-normal limit

$$\omega_1 = \omega' + \lambda$$
$$\omega_3 = \omega' - \lambda \tag{3.30}$$
$$x_{11} = x_{33} = \frac{x_{13}}{4} = \frac{-\omega x}{2}$$

It is also possible to show by comparing the predictions of Eq. (3.29) with the eigenvalues of $H^{\pm}(1)$ and $H^{-}(2)$ and the trace of $H^{+}(2)$ that for arbitrary $\lambda/\omega x$ ratios

$$\omega' = \frac{\omega_1 + \omega_3}{2}$$
$$\omega x = \frac{-(x_{11} + x_{33})}{2} - \frac{x_{13}}{4} \tag{3.31}$$
$$\lambda = \frac{\omega_1 - \omega_3}{2} + x_{11} - x_{33}$$

provided

$$\Delta = x_{11} + x_{33} - \frac{x_{13}}{2} = 0 \tag{3.32}$$

The extent to which Eq. (3.32) is satisfied is a measure of the contribution of pure bond anharmonicity to the normal mode anharmonic parameters. Table III shows that Eqs. (3.31) and (3.32) are at least moderately well satisfied for the molecules listed, as can be seen from the experimental data.[111-114] Hence, as suggested by Halonen et al.,[80] Eqs. (3.31) may be employed to determine ω', ωx, and λ for use with Eqs. (3.26). This approach was found to give a much better fit to the spectrum of C_2H_2 than that using Eq. (3.29), despite a reduction from five to three adjustable parameters.

Finally, it is of interest that the eigenvalues of the matrices in Eqs. (3.26) are found to be identical with those derived by conventional techniques,[110] provided the contributions of off-diagonal Darling–Dennison[115] couplings are included.[116] Specifically the effective Hamiltonian may be written

$$
\frac{H}{hc} = \tfrac{1}{2}\omega_1\left(p_1^2 + q_1^2 \right) + \tfrac{1}{2}\omega_3\left(p_3^2 + q_3^2 \right) + \tfrac{1}{4}x_{11}\left(p_1^2 + q_1^2 \right)^2
$$
$$
+ \tfrac{1}{4}x_{13}\left(p_1^2 + q_1^2 \right)\left(p_3^2 + q_3^2 \right) + \tfrac{1}{4}x_{33}\left(p_3^2 + q_3^2 \right)^2
$$
$$
+ \tfrac{1}{8}\gamma\left(q_{1+}^2 q_{3-}^2 + q_{1-}^2 q_{3+}^2 \right) \tag{3.33}
$$

where

$$
q_{r\pm} = q_r \mp ip_r
$$

and

$$
q_r = \left(\frac{2\pi c \omega_r}{\hbar} \right)^{1/2} Q_r \tag{3.34}
$$

is a dimensionless normal coordinate and p_r its conjugate momentum operator. The parameters ω_1, ω_3, x_{11}, x_{13}, and x_{33} and the Darling–Dennison

TABLE III
Anharmonic Parameters and the Local Mode Defect Δ

	x_{11} (cm^{-1})	x_{13} (cm^{-1})	x_{33} (cm^{-1})	Δ (cm^{-1})
H_2O^a	-42.576	-165.824	-47.566	-7.23
D_2O^a	-22.58	-87.15	-26.92	5.93
$C_2H_2^b$	-18.57	-102.39	-30.95	-1.68
$C_2D_2^b$	-10.87	-47.26	-14.51	1.75
SO_2^c	-3.99	-13.71	-5.17	-2.31
H_2S^d	-25.1	-94.7	-24.0	-1.8

[a] From Ref. 111.
[b] From Ref. 112.
[c] From Ref. 113.
[d] From Ref. 114.

constant γ are then related to ω', ωx and λ by the equations

$$\omega_1 = \omega' + \lambda$$
$$\omega_3 = \omega' - \lambda$$
$$x_{11} = x_{33} = -\tfrac{1}{2}\omega x \tag{3.35}$$
$$x_{13} = 2\gamma = -2\omega x$$

These results are obtained by constructing the appropriate hamiltonian matrices in the harmonic basis for each overtone manifold separately and comparing the coupling matrices with those in Eq. (3.26). These new matrices $E^{(\pm)}(v)$ are given below up to $v = 5$ (energies have been measured from the zero point level as before)

$$E^{(\pm)}(1) = \omega' - 2\omega x \pm \lambda$$

$$E^{(+)}(2) = \begin{bmatrix} 2\omega' - 5\omega x + 2\lambda & -\omega x \\ -\omega x & 2\omega' - 5\omega x - 2\lambda \end{bmatrix}$$

$$E^{(-)}(2) = 2\omega' - 6\omega x$$

$$E^{(\pm)}(3) = \begin{bmatrix} 3\omega' - 9\omega x \pm 3\lambda & -\sqrt{3}\,\omega x \\ -\sqrt{3}\,\omega x & 3\omega' - 11\omega x \mp \lambda \end{bmatrix}$$

$$E^{(+)}(4) = \begin{bmatrix} 4\omega' - 14\omega x + 4\lambda & -\sqrt{6}\,\omega x & 0 \\ -\sqrt{6}\,\omega x & 4\omega' - 18\omega x & -\sqrt{6}\,\omega x \\ 0 & -\sqrt{6}\,\omega x & 4\omega' - 14\omega x - 4\lambda \end{bmatrix} \tag{3.36}$$

$$E^{(-)}(4) = \begin{bmatrix} 4\omega' - 17\omega x + 2\lambda & -3\omega x \\ -3\omega x & 4\omega' - 17\omega x - 2\lambda \end{bmatrix}$$

$$E^{(\pm)}(5) = \begin{bmatrix} 5\omega' - 20\omega x \pm 5\lambda & -\sqrt{10}\,\omega x & 0 \\ -\sqrt{10}\,\omega x & 5\omega' - 26\omega x \pm \lambda & -3\sqrt{2}\,\omega x \\ 0 & -3\sqrt{2}\,\omega x & 5\omega' - 24\omega x \mp 3\lambda \end{bmatrix}.$$

It is particularly interesting to see that these two different approaches lead to the same eigenvalues without the inclusion of cubic terms in Eq. (3.33). The harmonically coupled anharmonic oscillator model is probably more illuminating from the physical point of view, but this second model has some great advantages: it is easy to include the bending vibration and the rotational motions in it because the whole machinery of the conventional vibration–rotation theory is readily usable.

IV. ABSORPTION INTENSITIES

Stannard et al.[86] identify three features that contribute to the intensities of overtone and combination bands in a local mode model: (i) electrical anharmonicity, (ii) potential anharmonicity, and (iii) vibrational coupling. The first relates to the form of the dipole function $\mu(\mathbf{R})$ in the transition moment integral

$$\mu_{if} = \langle i|\mu(\mathbf{R})|f \rangle \tag{4.1}$$

The second relates to basis functions; the third, to the coefficients of these functions in the initial and final states.

A common approximation[52,56,61] has been to represent $\mu(R)$ as a sum of bond dipole contributions, so that for two active bonds

$$\mu(R) = \sum_{\nu} \mu_{\nu}(R_{\nu})\mathbf{e}_{\nu}, \qquad \nu = a, b \tag{4.2}$$

where $\mathbf{e}_a$ and $\mathbf{e}_b$ are unit vectors directed along the bonds. One may also assume with confidence that the ground state $|i\rangle$ is well represented by the simple local mode product state $|00\rangle$ (but see Refs. 53 and 85), while

$$|f\rangle = \sum_{n,m} c_{nm}^{f}|n \quad m\rangle \tag{4.3}$$

Thus

$$\mu_{if} = \sum_{n} c_{n0}^{f}\langle 0|\mu_a|n\rangle + \sum_{m} c_{0m}^{f}\langle 0|\mu_b|m\rangle \tag{4.4}$$

because

$$\langle 0|m\rangle = \langle 0|n\rangle = 0 \tag{4.5}$$

Hence in the bond dipole approximation the magnitude of the transition moment from the ground state to a given overtone level depends on the admixture coefficients c_{n0}^{f} and c_{m0}^{f} relevant only to the singly excited bond oscillator functions $|n0\rangle$ and $|0m\rangle$, and these are, according to the previous discussion, just the states that become progressively decoupled from the rest of the manifold as n or m increases. Consequently the dominant intensity in any manifold will become progressively concentrated into the lowest energy component of the manifold, an effect demonstrated by most local mode calculations,[21,56,61,79] and one which is also confirmed by experiment.[1,6,20,87]

Note, however, that it depends both on anharmonic decoupling arising from the dynamics and on the assumption of a bond dipole model.

In assessing the relative intensities of different bands in a given manifold due to admixture with the doorway states $|n0\rangle$ and $|0m\rangle$, it is convenient to distinguish between the effects of intermanifold and intramanifold coupling. In the former the state $|f\rangle$ belongs to a different manifold from the admixed state $|n0\rangle$ and the coefficient in c_{n0}^f is normally much smaller than those arising from coupling within the manifold. Hence it is tempting to ignore the intermanifold terms in Eq. (4.4) because the relative intensities for transitions to different components of the manifold are determined by the relative values of $|c_{n0}^f|^2$ or $|c_{0m}^f|^2$, the transition moments $\langle 0|\mu|n\rangle$ and $\langle 0|\mu|m\rangle$ being the same for all transitions. Calculations of this type are very revealing,[52,79] but they ignore the fact that, while the intermanifold admixture coefficients may be small, the transition moment $\langle 0|\mu|n\rangle$ to a lower manifold may be much larger than that to the manifold in question. This has particular relevance to the $v = 3$ manifold because the bilinear terms $p_a p_b$ and $q_a q_b$ in Eq. (3.4) couple the fundamental level $|01\rangle$, for example, not only to $|10\rangle$ (as discussed in Section III) but also to $|12\rangle$, and the transition moment $\langle 0|\mu|1\rangle$ will normally be much larger than $\langle 0|\mu|3\rangle$. This complicates the calculation of relative intensities in the $v = 3$ manifold. On the other hand, as discussed by Halonen and Child,[56] it offers the possibility of independent assessments of $\langle 0|\mu|1\rangle$ and $\langle 0|\mu|3\rangle$ by intensity measurements in a narrow frequency region, because the coefficients c_{0n}^f etc. may be deduced from a fit to the eigenvalue spectrum, as discussed for individual molecules in the following sections.

We turn now to the calculation of the dipole matrix elements $\langle 0|\mu|n\rangle$, which raises questions about the relative importance of electrical and potential anharmonicity. The effect of the latter alone may be assessed by adopting a linear dipole assumption for a given bond $R = R_\nu$, with equilibrium length R_e

$$\mu(R) = \mu_0 + \mu_1(R - R_e) \tag{4.6}$$

in which case, with $r = R - R_e$, it follows from Eqs. (3.12) and (3.17) that

$$\langle n|\mu|0\rangle = \mu_1\langle n|R - R_e|0\rangle = (-1)^{n+1}\mu_1 a^{-2} N_0 N_n\left[\frac{\Gamma(k-n)}{n(k-n-1)}\right]$$

$$\overset{k \gg 1}{\sim} (-1)^{n+1}\mu_1\left(\frac{\hbar^2 g_{aa}}{4k_{aa}}\right)^{1/4}\left[\frac{k^{1-n}n!}{n^2}\right]^{1/2} \tag{4.7}$$

Hence $\langle n|\mu|0\rangle$ decreases monotonically with increasing n; for $k = 50$,

$|\langle n|\mu|0\rangle|$ is found to fall from the value at $n = 1$ by successive factors of approximately 0.1 at $n = 2, 5, 7$, and 10.

The contribution of electrical anharmonicity depends, of course, on the assumed form of $\mu(R)$. One approach is to add further terms in a Taylor expansion about R_e,[21,83] calculating the matrix elements by the methods given by Sage[51] or discussed by Carney et al.[49] It is also possible to perform a Taylor expansion in powers of the Morse variable

$$y = 1 - \exp[-a(R - R_e)] \tag{4.8}$$

rather than in powers of $(R - R_e)$ itself, because the necessary matrix elements have also been evaluated.[52,117] The most convenient compact form with the correct limiting behavior as $R \to 0$ and $R \to \infty$ is, however,

$$\mu(R) = \mu_0 R \exp(-\gamma a R) \tag{4.9}$$

for which, according to Schek et al.[52] and Halonen and Child,[56]

$$\langle n|\mu(R)|0\rangle = N_n N_0 \mu_0 a^{-2} k^{-\gamma} e^{-\gamma a R_e} \binom{-\gamma}{n} \Gamma(k + \gamma - n - 1) f$$
$$f = a R_e + \ln k - [\Phi(1 - \gamma - n) - \Phi(1 - \gamma) + \Phi(k + \gamma - n - 1)] \tag{4.10}$$

As before, $\Phi(z)$ is the digamma function.[108] This may be compared with the result for a linear dipole function by scaling $\mu(R)$ to have unit slope at $R = R_e$ and then considering the ratio

$$Q(n) = \frac{\langle n|R \exp(-\gamma a R)|0\rangle}{(1 - \gamma a R_e)\exp(-\gamma a R_e)\langle n|R - R_e|0\rangle}$$
$$= (-1)^{n+1}\left(\frac{k^{-\gamma}}{1 - \gamma a R_e}\right)\binom{-\gamma}{n}\frac{\Gamma(k + \gamma - n - 1)}{\Gamma(k - n - 1)} nf \tag{4.11}$$

The general behavior of this ratio for large k values may be assessed by means of the approximations

$$\frac{\Gamma(k + \gamma - n - 1)}{\Gamma(k - n - 1)} \simeq k^{\gamma} \tag{4.12}$$

and

$$\Phi(k + \gamma - n - 1) \simeq \ln k \tag{4.13}$$

both of which are derivable from Stirling's approximation[108] and by means of the identity[108]

$$\Phi(1-\gamma-n)-\Phi(1-\gamma)=\sum_{r=0}^{n-1}(\gamma+r)^{-1} \qquad (4.14)$$

It then follows that

$$Q(n)=\frac{(\gamma+1)(\gamma+2)\cdots(\gamma+n-1)}{(n-1)!}\left[\frac{\displaystyle\sum_{r=0}^{n-1}[\gamma/(\gamma+r)]-\gamma a R_e}{1-\gamma a R_e}\right] \qquad (4.15)$$

which reduces as expected to $Q=1$ when $\gamma=0$.

Figure 10, which contains information similar to that in Fig. 1 of Schek et al.,[52] shows the variation of this electrical anharmonicity ratio with n, for various values of γ when $aR_e=2$, as roughly applicable to naphthalene.[52] Here $Q(n)$ must of course be multiplied by the linear dipole result given by

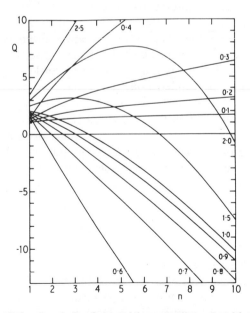

Fig. 10. Variation of the electrical anharmonicity parameter, Q, with final state quantum number, n, at different values of γ, when $aR_e=2$.

(4.7) to describe the full variation of $\langle n|\mu|0\rangle$. The main feature of interest in Fig. 10 is the change in behavior on passing from $\gamma < 0.5$ to $\gamma > 0.5$, the significance of $\gamma = 0.5$ being that this is the value for which R_e coincides with the maximum in the dipole function—hence $\mu'(R_e) = 0$ and $Q(n) = \infty$ for $n \geq 2$; this is why no curve is shown for $\gamma = 0.5$. It appears from the behavior in Fig. 10 that the electrical anharmonicity introduced by Eq. (4.9) makes little contribution to $\langle 0|\mu|n\rangle$ for $\gamma < 0.1$, and that for $0.1 < \gamma < 0.5$, when $R_e < R_{max}$, the electrically anharmonic terms act in the same sense as those arising from potential anharmonicity. The resulting electrical enhancement factor therefore rises with increasing γ for any given n, and also generally increases with n. For $\gamma > 0.5$, on the other hand, when $R_e > R_{max}$, the signs of successive derivatives $\mu^{(n)}(R_e)$ no longer vary uniformly with n, and this may lead to exact cancellation of different contributions to $\langle 0|\mu|n\rangle$, for certain (γ, n) contributions; for example, $\langle 0|\mu|2\rangle \simeq 0$ for $\gamma = 0.7$. Evidence of such cancellations is seen in the absorption intensity calculations for H_2O by Lawton and Child[61] and for CH_4 by Halonen and Child.[56]

This discussion is, of course, open to serious question in relation to the bond dipole approximation and to the specific form chosen to model the function in Eq. (4.9). Some points are clear, however. First, the bond dipole assumption leads to predominant absorption intensity in the most anharmonic (lowest frequence) bands associated with any given manifold. Secondly, the intensity variation from one such band to another in principle provides information on the form of the dipole function, which would be particularly clear-cut if there were a low-intensity anomaly at a particular n value, attributable in the language of the present discussion to cancellation within $Q(n)$. A third point is that it would be misleading to discount the contribution of potential anharmonicity to the overtone intensity because Fig. 10 shows that $|Q(n)|^2 \leq 10$ for $n = 1$–5 and $0 < \gamma < 0.2$ and $0.9 < \gamma < 1.5$, whereas the contribution for zero electrical anharmonicity $\gamma = 0$ would be $Q(n) = 1$ for all n. Finally, it is fruitful in any discussion to distinguish the effects on the intensity of vibrational coupling within a given overtone manifold from coupling between manifolds. The former gives rise to intensity borrowing from $\langle 0|\mu|n\rangle$; the latter, from $\langle 0|\mu|n \pm 2\rangle$.

V. XY$_2$ SYSTEMS

The notation XY$_2$ adopted here is intended to indicate two coupled oscillators, with Y normally an H or D atom but with X either an atom or a molecular fragment. The general structure of the theory for molecules of this type has been outlined together with a local mode to normal mode correlation diagram in section 3. The specific calculations reported below refer to H_2O,[54,55,61,62,86] C_2H_2, C_2D_2,[80,81] CH_2Z_2 (Z = Cl, Br, or I),[106] H_2S,[118] SO_2.[54,55]

A. Water

Water is the obvious prototype molecule for this study both because of its small size and because there are extensive experimental data[64-76] for comparison with theoretical results.

Many calculations have been reported in the stretching only approximation,[42,54,55,61,119] using either Morse-based potentials with[55] or without[42,54,119] potential coupling or more complicated empirical potentials.[61] Attempts have also been made to include the effects of the bending vibration.[43,62,86,118,119] The aims have been to understand and to predict the eigenvalue structure, to analyze the wave functions in relation to their local mode or normal coordinate character, and to calculate absorption intensities.

Sample results for selected eigenvalues are compared with the experimental term values in Table IV. The first column gives the conventional normal coordinate quantum numbers (v_1, v_3), and the second the more revealing local mode labels ($nm^\pm$) suggested by the form of wave function in Eq. (3.21). As predicted in Sections II and III, the experimental local mode doublet splittings $\Delta G(nm)$ decrease dramatically as the larger quantum number m increases at fixed n. These results refer to the lowest bending state, $v_2 = 0$, but the same pattern is repeated with remarkably little change for $v_2 = 1$ and $v_2 = 2$.[62,86]

The three calculations have been chosen to illustrate the influence of three different potential surfaces: LC[62] and SEG[86] are three-dimensional calculations on the Sorbie–Murrell[77] and a hybrid Sorbie–Murrell–quartic surface, respectively, while WHR[54] employs a bond diagonal Morse oscillator potential in the stretching-only approximations. It is noticeable that the absence of potential coupling in WHR leads to a fundamental splitting, $\Delta G(01) = 53.5$ cm^{-1}, which is substantially smaller than the experimental (98.8 cm^{-1}), LC (102.2 cm^{-1}) and SEG (101 cm^{-1}) values. The general success of the two-parameter WHR potential is, however, remarkable. Overall the LC and WHR results are roughly comparable in accuracy with rms deviations from the experimental values of 28 and 30 cm^{-1}, respectively, but the SEG rms deviation of 70 cm^{-1} is significantly larger. This indicates that the inclusion of quartic coupling terms in the SEG potential gives a relatively poor picture of the forces in the molecule.

Analyses of the nature of the vibrational eigenfunctions also naturally vary with the form of the potential. Møller and Mortensen[119] show both for a Morse-type potential and for a quartic potential that the symmetry adapted

TABLE IV
Experimental and Calculated Term Values for H_2O in the
$v_2 = 0$ Bending State

Normal v_1	v_3	Local nm^+	Expt[a] G (cm^{-1})	Expt[a] ΔG (cm^{-1})	LC[a] G (cm^{-1})	SEG[b] G (cm^{-1})	WHR[c] G (cm^{-1})
1	0	01^+	3657.1	98.8	3662.7	3674	3673.5
0	1	01^-	3755.9		3764.9	3775	3727
2	0	02^+	7201.5	48.3	7206.0	7214	7223
1	1	02^-	7249.8		7257.0	7288	7240
0	2	11	7445.0		7462.5	7485	7416
3	0	03^+	10599.7	13.7	10593.6	10618	10608.5
2	1	03^-	10613.4		10608.6	10669	10611.3
1	2	12^+	10868.9	163.5	10877.9	10902	10896
0	3	12^-	11032.4		11054.6	11112	10994
2	2	04^+	13828.3	2.6	13799.7	13888	13826.1
3	1	04^-	13830.9		13802.3	13921	13826.5
4	0	13^+	14221.1	97.7	14213.7	14206	14282
1	3	13^-	14318.8		14322.8	14436	14324
0	4	22	14536.9		14560.2	14679	14513
3	2	05^+	16898.4	0.4	16830.0	16979	16880.6
4	1	05^-	16898.8		16830.1	16987	16880.7
5	0	14^+	17458.2	37.3	17429.3	17424	17517
2	3	14^-	17495.5		17473.1	17565	17526
1	4	23^+	17748.1		17746.7	17855	17803
0	5	23^-	—		17970.9	18183	17934
rms deviation (cm^{-1})					27.5	70	30

[a] From Ref. 62.
[b] From Ref. 86.
[c] From Ref. 54.

local mode representation, as in Eq. (3.21), gives a more compact description of the wave functions than does the conventional normal coordinate representation. Similarly Lawton and Child[61] were only able to obtain fully converged results in the local mode representation. This does not mean, however, that the wave function is always represented by a single local mode structure.

Such strict localization always occurs for $v_1 + v_3 = 1$, but for the trivial reason that both local mode and normal coordinate wave functions are close approximations to the $v = 1$ state of a degenerate harmonic oscillator. Deviations in the local mode case depend on the ratio $(\omega x / \omega)$, and those in the normal coordinate representation on $|(\omega_1 - \omega_3)/\bar{\omega}|$; both these ratios are small.

At higher levels there is a general increase of intermanifold mixing as the number of states per manifold increases, followed by the development of increasing localization in the lowest components of the manifold as the local mode doubling becomes small compared with the fundamental splitting. These trends are more strongly marked for the Sorbie–Murrell potential[62] than for the more strongly coupled SEG potential.[86]

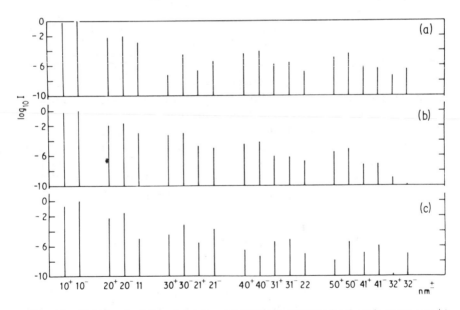

Fig. 11.   Calculated relative absorption intensities to different local mode states $|nm^\pm\rangle$ obtained by (a) Watson et al.[54] (WHR); (b) Lawton and Child[61] (LC1); and (c) Standard et al.[86] (SEG).

Turning to the question of absorption intensities, Fig. 11 gives results for the SEG and WHR calculations reported in Table IV and for a stretching-only calculation, LC1,[61] on the Sorbie–Murrell potential. Detailed comparison is difficult because different dipole functions were employed as well as different potentials: WHR assumes linear bond dipoles; LC1 assumes an $R \exp(-\gamma aR)$ bond dipole combination; while SEG adopts an experimentally based multiterm expansion including both single-bond and crossed-bond terms. There is however a striking similarity between the WHR and LC1 calculations, particularly if the $v = 3$ manifold is omitted and the discussion in Section IV shows that interference between vibrational interaction, electrical and potential anharmonicity may be specially severe in this case. The general trend is that the intensity of the lowest frequency band in each manifold is typically two orders of magnitude larger than any other band. The main effect of the electrical anharmonicity introduced in LC1 appears to be to decrease the rate of falloff in intensity with frequency, but this does not have a major influence until $v = 5$. The value of the comparison between Fig. 11c and the other two sets of results is that the relative intensities within any manifold are sufficiently different that a comparison with experiment might give clear evidence for or against the bond dipole approximation.

B.   Acetylene

Acetylene was taken by Hayward and Henry[2] as one of the molecules in which the most prominent high-frequency bands in the CH stretching progression could be fitted to the Morse eigenvalue relation. Subsequent calculations[55,80] identify these bands with transitions to the lowest frequency components of successive manifolds that show the typical decrease in local mode splitting indicative of progressively complete decoupling from the rest of the manifold. Another point of interest is that the qualitative behavior of C_2D_2 differs considerably from that of C_2H_2 in that both calculated[55,80] and experimental[64,120–132] spectra are more nearly normal in character. This is attributed to the dominant influence of second-order inter-CH or -CD coupling via the C—C bond, the difference between C_2H_2 and C_2D_2 being that the frequency of the C—C vibrations is more nearly resonant with that of CD than of CH.

The most extensive calculations have been performed by Halonen et al.[80] using a Morse potential model for the CH or CD oscillators and a series expansion for the C—C bond potential. Interbond coupling was introduced via the kinetic energy operator and a bilinear CH/CC potential coupling term. Bending motions were ignored except to the extent that slightly different parameter values were allowed for the CH and CD Morse potential functions. The results obtained by a least squares fit to the experimental data[64,120-132] gave the ratios $(\omega_{CH}/\omega_{CD})^2 = 1.850$ and $\omega x_{CH}/\omega x_{CD} = 1.831$, which may be compared with the reduced mass ratio $\mu_{CD}/\mu_{CH} = 1.857$.[80] These small discrepancies are plausibly attributed to weak mixing with the bending modes.

Table V gives a comparison between calculated and experimental term values for C_2H_2 and C_2D_2.[80] The general level of agreement is seen to be excellent: in all 21 levels for C_2H_2 and 20 levels for C_2D_2 were shown[80] to be well reproduced in each case by means of a six-parameter potential, which is transferable to the mixed isotopic species CHCD.[81]

Caution is suggested in assessing the reliability of the potential function, however, because Lehmann et al.[133] have recently shown that the predicted vibrational state dependence of the rotational constant is systematically underestimated by 5–10%, which is outside the limits of experimental uncertainty. Work on rotational analysis from a local mode viewpoint has also been reported by Henry et al.[15]

TABLE V

Experimental and Calculated[80] Stretching Eigenvalues for $|v_1 0 v_3 0^0 0^0\rangle$ States of C_2H_2 and C_2D_2

Normal		Local	C_2H_2		C_2D_2	
v_1	v_3	$nm^\pm$	ν_{obs} (cm^{-1})	ν_{calc} (cm^{-1})	ν_{obs} (cm^{-1})	ν_{calc} (cm^{-1})
1	0	01$^+$	3372.9	3371.5	2705.2	2704.1
0	1	01$^-$	3294.8	3290.0	2439.2	2439.9
0	2	02$^+$	6502.3	6510.5	—	4850.0
1	1	02$^-$	6556.5	6556.6	5097.2	5097.0
2	0	11	6709.0	6707.1	—	5385.1
1	2	03$^+$	—	9660.9	—	7460.0
0	3	03$^-$	9639.9	9643.8	7230.2	7230.3
3	0	12$^+$	—	9989.0	—	8043.1
2	1	12$^-$	9835.2	9835.8	7733.9	7733.6
0	4	04$^+$	—	12670.7	—	9581.0
1	3	04$^-$	12675.7	12674.6	9794.1	9793.5
2	2	13$^+$	—	12921.4		10052.9
3	1	13$^-$	13033.3	13030.5	10347.9	10349.0
4	0	22	—	13233.8		10678.1
1	4	05$^+$	—	15591.5		12097.9
0	5	05$^-$	15600.2	15590.9	11905.3	11902.2
3	2	14$^+$	—	16007.1		12626.9
2	3	14$^-$	15948.5	15948.7	12344.5	12343.3
5	0	23$^+$	—	16408.2		13291.0
4	1	23$^-$	—	16199.6	—	12943.2

C.   Other Molecules

Local mode calculations for molecules of XY_2 type have also been reported for dihalomethanes CH_2Cl_2, CH_2Br_2 and CH_2I_2,[41,106,134] H_2S and D_2S,[118] and SO_2.[54,55] Comparisons between theory and experiment are given in Table VI for CH_2Cl_2[106] and SO_2.[54] The former are included to emphasize the role of the dihalomethane spectra in the development of local mode concepts[35] and also to demonstrate the success of the intramanifold coupling model, based on Eqs. (3.26) for these molecules. The interest in the case of SO_2 is that, although the molecule lies toward the extreme normal mode limit in Fig. 9, the calculations can be converged in the local mode representation. Secondly, the experimental stretching eigenvalues are very well reproduced by a two-parameter-bond Morse oscillator potential. This shows, in contrast to the case of H_2O, that the overwhelming coupling mechanism is momentum transfer through the S atom.

TABLE VI

Experimental Calculated Stretching Eigenvalues for CH_2Cl_2[106] and SO_2[54]

Local $nm^{\pm}$	CH_2Cl_2		SO_2	
	ν_{obs} (cm^{-1})	ν_{calc} (cm^{-1})	ν_{obs} (cm^{-1})	ν_{calc} (cm^{-1})
01^+	2986	2986	1151	1152
01^-	3048	3048	1362	1359
02^+	5903	5882	2296	2298
02^-	5922	5907	2500	2498
11	6069	6058	2715	2708
03^+	8656	8656	3431	3439
03^-	8656	8661	3630	3631
12^+	8903	8875	3838	3835
12^-	8998	8994	4054	4049
04^+	11301	11296	4560	4573
04^-	11301	11297	4751	4757
13^+	11634	11635	—	4954
13^-	11700	11697	5166	5162
22	—	11876	—	5380
05^+	13831	13808	—	5701
05^-	13831	13808	5872	5877
14^+	—	—	—	6068
14^-	—	—	—	6269
23^+	—	—	6489	6481
23^-	—	—	6689	6702

VI. AH₃ SYSTEMS

The model introduced for two coupled equivalent oscillators in Eqs. (3.1)–(3.5) is readily generalized to larger systems. Two types of generalization to three oscillators have been reported.

The first is to adjust the potential parameters by means of a least squares fit to experimental data and hence to predict the absorption frequencies and intensities of unobserved bands. Such calculations have been reported for NH_3,[54] CH_3Cl and CD_3Cl,[135] and CH_3D, CHD_3, SiH_3D, and $SiHD_3$.[78] Comparisons between calculated and experimental XH_3 or XD_3 stretching term values for NH_3 and CH_3D are given in Table VII.[54,78] Estimated ab-

TABLE VII
Calculated and Experimental Term Values for NH_3 and CH_3D. Most of the Assignments Are Tentative

v	Γ		NH_3[a]		CH_3D[b]	
			ν_{obs} (cm^{-1})	ν_{calc} (cm^{-1})	ν_{obs} (cm^{-1})	ν_{calc} (cm^{-1})
1	A_1	ν_1	3337	3338.5	ν_1	2970	2964.9
	E	ν_3	3443	3441.5	ν_4	3016.6	3003.6
2	A_1	$2\nu_1$	—	6600	$2\nu_1$	—	5853.1
	E	$\nu_1+\nu_3$	6608	6654	$\nu_1+\nu_4$	5856.1	5864.7
	A_1	$2\nu_3$	6850	6806	$2\nu_4$	5980.4	5969.8
	E		6850	6860		6022.0	6000.9
3	A_1	$3\nu_1$	9760	9737	$3\nu_1$	—	8620.4
	E	$2\nu_1+\nu_3$	9760	9750	$2\nu_1+\nu_4$	—	8622.0
	E	$\nu_1+2\nu_3$	10105	10133	$\nu_1+2\nu_4$	—	8824.2
	A_1	—	—	—	$3\nu_4$	9020.9	8995.5
4	A_1	$4\nu_1$	12609	12705	$4\nu_1$	11280	11265.4
	E	$3\nu_1+\nu_3$	12609	12707	$3\nu_1+\nu_4$	—	11265.5
	A_1	$2\nu_1+2\nu_3$	—	13062	$2\nu_1+2\nu_4$	11600	11602.2
5	A_1	$5\nu_1$	15440	15514.0	$5\nu_1$	13753	13793.6
	E	$4\nu_1+\nu_3$	15440	15514.2	$4\nu_1+\nu_4$	—	13793.6
6	A_1	$6\nu_1$	18150	18172.25	$6\nu_1$	—	16207.2
	E	$5\nu_1+\nu_3$	18150	18172.27	$5\nu_1+\nu_4$	—	16207.2
7	A_1	$7\nu_1$	20645	20677.65	$7\nu_1$	—	18514.1
	E	$6\nu_1+\nu_3$	20645	20677.65	$6\nu_1+\nu_4$	—	18514.1

[a] From Ref. 54.
[b] From Ref. 78.

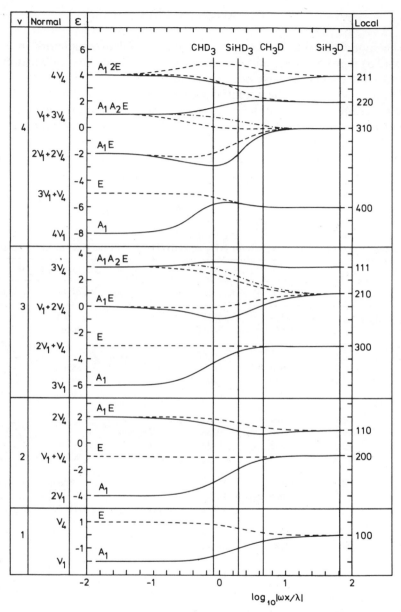

Fig. 12. The normal mode to local mode correlation diagram for three equivalent anharmonic oscillators. (Taken from Halonen and Child,[78] with permission.) The label ν_1 should be replaced by ν_2 in the case of CHD_3 and $SiHD_3$.

sorption intensities for NH_3 were also calculated[54] in the linear bond dipole approximation. An early local mode fit[39] to the spectrum of NH_3 may also be cited. The calculations of Halonen and Child[78] include term values for the XD vibrations in XH_3D and for the XH vibrations in XHD_3 in addition to those given in Table VII. These were particularly valuable in the case of CHD_3 because the observed progression in the CH vibration ν_1 was used together with the available data on the CD_3 vibrations ν_2 and ν_4 to determine the potential function for both CH_3D and CHD_3. Strong anharmonic resonances in CH_3D were found to complicate the direct optimization procedure for this molecule.[136]

The second type of calculation is an extension of the intramanifold coupling approximation represented by Eqs. (3.26). The extended matrices required for the XY_3 systems are given in the Appendix. Figure 12 shows the resulting correlation diagram, with the reduced eigenvalues for each manifold with mean energy $\bar{E}$,

$$\varepsilon = \frac{E - \bar{E}}{\left[\lambda^2 + (\omega x)^2\right]^{1/2}} \tag{6.1}$$

plotted as a function of $\log|\omega x/\lambda|$. Here $\lambda < 0$ for all the molecules illustrated; NH_3 would lie at approximately $\log(-\omega x/\lambda) = 0.32$, corresponding to $\omega' \simeq 3550$ cm^{-1}, $\omega x \simeq 75$ cm^{-1}, and $\lambda = -36$ cm^{-1}.

VII. AH_4 SYSTEMS

Calculations of the types reported for XY_3 systems have also been extended to the tetrahedral hydrides and their deuterium substituents.[56,135,137,138]

The local mode to normal mode correlation diagram, for which the necessary coupling matrices are given in the Appendix, is shown in Fig. 13. Here it is of particular interest that SiH_4 lies very close to the local mode limit (as does GeH_4[56]), while the behavior of SiD_4 (and GeD_4) is more intermediate in character. A similar change is also apparent in passing from SiH_3D to $SiHD_3$ in Fig. 12. The origin of this difference may be traced[56] to a near cancellation between potential and kinetic contributions to the interbond coupling parameter λ in SiH_4, as expressed by Eq. (3.23) when $\alpha \simeq -\beta$ and $\alpha_\mu \simeq \beta_\mu \simeq 0$; the cancellation is much less complete for SiD_4, SiD_3H, and GeD_4 due to the enhancement of $|\alpha/\beta|$ by a factor of μ_{XD}/μ_{XH}.

The molecule of greatest interest is CH_4, for which the first recent detailed calculation was performed by Bron and Wallace.[135] Both bending and stretching modes were included in the model, but results were reported for

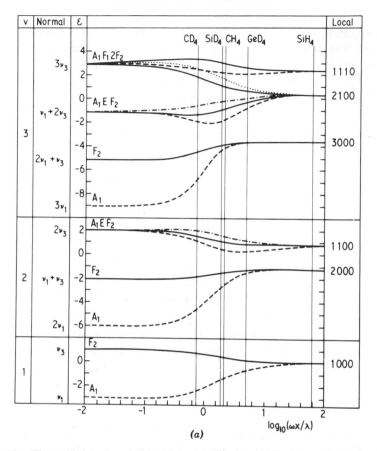

Fig. 13.   The normal mode to local mode correlation diagram for four equivalently coupled anharmonic oscillators (T_d): (a) $v = 1-3$ and (b) $v = 4$. (Taken from Halonen and Child,[56] with permission.)

CH$_4$ and CD$_4$ only up to the first overtone and binary combination levels. Halonen and Child[56] performed stretching-only calculations with reported converged eigenvalues up to the fifth excitation manifold, and more recently Halonen[138] has extended these calculations including all stretching states up to the twelfth manifold by using symmetrized basis functions. The potential functions used by Halonen were the same as in the earlier paper by Halonen and Child.[56] There is evidence, however, of considerable interaction with the bending modes because the best fit interaction force constant $f_{rr'}$ for CH$_4$ (2140 cm^{-1}Å^{-2}) differs considerably from that for CD$_4$ (4589 cm^{-1}Å^{-2}).[56]

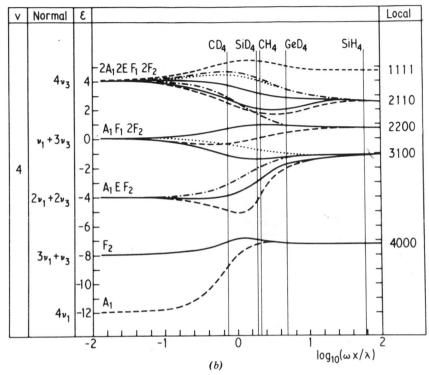

Fig. 13. (*Continued*).

This interaction is taken into account by Abram et al.[137] by an extension of Eqs. (3.22)–(3.23) to include intermanifold as well as intramanifold coupling. Results are reported for CH_4 but not CD_4, so the validity of this perturbation approximation cannot be fully assessed.

Table VIII compares the results obtained by Halonen[138] and Abram et al.[137] with the experimental data[56,87,139–142] on CH_4 and SiH_4.

It is also of interest that Halonen and Child[56] and Abram et al.[137] give graphic predictions of the relative intensities of different bands: Abram et al. adopted an unspecified bond dipole approximation and neglect intermanifold coupling contributions to the intensity (see Section IV). This enables them to deduce relative intensities within any manifold from the coefficient of the $|v000; F_2\rangle$ component of the eigenstate. Halonen and Child show, however, by use of the $R\exp(-a\gamma R)$ bond dipole form, that the relative intensities, particularly within the $v = 3$ manifold, are sensitive to this intermanifold coupling effect.

TABLE VIII
Observed and Calculated Stretching Vibrational Term Values for CH_4 and SiH_4[a]

v	Γ	Normal	CH$_4$			SiH$_4$		Local
			ν_{obs} (cm^{-1})[b]	ν_{calc} (cm^{-1})[c]	ν_{calc} (cm^{-1})[d]	ν_{obs} (cm^{-1})[e]	ν_{calc} (cm^{-1})[c]	
1	A_1	ν_1	2916.47	2916.4	2918	2186.87	2186.8⎫	
	F_2	ν_3	3019.49	3021.0	3019	2189.19	2189.0⎭	1000
2	A_1	$2\nu_1$		5791.6	5793	4308.38	4308.2⎫	
	F_2	$\nu_1+\nu_3$	5861	5856.4	5851	4309.36	4309.3⎭	2000
	F_2	$2\nu_3$	6004.65	6010.2	6006	4378.38	4378.5	1100
3	F_2	$2\nu_1+\nu_3$	8604[f]	8612.5	8601		6361.9	3000
	F_2	$\nu_1+2\nu_3$	8807	8807.8	8802		6499.6	2100
	F_2	$3\nu_3$	8900	8909.5	8904		6501.5⎫	
	F_2	$3\nu_3$	9045.92	9041.7	9034		6571.0⎭	1110
4	F_2	$3\nu_1+\nu_3$	11270	11262.5	11240	8349.3	8347.0	4000
	F_2	$2\nu_1+2\nu_3$	(11620)[g]	11553.3	11539		8554.3	3100
	F_2	$4\nu_3$[h]	(11885)	11907.2	11883		8695.8	2110
5	F_2	$4\nu_1+\nu_3$	13790	13796.4	13760	10269	10264.3	5000
	F_2	$3\nu_1+2\nu_3$	14220	14206.9	14211		10541.3	4100
	F_2	$5\nu_3$[i]	(14640)	14641.1	14618		10751.0	3110
6	F_2	$5\nu_1+\nu_3$	16160	16213.3			12113.7	6000
	F_2	$4\nu_1+2\nu_3$	16740	16745.1			12460.5	5100
7	F_2	$6\nu_1+\nu_3$	18420	18513.5			13895.4	7000
	F_2	$5\nu_1+2\nu_3$	19120	19166.0			14312.0	6100
8	F_2	$7\nu_1+\nu_3$	20600	20697.4			15609.3	8000
9	F_2	$8\nu_1+\nu_3$	22660	22765.3			17255.4	9000

[a]Assignments in parentheses are only tentative. Qualitative intensity calculations have been used as one aid in the assignment.
[b]From Refs. 56 and 87.
[c]From Ref. 138.
[d]From Ref. 137.
[e]From Refs. 56 and 142.
[f]Assigned as $2\nu_3+2\nu_4$ in Ref. 64.
[g]Assigned as $\nu_1+3\nu_3$ in Ref. 56.
[h]Higher frequency F_2 in the $4\nu_3$ manifold.
[i]Lowest frequency F_2 assignment in the $5\nu_3$ manifold.

VIII. XY_6 SYSTEMS

Local mode behavior has been adduced to explain the structure of the $n\nu_3$ manifolds of SF_6,[143,144] but in a different sense from that employed in this article. Here the motion is treated as local if the bond anharmonicity is sufficiently strong to decouple the vibrations of one bond from another. Thus the local motions are individual bond vibrations. This situation certainly does not apply to SF_6, for which the spread of the stretching fundamentals is roughly 300 cm^{-1},[143,145,146] while the bond anharmonicity is estimated as only 3 cm^{-1}.[82] It has been argued, however, that within an overall normal coordinate picture there is a Cartesian localization of the ν_3 motion.[143,144] This means that the anharmonicity of the ν_3 vibration results in zeroth-order splittings between the component $|v_{3x}v_{3y}v_{3z}\rangle$ levels for given $v_3 = v_{3x} + v_{3y} + v_{3z}$. For example, the $|200\rangle$ level lies below the $|110\rangle$ level but would be coupled to it by the ξ_3 Coriolis operator. This Cartesian localization is therefore revealing for interpretation of the rotational structure.

There are molecules other than SF_6 for which the mass ratio m_x/m_y is larger, but even so a normal coordinate description of the motion will usually be most appropriate. Nevertheless, it is interesting to see to what extent a bilinearly coupled Morse oscillator model for the stretching vibrations is adequate and what predictions can be made with this model. Such calculations have been reported for SF_6,[82] $^{184}WF_6$,[82] $^{235}UF_6$,[82,147] and $^{238}UF_6$.[82,147] Results are compared with the available spectroscopic information in Table IX, and the general level of agreement is seen to be excellent. The two calculations for $^{238}UF_6$ differ in that the four potential parameters were determined by a full least squares fit in one case[82] and by a fit to the lowest frequency levels in the other.[147] It is therefore not surprising that better agreement is obtained in the first case.

It is noticeable that the relative values of the fundamental frequencies, which affect the nature of the whole spectrum, vary from one molecule to another. These changes can be understood within the intramanifold coupling approximation in terms of the relative strengths of the bond anharmonicity and the adjacent and opposite interbond coupling terms. Table X gives the values of the fitted bond anharmonicity ωx and parameters f_1 and f_2 for bilinear potential coupling between adjacent and opposite bonds, respectively.[82] There is also a term in the kinetic energy operator that couples opposite bonds and has a sign opposite to that of the potential coupling term. The resulting overall coupling parameters λ_1 and λ_2, analogous to λ in Eq. (3.23), are also given in Table X; it is seen that λ_1 is roughly constant from SF_6 to UF_6, whereas λ_2 varies markedly from molecule to molecule due to the change in mass ratio m_F/m_x.

TABLE IX
Observed and Calculated Term Values for SF$_6$, WF$_6$, and UF$_6$[a]

v		Γ	^{32}SF$_6$		^{184}WF$_6$		^{238}UF$_6$		
			ν_{obs} (cm^{-1})[b]	ν_{calc} (cm^{-1})[b]	ν_{obs} (cm^{-1})[b]	ν_{calc} (cm^{-1})[b]	ν_{obs} (cm^{-1})[b]	ν_{calc} (cm^{-1})[b]	ν_{calc} (cm^{-1})[c]
1	ν_1	A_{1g}	774.54	770.9	772.1	771.8	667.1	665.9	658
	ν_2	E_g	643.35	643.6	678.2	677.4	534.1	533.5	534
	ν_3	F_{1u}	948.10	947.2	712.4	712.3	625.5	625.5	626
2	$2\nu_1$	A_{1g}	—	1542.0	—	1543.4	1331.9	1331.6	1344
	$2\nu_2$	A_{1g}	—	1286.4	1354	1353.9	1066.3	1066.0	1072
		E_g	—	1287.1	1354	1354.5	1066.3	1066.7	1073
	$2\nu_3$	E_g	1891.60	1889.6	1422.4	1422.4	—	1248.9	1255
		A_{1g}	1889.05	1889.6	1422.4	1422.5	—	1248.9	1255
		F_{2g}	1896.53	1895.6	1422.4	1425.0	—	1251.6	1259
	$\nu_1+\nu_2$	E_g	—	1413.7	—	1448.5	1197	1198.2	1204
	$\nu_1+\nu_3$	F_{1u}	1719.59	1717.2	1482.8	1483.0	1290.9	1290.7	1297
	$\nu_2+\nu_3$	F_{1u}	1588.1	1587.6	1387.1	1387.0	1156.9	1156.7	1161
3	$3\nu_3$	F_{1u}	2827.55	2827.6	—	2130.0	(1874.6)	1870.1	1875
		F_{2u}	2840.35	2839.3	—	2135.3	—	1875.6	1882
		F_{1u}	2839.04	2839.5	—	2135.5	1874.6	1875.6	1882
		A_{2u}	2845.28	2845.3	—	2138.2	—	1878.3	1885
	$2\nu_1+\nu_3$	F_{1u}	2489.47	2488.4	—	2253.6	1955	1955.6	1966
	$2\nu_2+\nu_3$	F_{1u}	2227.5	2227.6	—	2060.8	1687.5	1687.1	1689
		F_{1u}	(2227.5)	2235.7	—	2067.3	(1687.5)	1693.1	1699
	$\nu_1+\nu_2+\nu_3$	F_{1u}	—	2357.1	—	2157.2	1821	1820.7	1824

[a] Values in parentheses are alternative assignments.
[b] From Ref. 82.
[c] From Ref. 147.

TABLE X
Potential Parameters and Effective Coupling Strengths

	SF_6	$^{184}WF_6$	$^{238}UF_6$
ω (cm^{-1})	843.9	717.6	606.3
ωx (cm^{-1})	2.96	2.10	1.74
$10^{-4}f_1$ (cm^{-1}Å^{-2})	1.695	1.289	1.479
$10^{-4}f_2$ (cm^{-1}Å^{-2})	1.790	2.370	0.0
λ_1 (cm^{-1})	21.2	15.7	22.2
λ_2 (cm^{-1})	-130.6	-1.5	-23.5

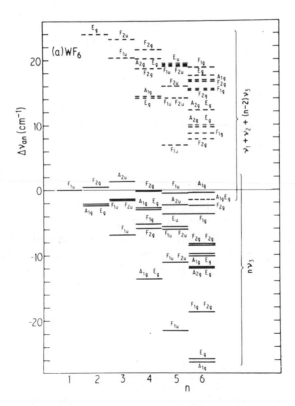

Fig. 14. Anharmonic energies, $\nu - n\nu_3$, for the $n = 1$–6 ν_3 manifolds of (a) WF$_6$ and (b) UF$_6$. (Taken from Halonen and Child,[82] with permission.)

47

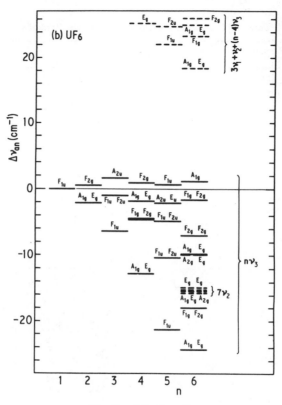

Fig. 14. (*Continued*).

The presence of two coupling parameters, λ_1 and λ_2, for adjacent and axial coupling, respectively, means that the correlation diagram analogous to Figs. 12 and 13 is three dimensional. The necessary coupling matrices for $v = 1$–3 are given in the Appendix (at the end of this chapter), and a section for $\lambda_1 = -\lambda_2$ roughly applicable to UF_6 has been given by Halonen and Child.[82]

A further outcome of the calculations on WF_6 and UF_6[82] was the predicted pattern on $n\nu_3$ overtone levels up to $n = 6$, as summarized in Fig. 14. These are believed to be converged to within 1 cm^{-1}, but similar convergence could be obtained for SF_6 only up to $n = 4$ due to the very strong kinetic energy coupling in this molecule.[82] The point of particular interest for the theory of multiphoton dissociation[148–157] is that the $n\nu_3$ ladder alone appears to contain levels in resonance to within 1 cm^{-1} of n quanta of the fundamental. This is in marked contrast to the situation for SF_6 where a transition from the $n\nu_3$ to the $(n-1)\nu_3 + \nu_2 + \nu_6$ ladder is required after $n = 4$–6 in order to maintain a roughly equal level spacing.[153,156]

TABLE XI

Observed[a] and Calculated Vibrational Term Values and Absorption Intensities for Liquid C_6H_6 and C_6D_6[b]

Local							C_6H_6					C_6D_6			
									BA[c]	H[d]				H[d]	
n_1	n_2	n_3	n_4	n_5	n_6	Γ	ν_{obs} (cm^{-1})	I_{obs}	ν_{calc} (cm^{-1})	ν_{calc} (cm^{-1})	I_{calc}	ν_{obs} (cm^{-1})	I_{obs}	ν_{calc} (cm^{-1})	I_{calc}
1	0	0	0	0	0	B_{1u}	(3048)[e]			3029.1		(2275)[e]		2256.0	
						E_{2g}	3048			3032.6		2266		2265.5	
						E_{1u}	3036[f]	[1]	3059	3039.8	[1]	2280[g]	[1]	2284.4	[1]
						A_{1g}	3062			3043.3		2293		2293.9	
2	0	0	0	0	0	E_{1u}	5957	2.1×10^{-2}	5963	5955.6	3.3×10^{-2}	4484	5.2×10^{-2}	4477.2	2.2×10^{-2}
3	0	0	0	0	0	E_{1u}	8763	1.7×10^{-3}	8768	8760.1	1.7×10^{-3}	6623	1.3×10^{-3}	6627.3	9.3×10^{-4}
4	0	0	0	0	0	E_{1u}	11442	1.3×10^{-4}	11456	11448.4	1.2×10^{-4}	8710	7.0×10^{-5}	8709.8	5.0×10^{-5}
5	0	0	0	0	0	E_{1u}	14024	1.3×10^{-5}	14015	14021.1	1.1×10^{-5}	10730	3.3×10^{-6}	10729.5	3.4×10^{-6}
6	0	0	0	0	0	E_{1u}	16470	—	16477	—	—	—	—	—	—

[a] From Refs. 1 and 161.
[b] The local mode combinations bands are not given because these assignments are only tentative (see Refs. 21 and 161).
[c] From Ref. 21.
[d] From Ref. 79.
[e] Uncertain because not a direct observation (Refs. 160 and 161).
[f] Strong Fermi resonance. The unperturbed value is 3053 cm^{-1} (from Ref. 161).
[g] Fermi resonance. The unperturbed value is 2276 cm^{-1} (from Ref. 161).

IX. BENZENE

The CH stretching spectrum of benzene is of great current interest because (i) the strongest overtone bands were among the first to be fitted to a Morse progression,[1,17,37,38] (ii) these bands suffer only a proportional change in intensity when the molecule is partially deuterated,[19] and (iii) the band width appears to decrease on excitation to higher overtone manifolds.[25] The first two of these observations are taken as indicative that the CH oscillators are essentially uncoupled. The third has given rise to much discussion of the possible kinetic consequences of local mode behavior,[25,28,88,95,97] but this is outside the scope of the present review.

Detailed local mode style calculations of the overtone term values and electromagnetic absorption intensities have been given by several authors.[21,79,83] Table XI gives results from two papers[21,79] that include interbond coupling and hence include a mechanism for borrowing intensity within a given overtone manifold. Burberry and Albrecht[21] include a potential coupling term from the normal coordinate force field of Whiffen,[158] although comparison with other force field calculations[159,160] shows this term to be poorly defined. Halonen,[79] on the other hand, includes only the mass determined second-order kinetic coupling via the C—C stretching modes. The most direct test of each model is to calculate the spread of the fundamental term values of different symmetries. This information is given only by Halonen,[79] who correctly predicts the spread both for C_6H_6 (if the unperturbed E_{1u} fundamental frequency is used) and C_6D_6 but finds that the calculated fundamental term values for C_6H_6 are systematically low, due to Fermi resonance with bending vibrations.[161] Both calculations obtain good agreement with the observed overtone term values and absorption intensities. Results are given in Table XI for liquid C_6H_6 and C_6D_6.

APPENDIX:
SYMMETRIZED EFFECTIVE HAMILTONIAN MATRICES

Symmetrized effective Hamiltonian matrices for AX_3, AX_4, and AX_6 systems. The symmetrized bra labels $\langle abc\ldots;\ \Gamma|$ are indicated by rows. These labels also apply to the corresponding column.

$AX_3(C_{3v})$:

$v=1$　　$\langle 100;\ A_1|\ \omega'-2\omega x+2\lambda$

　　　　$\langle 100;\ E|\ \omega'-2\omega x-\lambda$

$v=2$　$\begin{matrix}\langle 200;\ A_1| \\ \langle 110;\ A_1|\end{matrix}\begin{bmatrix} 2\omega'-6\omega x & 2\sqrt{2}\lambda \\ 2\sqrt{2}\lambda & 2\omega'-4\omega x+2\lambda \end{bmatrix}$

$$
\begin{array}{l}
\langle 200;\,E| \\
\langle 110;\,E|
\end{array}
\begin{bmatrix}
2\omega' - 6\omega x & -\sqrt{2}\,\lambda \\
-\sqrt{2}\,\lambda & 2\omega' - 4\omega x - \lambda
\end{bmatrix}
$$

$v = 3$
$$
\begin{array}{l}
\langle 300;\,A_1| \\
\langle 210;\,A_1| \\
\langle 111;\,A_1|
\end{array}
\begin{bmatrix}
3\omega' - 12\omega x & \sqrt{6}\,\lambda & 0 \\
\sqrt{6}\,\lambda & 3\omega' - 8\omega x + 3\lambda & 2\sqrt{3}\,\lambda \\
0 & 2\sqrt{3}\,\lambda & 3\omega' - 6\omega x
\end{bmatrix}
$$

$\langle 210;\,A_2|\ 3\omega' - 8\omega x - 3\lambda$

$$
\begin{array}{l}
\langle 300;\,E| \\
\langle 210;\,1E| \\
\langle 210;\,2E|
\end{array}
\begin{bmatrix}
3\omega' - 12\omega x & \sqrt{6}\,\lambda & 0 \\
\sqrt{6}\,\lambda & 3\omega' - 8\omega x & -\sqrt{3}\,\lambda \\
0 & -\sqrt{3}\,\lambda & 3\omega' - 8\omega x
\end{bmatrix}
$$

$v = 4$
$$
\begin{array}{l}
\langle 400;\,A_1| \\
\langle 310;\,A_1| \\
\langle 220;\,A_1| \\
\langle 211;\,A_1|
\end{array}
\begin{bmatrix}
4\omega' - 20\omega x & 2\sqrt{2}\,\lambda & 0 & 0 \\
2\sqrt{2}\,\lambda & 4\omega' - 14\omega x + \lambda & 2\sqrt{3}\,\lambda & \sqrt{6}\,\lambda \\
0 & 2\sqrt{3}\,\lambda & 4\omega' - 12\omega x & 2\sqrt{2}\,\lambda \\
0 & \sqrt{6}\,\lambda & 2\sqrt{2}\,\lambda & 4\omega' - 10\omega x + 4\lambda
\end{bmatrix}
$$

$\langle 310;\,A_2|\ 4\omega' - 14\omega x - \lambda$

$$
\begin{array}{l}
\langle 400;\,E| \\
\langle 310;\,1E| \\
\langle 310;\,2E| \\
\langle 220;\,E| \\
\langle 211;\,E|
\end{array}
\begin{bmatrix}
4\omega' - 20\omega x & 2\sqrt{2}\,\lambda & 0 & 0 & 0 \\
2\sqrt{2}\,\lambda & 4\omega' - 14\omega x + \lambda & 0 & -\sqrt{3}\,\lambda & \sqrt{6}\,\lambda \\
0 & 0 & 4\omega' - 14\omega x - \lambda & 3\lambda & 0 \\
0 & -\sqrt{3}\,\lambda & 3\lambda & 4\omega' - 12\omega x & -\sqrt{2}\,\lambda \\
0 & \sqrt{6}\,\lambda & 0 & -\sqrt{2}\,\lambda & 4\omega' - 10\omega x - 2\lambda
\end{bmatrix}
$$

AX$_4$(T_d):

$v = 1$ $\langle 1000;\,A_1|\ \omega' - 2\omega x + 3\lambda$

        $\langle 1000;\,F_2|\ \omega' - 2\omega x - \lambda$

$v = 2$
$$
\begin{array}{l}
\langle 2000;\,A_1| \\
\langle 1100;\,A_1|
\end{array}
\begin{bmatrix}
2\omega' - 6\omega x & 2\sqrt{3}\,\lambda \\
2\sqrt{3}\,\lambda & \cdot \quad 2\omega' - 4\omega x + 4\lambda
\end{bmatrix}
$$

$\langle 1100;\,E|\ 2\omega' - 4\omega x - 2\lambda$

$$
\begin{array}{l}
\langle 2000;\,F_2| \\
\langle 1100;\,F_2|
\end{array}
\begin{bmatrix}
2\omega' - 6\omega x & 2\lambda \\
2\lambda & 2\omega' - 4\omega x
\end{bmatrix}
$$

$v = 3$
$$
\begin{array}{l}
\langle 3000;\,A_1| \\
\langle 2100;\,A_1| \\
\langle 1110;\,A_1|
\end{array}
\begin{bmatrix}
3\omega' - 12\omega x & 3\lambda & 0 \\
3\lambda & 3\omega' - 8\omega x + 4\lambda & 2\sqrt{6}\,\lambda \\
0 & 2\sqrt{6}\,\lambda & 3\omega' - 6\omega x + 3\lambda
\end{bmatrix}
$$

$\langle 2100;\,E|\ 3\omega' - 8\omega x + \lambda$

$\langle 2100;\,F_1|\ 3\omega' - 8\omega x - 3\lambda$

$$
\begin{array}{l}
\langle 3000;\,F_2| \\
\langle 2100;\,1F_2| \\
\langle 2100;\,2F_2| \\
\langle 1110;\,F_2|
\end{array}
\begin{bmatrix}
3\omega'-12\omega x & \sqrt{6}\lambda & \sqrt{3}\lambda & 0 \\
\sqrt{6}\lambda & 3\omega'-8\omega x-\lambda & \sqrt{2}\lambda & 0 \\
\sqrt{3}\lambda & \sqrt{2}\lambda & 3\omega'-8\omega x+2\lambda & -2\sqrt{2}\lambda \\
0 & 0 & -2\sqrt{2}\lambda & 3\omega'-6\omega x-\lambda
\end{bmatrix}
$$

$v=4$

$$
\begin{array}{l}
\langle 4000;\,A_1| \\
\langle 3100;\,A_1| \\
\langle 2200;\,A_1| \\
\langle 2110;\,A_1| \\
\langle 1111;\,A_1|
\end{array}
\begin{bmatrix}
4\omega'-20\omega x & 2\sqrt{3}\lambda & 0 & 0 & 0 \\
2\sqrt{3}\lambda & 4\omega'-14\omega x+2\lambda & 2\sqrt{3}\lambda & 2\sqrt{3}\lambda & 0 \\
0 & 2\sqrt{3}\lambda & 4\omega'-12\omega x & 4\lambda & 0 \\
0 & 2\sqrt{3}\lambda & 4\lambda & 4\omega'-10\omega x+6\lambda & 2\sqrt{6}\lambda \\
0 & 0 & 0 & 2\sqrt{6}\lambda & 4\omega'-8\omega x
\end{bmatrix}
$$

$$
\begin{array}{l}
\langle 3100;\,E| \\
\langle 2200;\,E| \\
\langle 2110;\,E|
\end{array}
\begin{bmatrix}
4\omega'-14\omega x-\lambda & 2\sqrt{3}\lambda & -\sqrt{3}\lambda \\
2\sqrt{3}\lambda & 4\omega'-12\omega x & -2\lambda \\
-\sqrt{3}\lambda & -2\lambda & 4\omega'-10\omega x-3\lambda
\end{bmatrix}
$$

$$
\begin{array}{l}
\langle 3100;\,F_1| \\
\langle 2110;\,F_1|
\end{array}
\begin{bmatrix}
4\omega'-14\omega x-\lambda & -\sqrt{3}\lambda \\
-\sqrt{3}\lambda & 4\omega'-10\omega x-3\lambda
\end{bmatrix}
$$

$$
\begin{array}{l}
\langle 4000;\,F_2| \\
\langle 3100;\,1F_2| \\
\langle 3100;\,2F_2| \\
\langle 2110;\,1F_2| \\
\langle 2110;\,2F_2| \\
\langle 2200;\,F_2|
\end{array}
\begin{bmatrix}
4\omega'-20\omega x & 2\sqrt{2}\lambda & 2\lambda & 0 & 0 & 0 \\
2\sqrt{2}\lambda & 4\omega'-14\omega x+\lambda & \sqrt{2}\lambda & \sqrt{3}\lambda & \sqrt{6}\lambda & 0 \\
2\lambda & \sqrt{2}\lambda & 4\omega'-14\omega x & \sqrt{6}\lambda & 0 & 2\sqrt{3}\lambda \\
0 & \sqrt{3}\lambda & \sqrt{6}\lambda & 4\omega'-10\omega x+3\lambda & -\sqrt{2}\lambda & 2\sqrt{2}\lambda \\
0 & \sqrt{6}\lambda & 0 & -\sqrt{2}\lambda & 4\omega'-10\omega x & 0 \\
0 & 0 & 2\sqrt{3}\lambda & 2\sqrt{2}\lambda & 0 & 4\omega'-12\omega x
\end{bmatrix}
$$

$AX_6(O_h)$:

$v=1$ $\langle 100000;\,A_{1g}|\ \omega'-2\omega x+4\lambda_1+\lambda_2$

$\langle 100000;\,E_g|\ \omega'-2\omega x-2\lambda_1+\lambda_2$

$\langle 100000;\,F_{1u}|\ \omega'-2\omega x-\lambda_2$

$$
v=2 \quad
\begin{array}{l}
\langle 200000;\,A_{1g}| \\
\langle 110000;\,A_{1g}| \\
\langle 100100;\,A_{1g}|
\end{array}
\begin{bmatrix}
2\omega'-6\omega x & 4\lambda_1 & 2\lambda_2 \\
4\lambda_1 & 2\omega'-4\omega x+4\lambda_1+2\lambda_2 & 4\lambda_1 \\
2\lambda_2 & 4\lambda_1 & 2\omega'-4\omega x
\end{bmatrix}
$$

$$
\begin{array}{l}
\langle 200000;\,E_g| \\
\langle 110000;\,E_g| \\
\langle 100100;\,E_g|
\end{array}
\begin{bmatrix}
2\omega'-6\omega x & 2\lambda_1 & 2\lambda_2 \\
2\lambda_1 & 2\omega'-4\omega x-2\lambda_1+2\lambda_2 & 2\lambda_1 \\
2\lambda_2 & 2\lambda_1 & 2\omega'-4\omega x
\end{bmatrix}
$$

$\langle 110000;\,F_{2g}|\ 2\omega'-4\omega x-2\lambda_2$

$$
\begin{array}{l}
\langle 200000;\,F_{1u}| \\
\langle 110000;\,F_{1u}|
\end{array}
\begin{bmatrix}
2\omega'-6\omega x & 2\sqrt{2}\lambda_1 \\
2\sqrt{2}\lambda_1 & 2\omega'-4\omega x+2\lambda_1
\end{bmatrix}
$$

$\langle 110000;\,F_{2u}|\ 2\omega'-4\omega x-2\lambda_1$

$v = 3$

$$\langle 300000; A_{1g}| \quad \begin{bmatrix} 3\omega' - 12\omega x & 2\sqrt{3}\lambda_1 & \sqrt{3}\lambda_2 & 0 & 0 \\ 2\sqrt{3}\lambda_1 & 3\omega' - 8\omega x + 4\lambda_1 + \lambda_2 & 2\lambda_1 & 2\sqrt{6}\lambda_1 & 2\lambda_1 + 2\lambda_2 \\ \sqrt{3}\lambda_2 & 2\lambda_1 & 3\omega' - 8\omega x + 2\lambda_2 & 0 & 4\lambda_1 \\ 0 & 2\sqrt{6}\lambda_1 & 0 & 3\omega' - 6\omega x + 3\lambda_2 & 2\sqrt{6}\lambda_1 \\ 0 & 2\lambda_1 + 2\lambda_2 & 4\lambda_1 & 2\sqrt{6}\lambda_1 & 3\omega' - 6\omega x + 4\lambda_1 + \lambda_2 \end{bmatrix}$$

with rows $\langle 300000; A_{1g}|$, $\langle 210000; A_{1g}|$, $\langle 200100; A_{1g}|$, $\langle 111000; A_{1g}|$, $\langle 110100; A_{1g}|$

$$\langle 210000; A_{2g}| \quad \begin{bmatrix} 3\omega' - 8\omega x - 4\lambda_1 + \lambda_2 & -2\lambda_1 + 2\lambda_2 \\ -2\lambda_1 + 2\lambda_2 & 3\omega' - 6\omega x - 4\lambda_1 + \lambda_2 \end{bmatrix}$$

with rows $\langle 210000; A_{2g}|$, $\langle 110100; A_{2g}|$

$$\langle 300000; E_g| \quad \begin{bmatrix} 3\omega' - 12\omega x & 2\sqrt{3}\lambda_1 & 0 & \sqrt{3}\lambda_2 & 0 & 0 \\ 2\sqrt{3}\lambda_1 & 3\omega' - 8\omega x + \lambda_1 + \lambda_2 & -\sqrt{3}\lambda_1 & 2\lambda_1 & -\lambda_1 - \lambda_2 & \sqrt{3}(-\lambda_1 + \lambda_2) \\ 0 & -\sqrt{3}\lambda_1 & 3\omega' - 8\omega x - \lambda_1 + \lambda_2 & 0 & \sqrt{3}(\lambda_1 + \lambda_2) & -\lambda_1 + \lambda_2 \\ \sqrt{3}\lambda_2 & 2\lambda_1 & 0 & 3\omega' - 8\omega x + 2\lambda_2 & -2\lambda_1 & 2\sqrt{3}\lambda_1 \\ 0 & -\lambda_1 - \lambda_2 & \sqrt{3}(\lambda_1 + \lambda_2) & -2\lambda_1 & 3\omega' - 6\omega x + \lambda_1 + \lambda_2 & -\sqrt{3}\lambda_1 \\ 0 & \sqrt{3}(-\lambda_1 + \lambda_2) & -\lambda_1 + \lambda_2 & 2\sqrt{3}\lambda_1 & -\sqrt{3}\lambda_1 & 3\omega' - 6\omega x - \lambda_1 + \lambda_2 \end{bmatrix}$$

with rows $\langle 300000; E_g|$, $\langle 210000; 1E_g|$, $\langle 210000; 2E_g|$, $\langle 200100; E_g|$, $\langle 110100; 1E_g|$, $\langle 110100; 2E_g|$

$$\langle 210000; F_{1g}| \quad \begin{bmatrix} 3\omega' - 8\omega x - 2\lambda_1 - \lambda_2 \end{bmatrix}$$

$$\langle 210000; F_{2g}| \quad \begin{bmatrix} 3\omega' - 8\omega x + 2\lambda_1 - \lambda_2 & 2\sqrt{2}\lambda_1 \\ 2\sqrt{2}\lambda_1 & 3\omega' - 6\omega x - \lambda_2 \end{bmatrix}$$

with rows $\langle 210000; F_{2g}|$, $\langle 111000; F_{2g}|$

$$\langle 111000; A_{2u}| \quad \begin{bmatrix} 3\omega' - 6\omega x - 3\lambda_2 \end{bmatrix}$$

$$\langle 300000; F_{1u}| \quad \begin{bmatrix} 3\omega' - 12\omega x & 2\sqrt{3}\lambda_1 & 0 & \sqrt{3}\lambda_2 & 0 & 0 \\ 2\sqrt{3}\lambda_1 & 3\omega' - 8\omega x + 2\lambda_1 + \lambda_2 & 2\lambda_1 & 2\lambda_1 & 2\sqrt{2}\lambda_1 & 2\lambda_1 \\ 0 & 2\lambda_1 & 3\omega' - 8\omega x - \lambda_2 & 0 & 2\sqrt{2}\lambda_1 & 2\lambda_2 \\ \sqrt{3}\lambda_2 & 2\lambda_1 & 0 & 3\omega' - 8\omega x - 2\lambda_2 & 0 & 0 \\ 0 & 2\sqrt{2}\lambda_1 & 2\sqrt{2}\lambda_1 & 0 & 3\omega' - 6\omega x + \lambda_2 & 2\sqrt{2}\lambda_1 \\ 0 & 2\lambda_1 & 2\lambda_2 & 0 & 2\sqrt{2}\lambda_1 & 3\omega' - 6\omega x - \lambda_2 \end{bmatrix}$$

with rows $\langle 300000; F_{1u}|$, $\langle 210000; 1F_{1u}|$, $\langle 210000; 2F_{1u}|$, $\langle 200100; F_{1u}|$, $\langle 111000; F_{1u}|$, $\langle 110100; F_{1u}|$

$$\langle 210000; 1F_{2u}| \quad \begin{bmatrix} 3\omega' - 8\omega x - 2\lambda_1 + \lambda_2 & 2\lambda_1 & 2\lambda_1 \\ 2\lambda_1 & 3\omega' - 8\omega x - \lambda_2 & 2\lambda_2 \\ 2\lambda_1 & 2\lambda_2 & 3\omega' - 6\omega x - \lambda_2 \end{bmatrix}$$

with rows $\langle 210000; 1F_{2u}|$, $\langle 210000; 2F_{2u}|$, $\langle 110100; F_{2u}|$

Acknowledgments

The authors are grateful for helpful discussions with Dr. A. G. Robiette and for comments on the manuscript by Dr. S. K. Gray and Mr. P. Coveney. One author (L.H.) acknowledges financial support by the U.K. Science and Engineering Research Council and the Academy of Finland.

REFERENCES

1. T. E. Martin and A. H. Kalantar, *J. Chem. Phys.* **49**, 235 (1968).
2. R. J. Hayward and B. R. Henry, *Chem. Phys.* **12**, 387 (1976).
3. W. R. A. Greenlay and B. R. Henry, *Chem. Phys. Lett.* **53**, 325 (1978).
4. B. R. Henry and R. J. D. Miller, *Chem. Phys. Lett.* **60**, 81 (1978).
5. W. R. A. Greenlay and B. R. Henry, *J. Chem. Phys.* **69**, 82 (1978).
6. J. W. Perry and A. H. Zewail, *J. Chem. Phys.* **70**, 582 (1979).
7. J. W. Perry and A. H. Zewail, *Chem. Phys. Lett.* **65**, 31 (1979).
8. A. H. Zewail and D. J. Diestler, *Chem. Phys. Lett.* **65**, 37 (1979).
9. B. R. Henry and J. A. Thomson, *Chem. Phys. Lett.* **69**, 275 (1980).
10. B. R. Henry and M. A. Mohammadi, *Chem. Phys. Lett.* **75**, 99 (1980).
11. J. W. Perry and A. H. Zewail, *J. Phys. Chem.* **85**, 933 (1981).
12. B. R. Henry and M. A. Mohammadi, *Chem. Phys.* **55**, 385 (1981).
13. R. Nakagi and I. Hanazaki, *Chem. Phys. Lett.* **83**, 512 (1981).
14. R. Nakagi and I. Hanazaki, *Chem. Phys.* **72**, 93 (1982).
15. B. R. Henry, M. A. Mohammadi, and A. W. Tarr, *J. Chem. Phys.* **77**, 3295 (1982).
16. M. E. Long, R. L. Swofford, and A. C. Albrecht, *Science* **191**, 183 (1976).
17. R. L. Swofford, M. E. Long, and A. C. Albrecht, *J. Chem. Phys.* **65**, 179 (1976).
18. R. L. Swofford, M. E. Long, M. S. Burberry and A. C. Albrecht, *J. Chem. Phys.* **66**, 664 (1977).
19. R. L. Swofford, M. S. Burberry, J. A. Morrell, and A. C. Albrecht, *J. Chem. Phys.* **66**, 5245 (1977).
20. M. S. Burberry, J. A. Morrell, A. C. Albrecht, and R. L. Swofford, *J. Chem. Phys.* **70**, 5522 (1979).
21. M. S. Burberry and A. C. Albrecht, *J. Chem. Phys.* **71**, 4631 (1979).
22. H. L. Fang and R. L. Swofford, *J. Chem. Phys.* **72**, 6382 (1980).
23. H. L. Fang and R. L. Swofford, *J. Chem. Phys.* **73**, 2607 (1980).
24. K. V. Reddy and M. J. Berry, *Faraday Discuss. Chem. Soc.* **67**, 188 (1979).
25. R. G. Bray and M. J. Berry, *J. Chem. Phys.* **71**, 4909 (1979).
26. C. K. N. Patel and A. C. Tam, *Chem. Phys. Lett.* **62**, 511 (1979).
27. C. K. N. Patel, A. C. Tam, and R. J. Kerl, *J. Chem. Phys.* **71**, 1470 (1979).
28. K. V. Reddy, D. F. Heller, and M. J. Berry, *J. Chem. Phys.* **76**, 2814 (1982).
29. R. F. Menefee, R. R. Hall, and M. J. Berry, *Appl. Phys.* **B28**, 121 (1982).
30. J. S. Wong and C. B. Moore, *J. Chem. Phys.* **77**, 603 (1982).
31. J. S. Wong and C. B. Moore, in *Frontiers of Chemistry* (International Union of Pure and Applied Chemistry), K. J. Laidler, ed., Pergamon Press, Elmsford, N.Y., 1982, p. 353.

32. K. K. Lehmann, G. J. Scherer, and W. Klemperer, *J. Chem. Phys.* **77**, 2853 (1982).

33. B. R. Henry, *J. Phys. Chem.* **80**, 2160 (1976).

34. B. R. Henry, *Acc. Chem. Res.* **10**, 207 (1977).

35. B. R. Henry, in *Vibrational Spectra and Structure*, J. R. Durig, ed., Vol. 10, Elsevier, New York, 1981.

36. M. L. Sage and J. Jortner, *Advan. Chem. Phys.* **47**, Pt. 1, 293 (1981).

37. B. R. Henry and W. Siebrand, *J. Chem. Phys.* **49**, 5369 (1968).

38. R. J. Hayward, B. R. Henry, and W. Siebrand, *J. Mol. Spectrosc.* **46**, 207 (1973).

39. R. J. Hayward and B. R. Henry, *J. Mol. Spectrosc.* **50**, 58 (1974).

40. R. J. Hayward and B. R. Henry, *J. Mol. Spectrosc.* **57**, 221 (1975).

41. B. R. Henry and I.-F. Hung, *Chem. Phys.* **29**, 465 (1978).

42. R. Wallace, *Chem. Phys.* **11**, 189 (1975).

43. M. L. Elert, P. R. Stannard, and W. M. Gelbart, *J. Chem. Phys.* **67**, 5395 (1977).

44. P. M. Morse, *Phys. Rev.* **34**, 57 (1929).

45. D. ter Haar, *Phys. Rev.* **70**, 222 (1946).

46. L. Infeld and T. E. Hull, *Rev. Mod. Phys.* **23**, 21 (1951).

47. E. M. Greenawalt and A. S. Dickinson, *J. Mol. Spectrosc.* **30**, 427 (1969).

48. G. D. Carney and R. N. Porter, *J. Chem. Phys.* **65**, 3547 (1976).

49. G. D. Carney, L. L. Sprandel, and C. W. Kern, *Advan. Chem. Phys.* **37**, 305 (1978).

50. R. Wallace, *Chem. Phys. Lett.* **37**, 115 (1976).

51. M. L. Sage, *Chem. Phys.* **35**, 375 (1978).

52. I. Schek, J. Jortner, and M. L. Sage, *Chem. Phys. Lett.* **64**, 209 (1979).

53. M. L. Sage and J. A. Williams, *J. Chem. Phys.* **78**, 1348 (1983).

54. I. A. Watson, B. R. Henry, and I. G. Ross., *Spectrochim. Acta* **37A**, 857 (1981).

55. M. S. Child and R. T. Lawton, *Faraday Discuss. Chem. Soc.* **71**, 273 (1981).

56. L. Halonen and M. S. Child, *Mol. Phys.* **46**, 239 (1982).

57. R. T. Lawton and M. S. Child, *Mol. Phys.* **37**, 1799 (1979).

58. R. T. Lawton and M. S. Child, *Mol. Phys.* **44**, 709 (1981).

59. C. Jaffé and P. Brumer, *J. Chem. Phys.* **73**, 5646 (1980).

60. E. L. Sibert, W. P. Reinhardt, and J. T. Hynes, *J. Chem. Phys.* **77**, 3583 (1982).

61. R. T. Lawton and M. S. Child, *Mol. Phys.* **40**, 773 (1980).

62. M. S. Child and R. T. Lawton, *Chem. Phys. Lett.* **87**, 217 (1982).

63. E. L. Sibert, J. T. Hynes, and W. P. Reinhardt, *J. Chem. Phys.* **77**, 3595 (1982).

64. G. Herzberg, *Infrared and Raman Spectra*, Van Nostrand, New York, 1945.

65. J. W. Swensson, W. S. Benedict, L. Delbouille, and G. Roland, *Mém. Soc. Roy. Sci. Liège*, Spec. *Vol.* **5** (1970).

66. J.-M. Flaud, C. Camy-Peyret, and A. Valentin, *J. Phys.* (Paris) **33**, 741 (1972).

67. J.-M. Flaud, and C. Camy-Peyret, *Mol. Phys.* **26**, 811 (1973).

68. C. Camy-Peyret and J.-M. Flaud, *Spectrochim. Acta.* **29A**, 1711 (1973).

69. J.-M. Flaud, C. Camy-Peyret, and J.-P. Maillard, *Mol. Phys.* **32**, 499 (1976).

70. C. Camy-Peyret, J.-M. Flaud, J.-P. Maillard, and G. Guelachvili, *Mol. Phys.* **33**, 1641 (1977).

71. L. S. Rothman, *Appl. Opt.* **17**, 3517 (1978).

72. J.-M. Flaud, C. Camy-Peyret, K. N. Rao, D.-W. Chen, Y.-S. Hoh, and J.-P. Maillard, *J. Mol. Spectrosc.* **75**, 339 (1979).

73. C. Camy-Peyret, J.-M. Flaud, and J.-P. Maillard, *J. Phys.* (Paris) **41**, L23 (1980).

74. J.-M. Flaud, and C. Camy-Peyret, *J. Mol. Spectrosc.* **51**, 142 (1974).

75. C. Camy-Peyret and J.-M. Flaud, *J. Mol. Spectrosc.* **59**, 327 (1976).

76. A. Rajaratnam and K. T. Lua, *J. Phys.* **B15**, 3615 (1982).

77. K. S. Sorbie and J. N. Murrell, *Mol. Phys.* **29**, 1387 (1975).

78. L. Halonen and M. S. Child, *J. Chem. Phys.* **79**, 4355 (1983).

79. L. Halonen, *Chem. Phys. Lett.* **87**, 221 (1982).

80. L. Halonen, M. S. Child, and S. Carter, *Mol. Phys.* **47**, 1097 (1982).

81. L. Halonen, D. W. Noid, and M. S. Child, *J. Chem. Phys.*, **78**, 2803 (1983).

82. L. Halonen and M. S. Child, *J. Chem. Phys.* **79**, 559 (1983).

83. M. S. Burberry and A. C. Albrecht, *J. Chem. Phys.* **70**, 147 (1979).

84. M. S. Burberry and A. C. Albrecht, *J. Chem. Phys.* **71**, 4768 (1979).

85. M. L. Sage and J. Jortner, *Chem. Phys. Lett.* **79**, 9 (1981).

86. P. R. Stannard, M. L. Elert, and W. M. Gelbart, *J. Chem. Phys.* **74**, 6050 (1981).

87. L. P. Giver, *J. Quant. Spectrosc. Radiat. Transfer* **19**, 311 (1978).

88. M. L. Sage and J. Jortner, *Chem. Phys. Lett.* **62**, 451 (1978).

89. P. J. Nagy and W. L. Hase, *Chem. Phys. Lett.* **54**, 73 (1978).

90. G. M. Korenowski and A. C. Albrecht, *Chem. Phys.* **38**, 239 (1979).

91. D. F. Heller and S. Mukamel, *J. Chem. Phys.* **70**, 463 (1979).

92. K. V. Reddy and M. J. Berry, *Chem. Phys. Lett.* **52**, 111 (1977).

93. K. V. Reddy and M. J. Berry, *Chem. Phys. Lett.* **66**, 223 (1979).

94. K. V. Reddy and M. J. Berry, *Chem. Phys. Lett.* **72**, 29 (1980).

95. P. R. Stannard and W. M. Gelbart, *J. Phys. Chem.* **85**, 3592 (1981).

96. M. L. Sage, *Chem. Phys.* **72**, 249 (1982).

97. E. L. Sibert, W. P. Reinhardt, and J. T. Hynes, *Chem. Phys. Lett.* **92**, 455 (1982).

98. D. W. Chandler, W. E. Farneth, and R. N. Zare, *J. Chem. Phys.* **77**, 4447 (1982).

99. V. Lopez and R. A. Marcus, *Chem. Phys. Lett.* **93**, 232 (1982).

100. R. J. Wolf and W. L. Hase, *J. Chem. Phys.* **73**, 3779 (1980).

101. J. T. Muckermann, R. T. Lawton, and M. S. Child, to be published.

102. H. Goldstein, *Classical Mechanics*, Addison-Wesley, Reading, Mass., 1972.

103. M. Born, *The Mechanics of the Atom*, Bell, London, 1927.

104. B. V. Chirikov, *Phys. Rep.* **52**, 263 (1979).

105. *NBS Appl. Math. Ser.* **59**, 2nd ed. (1966).

106. O. S. Mortensen, B. R. Henry, and M. A. Mohammadi, *J. Chem. Phys.* **75**, 4800 (1981).

107. M. A. Mohammadi, *Chem. Phys.* **66**, 153 (1982).

108. M. Abramowitz and I. A. Stegun, *Handbook of Mathematical Functions*, Dover, New York, 1965.

109. E. B. Wilson, J. C. Decius, and P. C. Cross, *Molecular Vibrations*, McGraw-Hill, New York, 1955.

110. I. M. Mills, in *Molecular Spectroscopy: Modern Research*, K. N. Rao and C. W. Matthews, eds., Vol. I, Academic Press, New York, 1972.

111. A. R. Hoy, I. M. Mills, and G. Strey, *Mol. Phys.* **24**, 1265 (1972).

112. G. Strey and I. M. Mills, *J. Mol. Spectrosc.* **59**, 103 (1976).

113. R. D. Shelton, A. H. Nielsen, and W. H. Fletcher, *J. Chem. Phys.* **21**, 2178 (1953).

114. K. Kuchitsu and Y. Morino, *Bull. Chem. Soc. Japan* **38**, 814 (1965).

115. B. T. Darling and D. M. Dennison, *Phys. Rev.* **57**, 128 (1940).

116. I. M. Mills and A. G. Robiette, private communication (1982).

117. J. N. Huffaker and L. Binh Tran, *J. Chem. Phys.* **76**, 3838 (1982).

118. J. Bron and R. Wallace, *Can. J. Chem.* **56**, 2167 (1978).

119. H. S. Møller and O. S. Mortensen, *Chem. Phys. Lett.* **66**, 539 (1979).

120. T. A. Wiggins, E. K. Plyler, and E. D. Tidwell, *J. Opt. Soc. Amer.* **51**, 1219 (1961).

121. E. K. Plyler, E. D. Tidwell, and T. A. Wiggins, *J. Opt. Soc. Amer.* **53**, 589 (1963).

122. W. J. Lafferty and R. J. Thibault, *J. Mol. Spectrosc.* **14**, 79 (1964).

123. E. Kostyk and H. L. Welsh, *Can. J. Phys.* **58**, 534 (1980).

124. E. Kostyk and H. L. Welsh, *Can. J. Phys.* **58**, 912 (1980).

125. I. Suzuki and J. Overend, *Spectrochim. Acta* **25A**, 977 (1969).

126. A. Baldacci, S. Ghersetti, S. C. Hurlock, and K. N. Rao, *J. Mol. Spectrosc.* **42**, 317 (1972).

127. B. D. Saskena, *J. Chem. Phys.* **20**, 95 (1952).

128. S. Ghersetti, J. Pliva, and K. N. Rao, *J. Mol. Spectrosc.* **38**, 53 (1971).

129. K. F. Palmer, M. E. Michelson, and K. N. Rao, *J. Mol. Spectrosc.* **44**, 131 (1972).

130. S. Ghersetti, J. E. Adams, and K. N. Rao, *J. Mol. Spectrosc.* **64**, 157 (1977).

131. A. Baldacci, S. Ghersetti, and K. N. Rao, *J. Mol. Spectrosc.* **68**, 183 (1977).

132. G. J. Scherer, K. K. Lehmann, and W. Klemperer, *J. Chem. Phys.* **78**, 2817 (1983).

133. K. K. Lehmann, G. J. Scherer, and W. Klemperer, *J. Chem. Phys.* **79**, 1369 (1983).

134. R. Wallace and A. A. Wu, *Chem. Phys.* **39**, 221 (1979).

135. J. Bron and R. Wallace, *Can. J. Chem.* **57**, 2321 (1979).

136. J. K. Wilmshurst and H. J. Bernstein, *Can. J. Chem.* **35**, 226 (1957).

137. I. Abram, A. de Martino, and R. Frey, *J. Chem. Phys.* **76**, 5727 (1983).

138. L. Halonen, unpublished results.

139. B. L. Lutz, T. Owen, and R. D. Cess, *Astrophys. J.* **203**, 541 (1976).

140. K. A. Dick and U. Fink, *J. Quant. Spectrosc. Radiat. Transfer* **18**, 433 (1977).

141. K. R. Ramaprasad, J. Caldwell, and D. S. McClure, *Icarus* **35**, 400 (1978).

142. J. Bardwell and G. Herzberg, *Astrophys. J.* **117**, 462 (1953).

143. C. W. Patterson, B. J. Krohn, and A. S. Pine, *Opt. Lett.* **6**, 39 (1981).

144. D. P. Hodgkinson, private communication (1982).

145. A. Aboumajd, H. Berger, and R. Saint-Loup, *J. Mol. Spectrosc.* **78**, 486 (1979).

146. P. Esherick and A. Owyoung, *J. Mol. Spectrosc.* **92**, 162 (1982).

147. J. Bron and R. Wallace, *J. Chem. Soc., Faraday Trans.*, *II* **74**, 611 (1978).

148. J. Jortner, R. D. Levine, and S. A. Rice, eds., *Advan. Chem. Phys.* **47**: "Photoselective Chemistry," Pts. 1 and 2 (1981).

149. C. D. Cantrell and H. W. Galbraith, *Opt. Commun.* **18**, 513 (1976).

150. C. D. Cantrell and H. W. Galbraith, *Opt. Commun.* **21**, 374 (1977).

151. W. Fuss, *Chem. Phys. Lett.* **71**, 77 (1980).

152. H. W. Galbraith and J. R. Ackerhalt, *Chem. Phys. Lett.* **84**, 458 (1981).

153. D. P. Hodgkinson and A. G. Robiette, *Chem. Phys. Lett.* **82**, 193 (1981).

154. D. P. Hodgkinson, A. J. Taylor, and A. G. Robiette, *J. Phys.* **B14**, 1803 (1981).

155. C. di Lauro and F. Lattanzi, *Chem. Phys.* **71**, 233 (1982).

156. C. di Lauro, F. Lattanzi, and G. Sanna, *Chem. Phys.* **73**, 215 (1982).

157. A. J. Taylor, D. P. Hodgkinson, and A. G. Robiette, *Opt. Commun.* **41**, 320 (1982).

158. D. H. Whiffen, *Phil. Trans. Roy. Soc. London*, Ser. A **248**, 131 (1955).

159. J. C. Duinker and I. M. Mills, *Spectrochim. Acta* **24A**, 417 (1968).

160. P. Pulay, G. Fogarasi, and J. E. Boggs, *J. Chem. Phys.* **74**, 3999 (1981).

161. S. Brodersen and A. Langseth, *Kgl. Danske Videnskab. Selskab, Mat. Fys. Skrifter* **1**, 1 (1956).

MULTIMODE MOLECULAR DYNAMICS BEYOND THE BORN–OPPENHEIMER APPROXIMATION

H. KÖPPEL, W. DOMCKE, AND L. S. CEDERBAUM

Department of Theoretical Chemistry
Institute for Physical Chemistry, University of Heidelberg
Heidelberg, Federal Republic of Germany

CONTENTS

I. INTRODUCTION

The Born–Oppenheimer adiabatic approximation[1-3] represents one of the cornerstones of molecular physics and chemistry. It allows the calculation of dynamical processes in molecules to be divided into two stages. In the first stage, the electronic problem is solved keeping the atomic nuclei fixed in space. The calculation of electronic energies and wave functions for fixed nuclei has been developed to a high degree of sophistication and constitutes the core of modern quantum chemistry. In the second stage, the nuclear dynamics on a given predetermined electronic potential energy surface is treated.

The bulk of chemical processes can be rationalized by considering the dynamics on a single electronic energy surface, in most cases the surface of the electronic ground state of the system. Specific fascinating theoretical aspects of this problem are, for example, the local versus normal mode description of molecular vibrations,[4] the regular or chaotic behavior of multimode dynamics,[5] the decay of excited vibrational levels in polyatomic molecules[6] and the calculation of reaction cross sections in atom–molecule collisions.[7] Another class of important and interesting phenomena, which is the subject of this article, is associated with dynamical processes that are not confined to a single electronic surface. As is well known,[1-3] the Born–Oppenheimer approximation is based on the fact that the spacing of electronic eigenvalues is generally large compared to typical spacings associated with nuclear motion. When this condition is violated, the residual coupling via the nuclear kinetic energy operator causes transitions between the adiabatic electronic states.[1-3] Typical phenomena associated with a violation of the Born–Oppenheimer separation of electronic and nuclear motions are inelastic atom–atom and atom–molecule collisions[8] and the radiationless decay of excited electronic states.[9] Usually only two or at most a few electronic states are close in energy, and this only in a limited range of the nuclear coordinate space. Therefore it is not necessary to abandon the Born–Oppenheimer approximation altogether; it suffices to develop efficient methods to deal with the near-degeneracy of electronic states in localized regions of coordinate space. The most important and fruitful concept to handle non-Born–Oppenheimer phenomena is the introduction of suitable "diabatic" electronic states that may cross as a function of the internuclear distance(s), whereas the adiabatic electronic states are subject to the noncrossing rule.[10-14]

Naturally, the investigation of nonadiabatic effects and the quantitative assessment of the accuracy of the adiabatic approximation were initially confined to simple diatomic molecules and atom–atom collisions. Kolos and Wolniewicz carried out an exhaustive study of the accuracy of the adiabatic approximation for the low-lying electronic states of H_2^+ and H_2.[15] They showed that the error in the vibrational energy levels originating from the adiabatic approximation is typically a few wave numbers and should be correspondingly smaller for heavier nuclei. For highly excited electronic states of diatomic molecules the non-Born–Oppenheimer perturbations are usually stronger. Their consideration is essential for the analysis of optical spectra in the VUV (vacuum ultraviolet) region.[16] Another common non-Born–Oppenheimer phenomenon in diatomic molecules is predissociation.[17] In the field of atomic collisions the adiabatic electronic representation plays a less central role than in spectroscopy, since the nonadiabatic coupling terms can be made arbitrarily large by increasing the relative velocity of the collision partners. (For a review of the theory of nonadiabatic transitions in atom–atom collisions see, e.g., Ref. 18.)

The situation is necessarily less transparent for polyatomic molecules on which this article focuses. The existence of several vibrational degrees of freedom allows for completely new types of phenomena that have no counterpart in diatomic molecules. Therefore the experience gained from the study of diatomic molecules should not be simply extrapolated to polyatomic systems. To stress and exemplify this point is the main intention of the present work. First of all, the probability that two given electronic states approach each other closely in energy can be expected to increase with the number of vibrational modes. Therefore, near-degeneracies of electronic states are expected to be more common in polyatomic systems than in diatomic molecules. Moreover, the noncrossing rule, which prevails in diatomic systems, is relaxed when more than two vibrational degrees of freedom are available.[19–23] Already in a two-mode system two electronic states of the same symmetry may become exactly degenerate at a certain point of the two-dimensional coordinate space, forming a so-called conical intersection.[23–26] A conical intersection necessarily causes a complete breakdown of the Born–Oppenheimer approximation. A well-known special case of a conical intersection is the Jahn–Teller effect, in particular the so-called $E \times \varepsilon$ Jahn–Teller effect involving the splitting of a doubly degenerate electronic state by a doubly degenerate vibrational mode.[12,27–29] The energy levels of Jahn–Teller coupled states have been extensively investigated both experimentally and theoretically.[28,29] The spectroscopic consequences of more general types of conical intersections in molecules have been investigated more recently.[30,31] Conical intersections also provide an important pathway for unimolecular decay[32] and atom–molecule reaction processes.[33,34]

In contrast to diatomic molecules, where the energies and wave functions of the participating electronic states can be obtained from accurate first-principles calculations, one has to resort to simplified models when considering dynamical processes in polyatomic molecules. Even the straightforward pointwise determination of the multidimensional electronic potential hypersurfaces using quantum chemical ab initio methods—not to mention the treatment of the quantum mechanical nuclear dynamics in many dimensions—would require a prohibitive amount of computing time, except perhaps the smallest three- and four-atomic systems. For this reason the construction of simple model Hamiltonians, which are suitable to describe the vibronic structure of electronic spectra of polyatomic molecules, has a long tradition. To render the problem tractable, one usually restricts the number of interacting electronic states and participating vibrational modes to a minimum, typically *two* states and *one* mode. In addition, one adopts simplifying approximations such as the harmonic approximation for the (adiabatic or diabatic) potential energy curves. Since a comprehensive survey of the vast amount of existing literature is not the intention of this review, we confine ourselves here to citing Refs. 35–45 as representative examples. Model Hamiltonians of this type facilitate the understanding and qualitative description of several interesting effects, such as the symmetry lowering of molecules in excited or ionic states and the associated strong excitation of non-totally symmetric modes in optical absorption or photoelectron spectra and the appearance of optically forbidden bands in absorption or emission spectra. Many spectroscopic phenomena have been satisfactorily explained in this way (see, for example, Refs. 46–53).

The emphasis of the present review is placed, as stated in the title, on the multimode aspects of non-Born–Oppenheimer effects, i.e., on those phenomena that cannot be explained by considering the coupling of two or more electronic states via a single vibrational mode or several modes, which are considered as independent. The coupling of electronic states via several modes has the characteristic property that the resulting vibronic spectrum is not simply a convolution of progressions corresponding to the individual modes. Therefore the inclusion of several coupling modes is a nontrivial step, in contrast to the excitation of totally symmetric modes on a single surface, where the convolution approximation prevails in lowest order. An example of the nonseparability of coupling modes is the Jahn–Teller effect with several active vibrational modes.[54-57] The present survey includes, in particular, the general conical intersection problem as a typical multimode effect that is not present in diatomic molecules. In its simplest form, a conical intersection may be viewed as two nondegenerate electronic states of different symmetry interacting through a single nontotally symmetric mode while an additional totally symmetric mode modulates the energy separation of the interacting states.[24-26] Since the coupling depends sensitively on the energy separation, the vibronic interaction leads indirectly to an intricate coupling of the two modes. The problem may be arbitrarily increased in complexity by including more than just two modes. It seems to us that the nontrivial multimode aspects of the vibronic coupling problem have received relatively little attention so far, and we shall attempt here to give a fairly complete account of the existing literature on multimode vibronic coupling. Multimode aspects of resonance Raman scattering have recently been reviewed by Champion and Albrecht.[58]

To provide a solid theoretical foundation for the ensuing discussion of specific examples, we review the general theory of vibronic coupling. The basic concepts such as the adiabatic and diabatic representations and the adiabatic and Franck–Condon approximations and their properties are discussed. The diabatic representation provides the basis for the systematic construction of simple model Hamiltonians. We establish the connection between the parameters of model Hamiltonians and the data obtainable from ab initio electronic structure calculations. The discussion is confined to bound nuclear motion, that is, reactions and unimolecular decay processes are excluded. We are thus mainly interested in phenomena that occur below the lowest dissociation threshold. The phenomena considered cover spectroscopy, in particular optical and photoelectron spectroscopy, and radiationless decay as a typical non-Born–Oppenheimer effect. Despite the simplicity of the models, their solution is already a formidable task due to the nonseparability of the vibrational modes. Perturbation theory is *not* applicable in most cases of interest. Therefore, one has to resort to numerical methods, that is, the numerical diagonalization of large secular matrices. The exact numerical calculation of spectra and time-dependent occupation probabilities provides the basis for a critical assessment of genuine multimode effects in spectroscopy and radiationless decay.

Any approach based on simplified models derives its justification from the comparison of theory and experiment and the a posteriori verification of the adopted model Hamiltonian. Therefore a large part of the present work is devoted to the analysis of electronic spectra and their theoretical reproduction. The ab initio calculation of the parameters entering the model Hamiltonian is a crucial ingredient of our approach. It removes possible ambiguities of the fitting procedure and opens the way to the analysis of very complex spectra. The comparison of model parameters determined by fitting experimental data with independently obtained ab initio values provides a convincing justification of the adequacy of the model. The restriction to rather small molecules with partly well-resolved spectra allows for a serious check of the validity of the models employed (with respect to the number of states included, the number of modes included, the choice of parameters, etc.).

We shall consider in some detail the photoelectron spectra of ethylene and the cumulene series. The cumulene radical cations represent a family of molecules which nicely exhibit a common type of vibronic coupling activity. The photoelectron spectra of BF_3 and HCN are considered as examples for multimode vibronic coupling involving degenerate coupling modes. In all cases considered, the vibronic structure of the experimental spectra can be reproduced nearly quantitatively. We shall also discuss the optical absorption spectrum of NO_2, the analysis of which is a notoriously difficult problem and is thus of particular theoretical interest. Our calculations give a good qualitative account of the highly irregular vibronic fine structure, the high density of vibronic levels, and the quenching of the radiative width observed experimentally.[59] In all these examples the consideration of multimode effects is essential to explain the observations.

Vibronic coupling via several modes can lead to extremely complex vibronic line spectra, expecially when a conical intersection of the adiabatic potential surfaces is involved. Some examples to be discussed will exhibit such high level densities that the assignment and analysis of individual spectral lines become rather meaningless. A more schematic description of the spectrum is then desirable. For example, one may confine oneself to consider only the envelope of the dense manifold of spectral lines, which corresponds to considering only the short-time behavior of the system. An alternative approach is the statistical analysis of the energy level pattern and the intensity distribution. Statistical aspects of the multimode vibronic coupling problem will be touched on, citing work on $C_2H_4^+$ and NO_2 as prototypical.

When one considers the literature on vibronic coupling in polyatomic molecules, vibronic effects in spectroscopy and the radiationless decay of excited electronic states appear as largely independent phenomena. This artificial separation is due to the fact that spectroscopic calculations are usually restricted to very few modes. In the theory of radiationless relaxation in large polyatomic molecules, on the other hand, one adopts a statistical description and includes many modes in a cursory manner.[60] The consideration of vibronic coupling models with several active modes (say, three to five, depending on the coupling strength) enables us to fill the existing gap between the detailed spectroscopic calculations and phenomenological relaxation theories. It will be shown that already for a three-mode model the level density can become sufficiently high so that irreversible behavior develops. Multimode vibronic coupling thus provides the bridge between spectroscopy and radiationless relaxation phenomena.

II. THEORY OF MULTIMODE VIBRONIC COUPLING EFFECTS

A. The Hamiltonian

1. General Remarks

We consider a molecule described by the Hamiltonian

$$H = T_e + T_N + U(\mathbf{r}, \mathbf{Q}) \tag{2.1}$$

where T_e and T_N are the operators of the kinetic energy of the electrons and nuclei, respectively, and $U(\mathbf{r}, \mathbf{Q})$ is the total potential energy of the electrons and nuclei. The vector $\mathbf{r}$ denotes the set of electronic coordinates and the vector $\mathbf{Q} = (Q_1, Q_2, \ldots, Q_M)$ stands for the nuclear coordinates describing the displacements from a reference configuration. If we put T_N equal to zero, H describes the electronic motion in a molecule with fixed nuclei. The orthonormal electronic wave functions $\Phi_n(\mathbf{r}, \mathbf{Q})$ and energies $V_n(\mathbf{Q})$ defined by

$$\left[H_e - V_n(\mathbf{Q}) \right] \Phi_n(\mathbf{r}, \mathbf{Q}) = 0 \qquad (2.2a)$$

$$H_e = T_e + U(\mathbf{r}, \mathbf{Q}) \qquad (2.2b)$$

depend on the nuclear geometry as parameters. They are the usual Born–Oppenheimer electronic states and potential energy surfaces. The exact eigenstates of the system can be expanded in the Born–Oppenheimer electronic states

$$\Psi(\mathbf{r}, \mathbf{Q}) = \sum_n \chi_n(\mathbf{Q}) \Phi_n(\mathbf{r}, \mathbf{Q}) \qquad (2.3)$$

Inserting this *ansatz* into the Schrödinger equation

$$(H - E)\Psi(\mathbf{r}, \mathbf{Q}) = 0 \qquad (2.4)$$

one readily obtains[2] the following set of coupled equations for the expansion coefficients in Eq. (2.3)

$$\left[T_N + V_n(\mathbf{Q}) - E \right] \chi_n(\mathbf{Q}) = \sum_m \hat{\Lambda}_{nm} \chi_m(\mathbf{Q}) \qquad (2.5)$$

The nonadiabatic operators $\hat{\Lambda}_{nm}$ —as we shall call them—are given by

$$\hat{\Lambda}_{nm} = - \int d\mathbf{r} \, \Phi_n^* [T_N, \Phi_m] \qquad (2.6)$$

where $[A, B] = AB - BA$.

It is useful to decompose each nonadiabatic operator into a differential operator and a c-number in $\mathbf{Q}$-space. Choosing the nuclear kinetic energy operator to be of the general form[61]

$$T_N = - \sum_{i, j = 1}^{M} \frac{\partial}{\partial Q_i} \alpha_{ij}(\mathbf{Q}) \frac{\partial}{\partial Q_j} \qquad (2.7)$$

the decomposition of the nonadiabatic operators reads

$$\hat{\Lambda}_{nm} = \sum_{i=1}^{M} F_{nm}^{(i)} \frac{\partial}{\partial Q_i} - G_{nm} \tag{2.8}$$

The quantities $F_{nm}^{(i)}$ and G_{nm} are functions of $\mathbf{Q}$ and are given by the following expressions

$$F_{nm}^{(i)} = 2 \sum_{j=1}^{M} \int d\mathbf{r} \, \Phi_n^* \alpha_{ij} \frac{\partial \Phi_m}{\partial Q_j} \tag{2.9a}$$

$$G_{nm} = \int d\mathbf{r} \, \Phi_n^* (T_N \Phi_m) \tag{2.9b}$$

We introduce matrices $\hat{\Lambda}$, $\mathbf{G}$, and $\mathbf{F}^{(i)}$ with elements $\hat{\Lambda}_{nm}$, G_{nm}, and $F_{nm}^{(i)}$, respectively. It is clear that since $\hat{\Lambda}$ is a hermitian operator the matrix $\mathbf{G}$ is nonhermitian and $\mathbf{F}^{(i)}$ is antihermitian; that is,

$$\mathbf{F}^\dagger = -\mathbf{F} \tag{2.10}$$

where the superscript (i) has been omitted.

We now return to the fundamental set of equations given in Eq. (2.5). This set of coupled equations can be rewritten as a matrix Schrödinger equation

$$(\mathcal{H} - E\mathbf{1})\chi = 0 \tag{2.11}$$

with the matrix Hamiltonian

$$\mathcal{H} = T_N \mathbf{1} + \mathbf{V}(\mathbf{Q}) - \hat{\Lambda} \tag{2.12}$$

which describes the nuclear motion in the manifold of electronic states. Here χ is a column vector with elements χ_n; $\mathbf{1}$ is the unit matrix, and $\mathbf{V}(\mathbf{Q}) = \{V_n(\mathbf{Q})\delta_{nm}\}$ is the diagonal matrix of electronic energies.

An alternative derivation of a matrix Schrödinger equation for the nuclear motion is obtained by expanding the exact eigenstate of H in electronic states $\Phi_n(\mathbf{r}, \mathbf{Q}_0)$ determined at a fixed nuclear configuration $\mathbf{Q}_0$:

$$\Psi(\mathbf{r}, \mathbf{Q}) = \sum_n \chi_n(\mathbf{Q}) \Phi_n(\mathbf{r}, \mathbf{Q}_0) \tag{2.13}$$

Of course, the $\chi_n(\mathbf{Q})$ in Eqs. (2.3) and (2.13) are not identical: they are merely expansion coefficients to be determined. The latter expansion leads via Eq. (2.11) to a matrix Schrödinger equation for the expansion coefficients. The corresponding matrix Hamiltonian reads:

$$\mathcal{H} = T_N \mathbf{1} + \mathbf{U}(\mathbf{Q}) \tag{2.14}$$

where the elements of the potential matrix $\mathbf{U}(\mathbf{Q})$ are given by

$$U_{nm}(\mathbf{Q}) = \int d\mathbf{r} \, \Phi_n^*(\mathbf{r}, \mathbf{Q}_0) \, H_e \Phi_m(\mathbf{r}, \mathbf{Q}_0) \tag{2.15}$$

and the electronic Hamiltonian H_e has been defined by Eq. (2.2b). As long as the basis sets $\Phi(\mathbf{r}, \mathbf{Q})$ and $\Phi(\mathbf{r}, \mathbf{Q}_0)$ are complete in the electronic space, the matrix Schrödinger equation (2.11) is exact for both the Hamiltonians (2.12) and (2.14) independently of the choice of $\mathbf{Q}_0$. These two matrix Hamiltonians are merely two different representations of the same operator.

2. The Adiabatic Approximation

The *adiabatic approximation*[2] (or Born–Oppenheimer adiabatic approximation, in the nomenclature of Ballhausen and Hansen[3]) is obtained by neglecting the nonadiabatic operator $\hat{\Lambda}$ in Eq. (2.12). The adiabatic approximation is based on the assumption that the kinetic energy operator of the nuclei can be considered as a small perturbation of the electronic motion. It has been advocated (see, e.g., Refs. 3 and 15) that a more useful approximation is obtained by neglecting only the off-diagonal terms of the nonadiabatic operator. In this case the adiabatic electronic wave functions Φ_n remain unchanged and the corresponding potential energy surfaces are slightly redefined to include the matrix elements G_{nn} given in Eq. (2.9b), that is, $V_n \to V_n + G_{nn}$. In the adiabatic approximation the matrix Hamiltonian $\mathcal{H}$ becomes diagonal and the total wave function (2.3) becomes a product of a nuclear and electronic wave function

$$\Psi(\mathbf{r}, \mathbf{Q}) = \chi_n(\mathbf{Q}) \Phi_n(\mathbf{r}, \mathbf{Q}) \tag{2.16}$$

The nuclear motion can be thought of as proceeding on the potential energy surface $V_n(\mathbf{Q})$ of a given electronic state characterized by the index n. On the other hand, if we put the matrix Hamiltonian (2.14) to be diagonal, the total wave function becomes

$$\Psi(\mathbf{r}, \mathbf{Q}) = \chi_n(\mathbf{Q}) \Phi_n(\mathbf{r}, \mathbf{Q}_0) \tag{2.17}$$

and the corresponding potential energy surface is $U_{nn}(\mathbf{Q})$. This approximation is termed the *crude adiabatic approximation*.[3,12]

Although the adiabatic approximation is a very appealing and often useful approach, it may fail in many cases, especially when potential energy surfaces of different electronic states are energetically close together. In these cases the nonadiabatic operators $\hat{\Lambda}_{nm}$ cannot be neglected in the Hamiltonian (2.12) for those electronic indices n and m which belong to the manifold of closely lying electronic states. These electronic states are now vibronically coupled via the quantities $\hat{\Lambda}_{nm}$. In practical applications we are not interested in solving the vibronic coupling problem [Eq. (2.11)] for all electronic states simultaneously. We rather aim at solving the vibronic problem for a subset (α) of electronic states that are vibronically decoupled from all other electronic states. Of course, the electronic states of the latter group, which we denote by (β), may also interact vibronically with each other. The adiabatic approximation is thus generalized to apply to groups of electronic states instead of being applied to individual states. More precisely, those nonadiabatic operators $\hat{\Lambda}_{nn'}$ are neglected for which one of the electronic indices n and n' belongs to the subset (α) and the other to complementary subset (β). The matrix Schrödinger equation (2.11) and the matrix Hamiltonian (2.12) still apply, but are now restricted to the members of the vibronically coupled subset (α).

3.   The Diabatic Basis

If we also restrict the matrix Hamiltonian (2.14) to a subset (α) of electronic states, the resulting Hamiltonians (2.12) and (2.14) represented in the adiabatic and crude adiabatic basis, respectively, are not equivalent anymore. Within this subset (α) the crude adiabatic states $\Phi_n(\mathbf{r}, \mathbf{Q}_0)$ cannot be expanded in the adiabatic states $\Phi_n(\mathbf{r}, \mathbf{Q})$ and vice versa. Clearly, the total wave function $\Psi(\mathbf{r}, \mathbf{Q})$ is better approximated using the adiabatic than the crude adiabatic electronic wave functions[62] and, hence, the matrix Hamiltonian (2.12) in the subset (α) is superior to the matrix Hamiltonian (2.14). Nevertheless, many authors prefer the latter Hamiltonian as a starting point of their investigations. The main reason for this choice is the complicated behavior of the nonadiabatic operators, which reflects the fast changes of the adiabatic electronic states with the nuclear coordinates in the vicinity of avoided crossings or conical intersections[23] of potential energy surfaces. These fast changes of the adiabatic electronic wave functions complicate considerably the solution of the Schrödinger equation. Furthermore, the ab initio determination of the nonadiabatic operators $\hat{\Lambda}_{nm}$ is generally tedious, especially for polyatomic molecules where several nuclear coordinates are involved. These difficulties should not make us forget the serious deficiencies of the crude adiabatic basis used to construct the Hamiltonian (2.14) for the subset (α) of the vibronically coupled electronic states. To overcome these deficiencies we may either increase considerably the number of basis functions $\Phi_n(\mathbf{r}, \mathbf{Q}_0)$, which is an inefficient and inelegant procedure, or replace the crude adiabatic functions by new functions $\phi_n(\mathbf{r}, \mathbf{Q})$ that are smooth and slowly varying functions of the nuclear coordinates and correspond to potential energy surfaces which may cross at the avoided crossings of the adiabatic potential energy surfaces. Such *diabatic*[11,14,63] electronic states can be constructed in several ways and are considered in the following discussion. The concept of diabatic states has been found to be useful in many fields ranging from atom–atom collisions to spectroscopy, and the problem of their construction has attracted numerous scientists. Therefore, here we can mention only some recent developments.

For simplicity, let us assume the subset (α) to contain only two vibronically interacting electronic states denoted by Φ_1 and Φ_2. In the adiabatic representation the matrix Hamiltonian describing the nuclear motion in these two states reads:

$$\mathcal{H} = T_N \mathbf{1} + \begin{bmatrix} V_1(\mathbf{Q}) - \hat{\Lambda}_{11} & -\hat{\Lambda}_{12} \\ -\hat{\Lambda}_{21} & V_2(\mathbf{Q}) - \hat{\Lambda}_{22} \end{bmatrix} \qquad (2.18)$$

New electronic states are introduced by the following orthogonal transformation

$$\begin{bmatrix} \phi_1(\mathbf{r}, \mathbf{Q}) \\ \phi_2(\mathbf{r}, \mathbf{Q}) \end{bmatrix} = \begin{bmatrix} \cos[a(\mathbf{Q})] & -\sin[a(\mathbf{Q})] \\ \sin[a(\mathbf{Q})] & \cos[a(\mathbf{Q})] \end{bmatrix} \begin{bmatrix} \Phi_1(\mathbf{r}, \mathbf{Q}) \\ \Phi_2(\mathbf{r}, \mathbf{Q}) \end{bmatrix} \qquad (2.19)$$

Using this basis to represent the matrix Hamiltonian, we arrive at a new "nonadiabatic" operator ($n, m = 1, 2$)

$$\lambda_{nm} = \sum_{i=1}^{M} f_{nm}^{(i)} \frac{\partial}{\partial Q_i} - g_{nm} \qquad (2.20a)$$

where

$$f_{nm}^{(i)} = 2 \sum_{j=1}^{M} \int d\mathbf{r} \, \phi_n^* \alpha_{ij} \frac{\partial \phi_m}{\partial Q_j} \qquad (2.20b)$$

and the less important quantity g_{nm} is determined straightforwardly from Eq. (2.9b). In our subset of two electronic states only the elements $f_{12}^{(i)} (= -f_{21}^{(i)})$ are different from zero.

The main idea behind the introduction of diabatic states is to choose the free parameter $a(\mathbf{Q})$ in the transformation (2.19) in such a way that the magnitude of f_{12} becomes as small as possible. Following Smith,[14] Baer,[64] and others[65] we use the transformation (2.19) to rewrite the expression for

f_{12}:

$$f_{12}^{(i)} = F_{12}^{(i)} + 2 \sum_{j=1}^{M} \alpha_{ij} \frac{\partial a(\mathbf{Q})}{\partial Q_j} \tag{2.21}$$

The superscript i runs over all nuclear coordinates Q_i, for $i = 1, 2, \ldots, M$. In many cases the kinematic matrix $\boldsymbol{\alpha} = \{\alpha_{ij}\}$ is diagonal and Eq. (2.21) is especially simple. Obviously, the best choice of $a(\mathbf{Q})$ is to make the non-adiabatic coupling elements $f_{12}^{(i)}$ vanish, i.e.,

$$F_{12}^{(i)}(\mathbf{Q}) = -2 \sum_{j=1}^{M} \alpha_{ij} \frac{\partial a(\mathbf{Q})}{\partial Q_j} \tag{2.22}$$

This set of differential equations has been discussed by several authors.[64-66] Baer[64,67] has proposed a scheme to solve approximately this set of differential equations for systems with several nuclear degrees of freedom and any number of vibronically interacting electronic states. In this scheme one assumes a distinct nuclear configuration $\mathbf{Q}_0$ at which the transformation from adiabatic to diabatic states is particularly simple and the solution of the set of differential equations is subsequently carried out in a propagative way. The scheme has been applied in several cases.[68] It has been pointed out by Mead and Truhlar[66] that for a polyatomic molecule Eq. (2.22) will have, strictly speaking, no solution. Furthermore, they stress that Baer's scheme and related schemes do have a strict solution only becuase they have approximated the nonadiabatic coupling using the diatomics-in-molecules approximation.[69] Consequently, the exact coupling elements f_{nm} are not fully transformed away by introducing diabatic states via an orthogonal transformation as in Eq. (2.19) for a subset of interacting electronic states. In most cases of practical importance it is clear, however, that the largest part of the coupling can be transformed away (see, e.g., Refs. 70 and 66). From the above the following heuristic definition of diabatic states emerges: a basis is considered as diabatic if Eq. (2.22) is fulfilled to a sufficiently good approximation. The extension of the discussion to an arbitrary number of vibronically interacting electronic states is obvious.

Other ways to determine a diabatic basis have been developed in the literature. They can be divided into two groups. In the first group, which includes Baer's method,[64] it is assumed that the nonadiabatic coupling elements in Eq. (2.9) are available. Consequently, these methods concentrate on the solution of differential equations in several dimensions. As already mentioned, the ab initio determination of the nonadiabatic operator for a polyatomic molecule is a tedious task. Nevertheless, some results have been reported in the literature (see, e.g., Refs. 71 and 72) and more are to be expected. The methods of the second group have in common that they all attempt to construct the diabatic basis without using the nonadiabatic coupling elements. Werner and Meyer,[73] for example, define a diabatic basis for the two lowest $^1\Sigma^+$ states of LiF as the one which diagonalizes the corresponding dipole operator. A more general method has been proposed by Hendekovic.[74] According to his definition, the diabatic basis maximizes the sum of the squared occupation probabilities of natural spin orbitals. A method which is in principle rigorous has been discussed by Özkan and Goodman.[75] Starting from the matrix Hamiltonian (2.14) in a formally complete crude adiabatic basis, they solve for the Schrödinger equation of the states of subset (α) using partitioning techniques. The idea is to treat the coupling between the two subsets (α) and (β) by perturbation methods while fully taking into account interactions within the primary subset (α) of vibronically coupled states. Following the same idea, we propose instead to construct an effective matrix Hamiltonian for the states of subset (α) using the quasi-degenerate perturbation theory.[76,77] The effective matrix Hamiltonian in the subset (α) then reads:

$$\mathscr{H} = \{\mathscr{H}_{ij}\}; \qquad i, j \in (\alpha)$$

$$\mathscr{H}_{ij} = T_N \delta_{ij} + U_{ij}(\mathbf{Q}) + \frac{1}{2} \sum_{n \in (\beta)} \left[\frac{U_{in}U_{nj}}{U_{jj} - U_{nn}} + \frac{U_{in}U_{nj}}{U_{ii} - U_{nn}} \right] + \cdots \qquad (2.23)$$

The states of the subset (α) are assumed to be energetically well separated from the states of subset (β). Remembering, furthermore, that $\mathbf{U}(\mathbf{Q})$ is diagonal at $\mathbf{Q}_0$, the perturbation series in Eq. (2.23) should converge fast for nuclear configurations not too far from the reference configuration $\mathbf{Q}_0$. Formally, $\mathscr{H}$ in Eq. (2.23) can serve to rigorously define diabatic states.

4. A Useful Approximate Hamiltonian

We assume that a diabatic basis has been obtained for the subset (α) of vibronically interacting electronic states introduced above. In this basis the matrix Hamiltonian is given by

$$\mathscr{H} = T_N \mathbf{1} + \mathbf{W}(\mathbf{Q}) \qquad (2.24a)$$

where the matrix elements of the potential matrix $\mathbf{W}(\mathbf{Q})$ read*

$$W_{nm}(\mathbf{Q}) = \int d\mathbf{r}\, \phi_n^*(\mathbf{r}, \mathbf{Q})\, H_e \phi_m(\mathbf{r}, \mathbf{Q}) \qquad (2.24b)$$

*If necessary, the elements g_{nn} in Eq. (2.20a) can be included in the definition of W_{nn}.

and the $\phi_n(\mathbf{r}, \mathbf{Q})$ are diabatic wave functions with $n \in (\alpha)$. For a polyatomic molecule the accurate solution of the matrix Schrödinger equation (2.11) using the matrix Hamiltonian (2.24) generally requires an extreme effort. Therefore, approximations must be introduced to make the problem amenable to practical computations. We shall see later that, in general, the matrix Schrödinger equation (2.11) cannot be attacked by perturbation methods. Consequently, we do not attempt here an approximate solution of Eq. (2.11) with the full Hamiltonian (2.24), but rather aim at the full solution of the Schrödinger equation using an approximate matrix Hamiltonian. The probably simplest approximate Hamiltonian which correctly describes the multi-mode molecular dynamics beyond the adiabatic approximation is discussed next. This Hamiltonian will be found to include as special cases the Hamiltonians successfully used to analyze the Jahn–Teller effect[29,78] and the dimer problem.[35,36]

Let us assume that the elements of the potential energy matrix $\mathbf{W}(\mathbf{Q})$ in Eq. (2.24) are slowly varying functions of $\mathbf{Q}$. Thus, we may expand $\mathbf{W} - V_0 \mathbf{1}$ about a reference nuclear configuration $\mathbf{Q}_0$. The choice of V_0 depends on the application in mind (see below). Without loss of generality $\mathbf{Q}_0$ is chosen to be the configuration used in the construction of the diabatic basis out of the adiabatic one; that is, the adiabatic and diabatic states are identical at $\mathbf{Q}_0$ (see Section II.A.3). Expanding up to the linear term in $\mathbf{Q}$, we obtain the following expressions for the elements of the matrix Hamiltonian:

$$
\begin{aligned}
\mathscr{H}_{nn} &= T_N + V_0(\mathbf{Q}) + E_n + \sum_s \kappa_s^{(n)} Q_s \\
\mathscr{H}_{nn'} &= \sum_s \lambda_s^{(n,n')} Q_s.
\end{aligned}
\tag{2.25}
$$

The energies E_n which appear in the diagonal of $\mathscr{H}$ are constants given by $W_{nn}(\mathbf{Q}_0)$. We call the quantities $\kappa_s^{(n)}$ and $\lambda_s^{(n,n')}$ *intrastate* and *interstate* electron-vibrational coupling constants, respectively.

To be more specific, we consider the application of the Hamiltonian (2.25) in spectroscopy. The molecule is initially in its ground electronic and vibrational state and is optically (or otherwise) excited into a manifold of vibronically interacting electronic states. To calculate the excitation spectrum we use the Hamiltonian (2.25). It is natural to choose in this case the reference geometry $\mathbf{Q}_0$ to be the equilibrium geometry of the molecule in its ground state and $V_0(\mathbf{Q})$ to be the potential energy surface of this state. The quantity E_n is the vertical energy difference at $\mathbf{Q}_0$ between the ground state potential energy surface and the corresponding surface of the nth member of the vibronically interacting electronic manifold. The coupling constants merely reflect the $\mathbf{Q}$-dependent changes in the Hamiltonian due to the excitation of the molecule. Introducing the auxiliary quantities

$$x_{snm}(\mathbf{Q}) = \int d\mathbf{r}\, \phi_n^* \frac{\partial U(\mathbf{r},\mathbf{Q})}{\partial Q_s} \phi_m \tag{2.26a}$$

$$y_{snm}(\mathbf{Q}) = \int d\mathbf{r}\, \phi_n^* \frac{\partial \phi_m}{\partial Q_s} \tag{2.26b}$$

the coupling constants simply read

$$\kappa_s^{(n)} = x_{snn}(\mathbf{Q}_0) \tag{2.27a}$$

$$\lambda_s^{(n,m)} = x_{snm}(\mathbf{Q}_0) + (E_n - E_m)\, y_{snm}(\mathbf{Q}_0) \tag{2.27b}$$

If the potential energy surfaces $V_n(\mathbf{Q})$ and $V_m(\mathbf{Q})$ happen to intersect at $\mathbf{Q}_0$, the second term in the expression (2.27b) for the interstate coupling constant vanishes. This second term should be negligible in most cases, since the diabatic states minimize the elements f_{nm} in Eq. (2.20b).

The nonvanishing interstate coupling constants $\lambda_s^{(n,m)}$ are those for which the product of the irreducible representations of ϕ_n and ϕ_m and of the nuclear coordinate Q_s contains the totally symmetric representation Γ_A:

$$\Gamma_n \times \Gamma_{Q_s} \times \Gamma_m \supset \Gamma_A \tag{2.28a}$$

The analogous condition for the intrastate coupling constants $\kappa_s^{(n)}$ is

$$\Gamma_n \times \Gamma_{Q_s} \times \Gamma_n \supset \Gamma_A \tag{2.28b}$$

Obviously all totally symmetric modes can couple to the electronic motion, emphasizing the important role of these modes in the vibronic coupling problem.

In most applications we have done so far (see Sections V to VII) the coupling constants have been determined by ab initio methods. To describe briefly how the calculations are carried out, we restrict ourselves to the typical example of two nondegenerate vibronically interacting electronic states, ϕ_1 and ϕ_2, of different spatial symmetry. Following the selection rules (2.28) for this example, only the totally symmetric modes give rise to nonzero intrastate coupling constants and only non-totally symmetric modes to nonzero interstate coupling constants. The former constants are obtained from

$$\kappa_s^{(n)} = \left(\frac{\partial V_n(\mathbf{Q})}{\partial Q_s} \right)_{\mathbf{Q}_0}; \qquad n = 1, 2 \tag{2.29a}$$

The interstate coupling constants determine the repulsion of the adiabatic potential energy surfaces $V_1(\mathbf{Q})$ and $V_2(\mathbf{Q})$ in the vicinity of the reference geometry $\mathbf{Q}_0$. These constants are deduced from a comparison of the repulsion of the ab initio computed surfaces with the repulsion of the surfaces that result from the approximate Hamiltonian. The interstate coupling constant $\lambda_s^{(1,2)}$ is sensitive to the vertical energy difference between the surfaces. Consequently, the ab initio method used to compute these potential energy surfaces should be able to describe both surfaces to the same accuracy. (For more details see Refs. 79 and 80). Note that $\lambda_s^{(1,2)}$ can also be determined by differentiation of the square of the repulsion energy with respect to the non-totally symmetric nuclear coordinate Q_s. A useful formula is[81]

$$\lambda_s^{(1,2)} = \left\{ \frac{1}{8} \frac{\partial^2}{\partial Q_s^2} |V_1(\mathbf{Q}) - V_2(\mathbf{Q})|^2 \right\}_{\mathbf{Q}_0}^{1/2} \qquad (2.29b)$$

To conclude this section let us briefly discuss the range of applicability of the matrix Hamiltonian (2.25). In spectroscopy, for example, the Franck–Condon zone plays a crucial role in the explanation of the spectrum. The Franck–Condon zone is defined as that region in nuclear coordinate space where the absolute value of the initial state vibrational wave function is appreciable. Using a moment analysis, it has been shown[82,83] that the gross features of the spectrum solely depend on the properties of the final states potential energy surfaces *within* the Franck–Condon zone. Indeed, the approximate Hamiltonian (2.25) has been determined from the full Hamiltonian (2.24) by expanding the latter about the center of the Franck–Condon zone (if $\mathbf{Q}_0$ is chosen to be the equilibrium geometry of the molecule). Assuming the molecule to be initially in its vibrational ground state, the Franck–Condon zone is particularly narrow and the expansion is appropriate. Of course, the fine structure of the spectrum depends on the overall shape of the potential energy. In particular one should take care that possible critical points of the final state's potential energy surfaces (e.g., conical intersections) that are relevant for the spectrum are correctly described by Eq. (2.25) if they are located far outside the Franck–Condon zone. When the ground state potential energy surface $V_0(\mathbf{Q})$ is harmonic in the vicinity of the Franck–Condon zone, it is useful to define the Q_s as the normal coordinates* of the molecules in its ground state and write the following for the

---

*For convenience we use here dimensionless normal coordinates, which are obtained from those of Wilson et al.[61] by multiplying with $\sqrt{\omega}$.

ground state Hamiltonian H_0 ($\hbar = 1$ throughout):

$$H_0 \equiv T_N + V_0(\mathbf{Q}) = -\frac{1}{2} \sum_{s=1}^{M} \omega_s \frac{\partial^2}{\partial Q_s^2} + \frac{1}{2} \sum_{s=1}^{M} \omega_s Q_s^2 - \frac{1}{2} \sum_{s=1}^{M} \omega_s \quad (2.30)$$

Here the ω_s terms are the harmonic vibrational frequences. The vibrational ground state energy has been chosen to be the zero of the energy scale. Obviously, introducing the approximation (2.30) restricts the Hamiltonian (2.25) to the description of bound state problems. If problems involving a nuclear continuum are to be investigated, one should return to the full Hamiltonian (2.24) and introduce more appropriate approximations. Finally we mention that—as will become clear in Section II.B—the full solution of the Schrödinger equation with the approximate Hamiltonian in Eqs.(2.25) and (2.30) is already a cumbersome task when several nuclear degrees of freedom are involved.

B. The Excitation Spectrum

1. General Remarks

Assume that a molecule initially in the state Ψ_0 is excited by some mechanism described by the operator $\hat{T}$ into a manifold of vibronically interacting electronic states. According to Fermi's golden rule, the excitation spectrum is described by the function

$$P(E) = 2\pi \sum_{\nu} |\langle \Psi_0 | \hat{T} | \Psi_\nu \rangle|^2 \delta(E - E_\nu) \quad (2.31)$$

where E_ν denotes the energies of the final molecular vibronic states Ψ_ν. These vibronic states and energies emerge as solutions of the matrix Schrödinger equation for the subset (α) of interacting electronic states as discussed in the preceding sections. The energy of the molecule in its initial state Ψ_0 has been chosen as the zero of the energy scale. For example, when the spectrum (2.31) is an optical absorption spectrum, E is the photon energy and the operator $\hat{T}$ describes the interaction of the electrons with the electromagnetic field.

The excitation spectrum is easily rewritten in a more compact form that does not explicitly exhibit the final states and energies:

$$P(E) = \int dt\, e^{iEt} \langle \Psi_0 | \hat{T} e^{-iHt} \hat{T}^\dagger | \Psi_0 \rangle \quad (2.32)$$

We eliminate the electronic coordinates by representing the molecular

Hamiltonian H in the diabatic basis and expanding the exact initial state Ψ_0 in the same basis

$$H = \sum_{n, n'} |\phi_n\rangle \mathscr{H}_{nn'} \langle \phi_{n'}| \qquad (2.33a)$$

$$\Psi_0 = \sum_n \chi_{n0}(\mathbf{Q}) \phi_n(\mathbf{r}, \mathbf{Q}) \qquad (2.33b)$$

The excitation spectrum (2.32) now reads

$$P(E) = \sum_{n, m} \int dt\, e^{iEt} \langle \chi_{n0} | \mathbf{T}_n^\dagger e^{-i\mathscr{H}t} \mathbf{T}_m | \chi_{m0} \rangle \qquad (2.34)$$

where $\mathscr{H}$ is the matrix Hamiltonian describing the nuclear motion in the manifold (α) of vibronically interacting electronic states and $\mathbf{T}_n$ denotes the column vector of generalized oscillator strength amplitudes for excitations from the diabatic state ϕ_n to all the members of this subset (α); that is,

$$\mathbf{T}_n^\dagger = (\langle \phi_n | \hat{T} | \phi_0 \rangle, \langle \phi_n | \hat{T} | \phi_1 \rangle, \dots) \qquad (2.35)$$

Equation (2.34) gives the excitation spectrum within the context of the golden rule. No assumptions have been made so far about the initial state out of which the molecule is excited. It can be a member of the same subset (α) of vibronically interacting states as are the excited states. In this case Ψ_0 is an eigenstate of the matrix Hamiltonian $\mathscr{H}$ in Eq. (2.34), or, more precisely, the expansion coefficients χ_{n0} in Eq. (2.33b) are the elements of an eigenvector of $\mathscr{H}$ [see Eq. (2.11)]. Of course, Ψ_0 can also be an eigenstate of another matrix Hamiltonian describing the nuclear motion in an electronic manifold different from subset (α). This more complicated situation requires the solution of two matrix Schrödinger equations. In the majority of cases of interest the initial state is the molecular ground state. Furthermore, the ground state is usually energetically well separated from the excited states and does not interact with them vibronically. In other words, the adiabatic approximation applies to the ground state and the corresponding wave function can be written as a product of an electronic wave function and a nuclear wave function

$$\Psi_0(\mathbf{r}, \mathbf{Q}) = \phi_0(\mathbf{r}, \mathbf{Q}) \chi_{00}(\mathbf{Q}) \qquad (2.36)$$

It should be noted that there is no need to distinguish between the adiabatic and diabatic electronic state in this case ($\Phi_0 = \phi_0$). The use of the adiabatic approximation for the initial state simplifies the expression for the (electronic) excitation spectrum, which now reads

$$P(E) = \int dt \, e^{iEt} \langle 0|\tau^{\dagger} e^{-i\mathscr{H}t} \tau|0\rangle \tag{2.37}$$

where $|0\rangle$ is a shorthand notation for the initial nuclear state $|\chi_{00}\rangle$. If not otherwise stated, we assume for simplicity that $|0\rangle$ is the vibrational ground state. Here τ is a column vector of generalized oscillator strength amplitudes τ_n for the final electronic states

$$\tau_n^* = \langle \phi_0|\hat{T}|\phi_n\rangle \tag{2.38}$$

The amplitudes τ_n would have been strictly independent of the nuclear coordinates $\mathbf{Q}$ if $\mathscr{H}$ were represented in a crude adiabatic basis; since here we are using a diabatic basis, they depend weakly on $\mathbf{Q}$. This weak dependence on $\mathbf{Q}$ can safely be neglected for most purposes and should not be confused with the possibly strong $\mathbf{Q}$ dependence that is found when the Hamiltonian is represented in the adiabatic basis. Another point should be mentioned here. The oscillator strengths depend on several parameters. For example, they depend on the energy and polarization of the incident photon in the case of optical absorption and on the kinetic energy and ejection angle of the photoelectron in photoionization. These parameters, which are of minor importance here, are hidden in $\hat{T}$. When studying model applications and in the discussion of photoelectron spectra, we simply take the τ_n to be given constants. This approximation is accurate if the energy of the photoelectron is large enough compared to the width of the observed spectrum and the electron is not caught in a resonance or autoionizing state. In optical absorption spectra the oscillator strengths $|\tau_n|^2$ are proportional to the photon energy.

The excitation spectrum (2.37) exhibits a series of lines that are represented by δ functions. By formally performing the integration in Eq. (2.37) we obtain

$$P(E) = 2\pi\langle 0|\tau^{\dagger} \delta(E - \mathscr{H}) \tau|0\rangle \tag{2.39}$$

To account for the finite experimental resolution and for degrees of freedom not considered here (e.g., rotation), we convolute $P(E)$ with a Lorentzian $L(E)$ of fwhm (full width at half maximum) Γ:

$$L(E) = \frac{1}{\pi} \frac{\Gamma/2}{E^2 + (\Gamma/2)^2} \tag{2.40a}$$

The result, which we call the *envelope* of the spectrum, reads

$$P_\Gamma(E) = 2\langle 0|\tau^\dagger \frac{\Gamma/2}{(E - \mathscr{H})^2 + (\Gamma/2)^2} \tau|0\rangle \qquad (2.40b)$$

Clearly $P_\Gamma(E)$ approaches $P(E)$ when $\Gamma \to 0^+$. If Γ is of the size of or larger than the typical line spacings in the spectrum (2.39), $P_\Gamma(E)$ becomes the "band shape" of the spectrum. In some cases, for example, the Jahn–Teller effect, the band shape may show characteristic features that are observed in low-resolution experiments and can be used to recognize the underlying phenomenon. This point is addressed in Section IV. In all our applications the discrete line spectrum together with its envelope are shown.

Excitation spectra in some simple special cases are considered in the following discussion. The approximate matrix Hamiltonian introduced in Section II.A.4 is used. Hence, $|0\rangle$ is the ground state of a multidimensional harmonic oscillator. When all the interstate coupling constants are put equal to zero, all the nonadiabatic effects vanish and the spectrum is particularly simple. For each final electronic state ϕ_f, the spectrum is the well-known spectrum of a shifted multidimensional harmonic oscillator. All nuclear motions decouple and, consequently, the spectrum is a convolution of the spectra of the individual one-dimensional oscillators. Each of the latter spectra can be written as

$$P(E) = 2\pi|\tau_f|^2 \exp(-a_f) \sum_{n=0}^\infty \frac{a_f^n}{n!} \delta\left(E - E_f + a_f^2\omega - n\omega\right) \qquad (2.41)$$

which is a series of equidistant peaks weighted by a Poisson distribution with a Poisson parameter $a_f = \frac{1}{2}(\kappa_f/\omega)^2$. Another simplification arises if the intrastate coupling constants $\kappa_s^{(n)}$ are equal for all the diabatic states ϕ_n in $\mathscr{H}$ for one nuclear coordinate Q_s. This nuclear degree of freedom can then be treated separately. The final spectrum is a convolution of the spectrum of the shifted harmonic oscillator in Q_s and the combined spectrum of all the remaining nuclear coordinates.

To make clear what we mean when we state that a nuclear degree of freedom can be treated separately, we decompose the full matrix Hamiltonian $\mathscr{H}$ into two parts

$$\mathscr{H} = \mathscr{H}_s + \mathscr{H}_r \qquad (2.42a)$$

where $\mathscr{H}_s$ is a function of Q_s and $\partial/\partial Q_s$ and all the other nuclear coordinates and momenta are contained in $\mathscr{H}_r$. The mode "s" decouples from the other modes and can be treated separately if

$$[\mathscr{H}_s, \mathscr{H}_r] = 0 \tag{2.42b}$$

Then the potential energy surfaces decompose as $V_n = V_n(Q_s) + V_n(\mathbf{Q}_r)$. Furthermore, the excitation spectrum can be rewritten to give

$$P(E) = 2\pi\tau^\dagger \int dE' \langle 0_s | \delta(E' - \mathscr{H}_s) | 0_s \rangle \langle 0_r | \delta(E - E' - \mathscr{H}_r) | 0_r \rangle \tau \tag{2.42c}$$

where it is assumed that the τ's are constants and the initial vibrational state can be factorized according to $|0\rangle = |0_s\rangle |0_r\rangle$. For the simple case discussed above (the $\kappa_s^{(n)}$ are equal for all ϕ_n), for example, the vector $\tau^\dagger$ in Eq. (2.42c) commutes with $\mathscr{H}_s$ and we immediately obtain

$$P(E) = \frac{1}{2\pi} \int dE' \, P_s(E') P_r(E - E') \tag{2.42d}$$

Here $P_s(E)$ is identical with the Poisson spectrum in Eq. (2.41) when we put $\tau_f = 1$, $E_f = 0$, and $a_f = \frac{1}{2}(\kappa_s/\omega_s)^2$; $P_r(E)$ is the full excitation spectrum in the absence of the mode "s."

2. The Spectrum in the Adiabatic and Franck–Condon Approximations

We refer to excitation spectra (2.34) and (2.37) investigated in the preceding section as the *exact* excitation spectra. In this subsection we shall discuss the well-known adiabatic and Franck–Condon approximations to the exact spectra. In molecular spectroscopy these approximations are commonly used to interpret and to calculate electronic spectra. We shall concentrate on the spectrum (2.37), bearing in mind that the initial electronic state does not interact vibronically with the excited electronic states.

In the adiabatic approximation the nonadiabatic operators $\hat{\Lambda}_{nm}$ introduced in Section II.A are neglected. To make use of this definition we first diagonalize the potential part of the matrix Hamiltonian $\mathscr{H}$ appearing in the exact spectrum (2.37):

$$[\mathscr{H} - T_N \mathbf{1}] \mathbf{S}(\mathbf{Q}) = \mathbf{S}(\mathbf{Q}) \mathbf{V}(\mathbf{Q})$$
$$\mathbf{S}\mathbf{S}^\dagger = \mathbf{1} \tag{2.43}$$

Here the eigenvalue matrix $\mathbf{V}$ is the diagonal matrix of the adiabatic potential energy surfaces [see Eq. (2.12)] and $\mathbf{S}$ is the corresponding orthogonal eigenvector matrix. The latter matrix has a simple meaning too: it transforms the adiabatic states Φ_n to the diabatic states ϕ_n. Introducing the column vectors $\mathbf{\Phi}$ and ϕ of adiabatic and diabatic states in the primary subset (α) of vibronically interacting electronic states, we may write

$$\phi = \mathbf{S}\mathbf{\Phi} \qquad (2.44)$$

For the special case of two interacting states in subset (α), this transformation has been discussed in Section II.A.3. With the aid of the transformation matrix $\mathbf{S}$ we can easily rewrite the exact spectrum in Eq. (2.39) and obtain

$$P(E) = 2\pi \langle 0|\tilde{\tau}^\dagger \, \delta(E - \tilde{\mathscr{H}}) \, \tilde{\tau}|0\rangle \qquad (2.45a)$$

where the new oscillator strengths

$$\tilde{\tau} = \mathbf{S}^\dagger(\mathbf{Q})\tau \qquad (2.45b)$$

now refer to the adiabatic basis and thus may depend severely on the nuclear coordinates. The matrix Hamiltonian $\tilde{\mathscr{H}}$ appearing above

$$\tilde{\mathscr{H}} = \mathbf{S}^\dagger \mathscr{H} \mathbf{S} = T_N \mathbf{1} + \mathbf{V}(\mathbf{Q}) - \hat{\mathbf{\Lambda}} \qquad (2.45c)$$

$$\hat{\mathbf{\Lambda}} = -\mathbf{S}^\dagger [T_N, \mathbf{S}] \qquad (2.45d)$$

is the matrix Hamiltonian in the adiabatic basis and is, of course, identical to the one in Eq. (2.12) except that it is restricted here to the electronic states of subset (α).

The excitation spectrum in the adiabatic approximation—or briefly the adiabatic spectrum—is now determined from the exact spectrum in Eq. (2.45a) by neglecting the nonadiabatic operator $\hat{\mathbf{\Lambda}}$. The adiabatic spectrum reads

$$P_{ad}(E) = 2\pi \langle 0|\tilde{\tau}^\dagger \, \delta(E - T_N \mathbf{1} - \mathbf{V}) \, \tilde{\tau}|0\rangle \qquad (2.46)$$

Since the Hamiltonian $T_N \mathbf{1} + \mathbf{V}(\mathbf{Q})$ is diagonal, we can define vibrational states for each adiabatic electronic state Φ_n according to

$$[T_N + V_n(\mathbf{Q}) - E_{ni}]\chi_{ni} = 0 \qquad (2.47)$$

and calculate the adiabatic spectrum

$$P_{ad}(E) = 2\pi \sum_n \sum_i |\langle \chi_{ni}|\tilde{\tau}_n|0\rangle|^2 \delta(E - E_{ni}) \qquad (2.48)$$

The spectrum in the adiabatic approximation is thus given by a superposition of independent spectra resulting from the adiabatic potential energy surfaces $V_n(\mathbf{Q})$.

The excitation spectrum in the Franck–Condon approximation—or briefly the Franck–Condon spectrum—is obtained from the adiabatic spectrum by replacing the $\mathbf{Q}$-dependent oscillator strengths $\tilde{\tau}_n(\mathbf{Q})$ by $\tilde{\tau}_n(\mathbf{Q}_0)$; $\mathbf{Q}_0$ is commonly chosen to be the equilibrium geometry of the molecule in its ground state. If this nuclear configuration coincides with our choice of reference geometry for the construction of the diabatic basis from the adiabatic basis (see Section II.A.3), the $\tilde{\tau}_n(\mathbf{Q}_0)$ are essentially equal to the oscillator strengths τ_n in the diabatic basis. One obtains the following expression for the Franck–Condon spectrum

$$P_{FC}(E) = 2\pi \sum_n |\tilde{\tau}_n(\mathbf{Q}_0)|^2 \sum_i |\langle \chi_{ni}|0\rangle|^2 \delta(E - E_{ni}) \qquad (2.49)$$

The overlaps $|\langle \chi_{ni}|0\rangle|^2$ are the so-called Franck–Condon factors. The Franck–Condon spectrum is given by the superposition of independent Franck–Condon spectra, one for each electronic state. Each of these spectra has its own weight factor $|\tilde{\tau}_n(\mathbf{Q}_0)|^2$. The positions of the vibrational levels are the same in both the adiabatic and Franck–Condon spectra. The intensities of these levels are different in both spectra owing to the $\mathbf{Q}$ dependence of the oscillator strengths $\tilde{\tau}_n$ appearing in the adiabatic spectrum. This dependence accounts for the well-known intensity *borrowing effects* which are often calculated using the Herzberg–Teller expansion[84] of the electronic wave function. The expansion of the $\tilde{\tau}_n$ about $\mathbf{Q}_0$ can explain the appearance of forbidden electronic transitions and the excitation of single quanta of non-totally symmetric modes that cannot be accounted for in the Franck–Condon approximation.

For some purposes it is advantageous to eliminate the $\mathbf{Q}$ dependence of the oscillator strengths in the adiabatic spectrum. Using Eqs. (2.43) and (2.45b) we immediately arrive at the alternative expression for the adiabatic spectrum

$$P_{ad} = 2\pi \langle 0|\tau^\dagger \, \delta(E - \overline{\mathcal{H}}) \, \tau|0\rangle \qquad (2.50a)$$

where

$$\overline{\mathcal{H}} = \mathcal{H} + \mathbf{S}\hat{\Lambda}\mathbf{S}^\dagger \qquad (2.50b)$$

is the Hamiltonian $T_N \mathbf{1} + \mathbf{V}(\mathbf{Q})$ describing the nuclear motion in the adiabatic potential energy surfaces but represented in the diabatic basis. We have thus obtained the adiabatic spectrum by adding a term to the matrix Hamiltonian appearing in the exact spectrum (2.39). From Eq. (2.45d) we see that this term is simply given by $S[T_N, S^\dagger]$, which contains derivatives of the transformation matrix with respect to the nuclear coordinates

$$\frac{\partial S_{nm}}{\partial Q_s} = -\sum_q \langle \phi_n | \frac{\partial}{\partial Q_s} | \phi_q \rangle S_{qm} + \sum_q \langle \Phi_q | \frac{\partial}{\partial Q_s} | \Phi_m \rangle S_{nq} \qquad (2.51)$$

The first term on the right-hand side of this equation is negligible owing to the smoothness of the diabatic wave functions ϕ_i with respect to the nuclear coordinates. The second term on the right-hand side of Eq. (2.51) is more relevant and deserves special attention. Differentiating $\langle \Phi_q | H_e | \Phi_m \rangle$ with respect to Q_s one easily arrives at

$$\langle \Phi_q | \frac{\partial}{\partial Q_s} | \Phi_m \rangle = -\left[V_q(\mathbf{Q}) - V_m(\mathbf{Q}) \right]^{-1} \langle \Phi_q | \frac{\partial U(\mathbf{r}, \mathbf{Q})}{\partial Q_s} | \Phi_m \rangle \quad (2.52)$$

for $q \neq m$. At points of degeneracies of the potential energy surfaces V_q and V_m (e.g., at a conical intersection), the off-diagonal matrix elements of the first derivative and also of higher derivatives diverge because $\langle \Phi_q | \partial U / \partial Q_s | \Phi_m \rangle \neq 0$, in general.

The divergencies of the matrix Hamiltonian $\overline{\mathscr{H}}$ at intersections of potential energy surfaces have consequences on the excitation spectrum calculated in the adiabatic approximation. To examine their influence on the spectrum we briefly discuss the moments of the spectrum. It is well-known that the first few moments of a spectrum determine the gross features of this spectrum.[85] The kth moment of the spectrum $P(E)$ is defined by

$$M_k = \int_{-\infty}^{\infty} E^k P(E) \, dE \qquad (2.53)$$

To compute the moments of the exact and adiabatic spectra we use Eqs. (2.39) and (2.50) and obtain immediately

$$M_k = 2\pi \langle 0 | \tau^\dagger \, \mathscr{H}^k \, \tau | 0 \rangle \qquad (2.54a)$$

$$M_k^{ad} = 2\pi \langle 0 | \tau^\dagger \, \overline{\mathscr{H}}^k \, \tau | 0 \rangle \qquad (2.54b)$$

The zeroth moment M_0 determines the total intensity of the excitation spectrum

$$M_0 = M_0^{ad} = 2\pi \sum_n |\tau_n|^2 \tag{2.55}$$

The quotients M_1/M_0 and $(M_2 - M_1^2)/M_0$ determine the centroid and the width of the spectrum, respectively. Whereas these quantities are well defined for the exact and Franck–Condon spectrum, they—and all higher moments—diverge for the adiabatic spectrum in the presence of conical and other intersections. The centroid of the adiabatic spectrum thus lies at infinite energy, and the spectral width is infinite too. This is explicitly demonstrated for a Jahn–Teller situation in Section III.B. The divergence of the adiabatic spectral moments in the presence of conical intersections is an inevitable feature of any adiabatic approximation and reflects its pathological nature in the conical intersection situation. How this divergence affects the calculated adiabatic spectrum is studied in Section III.B.

It is important to note that the eigenvector matrix $\mathbf{S}$ is not uniquely determined by the requirement that it diagonalizes $(\mathcal{H} - T_N \mathbf{1})$. We may multiply each eigenvector (column of $\mathbf{S}$) by a phase factor which may be an arbitrary function of the nuclear coordinates. Consequently, there is an infinite manifold of adiabatic approximations which will yield different results especially when a conical intersection is present. This reflects the fact that the adiabatic approximation is an inherently inconsistent approximation. For the conventional choice[20,24] of the phase factor, $\mathbf{S}$ and thus the adiabatic electronic wave functions are not single-valued functions of the nuclear coordinates in a conical intersection situation. Another choice of the phase factor discussed in Ref. 30 and Section III leads to single-valued but discontinuous adiabatic electronic wave functions. These wave functions become continuous functions of the nuclear coordinates within the Franck–Condon zone when the intersection moves out of this zone. Mead and Truhlar[86] have used the freedom in the choice of the phase factor to make the adiabatic electronic wave functions single valued and continuous along any loop enclosing the point of intersection. This introduces diagonal elements of $\hat{\Lambda}$ which are not present in the conventional[20,24] choice of the phase factor and which have to be included in the calculation of the vibrational states. It is easily verified for the special case of the $E \times \varepsilon$ Jahn–Teller problem that this procedure yields identical results as obtained by choosing double-valued electronic wave functions and double-valued vibrational wave functions to make the total adiabatic wave function [Eq. (2.16)] single valued.[87] With these modified definitions of the adiabatic approximation, one still cannot eliminate the discontinuity of the adiabatic electronic wave functions at the point of intersection itself, and the adiabatic spectral moments diverge. We are not interested here in searching for the best adiabatic approximation in a conical intersection situation. We rather wish to point out that when the conventional approximation fails severely any other choice of the adiabatic approximation fails too.

3. Numerical Methods

In general, it is not possible to evaluate the excitation spectrum analytically, even when we are using the approximate Hamiltonian discussed in Section II.A.4. Moreover, perturbation theory fails severely in approximating the spectrum in situations where the potential energy surfaces intersect. Consequently, we must resort to numerical methods for computing the excitation spectrum. In what follows we shall discuss the numerical methods which have been used to calculate the exact, Franck–Condon, and adiabatic excitation spectra.

We start with the exact spectrum (2.37) and consider a manifold of N vibronically interacting electronic states, that is, where the Hamiltonian $\mathscr{H}$ is a $N \times N$ matrix operator. We may classify the N diabatic electronic states according to their irreducible representations in the symmetry point group of the molecule in its equilibrium geometry (each component of a degenerate electronic state is counted separately). As usually done for the Jahn–Teller problem[78] $\mathscr{H}$ is represented in a basis of the vibrational states $|n_1 n_2 \cdots n_M\rangle$ chosen to be the eigenstates of the nuclear Hamiltonian in the electronic ground state

$$[T_N + V_0(\mathbf{Q})]|\mathbf{n}\rangle = E_\mathbf{n}|\mathbf{n}\rangle \tag{2.56}$$

The n_s are vibrational quantum numbers and the compact notation $|\mathbf{n}\rangle = |n_1 n_2 \cdots n_M\rangle$ has been used. One obtains a supermatrix with $N \times N$ matrices $\mathscr{H}_{\mathbf{nm}}$ as elements

$$\mathscr{H}_{\mathbf{nm}} \equiv \langle \mathbf{n}|\mathscr{H}|\mathbf{m}\rangle \tag{2.57}$$

Fortunately, the supermatrix often decouples into several submatrices $\mathbf{M}_J$, each being specified by a *vibronic symmetry* index J. When the molecular point group is Abelian, the situation is particularly simple and J denotes irreducible representations Γ_J of the point group. Thus, the maximum number of submatrices $\mathbf{M}_J$ is the number of irreducible representations in the point group. Only those $\mathbf{M}_J$ play a role in the calculation of the spectrum for which at least one diabatic electronic state of the interacting manifold transforms as Γ_J. This reduces the number of submatrices to be diagonalized. For non-Abelian point groups the problem is somewhat more complicated, but the solution is still straightforward. For linear molecules, for example, the projection of the total electronic-vibrational angular momentum on the molecular axis must commute with $\mathscr{H}$. Consequently, the index

J specifies a vibronic angular momentum quantum number which takes the values $0, \pm 1, \pm 2, \ldots$.

To compute the excitation spectrum one has to diagonalize the relevant submatrices $\mathbf{M}_J$. The eigenvalues of those matrices give the positions of the vibronic lines in the spectrum and the corresponding eigenvectors together with the oscillator strength vector τ determine the intensity of these lines. Assuming that k diabatic electronic states $\phi_1, \phi_2, \ldots, \phi_k$ transform as Γ_J, the intensity of a line ν in the spectrum reads

$$I_\nu = \left| \sum_{i=1}^{k} \tau_i x_{i\nu} \right|^2 \qquad (2.58)$$

where $x_{i\nu}$ denotes the ith component of the νth eigenvector of $\mathbf{M}_J$. The problem simplifies and becomes more amenable to numerical calculations if each diabatic state transforms differently. The intensity of a line in the spectrum with vibronic symmetry J is then obtained by multiplying the square of the first element of the corresponding eigenvector of $\mathbf{M}_J$ by $|\tau_i|^2$, where ϕ_i is the diabatic state which transforms as Γ_J. Since the dependence of the τ's on the nuclear coordinates is, in general, negligible, we may introduce a Q-independent orthogonal transformation which transforms all but one of the τ_i in Eq. (2.58) to zero. Consequently, only a single element of each eigenvector will be needed in the computation of the spectral intensities.

The above submatrices $\mathbf{M}_J$ are particularly simple when the approximate matrix Hamiltonian $\mathscr{H}$ in Eqs. (2.25) and (2.30) is used. The matrix elements of $\mathscr{H}$ in the harmonic oscillator basis read

$$(\mathscr{H}_{ii})_{\mathbf{nm}} = \left[E_i + \sum_{s=1}^{M} \omega_s n_s \right] \delta_{\mathbf{nm}} + \sum_{s=1}^{M} \kappa_s^{(i)} f_{n_s m_s} \prod_{s' \neq s}^{M} \delta_{n_{s'} m_{s'}} \qquad (2.59a)$$

$$(\mathscr{H}_{ij})_{\mathbf{nm}} = \sum_{s=1}^{M} \lambda_s^{(i,j)} f_{n_s m_s} \prod_{s' \neq s}^{M} \delta_{n_{s'} m_{s'}}, \qquad i \neq j \qquad (2.59b)$$

where

$$f_{nm} = \left[\frac{m+1}{2} \right]^{1/2} \delta_{n, m+1} + \left[\frac{m}{2} \right]^{1/2} \delta_{n, m-1} \qquad (2.59c)$$

Obviously, $\mathscr{H}$ expressed in the basis of harmonic oscillator functions is a sparse matrix. The submatrices $\mathbf{M}_J$ can be arranged to be so-called *banded* matrices. For banded matrices there are efficient diagonalization procedures available.[88] The numerical effort in solving the secular equation is especially low when only one element of each eigenvector is computed.

As an example we briefly discuss a two-state two-mode problem in a molecule with an Abelian symmetry point group. Let ϕ_1 and ϕ_2 be the diabatic states and ω_a and ω_b be the two vibrational modes which transform as the irreducible representations Γ_1, Γ_2, Γ_a, and Γ_b, respectively; Γ_a is assumed to be the totally symmetric species. Further we have

$$\Gamma_1 \times \Gamma_2 = \Gamma_b$$

implying that the two diabatic electronic states are coupled vibronically by the non-totally symmetric mode ω_b. The matrix Hamiltonian of this problem takes on the appearance

$$\mathcal{H} = H_0 \mathbf{1} + \begin{pmatrix} E_1 + \kappa_a^{(1)} Q_a & \lambda_b^{(1,2)} Q_b \\ \lambda_b^{(1,2)} Q_b & E_2 + \kappa_a^{(2)} Q_a \end{pmatrix} \tag{2.60a}$$

where the initial state Hamiltonian reads

$$H_0 = \frac{1}{2} \omega_a \left[-\frac{\partial^2}{\partial Q_a^2} + Q_a^2 \right] + \frac{1}{2} \omega_b \left[-\frac{\partial^2}{\partial Q_b^2} + Q_b^2 \right] \tag{2.60b}$$

As discussed above, the solution of the matrix Schrödinger equation (2.11) requires the diagonalization of the two submatrices $\mathbf{M}_1$ and $\mathbf{M}_2$.

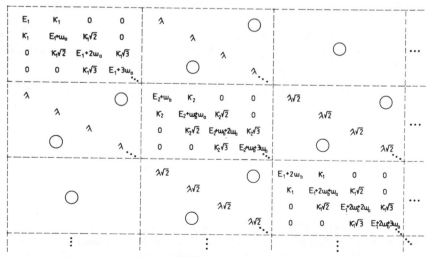

Fig. 1. The submatrix $\mathbf{M}_1$ of the two-state problem discussed in the text. To compare with the text, put $\kappa_1 \equiv \kappa_a^{(1)}$, $\kappa_2 \equiv \kappa_a^{(2)}$, and $\lambda \equiv \lambda_b^{(1,2)}$.

The submatrix M_1 is depicted in Fig. 1, and the submatrix M_2 is easily obtained from it by interchanging the indices "1" and "2" of the electronic states. Putting the interstate coupling constant λ equal to zero, M_1 decouples into a series of tridiagonal matrices. These matrices, out of which only the upper one in Fig. 1 is of relevance for the spectrum, describe shifted harmonic oscillators with frequency ω_a. The corresponding spectrum is given in Eq. (2.41). Neglecting, on the other hand, the intrastate coupling constants, M_1 can again be rearranged to be a tridiagonal matrix, but the spectrum can no longer be given in closed form except when $E_1 = E_2$. In the general case we must diagonalize a banded matrix in order to obtain the spectrum. A matrix with elements a_{ij} is termed banded when $a_{ij} = 0$ for $|i - j| > m$. The bandwidth of this matrix is $2m + 1$. When more than a single vibrational mode is considered, the bandwidth of M_1 depends on the number L_s of oscillator states $|n_s\rangle$ used as a basis in the expansion of $\mathcal{H}$. For M nuclear degrees of freedom being taken into account in the two-state problem, the bandwidth BW of M_1 simply reads

$$BW = 2 \prod_{s=1}^{M-1} L_s + 1 \tag{2.61}$$

where L_M is chosen to be the largest of all the L_s. The dimension of the matrix is, of course, given by the product of all the L_s. Since the numerical effort to diagonalize M_1 depends severely on the bandwidth, special care should be taken in the arrangement of this matrix. In practice we make an initial guess as to the numbers L_s according to the magnitude of the corresponding parameters $(\kappa_s/\omega_s)^2$ or $(\lambda_s/\omega_s)^2$ and subsequently vary these numbers to investigate convergence. It is important to rearrange the matrix M_1 in Fig. 1 in such a way that the matrix begins with the mode with the smallest L_s and ends with the mode Q_M. Otherwise, the resulting bandwidth might be larger than BW in Eq. (2.61).

The above discussion makes clear that the numerical effort to calculate the spectrum grows rapidly with the number of vibrational modes coupled to the electronic motion. In typical two-state two-mode problems with intermediate coupling ($|\kappa_s/\omega_s|, |\lambda_s/\omega_s| \gtrsim 1$), the dimension of the matrices M_J needed to obtain a converged spectrum is of the order of 500–1000. If we use standard diagonalization procedures especially adapted to banded matrices, the computation time amounts to about half an hour (IBM 370/168). In a typical two-state three-mode problem with intermediate coupling, the required dimensions rise to a few thousands and more. For the $\tilde{A}^2 B_{3g}$ state of the ethylene cation, for example, a dimension of 3600 is needed to secure the convergence of the spectrum which exhibits several hundred vibronic lines.[31] The computation of the spectrum then requires several hours of computer time. The situation becomes rather hopeless for two-state three-mode problems with strong coupling. In our approximate calculation of the absorption spectrum of NO_2, for instance, a matrix having a dimension of 18,630 has to be diagonalized.[89] This is far beyond the computational capabilities of standard diagonalization procedures. To overcome the problem, the Lanczos procedure has been used and is briefly discussed below.

In most standard diagonalization procedures the matrix to be diagonalized is first transformed into a tridiagonal form employing a sequence of Jacobi or other related rotations. Subsequently this tridiagonal matrix is diagonalized. Instead the Lanczos procedure[90] generates iteratively a sequence of tridiagonal matrices T_k, for $k = 1, 2, 3, \ldots$. Without going into detail here, we just mention that these T_k are the representations of the original matrix in the sequence of Krylov subspaces defined as span$\{Z, M_j Z, M_j^2 Z, \ldots, M_j^k Z\}$ where Z is a suitable test vector.[88,90] The main advantage of the Lanczos algorithm results from the fact that it can be directly applied to operators without constructing a matrix representation. Historically, the stability of the Lanczos algorithm was doubted, and a confident appropriation of the algorithm is only justified since the detailed analysis of Paige.[91] The Lanczos method plays a predominant role in all cases where many eigenvalues of large and sparse matrices are needed. Applications can be found, for instance, in nuclear shell model calculations[92] and in vibronic coupling problems.[93-95] The occurrence of complex symmetric matrices in physical problems such as the calculation of electron-spin resonance (ESR) and nuclear magnetic resonance (NMR) spectra[96,97] or the computation of the lifetimes of vibronic states[89] have led to a generalization of the Lanczos procedure by defining a pseudo-(complex-valued)scalar product. In the next subsection and in Section VII.A the problem of the radiative lifetime of vibronic states will be addressed.

The spectrum of the ethylene cation mentioned above has been recalculated using the Lanczos algorithm.[31] After 3600 iteration steps, the whole spectrum has been found to be identical within drawing accuracy with the one calculated using the standard procedure for banded matrices. The computation time is reduced by using the Lanczos algorithm by more than a factor of 10. With this algorithm even the aforementioned absorption spectrum of NO_2 has become amenable to numerical calculations.[89] This calculation requires, however, several hours of computer time. In all our calculations we have noticed that the band shape of the spectrum is excellently reproduced by the Lanczos method after only a moderate number of Lanczos iteration steps. This finding, which is of practical importance, is not at all surprising when we note the close connection of the Lanczos method to the method of moments in Hilbert space.[98,99]

The above discussion illustrates that the numerical effort involved in the calculation of the excitation spectrum of a polyatomic molecule is considerable. Even with the approximate Hamiltonian (2.25) a full quantum mechanical calculation of the spectrum is only possible when a few nuclear vibrations couple significantly to the electronic motion. It is clear that abandoning the approximate Hamiltonian and returning to the full one in Eq. (2.24) renders the situation quite hopeless except in some weak coupling cases.

We now turn to the computation of the adiabatic and Franck–Condon spectra given by Eqs. (2.48) and (2.49), respectively. In analogy to the computation of the exact spectrum calculated with the approximate Hamiltonian (2.25), we expand the Hamiltonian in the direct product harmonic oscillator basis. However, the Hamiltonian to be diagonalized is now $T_N \mathbf{1} + \mathbf{V}(\mathbf{Q})$, and the calculation can be carried out separately for each adiabatic electronic state Φ_i, for $i = 1, N$. To find the representation of $T_N + V_i$, we use matrix methods that are well known for the one-dimensional case.[100–102] We first diagonalize the nuclear coordinates Q_s in the harmonic oscillator basis $|n_s\rangle$ using L_s basis functions. Let the eigenvalues and the corresponding eigenvector elements be $q_{n_s}^{(s)}$ and $x_{n_s n_s'}^{(s)}$, respectively. We then evaluate for each adiabatic potential energy surface $V_i(\mathbf{Q}) = V_i(Q_1, Q_2, \ldots, Q_M)$ the quantities

$$ V_i\left(q_{n_1}^{(1)}, q_{n_2}^{(2)}, \ldots, q_{n_M}^{(M)}\right) - V_0\left(q_{n_1}^{(1)}, q_{n_2}^{(2)}, \ldots, q_{n_M}^{(M)}\right) $$

Considering these as the elements of a $L = L_1 \times L_2 \times \cdots \times L_M$ dimensional diagonal matrix, we transform this matrix back to the original harmonic oscillator representation via the L-dimensional matrix with elements $x_{n_1 n_1'}^{(1)} x_{n_2 n_2'}^{(2)} \cdots x_{n_M n_M'}^{(M)}$. Adding in the diagonal the elements of the unperturbed oscillator Hamiltonian

$$ \sum_{s=1}^{M} \omega_s \left(n_s + \tfrac{1}{2}\right) $$

we obtain the representation of the adiabatic Hamiltonians $T_N + V_i$ in the harmonic oscillator basis. Diagonalization of these matrices yields the vibrational energy levels E_{in} and wave functions χ_{in}, which completes the computation of the Franck–Condon spectrum. To determine the adiabatic spectrum we must, in addition, calculate matrix elements of the type $\langle 0 | S_{ij}(\mathbf{Q}) | \chi_{in} \rangle$. This is easily carried out once the S_{ij} are represented in the harmonic oscillator basis, which can be done as described above for the potential. As discussed in the preceding subsection, the S_{ij} depend on the choice of the phase factor. Our choice of this factor is discussed in Section III and Ref. 30.

The numerical effort needed to calculate the adiabatic spectrum of a multimode system is often larger than the one needed to compute the corresponding exact spectrum when the approximate Hamiltonian (2.25) is used. The matrices involved in the calculation of the adiabatic spectrum are usually full matrices, and the fast numerical procedures for sparse or banded matrices discussed in the context of the exact spectrum cannot be applied efficiently. Moreover, whereas only the first element of each eigenvector enters the calculation of the exact and Franck–Condon line intensities, all the elements are needed in the calculation of the adiabatic ones. We have computed the adiabatic spectrum in several cases mainly for the sake of discussion of the nonadiabatic effects.

C. Radiative and Nonradiative Decay

1. Introduction

The processes of radiative and nonradiative decay in polyatomic molecules have received much attention in the literature, both theoretical and experimental.[103-111] Fluorescence or phosphorescence is generally only observed from the lowest excited singlet or triplet states of neutral molecules, even if higher excited states have been populated initially (Kasha's rule).[112] Also, the majority of organic radical cations studied to date do not exhibit any detectable emission.[113] Many of these observations have to be attributed to fast or even ultrafast nonradiative decay from the higher excited electronic states to lower ones and/or to the ground state. In most theoretical investigations it was assumed that in zeroth-order the Born–Oppenheimer adiabatic separability between the electronic and nuclear vibrational motions can be invoked. In particular, the preparation and nonradiative decay of single Born–Oppenheimer vibrational levels via weak nonadiabatic coupling terms is considered. Although this assumption may hold in many cases, for example, for aromatic molecules, there is a vast number of counterexamples for which it does not apply. It is well known that the adiabatic approximation fails severely in situations where adiabatic potential energy surfaces intersect. An increasing number of ab initio calculations indicates that conical intersections are a common phenomenon in even small molecules[32,114-116] (see also Section V). There is also much experimental evidence[112,117] for ultrafast nonradiative decay, which in turn is an indication of a considerable nonadiabatic perturbation of the adiabatic vibrational levels.

In what follows we shall discuss the radiative and nonradiative decay of polyatomic molecules in situations where the adiabatic approximation breaks down due to strong nonadiabatic coupling terms (e.g., in conical intersection situations). To be specific, we consider the scheme depicted in Fig. 2.

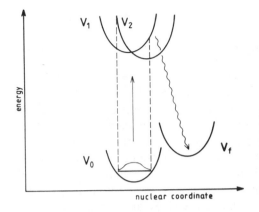

Fig. 2. Schematic representation of the excitation and decay process discussed in the text: V_0 and V_f are the adiabatic potential energy surfaces of the initial and final electronic states, respectively; V_1 and V_2 are the intersecting energy surfaces of the excited vibronically interacting electronic states. The Franck–Condon zone is indicated by the broken vertical lines.

The molecule initially in the electronic state Φ_0, which is usually the ground state, is excited into the manifold of vibronically interacting electronic (diabatic) states ϕ_1 and ϕ_2. Subsequently, the system decays via spontaneous emission to the final state (or states) Φ_f. This final electronic state may be identical with the initial state. It is assumed that the initial and final states do not interact vibronically with any state; that is, the adiabatic approximation is excellent for these states and there is no reason to distinguish between the adiabatic and diabatic picture ($\Phi_0 = \phi_0$ and $\Phi_f = \phi_f$). The excitation from Φ_0 to $\{\phi_1, \phi_2\}$ can take place via photons, electrons, or other particles as long as the excitation can be described by the golden rule. The situation applies to the excitation of neutral molecules, where usually $\phi_f = \Phi_0$, as well as to ionization where ϕ_1 and ϕ_2 are excited electronic states of the radical cation and Φ_f is usually the cationic ground state.

2. Radiative Decay of Individual Levels

For the discussion of radiative decay the molecular Hamiltonian H in Eq. (2.1) must be augmented by the Hamiltonian H_{rad} of the radiation field and the interaction between them

$$\tilde{H} = H + H_{rad} + H_{int} \qquad (2.62)$$

Since the interaction between radiation and matter is weak, we may, to a good approximation, describe the time evolution of nonstationary molecular states by the effective Hamiltonian[103,118,119]

$$H_{eff} = H - \frac{i}{2}\Gamma \qquad (2.63a)$$

where the operator Γ is defined through its matrix elements

$$\Gamma_{nm} = 2\pi \sum_f \sum_e \int d\Omega_k \langle \phi_n | H_{int} | \phi_f, \mathbf{ke} \rangle \langle \phi_f, \mathbf{ke} | H_{int} | \phi_m \rangle \rho(k) \quad (2.63b)$$

This expression for the radiative *damping matrix* $\Gamma = \{\Gamma_{nm}\}$ exhibits a sum over the final states as well as a sum over the polarization $\mathbf{e}$ and an integration over the directions Ω_k of the emitted photon with wave vector $\mathbf{k}$. Here $\rho(k)$ is the density of photon states; Γ_{nn} is just the definition of the golden rule radiative decay rate of the diabatic electronic state ϕ_n. Clearly, Γ as defined in Eq. (2.63b) is still a function of the nuclear coordinates, that is, $\Gamma = \Gamma(\mathbf{Q})$. Furthermore, the damping matrix depends on the photon energy. This latter dependence can be neglected when ϕ_1 and ϕ_2 are energetically well separated from the final states ϕ_f, which we have implicitly assumed in our excitation and decay process scheme shown in Fig. 2.

Usually the damping matrix is represented in the adiabatic basis. Consequently, its elements will strongly depend on the nuclear coordinates in conical intersection situations, for example, in NO_2 (see Sections III.B.2 and V.B). This property of the damping matrix highly complicates the calculation of the eigenvalues of H_{eff} when one is using an adiabatic basis. The calculation simplifies considerably, and more insight into the radiative decay is gained in the diabatic representation. The ϕ_i are smoothly varying functions of the nuclear coordinates, and for most practical purposes we may safely consider the elements Γ_{nm} of the damping matrix in Eq. (2.63b) as constant numbers. The matrix Hamiltonian $\mathscr{H}_{eff}$ describing the nuclear motion in the manifold of vibronically interacting diabatic states in the presence of the radiation field is now simply given by the expression

$$\mathscr{H}_{eff} = \mathscr{H} - \frac{i}{2}\Gamma \quad (2.64)$$

Since the damping matrix Γ in Eq. (2.64) is a constant matrix, we may diagonalize Γ by an orthogonal $\mathbf{Q}$-independent transformation Ω, thus defining a new matrix Hamiltonian and oscillator strengths

$$\overline{\mathscr{H}}_{eff} = \Omega \mathscr{H}_{eff} \Omega^\dagger \quad (2.65a)$$

$$\bar{\tau} = \Omega \tau \quad (2.65b)$$

For the two-state problem addressed in Fig. 2, the effective matrix Hamiltonian reads

$$\overline{\mathscr{H}}_{eff} = \begin{pmatrix} T_N + \overline{W}_{11}(\mathbf{Q}) & \overline{W}_{12}(\mathbf{Q}) \\ \overline{W}_{21}(\mathbf{Q}) & T_N + \overline{W}_{22}(\mathbf{Q}) \end{pmatrix} - \frac{i}{2}\begin{pmatrix} \Gamma_1 & 0 \\ 0 & \Gamma_2 \end{pmatrix} \quad (2.66)$$

Here, Γ_1 and Γ_2 are the eigenvalues of Γ and $\overline{W} = \Omega W \Omega^\dagger$, where W has been defined in Eq. (2.24). In several relevant situations Γ is a priori diagonal and, hence, $\overline{\mathscr{H}}_{\text{eff}} = \mathscr{H}_{\text{eff}}$. Due to selection rules the situation may occur that the final state ϕ_f is radiatively inaccessible from one of the vibronically interacting states, say, ϕ_2. Then, $\Gamma_{22} = \Gamma_{12} = \Gamma_{21} = 0$ and the damping matrix is diagonal. The damping matrix is also diagonal when ϕ_f can only be reached from ϕ_1 by spontaneous emission of photons of a specific polarization and from ϕ_2 by photons of a different polarization. Then, $\Gamma_{12} = \Gamma_{21} = 0$, but both Γ_{11} and Γ_{22} are different from zero.

Let us now turn to the discussion of the excitation spectrum in the presence of spontaneous emission. Here we should consult scattering theory rather than a spectral function theory, as done in Section II.B, and consider the physical process addressed in Fig. 2, that is, excitation followed by spontaneous emission, as a single process. Gel'mukhanov et al.[120,121] and Kaspar et al.[122] have investigated the influence of finite lifetime of electronic states on the vibrational structure of molecular electronic spectra. Extending Eq. (12) of Ref. 122 to the case of vibronic coupling discussed here, we obtain for the spectrum

$$P(E) = \langle 0 | \tau^\dagger \left[E - \mathscr{H}_{\text{eff}}^\dagger \right]^{-1} \Gamma \left[E - \mathscr{H}_{\text{eff}} \right]^{-1} \tau | 0 \rangle \qquad (2.67)$$

This positive semidefinite quantity can be rearranged using the operator identities

$$(A + iB)^{-1} - (A - iB)^{-1} = -2i(A + iB)^{-1} B (A - iB)^{-1} \qquad (2.68a)$$

and (with the imaginary parts of the eigenvalues of $A + iB$ having to be positive definite)

$$\int_0^\infty dt \, e^{i(A + iB)t} = i(A + iB)^{-1} \qquad (2.68b)$$

Putting $A = E - \mathscr{H}$ and $B = \Gamma/2$, we immediately arrive at the following expression for the excitation spectrum (Re = real part)

$$P(E) = 2 \operatorname{Re} \int_0^\infty dt \, e^{iEt} \langle 0 | \tau^\dagger \exp(-i \mathscr{H}_{\text{eff}} t) \, \tau | 0 \rangle \qquad (2.69)$$

To make contact with the excitation spectrum derived in Section II.B.1, not taking into account the interaction of radiation and matter, we let the damping matrix approach a positive infinitesimal

$$\frac{\Gamma}{2} \to \eta \mathbf{1} \qquad (2.70)$$

In this limit the excitation spectrum (2.69) is given by

$$P(E) = 2\,\mathrm{Re}\left\{ i\langle 0|\tau^\dagger \left[E - \mathscr{H} + i\eta \right]^{-1} \tau|0\rangle \right\} \qquad (2.71)$$

This equation can be further reduced using the well-known relation

$$\frac{1}{E - E_\nu + i\eta} = \mathscr{P}\frac{1}{E - E_\nu} - i\pi\delta(E - E_\nu) \qquad (2.72)$$

where $\mathscr{P}$ denotes the Cauchy principal value. The spectrum now takes on the appearance

$$P(E) = 2\pi\langle 0|\tau^\dagger \,\delta(E - \mathscr{H})\, \tau|0\rangle \qquad (2.73)$$

which is identical with the excitation spectrum derived in Section II.B.1 [see Eq. (2.39)]. Obviously, the scattering formula (2.67) reduces to the golden rule formula in the limit of an infinitesimal perturbation.

In Sections V and VII we shall present quantum mechanical calculations on small molecules exhibiting conical intersections (e.g., NO_2 and $C_2H_4^+$). Hence, we pay here special attention to a vibronic level density which is sufficiently small to ensure that

$$|E_\nu - E_\mu| \gg k_\nu, k_\mu, \qquad \nu \neq \mu \qquad (2.74)$$

in the absence of accidental degeneracies. The quantities appearing in Eq. (2.74) are related to the real and imaginary parts of the eigenvalues of $\mathscr{H}_{\mathrm{eff}}$

$$\mathscr{H}_{\mathrm{eff}}\chi_\nu = \varepsilon_\nu\chi_\nu \qquad (2.75a)$$

$$\varepsilon_\nu = E_\nu - \frac{i}{2}k_\nu \qquad (2.75b)$$

In typical situations, for instance NO_2, we have $k_\nu \approx 10^{-8}$ eV compared to the average next-neighbor distance d of vibronic levels of $d \approx 10^{-3}$ eV. Clearly, E_ν is very close to the corresponding eigenvalue of $\mathscr{H}$, and for all practical purposes we may take the eigenfunctions χ_ν also to be the eigenfunctions of $\mathscr{H}$ and vice versa [as long as Eq. (2.74) is valid].

Each eigenfunction of $\mathscr{H}$ is a two-component function

$$\chi_\nu = \begin{pmatrix} \chi_{1\nu} \\ \chi_{2\nu} \end{pmatrix} . \qquad (2.76)$$

referring to the diabatic wave functions ϕ_1 and ϕ_2. Inserting the closure relation into the expression (2.69) for the spectrum, we find that

$$P(E) = 2\pi \sum_{\nu} I_{\nu} L(E_{\nu}) \tag{2.77a}$$

where

$$I_{\nu} = |\langle \chi_{1\nu}|0\rangle \tau_1 + \langle \chi_{2\nu}|0\rangle \tau_2|^2 \tag{2.77b}$$

$$L(E_{\nu}) = \frac{1}{\pi} \frac{k_{\nu}/2}{(E - E_{\nu})^2 + (k_{\nu}/2)^2} \tag{2.77c}$$

This result for the excitation spectrum is physically appealing. The spectrum consists of a series of Lorentzian peaks of fwhm k_{ν} at the energies E_{ν}. The corresponding positive definite intensities I_{ν} gain contributions from both the diabatic states ϕ_1 and ϕ_2 and are subject to interference effects between these states. In case that Eq. (2.74) is violated for some vibronic states, Eq. (2.77) is not applicable at the corresponding energies. The resulting spectrum is then not correctly described by the superposition of Lorentzian peaks, and one should return to the original equation (2.69) or equivalently to (2.67) and recalculate the spectrum not using the assumption that the eigenfunctions of $\mathcal{H}$ are also the eigenfunctions of $\mathcal{H}_{\text{eff}}$.

As long as the inequality (2.74) is valid, we may easily calculate the fluorescence decay rate k_{ν} of the νth vibronic level

$$k_{\nu} = \Gamma_1 \langle \chi_{1\nu}|\chi_{1\nu}\rangle + \Gamma_2 \langle \chi_{2\nu}|\chi_{2\nu}\rangle \tag{2.78a}$$

For the sake of simplicity we have assumed the damping matrix to be diagonal; otherwise, χ_{ν} should be replaced by $\bar{\chi}_{\nu}$. It should be noted that

$$\langle \chi_{\nu}|\chi_{\nu}\rangle = \langle \chi_{1\nu}|\chi_{1\nu}\rangle + \langle \chi_{2\nu}|\chi_{2\nu}\rangle = 1 \tag{2.78b}$$

The decay rate k_{ν} is smaller than the larger of the two elements Γ_1 and Γ_2 of the damping matrix. Let us now discuss the situation where one of these elements vanishes and assume without loss of generality that $\Gamma_2 \neq 0$. In this situation the lifetime of the νth vibronic level, defined as the inverse decay rate k_{ν}^{-1}, is in the presence of vibronic coupling between the states always longer than the lifetime Γ_2^{-1} obtained in the absence of vibronic coupling. Remembering that the exact wave function of the νth vibronic level is given by

$$\Psi_{\nu} = \chi_{1\nu}\phi_1 + \chi_{2\nu}\phi_2 \tag{2.79a}$$

the quenching factor of the radiative rate reads

$$\frac{k_\nu}{\Gamma_2} = \langle \chi_{2\nu} | \chi_{2\nu} \rangle = \langle \Psi_\nu | \hat{P}_2 | \Psi_\nu \rangle \qquad (2.79b)$$

where $\hat{P}_2 = |\phi_2\rangle\langle\phi_2|$ is a purely electronic projector which projects on the electronic diabatic wave function ϕ_2. The quenching factors are the smaller the stronger the nonradiating electronic state ϕ_1 is intermixed in the exact wave function. Douglas[123] has argued that the anomalously long radiative lifetimes observed for some molecules (e.g., NO_2) arise from a mixing of the vibrational levels within one state or a mixing of the levels of two electronic states. Equation (2.79b) is merely the quantitative description of the latter mechanism proposed by Douglas. Section VII.A is devoted to the numerical computation and discussion of the quenching of the radiative rate in NO_2. It is found that Eq. (2.79b) indeed explains the observed anomalously long radiative lifetimes of NO_2. The Douglas effect in the limit of weak non-adiabatic perturbation has been discussed many times in the literature (see, e.g., Ref. 103).

3. The Time Evolution of Wave Packets

In the preceding subsection we have discussed the excitation spectrum in the presence of the radiation field and the radiative decay of individual vibronic levels. In the following discussion we investigate more global quantities such as the quantum yield, the autocorrelation function, and the occupation probability of a particular diabatic state, all as a function of time. To be specific, we again consider the scheme depicted in Fig. 2 and discussed in Section II.C.1. We do not intend to concentrate on the time evolution of individual or of a small group of coherently excited vibronic states. To simplify the discussion, we rather consider a "broad band" optical excitation which prepares the vibrational state $|0\rangle$ of the electronic ground state Φ_0 as the initial state at time $t = 0$ on a single diabatic potential energy surface which we put to correspond to ϕ_2. In other words, only the diabatic state ϕ_2 is optically accessible from the ground state, and the vibrational state $|0\rangle$ indicated in Fig. 2 is transferred vertically up the Franck–Condon zone to the upper potential energy surface. In the basis of the diabatic states ϕ_1 and ϕ_2, this initial wave packet takes on the appearance

$$|\psi(0)\rangle = \begin{pmatrix} 0 \\ |0\rangle \end{pmatrix} \qquad (2.80)$$

Unless otherwise stated we assume that $\Phi_0 = \Phi_f$. Consequently, only the vibrational levels of the state ϕ_2 can decay radiatively: $\Gamma_1 = 0$.

It is natural to start with the nondecay probability amplitude or autocorrelation function.[124,125] In the present situation this function reads

$$C(t) \equiv \langle \psi(0)|\psi(t)\rangle = \langle \psi(0)|\exp(-i\mathcal{H}_{eff}t)|\psi(0)\rangle \qquad (2.81)$$

where $C(t)$ represents the probability amplitude that the wave packet after time t is still the same as at starting time $t = 0$. When radiative processes are neglected, the autocorrelation function is proportional to the Fourier transform of the excitation spectrum in Eq. (2.37) for $\tau_1 = 0$. Obviously, a regular line spectrum leads to pronounced and regular recurrences of the modulus of the autocorrelation function, and an erratic irregular spectrum causes the $|C(t)|$ curve to be irregular and to fluctuate around a more or less small constant value C_0 given by

$$C_0^2 \equiv \sum_\nu |\langle \psi(0)|\Psi_\nu\rangle|^4 = \sum_\nu |\langle 0|\chi_{2\nu}\rangle|^4 \qquad (2.82)$$

A small value of C_0 has been associated with nonradiative decay processes[126] and $|C(t)|^2$ is often taken as a measure of these processes, especially when the prepared initial state $|\psi(0)\rangle$ is chosen as a vibrational eigenstate of the upper potential energy surface.[103,104] The latter choice of $\psi(0)$ is certainly a useful measure of the relaxation process in weak vibronic coupling situations. When the vibronic coupling becomes considerable, neither choice of the initially prepared state is adequate because the autocorrelation function itself ceases to be the appropriate quantity to describe the relaxation. This function describes the total relaxation of $\psi(0)$, that is, the decay to the non-radiating levels of the lower potential energy surface as well as to the other radiating vibrational levels of the upper surface. What we actually need to compute is the decay of $\psi(0)$ to the levels of the lower potential energy surface only. The appropriate quantity is the probability of occupying an *electronic* state rather than a *vibrational* level of this state. This probability $P_2(t)$ is discussed below.

Once the wave packet $\psi(0)$ is generated, the system begins to radiate and the self-overlap of $\psi(t)$ is less than unity for $t > 0$

$$\langle \psi(t)|\psi(t)\rangle = \langle \psi(0)|\exp(i\mathcal{H}_{eff}^\dagger t)\exp(-i\mathcal{H}_{eff}t)|\psi(0)\rangle < 1 \quad (2.83)$$

In the event that both diabatic states ϕ_1 and ϕ_2 have the same radiative decay rates, that is, $\Gamma_1 = \Gamma_2 = \Gamma$, the self-overlap of $\psi(t)$ becomes rather simple

$$\langle \psi(t)|\psi(t)\rangle = e^{-\Gamma t} \qquad (2.84)$$

Otherwise, when $\Gamma_1 = 0$ and $\Gamma_2 \neq 0$, we obtain a complicated expression. The self-overlap of $\psi(t)$ has a very simple physical interpretation: it is the probability of occupying the manifold of the vibronically interacting electronic states at time t. For large t the system has completely decayed to the ground state Φ_0 and the self-overlap of $\psi(t)$ approaches zero. With this interpretation it is clear that the quantum yield at time t can be expressed as

$$Y(t) = 1 - \langle \psi(t) | \psi(t) \rangle \tag{2.85}$$

Since only ϕ_2 is assumed to decay radiatively, the fraction of photons $dY(t)$ emitted in the time between t and $t + dt$ is proportional to the radiative decay rate of ϕ_2, to the probability $P_2(t)$ of occupying ϕ_2, and, of course, also to dt:

$$\frac{dY(t)}{dt} = \Gamma_2 P_2(t) \tag{2.86a}$$

The probability $P_2(t)$ is easily obtained either from its definition or by differentiating the quantum yield in Eq. (2.85) with respect to time. With the aid of Eq. (2.83) we immediately obtain

$$P_2(t) = \langle \psi(t) | \begin{pmatrix} 0 & 0 \\ 0 & 1 \end{pmatrix} | \psi(t) \rangle \tag{2.86b}$$

Here $\begin{pmatrix} 0 & 0 \\ 0 & 1 \end{pmatrix}$ is the matrix representation of the projector $\hat{P}_2 = |\phi_2\rangle\langle\phi_2|$ first introduced in Eq. (2.79); $\hat{P}_2$ can also be interpreted as the occupation number operator of the upper electronic state. Thus, $P_2(t)$ in Eq. (2.86b) equals the expectation value of this occupation number operator, taken with the full time-dependent wave function $\psi(t)$. Analogously, we may define the probability $P_1(t)$ of occupying the electronic state ϕ_1 and obtain the identity

$$1 - Y(t) = P_1(t) + P_2(t) \tag{2.87}$$

We shall now briefly discuss the above quantities, noting first that all the major formulas derived here are valid for any choice of the prepared initial state $\psi(0)$ and are not restricted to the choice in Eq. (2.80). This choice has only been taken to avoid the discussion and calculation (see Section VII.B) of probability functions for many individual vibrational levels. It can easily be seen from Eqs. (2.85) and (2.83) that the quantum yield at $t \to \infty$ is unity. This is not surprising, since the excited states ϕ_1 and ϕ_2 do not interact vibronically with the ground state Φ_0. Thus, strictly speaking we are not dealing here with nonradiative decay. Nevertheless, $P_2(t)$ [or $P_1(t)$] is an excellent and natural tool to describe the nonradiative decay from the state ϕ_2 to ϕ_1. We shall see in Section VII.B that irregular spectra which are obtained, for instance, in a conical intersection situation imply an ultrafast decaying function $P_2(t)$ associated with anomalously long radiative lifetimes. Therefore, it will last a "long" while, until the quantum yield is substantially different from zero, and thus $P_2(t)$ correctly explains the observations in short-time-scale experiments. The knowledge of $P_2(t)$ is also important for the interpretation of time-resolved experiments and when collisions enter the scene. In the latter case the collision-induced deactivation rates could be very different for ϕ_1 and ϕ_2 and the occupation probabilities of these states are of relevance.

In many experiments we encounter the situation that the ground state ϕ_0 and/or the final state ϕ_f is identical with the lower state ϕ_1 of the vibronically interacting manifold. For instance, in a photoionization experiment the cationic ground and first excited states may be vibronically coupled and the latter state decays radiatively to the first one. In these cases of radiative decay within the manifold of vibronically interacting states (internal emission) some of the results discussed above are not strictly valid because Γ cannot a priori be considered a constant matrix anymore. To arrive at the appropriate form of the radiative damping matrix one has to start from the explicit form of the interaction H_{int} between radiation and matter, Eq. (2.62). Then the radiative damping matrix for internal emission, $\Gamma = \Gamma^{(int)}$, can be derived without any ad hoc assumptions and with only very minor approximations.[302] Instead of going into the details we quote here only the final result[302]

$$\Gamma^{(int)} = 2\pi g_0^2 [V_2(\mathbf{Q}) - V_1(\mathbf{Q})][W_{22}(\mathbf{Q}) - W_{11}(\mathbf{Q})]^2 \hat{\mathbf{P}}_2 \qquad (2.88)$$

Here

$$g_0 = \text{const}\langle \phi_1 | \sum_i r_i | \phi_2 \rangle \qquad (2.89)$$

is the matrix element of the electronic dipole operator between the diabatic electronic states and $\hat{\mathbf{P}}_2$ is the projection operator $|\Phi_2\rangle\langle\Phi_2|$ on the upper adiabatic electronic state. Written in the diabatic electronic basis, $\hat{\mathbf{P}}_2$ takes the following appearance

$$\hat{\mathbf{P}}_2 = \mathbf{S}\begin{pmatrix} 0 & 0 \\ 0 & 1 \end{pmatrix}\mathbf{S}^+ = \tfrac{1}{2}\mathbf{1} + \frac{1/2}{V_2 - V_1}\begin{pmatrix} W_{11} - W_{22} & 2W_{12} \\ 2W_{12} & W_{22} - W_{11} \end{pmatrix} \qquad (2.90)$$

For the definition of $\mathbf{S} = \mathbf{S}(\mathbf{Q})$ see Eq. (2.43). As before, the time-resolved intensity of fluorescence $I(t)$ is given as the expectation value of Γ with the time-dependent wave function

$$I(t) = \frac{dY(t)}{dt} = \langle \psi(t)|\Gamma|\psi(t)\rangle \tag{2.91}$$

The above results should be compared with the previous Eq. (2.86) which related to the situation where the emission is to a third electronic state which is vibronically uncoupled from and energetically far below the vibronically interacting states (external emission). Here, the radiative damping matrix, $\Gamma = \Gamma^{(ext)}$, reads in the diabatic electronic basis

$$\Gamma^{(ext)} = \Gamma_2 \begin{pmatrix} 0 & 0 \\ 0 & 1 \end{pmatrix} \tag{2.92}$$

The different prefactors in Eqs. (2.88) and (2.92) are merely due to the instance that for external emission the potential energy differences of relevance are large and can be roughly taken as constant. More interesting is the projection operator involved in Γ. As has been argued above, for external emission the projection should refer to the diabatic electronic states. Why does it refer to the adiabatic basis in the case of internal emission? A simple heursitic argument, which is of course no substitute for the full calculation, proceeds as follows. For internal emission a radiative coupling between diabatic states of course implies such a coupling also between adiabatic states and vice versa. Energy conservation then makes it plausible that the occupation probability of the upper adiabatic state determines $I(t)$: only the adiabatic states have a definite energetic ordering and only the fraction of the wave packet being on the upper surface can radiate. Of course, in the adiabatic basis the dipole moment is a strongly varying function of $\mathbf{Q}$. Transforming the dipole moment function back to the diabatic basis, however, yields just the combination of energy prefactors displayed in Eq. (2.88).

In systems with internal emission the energy prefactors in (2.88) can certainly not be taken constant for quantitative purposes. However, the use of Eq. (2.92) instead of Eq. (2.88) greatly simplifies the numerical calculations, often rendering them possible at all. Therefore, in Section VII.A we shall treat the internal emission of some systems as if it were external.

To complete this section we shall now briefly explain how the numerical calculation of the occupation probability $P_2(t)$ is carried out. Numerical results on $C_2H_4^+$ are presented and discussed in Section VII.B (see also Ref. 127). The numerical procedures used to compute the eigenvalues and the first component of eigenvectors of $\mathscr{H}$ and $\mathscr{H}_{eff}$ have been discussed in Section II.B.3. The occupation probability $P_2(t)$ can be straightforwardly evaluated starting from Eq. (2.86b) if the secular problem of the matrix Hamiltonian is completely solved. In most cases we have been interested in, the complete calculation of all eigenvector components is a very difficult task in view of the large dimensions of the secular matrices (see Section II.B.3). In these cases we do not attempt such a solution, but rather compute $P_2(t)$ by numerical integration of the time-dependent Schrödinger equation, which directly yields the time-dependent wave function $\psi(t)$. Among the various available finite difference methods, we adopt a so-called predictor–corrector method[128] which is correct through fourth order in the interval length Δt. Since the formulas are straightforward extensions of equations quoted previously for the one-dimensional case,[128] we do not give more details here. The numerical effort is comparable to the successive evaluation of the time evolution operator $\exp(-i\mathscr{H}t)$ through fourth order in Δt in each interval.

III. THE TWO-STATE PROBLEM

In this section we shall analyze in detail the static and dynamic properties of a prototype model of multimode vibronic coupling effects.[30,75,81,129] The model consists of two nondegenerate electronic states, $|\phi_1\rangle$ and $|\phi_2\rangle$, of different spatial symmetry, which are coupled by a single nondegenerate, non-totally symmetric vibrational mode (the coupling mode) and whose energy separation $W_{22} - W_{11}$ (see Section II.A.4) depends on a single totally symmetric mode (the tuning mode). As discussed in Section II.A.4, the main simplifications of the model are the assumption of harmonic diabatic potentials and linear coupling to the vibrational modes. The model is representative for a wide class of vibronic coupling phenomena in molecules with (nontrivial) Abelian point groups.

It is our aim in this section to emphasize the nontrivial modification of the one-dimensional vibronic coupling problem by the totally symmetric tuning mode. In this sense the model represents multimode vibronic coupling effects in their simplest form. The true multimode problem, namely, vibronic coupling of two nondegenerate electronic states via many non-totally symmetric coupling modes, including many tuning modes, is discussed later in a separate subsection (III.C).

The matrix Hamiltonian describing the dynamics in the manifold of coupled electronic states reads[30,75,81,129,130]

$$\mathscr{H} = (T_N + V_0)\mathbf{1} + \begin{bmatrix} E_1 + \kappa_1 Q_g & \lambda Q_u \\ \lambda Q_u & E_2 + \kappa_2 Q_g \end{bmatrix} \tag{3.1a}$$

$$T_N = -\frac{1}{2}\omega_g \frac{\partial^2}{\partial Q_g^2} - \frac{1}{2}\omega_u \frac{\partial^2}{\partial Q_u^2} \tag{3.1b}$$

$$V_0 = \frac{1}{2}\omega_g Q_g^2 + \frac{1}{2}\omega_u Q_u^2 \tag{3.1c}$$

where we have introduced the simplifying notation $\kappa_1 \equiv \kappa_g^{(1)}$, $\kappa_2 \equiv \kappa_g^{(2)}$, $\lambda \equiv \lambda_u^{(1,2)}$ (see Section II.A.4). Here E_1 and E_2 (we assume $E_1 < E_2$ without loss of generality) are the excitation or ionization energies of the coupled electronic states at the reference geometry $\mathbf{Q} = 0$, where $\mathbf{Q}$ denotes collectively the set of nuclear coordinates (Q_g, Q_u); T_N is the nuclear kinetic energy operator and $V_0(\mathbf{Q})$ is the harmonic potential energy function of the molecule in its electronic ground state. The vibrational mode Q_u is assumed to transform according to a representation Γ_3 such that

$$\Gamma_1 \times \Gamma_2 \times \Gamma_3 \supset \Gamma_A \tag{3.2}$$

where Γ_1 and Γ_2 are the representations of the electronic states and Γ_A denotes the totally symmetric representation; Q_g is one of the totally symmetric modes of the molecule, that is, it transforms according to Γ_A. We use the indices g and u (*gerade*, *ungerade*) to distinguish clearly the totally symmetric tuning mode from the non-totally symmetric coupling mode.

We first concentrate on the static aspects of the problem, that is, the adiabatic potential energy surfaces and the dependence of the adiabatic electronic wave functions on the nuclear coordinates (Section III.A). We shall see that the Hamiltonian (3.1) represents the prototype model for a conical intersection. Section III.B is devoted to a detailed study of the nuclear dynamics described by this Hamiltonian.

A.   The Static Problem: Symmetry Breaking and Conical Intersection

1.   Adiabatic Potential Energy Surfaces

The concept of adiabatic electronic potential energy surfaces is of outstanding importance for the interpretation of all kinds of phenomena in molecular physics and chemistry. Therefore we shall consider in some detail the adiabatic potential energy surfaces associated with the Hamiltonian (3.1).

The adiabatic potential energy surfaces are obtained as the eigenvalues of $\mathscr{H} - T_N \mathbf{1}$, or, in other words, by diagonalizing $\mathscr{H}$ in the fixed-nuclei limit, $T_N \to 0$. Explicitly, we have

$$\mathbf{S}^\dagger(\mathscr{H} - T_N \mathbf{1})\mathbf{S} = \mathbf{V} \tag{3.3a}$$

$$\mathbf{V} = \begin{pmatrix} V_1(\mathbf{Q}) & 0 \\ 0 & V_2(\mathbf{Q}) \end{pmatrix} \tag{3.3b}$$

Here $V_1(\mathbf{Q})$ and $V_2(\mathbf{Q})$ are the adiabatic potential energy surfaces of the Hamiltonian (3.1). The unitary 2×2 matrix $\mathbf{S}$ transforms from the diabatic electronic basis, $|\phi_1\rangle$ and $|\phi_2\rangle$, to the adiabatic electronic basis, $|\Phi_1\rangle$ and $|\Phi_2\rangle$, via

$$\begin{pmatrix} |\Phi_1\rangle \\ |\Phi_2\rangle \end{pmatrix} = \mathbf{S}^\dagger \begin{pmatrix} |\phi_1\rangle \\ |\phi_2\rangle \end{pmatrix} \tag{3.4}$$

Here $\mathbf{S}$ is a function of all nuclear coordinates $\mathbf{Q}$ (see Section II.B.2).

For the explicit analysis of $\mathbf{V}$ and $\mathbf{S}$, it is convenient to rewrite $\mathcal{H}$ of Eq. (3.1) in the following form:

$$\mathcal{H} = H_0 \mathbf{1} + \begin{pmatrix} -d & c \\ c & d \end{pmatrix} \tag{3.5a}$$

$$H_0 = T_N + V_0 + \Sigma + \sigma Q_g \tag{3.5b}$$

$$\Sigma = \frac{E_1 + E_2}{2} \tag{3.5c}$$

$$\Delta = \frac{E_2 - E_1}{2} \tag{3.5d}$$

$$\sigma = \frac{\kappa_1 + \kappa_2}{2} \tag{3.5e}$$

$$\delta = \frac{\kappa_2 - \kappa_1}{2} \tag{3.5f}$$

$$d = \Delta + \delta Q_g \tag{3.5g}$$

$$c = \lambda Q_u \tag{3.5h}$$

The adiabatic potentials then read

$$V_{1,2}(\mathbf{Q}) = V_0(\mathbf{Q}) + \Sigma + \sigma Q_g \mp W \tag{3.6a}$$

$$W = (d^2 + c^2)^{1/2} \tag{3.6b}$$

and the transformation matrix $\mathbf{S}$ has the elements

$$S_{11} = S_{22} = 2^{-1/2} \left(1 + \frac{d}{W} \right)^{1/2} \mathrm{sgn}(Q_u) \tag{3.7a}$$

$$S_{12} = -S_{21} = 2^{-1/2} \left(1 - \frac{d}{W} \right)^{1/2} \tag{3.7b}$$

When the transformation $\mathbf{S}$, which diagonalizes $\mathcal{H} - T_N \mathbf{1}$, is applied to the full $\mathscr{H}$ of Eq. (3.1), one obtains (see Section II.B.2)

$$\mathscr{H}' = S^\dagger \mathscr{H} S = \mathscr{H}_{ad} - \hat{\Lambda} \tag{3.8a}$$

$$\mathscr{H}_{ad} = T_N \mathbf{1} + \mathbf{V} \tag{3.8b}$$

$$\hat{\Lambda} = -\sum_{\alpha = g, u} \frac{\omega_\alpha}{2} \begin{pmatrix} g_\alpha^2 & -\dfrac{\partial g_\alpha}{\partial Q_\alpha} - 2g_\alpha \dfrac{\partial}{\partial Q_\alpha} \\ \dfrac{\partial g_\alpha}{\partial Q_\alpha} + 2g_\alpha \dfrac{\partial}{\partial Q_\alpha} & g_\alpha^2 \end{pmatrix} \tag{3.8c}$$

$$g_g = \frac{-c\delta}{2(c^2 + d^2)} \tag{3.8d}$$

$$g_u = \frac{d\lambda}{2(c^2 + d^2)} \tag{3.8e}$$

In the adiabatic approximation, one replaces the full $\mathscr{H}'$ by $\mathscr{H}_{ad}$, which describes vibrational motion on the two separate potential energy surfaces $V_1(\mathbf{Q})$ and $V_2(\mathbf{Q})$. The nondiagonal 2×2 matrix $\hat{\Lambda}$ represents the nonadiabatic operator, that is, the vibronic coupling terms in the adiabatic representation.

We shall now devote some space to the discussion of the physical phenomena described by the Hamiltonian $\mathscr{H}'$ of Eq. (3.8). We first investigate the properties of the adiabatic potential energy surfaces $V_{1,2}(\mathbf{Q})$. Later we shall discuss the properties of the diabatic-to-adiabatic transformation matrix $\mathbf{S}$ and the nonadiabatic coupling terms (3.8c).

The lower adiabatic potential energy surface $V_1(\mathbf{Q})$ exhibits a characteristic and important phenomenon, namely, the breaking of the molecular symmetry. In the present context, symmetry breaking means that the minimum of the lower surface $V_1(\mathbf{Q})$ occurs at a nuclear geometry that is of lower symmetry than the equilibrium geometry of the molecule in its electronic ground state. The symmetry breaking is simply a consequence of the repulsion of the diabatic surfaces via the vibronic coupling. Therefore only the lower surface $V_1(\mathbf{Q})$ can develop new minima, while the upper surface $V_2(\mathbf{Q})$ can only become steeper.

Let us, for the moment, discard the tuning mode Q_g, that is, $\kappa_1 = \kappa_2 = 0$. It follows from Eq. (3.6) that the minimum of $V_1(\mathbf{Q})$ occurs at $Q_g^{(0)} = 0$ and

$$Q_u^{(0)} = \begin{cases} 0, & x \leq 1 \\ \pm \dfrac{\lambda}{\omega_u} \left[1 - \dfrac{1}{x^2} \right]^{1/2}, & x \geq 1 \end{cases} \tag{3.9a}$$

where we have introduced the dimensionless quantity

$$x = \frac{\lambda^2}{\omega_u \Delta} \tag{3.9b}$$

For $x > 1$ we have two equivalent minima with $Q_u^{(0)} \neq 0$ and the previous minimum $Q_u = 0$ is converted to a local maximum. The barrier height E_s, that is, the energy difference between $V_1(Q_u = 0)$ and $V_1(Q_u^{(0)})$, is given by

$$\frac{E_s}{\Delta} = \begin{cases} 0, & x \leq 1 \\ \dfrac{(1-x)^2}{2x}, & x \geq 1 \end{cases} \tag{3.10}$$

Here E_s may also be called the stabilization energy; it represents the lowering of the minimum of $V_1(\mathbf{Q})$ relative to the minimum in the absence of vibronic coupling ($\lambda = 0$). The simple functional dependence of the dimensionless stabilization energy E_s/Δ on the dimensionless parameter x is illustrated in Fig. 3. For $x < 1$ no symmetry breaking occurs. Just above the "threshold" $x = 1$, the stabilization energy increases quadratically with x, approaching a linear dependence on x for large x. The symmetry breaking by a single non-totally symmetric coupling mode is a well-known phenomenon and has been discussed, for example, in Refs. 35–40, 43, 44, and 49.

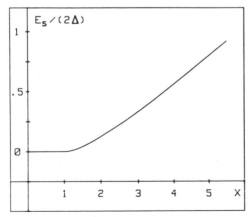

Fig. 3. The dependence of the stabilization energy E_s on the dimensionless quantity x, according to Eq. (3.10).

Including now the tuning mode Q_g, we obtain for the stationary point $\mathbf{Q}^{(0)}$ of $V_1(\mathbf{Q})$

$$Q_g^{(0)} = \frac{\Delta - F}{L - D} \frac{\delta}{\omega_g} - \frac{\sigma}{\omega_g} \qquad (3.11a)$$

$$Q_u^{(0)} = \pm \left(\frac{\lambda}{\omega_u}\right) \left[1 - \left(\frac{\Delta - F}{L - D}\right)^2\right]^{1/2} \qquad (3.11b)$$

where we have introduced the abbreviations

$$D = \frac{\delta^2}{\omega_g} = \frac{(\kappa_2 - \kappa_1)^2}{4\omega_g} \qquad (3.12a)$$

$$F = \frac{\delta\sigma}{\omega_g} = \frac{(\kappa_2 - \kappa_1)(\kappa_1 + \kappa_2)}{4\omega_g} \qquad (3.12b)$$

$$L = \frac{\lambda^2}{\omega_u} \qquad (3.12c)$$

The stabilization energy is

$$E_s = \frac{(\Delta + D - L - F)^2}{2(L - D)} \qquad (3.13)$$

The criteria for the existence of a stabilized minimum with $Q_u^{(0)} \neq 0$ are

$$\left|\frac{\Delta - F}{L - D}\right| < 1 \qquad (3.14a)$$

and

$$L > D \qquad (3.14b)$$

The additional condition (3.14b) indicates that there is a certain competition between the coupling mode and the tuning mode concerning the symmetry-breaking effect. If κ_1 and κ_2 have different signs, the symmetry-breaking effect of the coupling mode is reduced by the totally symmetric mode; that is, $Q_u^{(0)}$ and E_s are smaller than in the case $\kappa_1 = \kappa_2 = 0$. This is a consequence of the fact that the tuning tends to increase the separation of the interacting electronic states near the new equilibrium geometry $Q_g^{(0)}$. Examples of this quenching of the symmetry breaking effect will be presented in Sections V.A and V.D. If, on the other hand, κ_1 and κ_2 are of the same sign and $|\kappa_1| < |\kappa_2|$, the energy separation of the interacting states near the new minimum will decrease and the symmetry-breaking effect is enhanced by the tuning mode. If the additional condition (3.14b) is not fulfilled, the stationary points (3.11) represent saddle points rather than local minima.

Let us illustrate the results obtained so far by examining a typical case. As will be discussed in Section V.A, the nuclear dynamics in the two lowest electronic states of the butatriene cation, $C_4H_4^+$, is very well characterized by the Hamiltonian (3.1). The values of the parameters E_1, E_2, ω_g, ω_u, κ_1, κ_2 and λ pertinent to $C_4H_4^+$ are given later in Table IV (see Section V). Here we take the Hamiltonian (3.1) with the parameter values of $C_4H_4^+$ as a model of the vibronic coupling of two electronic states via a single coupling mode in the presence of a single tuning mode. Figure 4 shows a cross section through the adiabatic potential surfaces along the Q_g axis ($Q_u = 0$). Figure 5 shows a cross section along the Q_u axis for three representative values of Q_g. Figure 4 illustrates the tuning effect, that is, the dependence of the energy gap $W_{22} - W_{11}$ on the Q_g coordinate. The shape of the potential energy curve in the Q_u direction depends sensitively on this energy gap, as is seen in Fig. 5. In Fig. 5a ($Q_g = 0$) we see the double-minimum character of the lower potential curve. When Q_g increases, $W_{22} - W_{11}$ increases and the double-minimum becomes more shallow (see Fig. 5c). This illustrates the inhibition of the symmetry lowering effect by the tuning mode discussed above.

Going into the opposite direction of negative Q_g, the energy gap $W_{22} - W_{11}$ decreases until the two potential curves intersect at $Q_g = -0.8571$ (see Fig. 4). Figure 5b shows the cross section through the potential surfaces along

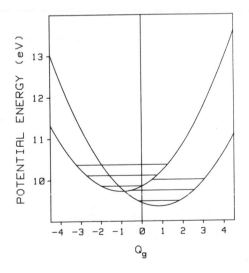

Fig. 4. Potential energy curves of the prototype conical intersection model along the Q_g direction for $Q_u = 0$. The electronic states cross at $Q_g = -0.8571$. The horizontal lines indicate the energy levels of the ν_g mode.

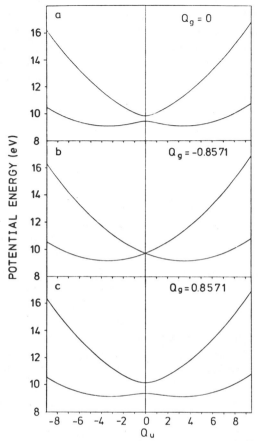

Fig. 5.   Adiabatic potential energy curves of the prototype conical intersection model along the non-totally symmetric coordinate Q_u for three characteristic values of the Q_g coordinate.

the Q_u direction for this particular value of Q_g. The potentials are seen to cross also in the Q_u direction. In other words, we have a degeneracy of V_1 and V_2 at $Q_g = -0.8571$, $Q_u = 0$, which is lifted in first order in both Q_g and Q_u. When considered as a surface in three-dimensional space, $V_1(Q_g, Q_u)$ and $V_2(Q_g, Q_u)$ form an elliptic double cone near the point of intersection. Therefore this particular phenomenon is called a conical intersection.[23–26] Figure 6 shows a perspective drawing of the adiabatic potential energy surfaces $V_{1,2}(Q_g, Q_u)$. The conical intersection is the single point where the upper surface V_2 touches the lower surface V_1. The double-minimum character of the lower surface is also visible. Clearly both adiabatic potential surfaces are strongly anharmonic, although they result from diabatic harmonic potentials (see Figs. 4 and 5b). Figure 6 illustrates the two main properties

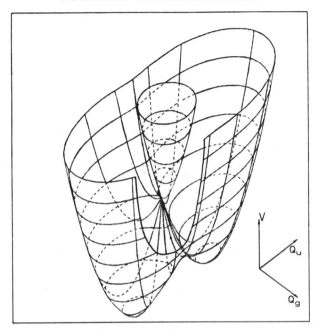

Fig. 6. Perspective view of the adiabatic potential energy surfaces of the prototype conical intersection model.

of adiabatic potential surfaces of vibronically interacting states, namely, the lowering of the symmetry of the equilibrium position of the lower surface and the existence of a conical intersection.

The existence of conical intersections of molecular potential energy surfaces has been predicted by Herzberg and Longuet-Higgins,[24] and others.[23,25,26] The well-known Jahn–Teller effect[27-29] is a special, highly symmetric case of a conical intersection situation. Indeed the two-state–two-mode model discussed here contains the Jahn–Teller effect as a special case. Putting $E_1 = E_2 = E$ and $\kappa_2 = -\kappa_1 = \kappa$, the Hamiltonian (3.1) describes the linear $E \times \beta$ Jahn–Teller effect.[28,29,131,132] When, in addition, $\omega_g = \omega_u = \omega$ and $\kappa = \lambda$, the Hamiltonian (3.1) represents the (linear) $E \times \varepsilon$ Jahn–Teller effect.[27-29] In the latter case the adiabatic potential energy surfaces take the simple form

$$V_{1,2} = \tfrac{1}{2}\omega\rho^2 + E \pm \lambda\rho \qquad (3.15)$$

where ρ is the radial coordinate defined by

$$Q_g = \rho\cos\phi, \qquad Q_u = \rho\sin\phi \qquad (3.16)$$

The potentials (3.15) are known as the "Mexican hat" potential energy surface of the $E \times \varepsilon$ Jahn–Teller problem. In the $E \times \beta$ Jahn–Teller problem the rotational symmetry with respect to the azimuthal angle ϕ is lost, since $\omega_g \neq \omega_u$ and $\kappa \neq \lambda$ in general. However, the intersection still occurs at $Q_g = Q_u = 0$, that is, at the center of the Franck–Condon zone.

The physical phenomena described by the Hamiltonian (3.1) are of a much wider variety than in the Jahn–Teller problem. The $E \times \varepsilon$ Jahn–Teller problem is characterized by the single dimensionless parameter κ/ω. In the $E \times \beta$ Jahn–Teller problem we have the three independent parameters ω_g/ω_u, κ, and λ. The more general two-dimensional conical intersection problem (3.1) is characterized by the five independent parameters ω_g/ω_u, $E_1 - E_2$, κ_1, κ_2, and λ. The range of spectroscopic phenomena described by this model for various choices of the parameters will be discussed below (Section III.B).

2. Adiabatic Electronic Wave Functions

Having obtained an overview over the shape of the adiabatic potential surfaces, we turn to a discussion of the **Q**-dependence of the corresponding electronic wave functions. This **Q**-dependence is described by the transformation matrix **S** of Eq. (3.7), since the dependence of the diabatic electronic wave functions on **Q** is weak by construction and therefore neglected (see Section II.A). The dependence of **S** on Q_u in the one-dimensional vibronic coupling case is well known[35,40-44] and need not be considered here. More interesting is the behavior of **S(Q)** in the vicinity of a conical intersection (see Section II.A,B). To get an idea of the qualitative behavior of S_{11} and S_{12} in the Q_g, Q_u plane, let us consider the $E \times \varepsilon$ Jahn–Teller limit $E_1 = E_2$, $\omega_g = \omega_u$, and $\kappa_1 = -\kappa_2 = \lambda$. Then Eq. (3.7) reduces to

$$S_{11} = S_{22} = \cos\left(\frac{\phi}{2}\right)$$
$$S_{12} = -S_{21} = \sin\left(\frac{\phi}{2}\right)$$

$$(3.17)$$

where ϕ is the azimuthal angle introduced in Eq. (3.16). It follows that the adiabatic electronic functions Φ_1 and Φ_2 change sign when the azimuthal angle increases by 2π radians. It has been shown by Herzberg and Longuet-Higgins[24] that this sign change is a general phenomenon when one transverses a closed loop around a conical intersection. The adiabatic electronic wave functions are thus double-valued functions in nuclear coordinate space. We can make them single-valued by drawing a cut, which connects the point of intersection with infinity, and allowing for a discontinuity of Φ_1 and Φ_2 on the cut.

As noted in Section II.B.2, the transformation matrix $\mathbf{S}$ is not uniquely determined by the requirement that it diagonalizes $\mathscr{H} - T_N \mathbf{1}$. We may multiply each eigenvector S_{1i}, S_{2i}, for $i = 1, 2$, by a phase factor which may be an arbitrary function of Q_g, Q_u. In the explicit calculations to be described below we take account of this freedom and replace S_{ij} of Eq. (3.7) by

$$S'_{ij} = S_{ij} \operatorname{sgn}(Q_u) \tag{3.18}$$

This has the advantage that the S'_{ij} reduce to the usual expressions[37-44] when considered as a function of Q_u alone; that is, S'_{11} is an even and S'_{12} an odd function of Q_u. Moreover, we shall be concerned with situations where the point of intersection, $Q_g^{(c)}$, lies on the negative Q_g axis (see Fig. 4). Using definition (3.18), we may cut the Q_g, Q_u plane along the Q_g axis between $Q_g^{(c)}$ and $-\infty$ and take S'_{12} to be discontinuous on this cut to get single-valued S'_{ij}. It follows that the S'_{ij} become continuous functions of Q_g, Q_u within the Franck–Condon zone when $Q_g^{(c)}$ moves out of the Franck–Condon zone. We can thus expect that the adiabatic approximation will produce reasonable results when the intersection point $Q_g^{(c)}$ is sufficiently outside the Franck–Condon zone.

Let us conclude the presentation of the static properties of the Hamiltonian (3.1) with a discussion of nonadiabatic coupling terms (3.8c). We consider again the $E \times \varepsilon$ Jahn–Teller limit to obtain a qualitative picture. In terms of the polar coordinates ρ, ϕ, the nonadiabatic operator becomes

$$\hat{\Lambda} = -\frac{\omega}{2\rho^2} \begin{pmatrix} \dfrac{1}{4} & \dfrac{\partial}{\partial \phi} \\[2ex] -\dfrac{\partial}{\partial \phi} & \dfrac{1}{4} \end{pmatrix} \tag{3.19}$$

Thus both the diagonal terms (which may be included as a correction to the adiabatic potentials[3]) as well as the nondiagonal coupling terms diverge as ρ^{-2} at the point of conical intersection, $\rho = 0$. This is the origin of the complete breakdown of the adiabatic approximation (where the nonadiabatic coupling terms are neglected) in the Jahn–Teller problem.[27-29] In the general two-state–two-mode problem represented by the Hamiltonian (3.1), $\hat{\Lambda}$ is a more complicated function of the coordinates and momenta but has the same type of singularity at the point of intersection. The elements g_g and g_u, which determine $\hat{\Lambda}$ according to Eq. (3.8), are shown in Fig. 7 for the $C_4H_4^+$ model introduced above. It can be seen that the behavior of these coupling functions agrees qualitatively with the behavior expected in the Jahn–Teller limit, namely, $g_g \sim \rho^{-1} \sin \phi$ and $g_u \sim \rho^{-1} \cos \phi$ [see Eqs.

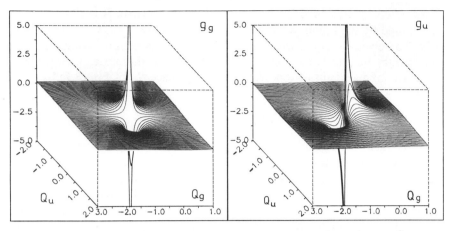

Fig. 7.   The elements g_g and g_u determining the nonadiabatic coupling operator $\hat{\Lambda}$ [see Eq. (3.8)] as a function of Q_g and Q_u for the prototype conical intersection model.

(3.8d, e)]. The nonadiabatic coupling terms of the generalized conical intersection model are thus as singular as in the Jahn–Teller case, and we should expect a complete breakdown of the adiabatic approximation for this model.

As discussed in Section II, there are two (in principle equivalent) approaches by which one can calculate exactly the nuclear dynamics in a certain manifold of interacting electronic states. In the adiabatic approach one uses adiabatic electronic basis states that render the electronic Hamiltonian diagonal but are coupled via the nuclear kinetic energy operator T_N. In the diabatic approach one chooses diabatic electronic basis states that render T_N essentially diagonal but are coupled via matrix elements of the electronic Hamiltonian. For the present model, we have given both representations. The original model Hamiltonian (3.1) is given in the diabatic representation. The adiabatic representation of the model is given by Eq. (3.8). It should be clear from the above discussion that the adiabatic approach has severe drawbacks in a conical intersection situation. The nonadiabatic coupling elements are principally singular at the point of intersection and thus difficult to handle numerically. In the diabatic approach, on the other hand, the coupling elements are well behaved and—as far as experience can tell—rather simple functions of the nuclear coordinates. For this reason all calculations of spectra discussed in the present work are performed using the diabatic approach. Nevertheless the adiabatic representation has its merits, say, for the qualitative interpretation of spectra. Another important aspect of the adiabatic representation is that the corresponding potential energies and electronic wave functions are directly obtained by standard fixed-nuclei quantum chemical ab initio calculations. Fischer[196] has recently proposed a canonical transformation of the model Hamiltonian (3.1) to a new basis which exhibits more explicitly than the diabatic basis the effects of anharmonicity caused by vibronic coupling but avoids the divergent coupling matrix elements of the adiabatic representation.

B. Study of the Nuclear Dynamics for a Typical Conical Intersection Problem

Spectroscopy is the most direct and sensitive tool for the investigation of molecular dynamics. In this section we shall discuss absorption spectra for the prototype conical intersection model defined by the Hamiltonian (3.1). We consider optical transitions from the noninteracting electronic ground state into the manifold of interacting states. Our intention in this section is to identify and interpret the impact of a conical intersection on the optical spectrum. The model Hamiltonian (3.1) describes a wide range of phenomena depending on the values of the five independent parameters $E_1 - E_2$, κ_1, κ_2, λ, and ω_g / ω_u. Therefore an exhaustive study of the spectra described by $\mathcal{H}$ of Eq. (3.1) is beyond the scope of the present work. We shall start with the parameter values introduced in the preceding section, for which we have a detailed knowledge of the adiabatic potentials (see Figs. 4, 5, and 6). Subsequently, we shall vary the vertical energy difference $E_2 - E_1$, which amounts to moving the conical intersection in coordinate space and in energy. The limit $\kappa_1 = \kappa_2 = 0$ is also considered to help in identification of multimode effects.

1. Computational Details

Calculation of the spectra is performed using the diabatic representation and has been described in Section II.B.3. The secular matrix of $\mathcal{H}$ in a direct product basis of harmonic oscillators is straightforwardly generated. For the examples given below, a dimension of 1128 for each of the two secular matrices (corresponding to vibronic g and u symmetry) was found to be sufficient to guarantee that the spectra are free of truncation errors.

To identify the nonadiabatic effects, we calculate the spectra also in the adiabatic and Franck–Condon approximations (see Section II.B.3). The adiabatic Hamiltonian $\mathcal{H}_{ad}$ of Eq. (3.8b) is represented in a direct product basis $|n_g, n_u\rangle$ of harmonic oscillator functions. For the present examples, 8(29) basis function for the $Q_g(Q_u)$ mode were found to be sufficient, giving a dimension of 232 for the secular matrices.

While the calculation of the spectrum in the Franck–Condon approximation (briefly called the Franck–Condon spectrum) is straightforward, the calculation of the adiabatic spectrum is somewhat tedious owing to the ill-behaved nature of the adiabatic approximation in the presence of a conical intersection. As discussed in the preceding section, the $S_{ij}(Q_g, Q_u)$ are necessarily discontinuous at the point of intersection and, depending on the choice of the phase factor, possibly discontinuous on a cut connecting the intersection point with infinity. These discontinuities of the $S_{ij}(Q_g, Q_u)$, which are unavoidable in the presence of a conical intersection, render the numerical calculation of the adiabatic spectrum rather difficult. As discussed in Section II.B.2, the discontinuities of the $S_{ij}(\mathbf{Q})$ cause the divergence of all but the zeroth moment of the adiabatic spectrum. The explicit calculation of the adiabatic moments (2.54b) is rather lengthy in the general case. To understand the principles, it suffices to consider again the $E \times \varepsilon$ Jahn–Teller limit $E_1 = E_2 = E$, $-\kappa_2 = \kappa_1 = \kappa = \lambda$, and $\omega_g = \omega_u = \omega$. It follows immediately that

$$M_1 = M_1^{FC} = E \qquad (3.20)$$

That is, the centroid of both the exact and the Franck–Condon spectrum is equal to the vertical electronic transition energy E. When computing the first moment of the adiabatic spectrum, we have to note that the kinetic energy operator T_N contained in $\mathscr{H}_{ad}$ acts on the $\mathbf{Q}$-dependence of $\mathbf{S}$ contained in $\tilde{\tau}$. A simple calculation using the Laplacian in polar coordinates gives

$$M_1^{ad} = E + \tfrac{1}{4}\omega\langle 0|\rho^{-2}|0\rangle \qquad (3.21)$$

The expectation value of ρ^{-2} is infinite. The centroid of the adiabatic spectrum thus lies at infinite energy. It follows that all higher moments of the adiabatic spectrum are infinite too. The divergence of the adiabatic spectral moments is a consequence of the discontinuity of $\mathbf{S}$ at the intersection point and reflects the ill-behaved nature of the adiabatic approximation in a conical intersection situation. When the point of intersection $Q_g^{(c)}$ moves out of the Franck–Condon zone, the divergence of the moments is damped exponentially with increasing $|Q_g^{(c)}|$, as is clear from the form of Eq. (3.21). Although, strictly speaking, the adiabatic spectral moments remain infinite for arbitrary $Q_g^{(c)}$, this no longer has observable consequences when $|Q_g^{(c)}| \gg 1$.

The divergence of the moments is of relevance for the numerical computation of the spectrum in the adiabatic approximation. Obviously a numerical calculation using an expansion in a finite basis set cannot yield infinite spectral moments. It follows that the numerically computed adiabatic spectrum cannot converge in a strict sense when the number of basis functions is increased, in contrast to the exact and Franck–Condon spectra, which have finite moments. A closer examination of the adiabatic spectra shows that they exhibit a tail of weak lines extending to very high energies. It is this tail which causes the divergence of the first and all higher moments. The numerical calculations show that apart from this tail the intensities of the lines in the adiabatic spectrum become stable with increasing number of basis functions. In this qualitative sense it is possible to compute the adiabatic spectrum with finite basis set expansion methods. When the point of intersection lies sufficiently outside the Franck–Condon zone, the total intensity contained in the high-energy tail becomes negligibly small and is no longer of relevance for the numerical calculations.

2. Discussion of the Spectra

Let us now discuss the results of the dynamical calculations. Figure 8 shows the spectra for the conical intersection example of the preceding subsection, obtained in the Franck–Condon, adiabatic, and exact vibronic calculations, respectively. The corresponding adiabatic potential energy surfaces are given in Figs. 4–6. The electronic transition matrix elements are taken to be equal: $\tau_1 = \tau_2 = 1$. All spectra are normalized such that the summed line intensity is equal to 2. The vertical lines show the actually calculated vibronic spectrum. Since experimental spectra always have a finite resolution, we have convoluted the line spectrum with a Lorentzian function of fwhm = 0.04 eV, representing the typical experimental resolution in a photoelectron spectrum or a low-resolution absorption spectrum.

Considering the potential energy surfaces of Fig. 6, the Franck–Condon spectrum (Fig. 8a) is relatively easy to interpret. The progression of lines centered at approximately 9.4 eV corresponds to the excitation of the Q_u mode in the lower state. The Q_u mode is strongly excited owing to the double-minimum character of the lower adiabatic surface. The intense line near 9.9 eV represents the vibrational ground level of the upper surface. Because of the conical shape of the upper adiabatic surface, the vibrational spacing is very large and the higher levels cannot be simply classified in terms of harmonic overtone and combination levels.

The adiabatic spectrum (Fig. 8b) differs from the Franck–Condon spectrum through the inclusion of intensity borrowing effects (see Section II.B.2). In the present definition of the adiabatic approximation, where the diagonal elements of $\hat{\Lambda}$ are not included in the adiabatic potentials, the line positions are the same in the adiabatic and Franck–Condon spectra. The intensity borrowing leads to the appearance of additional lines which are symmetry-

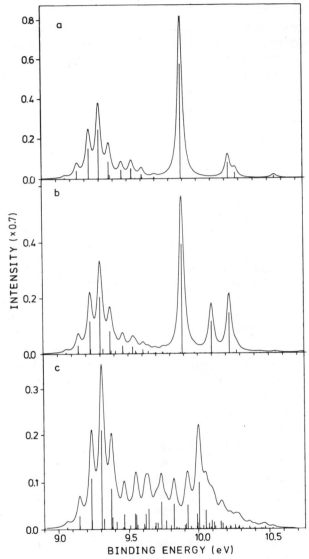

Fig. 8. Spectra for the prototype conical intersection model obtained in the Franck–Condon, adiabatic, and exact vibronic calculation, respectively, from top to bottom. The corresponding adiabatic potential energy surfaces are shown in Fig. 6.

116

forbidden in the Franck–Condon approximation. Inspection of Fig. 8 shows that intensity borrowing effects are rather weak in the lower electronic state but strong in the upper state. In Fig. 8b one can see the weak high-energy tail that causes the divergence of the spectral moments mentioned above.

Considering now the exact spectrum (Fig. 8c), it is immediately obvious that neither the Franck–Condon nor the adiabatic approximation give any reasonable description of the true vibronic spectrum. In the low-energy region, up to approximately 9.5 eV, the adiabatic spectrum agrees qualitatively with the exact spectrum (note the different ordinate scales in Figs. 8a, 8b, and 8c). Above 9.5 eV, however, there is no longer any correspondence between the two spectra. The exact spectrum exhibits a dense vibronic line structure. Most peaks of the envelope in Fig. 8c are accumulations of lines rather than single vibronic lines. The high-energy tail of the low-resolution spectrum is completely diffuse. It is particularly noteworthy that the vibrational energy levels of the upper adiabatic surface are not discernible in the exact spectrum. This proves the complete breakdown of the Born–Oppenheimer adiabatic approximation for the upper surface. In other words, the vibrational motion is not confined to the upper adiabatic electronic surface, but proceeds on both surfaces simultaneously owing to the strong nonadiabatic coupling. In the present example, the energetic position of the point of intersection is $E_c = 9.73$ eV. Our calculation shows that the breakdown of the adiabatic approximation starts a few tenths of an electron volt below the intersection and extends over the whole spectrum above.

Further evidence on the nature and importance of nonadiabatic effects is provided by the analysis of individual vibronic eigenstates. We have computed the full eigenvectors for all vibronic eigenstates in the spectrum of our prototype model (Fig. 8c). As expected, the vibronic states are a mixture of a vast number of electronic–vibrational basis states. To make contact with the usual picture based on the adiabatic approximation and to identify the nonadiabatic effects, we have reexpanded the exact vibronic eigenstates in terms of the vibrational states of the *adiabatic* electronic potential surfaces. For comparison, we have also reexpanded the exact vibronic eigenstates in terms of the vibrational states of the *diabatic* electronic potential surfaces. In Tables I and II we give these expansion coefficients for three vibronic states which correspond to prominent spectral lines in the high-energy part of Fig. 8c. It may be seen that the vibronic states are composed of a considerable number of vibrational levels, both in the diabatic and adiabatic representations. Summing over the various contributions, one finds that the vibronic states consist of an almost equal mixture of the diabatic electronic species. Concerning the adiabatic representation, even in the energy range of the upper potential energy surface the vibronic states contain predominantly the lower adiabatic electronic species. These findings corroborate the

TABLE I

Eigenvector Coefficients of Three Selected Vibronic Eigenstates in
Terms of the Diabatic Vibrational Wave Functions[a]

			E		
n_u	n_g	E_{diab}	9.733	10.007	10.054
3	1	9.895	0.158	⋯	⋯
1	2	9.970	⋯	−0.245	−0.136
5	1	10.077	−0.219	⋯	⋯
3	2	10.152	⋯	0.227	⋯
9	0	10.184	−0.159	⋯	⋯
1	3	10.228	⋯	⋯	0.257
7	1	10.259	⋯	0.154	⋯
13	0	10.549	0.201	⋯	⋯
15	0	10.711	⋯	−0.142	⋯
13	1	10.807	−0.134	0.141	⋯
17	0	10.913	−0.179	⋯	⋯
15	1	10.989	⋯	⋯	−0.180
13	2	11.064	⋯	⋯	0.150
19	0	11.096	−0.194	0.175	⋯
15	2	11.247	⋯	⋯	0.151
19	1	11.354	⋯	−0.154	⋯
0	0	9.724	0.365	−0.197	⋯
2	0	9.906	−0.272	⋯	−0.137
0	1	9.982	⋯	0.374	0.401
4	0	10.089	0.136	⋯	0.154
2	1	10.164	⋯	⋯	−0.196
4	1	10.347	⋯	⋯	0.201
8	0	10.454	−0.165	−0.155	⋯
8	2	10.969	⋯	⋯	−0.145

[a] The expansion coefficients are written as columns below the
corresponding vibronic energy E (entries smaller than 0.13 in absolute magnitude are suppressed). The vibrational wave functions
are characterized by the harmonic quantum numbers n_u and n_g
of the coupling and tuning modes, respectively, and their energy
E_{diab}. Note that an even value of n_u corresponds to the (vertically) upper diabatic state, while an odd value of n_u corresponds
to the (vertically) lower diabatic state. All energies are in electron
volts.

profound impact that the vibronic coupling and nonadiabatic interactions
exert on the nuclear motion in this example.

The data of Tables I and II shed some light on the feasibility of "decon-
volution" procedures which have been used to deperturb irregular vibronic
spectra.[137,138] The vibrational levels of the upper surface are taken to be embedded in and to interact with the manifold of excited vibrational levels of

TABLE II

Eigenvector Coefficients of Three Selected Vibronic Eigenstates in Terms of the Adiabatic Vibrational Wave Functions[a]

			E		
σ	n	E_{ad}	9.733	10.007	10.054
1	14	9.646	−0.127	· · ·	· · ·
1	15	9.695	0.190	· · ·	· · ·
1	17	9.746	0.859	· · ·	· · ·
1	19	9.783	0.254	· · ·	· · ·
1	23	9.877	0.127	−0.131	· · ·
1	26	9.964	· · ·	−0.153	· · ·
1	27	9.981	· · ·	0.748	−0.158
1	28	9.994	· · ·	−0.140	· · ·
1	29	10.012	· · ·	· · ·	0.440
1	31	10.074	· · ·	· · ·	−0.360
1	32	10.078	· · ·	· · ·	0.197
1	33	10.092	· · ·	0.231	−0.285
1	34	10.108	· · ·	· · ·	−0.464
1	35	10.156	· · ·	· · ·	0.131
1	38	10.210	· · ·	· · ·	0.121
1	39	10.214	· · ·	· · ·	0.164
1	45	10.324	· · ·	· · ·	−0.120
2	1	9.890	0.219	0.492	0.240
2	2	10.228	· · ·	−0.120	−0.210

[a] The expansion coefficients are written as columns below the corresponding vibronic energy E (entries smaller than 0.12 in absolute magnitude are suppressed). The adiabatic levels with $\sigma = 1$ ($\sigma = 2$) are those of the lower (upper) potential energy surface, n is a running index, and E_{ad} their energy. All energies are in electron volts.

the lower surface. In the simplest form, the levels of the upper surface are treated as independent; that is, the possible indirect coupling through the vibrational levels of the lower surface is ignored. This implies that the eigenvector of a given vibronic state should involve only a single vibrational level of the upper surface. We see from the two tables that this simple picture does not apply in the present example, especially in the diabatic representation. Rather, the strong vibronic coupling leads here also to a mixing of different vibrational levels of the (vertically) *upper* surface. In the adiabatic representation and for low energies, the vibronic eigenstates involve mainly a single vibrational level of the upper adiabatic surface (see the first column of Table II). This is caused by the fact that the spacings of the vibrational levels of the upper adiabatic surface are large and all these levels lie higher in energy than the vibronic eigenstates in question. For higher energies, such as in the third column of Table II, the vibronic eigenstates contain several vibrational levels of the upper adiabatic surface with notable weights. The mixing of different levels of the upper surface may give rise to interesting interference effects[139] when the vibrational levels involved have comparable Franck-Condon factors. For example, the intensity of the vibronic eigenstate at 10.054 eV is subject to constructive interference between the lowest vibrational levels of the upper adiabatic surface (at 9.89 and 10.23 eV). Similarly, in the diabatic representation, the vibronic eigenstate at 10.007 eV gains its intensity by constructive interference between the diabatic vibrational levels at 9.72 and 9.98 eV.

The complete breakdown of the adiabatic approximation is a genuine multimode effect. This is made explicit by Fig. 9, where the corresponding results for the one-dimensional vibronic coupling problem are shown. In these calculations only the coupling mode Q_u is considered. The corresponding adiabatic potential energy curves are given in Fig. 5a. Numerical studies of the one-dimensional $g-u$ vibronic coupling problem have been undertaken previously,[36,38-40,43] and the effects are well understood in principle. The numerical calculation of the spectra is of course much simpler in the one-dimensional case. We have included 100 harmonic oscillator basis functions in the exact vibronic calculation and 61 in the adiabatic and Franck-Condon approximations. The Franck-Condon spectrum (Fig. 9a) exhibits a regular progression in the lower electronic state reflecting the double-minimum shape of the potential. In the upper state the vibrational excitation is very weak, as expected from inspection of Fig. 5a. The adiabatic spectrum (Fig. 9b) shows that intensity borrowing effects are weak in the lower electronic state and moderate in the upper state. The exact vibronic calculation (Fig. 9c) exhibits a slight splitting of the intense line corresponding to the ground vibrational level of the upper adiabatic potential energy curve. Nonadiabatic effects are thus visible but do not lead to a complete disappearance of the lines in the adiabatic spectrum. The comparison of Figs. 8 and 9 illustrates the dramatic enhancement of the nonadiabatic coupling by the tuning mode Q_g. In other words, nonadiabatic effects are much more pronounced in the conical intersection problem than in the one-dimensional avoided crossing problem.

The model exhibits another interesting mode-mode coupling phenomenon, namely, mutual quenching of the excitation strengths of the two modes in the lower electronic state. This phenomenon can be fully understood already in the Franck-Condon approximation. A comparison of Fig. 8a with Fig. 9a shows that the excitation strength of the torsional mode is weaker in the two-dimensional calculation than in the one-dimensional calculation. The additional totally symmetric degree of freedom apparently "quenches" the

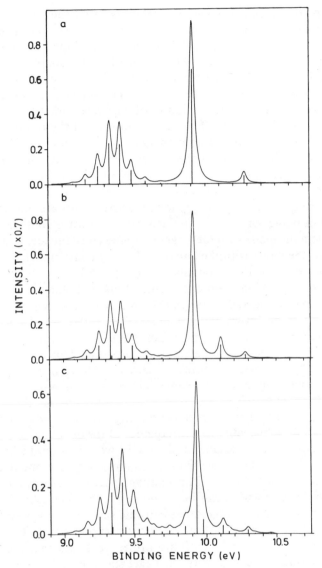

Fig. 9.   Spectra obtained by considering only the coupling mode and discarding the tuning mode
(Franck–Condon, adiabatic, and exact, respectively, from top to bottom).

121

excitation of the torsional mode. The effect occurs likewise for the totally symmetric mode. The excitation strength of the totally symmetric mode seen in Fig. 8a is smaller than expected from the value of the "Poisson parameter" $a = \frac{1}{2}(\kappa_1/\omega_g)^2 = 0.34$ (see Section II.B.1). The coupling mode thus quenches the excitation of the tuning mode. These trends can be qualitatively understood by looking at the adiabatic potential energy surface of Fig. 6 in more detail. The absolute minima of the lower adiabatic surface are found at $Q_u^{(0)} = \pm 3.42$, $Q_g^{(0)} = 0.10$. In the absence of the tuning mode, the minimum of the torsional potential energy curve is found at $Q_u^{(0)} = \pm 3.43$. In the absence of the coupling mode, on the other hand, the minimum of the lower potential energy curve is found at $Q_g^{(0)} = 0.82$. The shift of the minimum in the Q_g coordinate is thus strongly reduced by the coupling mode, explaining the quenching of the Q_g mode. These trends are not confined to the present example. One finds quite generally that for a nonseparable two-mode problem the mode with the larger coupling constant acts to quench the excitation of the mode with the smaller coupling constant. The nonseparability of the modes of Hamiltonians of the type (3.1) even in the adiabatic limit has been pointed out and analyzed in Refs. 133–136.

The example considered so far as the prototype of a conical intersection represents a case of "overlapping bands"; that is, the two electronic states are rather close in energy. To find a limitation of the validity of the adiabatic approximation is thus not altogether surprising. One could expect that the adiabatic approximation improves quickly as the vertical energy separation of the interacting electronic states increases. To investigate this point, we double the energy gap $E_2 - E_1$ in our prototype example, that is, $E_2 - E_1 = 0.8$ eV, leaving all other parameters unchanged. The vertical energy gap is now nearly 10 times the vibrational frequency of the coupling mode, and one would thus expect the nonadiabatic effect to be weak in accord with the usual arguments.[36,38-40,43] This expectation is confirmed by the one-dimensional calculation as shown in Fig. 10. The adiabatic potential energy curves corresponding to this calculation can be found in Fig. 5c. The vibrational levels corresponding to different adiabatic electronic states are now well separated in energy; that is, we are in the limit of nonoverlapping bands. It can be seen that the intensity borrowing effects are rather weak and that the adiabatic spectrum is virtually identical with the exact spectrum.

The spectra obtained in the two-mode calculation for $E_2 - E_1 = 0.8$ eV are shown in Fig. 11. Again, the bands corresponding to the two electronic states are clearly separated. Comparison of Fig. 11a with Fig. 11b shows that adiabatic intensity borrowing effects are very weak in the lower state and moderate in the upper state. Comparison of Fig. 11b with 11c shows that the lower band is accurately given by the adiabatic approximation, whereas it does not work at all for the upper band. Apparently, the nonadiabatic

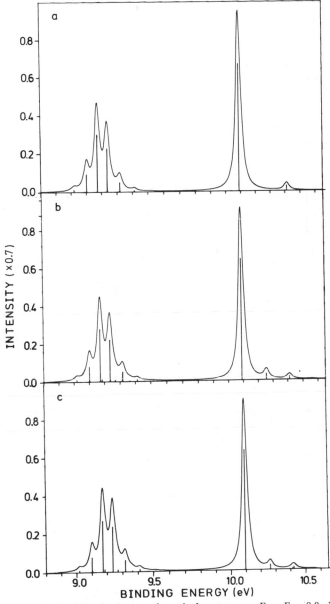

Fig. 10. Spectra obtained for the increased vertical energy gap $E_2 - E_1 = 0.8$ eV (twice the original value) in the one-dimensional calculation (Franck–Condon, adiabatic, and exact, respectively, from top to bottom).

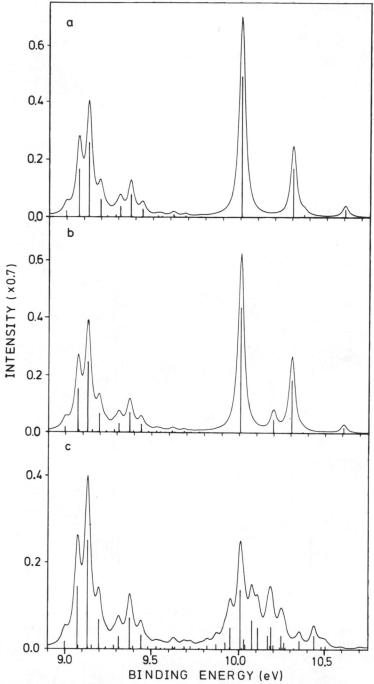

Fig. 11. Spectra obtained for $E_2 - E_1 = 0.8$ eV in the two-dimensional calculation (Franck–Condon, adiabatic, and exact, respectively, from top to bottom).

coupling is still strong enough to split the vibrational levels of the upper adiabatic surface and to redistribute the intensity completely. This result is in sharp contrast to the one-dimensional calculation and underlines once more the strong enhancement of nonadiabatic effects by the totally symmetric tuning mode.

Let us have a look at the corresponding adiabatic potential energy surfaces. For brevity we show in Fig. 12 only a cut through the surfaces along the Q_g axis ($Q_u = 0$). Since the coupling term λQ_u vanishes on the Q_g axis, these potentials are identical with the diabatic potentials of the model. The vibrational energy levels of the ν_g mode are indicated by the horizontal lines. The potential curves intersect at $Q_g^{(c)} = -1.7142$, which is well outside the Franck–Condon zone (defined as $|Q_g| < \sqrt{2}$). However, the point of intersection still occurs close to the minimum of the upper surface, at an energy $E_c = 9.99$ eV. All energy levels of the ν_g mode of the upper surface lie above the point of intersection. According to the classical picture, during each vibrational period the nuclear motion on the upper surface twice has to cross the point of intersection, where it is coupled with particular strength to the lower surface. This explains the complete failure of the adiabatic approximation.

When we once again double the vertical energy gap, that is, $E_2 - E_1 = 1.6$ eV, the point of intersection moves down the negative Q_g axis to $Q_g^{(c)} = -3.4284$ and moves up in energy to $E_c = 11.09$ eV. Figure 13 shows the corresponding diabatic (and adiabatic) potentials on the Q_g axis. It may be seen here that three vibrational levels of the upper potential lie below the point of intersection. Since the conical intersection now lies outside the classical

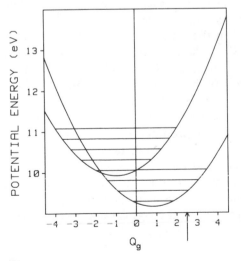

Fig. 12. Potential energy curves along the Q_g direction for $E_2 - E_1 = 0.8$ eV (twice the original value). The horizontal lines indicate the energy levels of the ν_g mode.

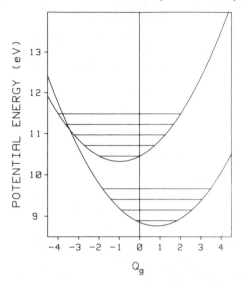

Fig. 13.   Potential energy curves along the Q_g direction for $E_2 - E_1 = 1.6$ eV (four times the original value). The horizontal lines indicate the energy levels of the ν_g mode.

turning points of these levels, one would expect that they should be well described in the adiabatic approximation. This expectation is confirmed by the numerical calculations shown in Fig. 14. Since for the lower band the adiabatic and exact spectra are identical to within drawing accuracy, only the upper band is shown. It can now be seen that the intensity borrowing effects (the difference between the adiabatic and the Franck–Condon spectrum) are extremely weak and that the envelope of the adiabatic spectrum agrees qualitatively with the envelope of the exact vibronic spectrum. However, we can still observe a splitting of the lines in the exact vibronic spectrum. The interpretation of this line splitting is as follows. The energy levels of the upper adiabatic surface are embedded in the rather dense manifold of vibrational levels of the lower surface. Since the conical intersection is situated outside the turning points of the nuclear motion for the vibrational levels of the upper adiabatic surface (see Fig. 13), the nonadiabatic coupling is rather weak, causing a significant mixing only for nearly degenerate levels. The resulting fine structure splitting can be interpreted as a "broadening" of the vibrational lines. The resulting "width" is proportional to the rate of nonradiative "decay" from the upper to the lower adiabatic surface. The relationship between conical intersections and radiationless decay is discussed in more detail in Section VII. Here we note only that even for energies below the point of intersection one obtains a significant splitting of the adiabatic energy levels (see Fig. 14c). When the vertical energy gap is further increased to $E_2 - E_1 = 2.0$ eV, the adiabatic spectrum becomes exact to

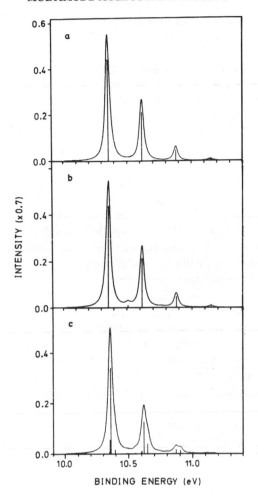

Fig. 14. Spectra obtained for $E_2 - E_1 = 1.6$ eV (four times the original value) in the two-dimensional calculation (Franck–Condon, adiabatic, and exact, respectively, from top to bottom).

within drawing accuracy; that is, the first three levels of the upper adiabatic surface are no longer affected by the conical intersection.

The foregoing examples have explicitly verified the transition from a conical intersection within the Franck–Condon zone, where the adiabatic approximation breaks down completely, to the case of a remote conical intersection, where the adiabatic picture becomes qualitatively valid. To complete the picture that has emerged so far, we next consider the fate of vibrational levels which lie energetically high above the conical intersection. In classical reasoning, the nuclei should have large kinetic energy when reaching the conical point. The nonadiabatic coupling should therefore be particularly strong and the nuclei should follow the diabatic surfaces to a good approximation. To what extent this qualitative argument is correct is shown by the following numerical experiment. We return to our model with $E_2 - E_1 = 0.8$ eV. To populate higher vibrational levels in the absorption spectrum, we shift the equilibrium geometry of the initial state of the optical transition from $Q_g = 0$ to $Q_g = 2.5$, indicated by the arrow in Fig. 12. It is clear from the Franck–Condon principle that now some 10 levels of the ν_g mode of the upper surface are populated with significant intensity. Figure 15 shows the resulting exact vibronic spectrum. Owing to the stronger vibrational excitation, a more extensive numerical calculation was necessary to generate this spectrum. We have included 40 basis functions of the ν_g mode and 50 basis functions of the ν_u mode, resulting in a secular matrix of dimension 2000. The resulting vibronic line structure is very complex over the whole range of the spectrum. Especially in the high-energy region, one observes a well-developed bunching of the vibronic lines, leading to a fairly regular progression of peaks in the envelope. The spacing of these peaks agrees with ω_g, the frequency of the Q_g mode on the diabatic surface. The line spacings on the upper adiabatic surface are significantly larger (see Fig. 11b). The development of a quasi-progression with spacing ω_g in the high-energy region of Fig. 15 illustrates the transition to the "diabatic limit," where the nuclear motion follows essentially the diabatic electronic surfaces. It can be seen from the numerical experiment shown in Fig. 15 that the diabatic limit is approached very slowly with increasing energy.

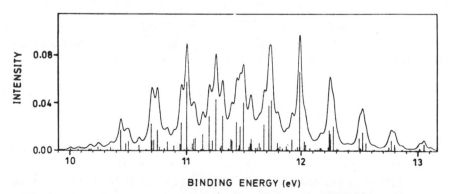

Fig. 15.   Exact vibronic spectrum obtained for $E_2 - E_1 = 0.8$ eV, assuming a shifted equilibrium geometry of the initial electronic state ($Q_g^{(0)} = 2.5$, indicated by an arrow in Fig. 12). This shift serves to populate higher vibrational levels of the upper electronic state.

Let us summarize the main results of these numerical studies of spectra for conically intersecting potentials. We have seen that vibrational energy levels sufficiently below the point of intersection approach the adiabatic limit, whereas energy levels significantly above the intersection approach the diabatic limit only very slowly. In a wide energy range around the energy of intersection the exact vibronic spectra are extremely complex. Whenever the conical intersection is energetically accessible, one observes a complete breakdown of the adiabatic approximation. The position of the intersection in coordinate space is comparatively irrelevant. These general statements are confirmed by a variety of additional model calculations and the comparison with experimental spectra, as discussed in Section V.

C. Inclusion of Additional Modes

Having discussed the prototype conical intersection problem involving one coupling mode and one tuning mode in some detail, we shall now give a brief overview of the trends expected when more coupling modes or tuning modes are active. Again we shall consider the static problem (potential energy surfaces and adiabatic electronic wave functions) and the dynamic problem (calculation of spectra) separately, starting with the static aspects.

The generalization of the model Hamiltonian (3.1) to the case of M coupling modes Q_{uj} and N tuning modes Q_{gi} is obvious:

$$\mathscr{H} = (T_N + V_0)\mathbf{1} + \begin{pmatrix} E_1 + \sum\limits_{i=1}^{N} \kappa_i^{(1)} Q_{gi} & \sum\limits_{j=1}^{M} \lambda_j Q_{uj} \\ \sum\limits_{j=1}^{M} \lambda_j Q_{uj} & E_2 + \sum\limits_{i=1}^{N} \kappa_i^{(2)} Q_{gi} \end{pmatrix} \tag{3.22a}$$

$$T_N = -\frac{1}{2} \sum_{i=1}^{N} \omega_{gi} \frac{\partial^2}{\partial Q_{gi}^2} - \frac{1}{2} \sum_{j=1}^{M} \omega_{uj} \frac{\partial^2}{\partial Q_{uj}^2} \tag{3.22b}$$

$$V_0 = \frac{1}{2} \sum_{i=1}^{N} \omega_{gi} Q_{gi}^2 + \frac{1}{2} \sum_{j=1}^{M} \omega_{uj} Q_{uj}^2 \tag{3.22c}$$

We assume thus that there exist several non-totally symmetric modes transforming according to the irreducible representation Γ_3 such that Eq. (3.2) is fulfilled. Of particular practical importance is the inclusion of several totally symmetric tuning modes, since even a medium-size molecule may have numerous totally symmetric modes.

For the discussion of the static aspects of the problem, it is convenient to use the notation introduced in Eq. (3.5) for the two-mode case. We write in obvious generalization of Eq. (3.5):

$$\mathscr{H} = H_0 \mathbf{1} + \begin{pmatrix} -d & c \\ c & d \end{pmatrix} \tag{3.23a}$$

$$H_0 = T_N + V_0 + \Sigma + \sum_{i=1}^{N} \sigma_i Q_{gi} \tag{3.23b}$$

$$\sigma_i = \frac{\kappa_i^{(1)} + \kappa_i^{(2)}}{2} \tag{3.23c}$$

$$\delta_i = \frac{\kappa_i^{(2)} - \kappa_i^{(1)}}{2} \tag{3.23d}$$

$$d = \Delta + \sum_{i=1}^{N} \delta_i Q_{gi} \tag{3.23e}$$

$$c = \sum_{j=1}^{M} \lambda_j Q_{uj} \tag{3.23f}$$

The constants Σ and Δ are defined in Eqs. (3.5c, d). The adiabatic potentials $V_1(\mathbf{Q})$ and $V_2(\mathbf{Q})$ are given by essentially the same expression as in Section III.A, namely,

$$V_{1,2}(\mathbf{Q}) = V_0(\mathbf{Q}) + \Sigma + \sum_{i=1}^{N} \sigma_i Q_{gi} \mp W \tag{3.24a}$$

$$W = (d^2 + c^2)^{1/2} \tag{3.24b}$$

The adiabatic eigenvector matrix $\mathbf{S}(\mathbf{Q})$ is given by Eq. (3.7). The nonadiabatic coupling matrix $\hat{\Lambda}$ is simply given by a sum of coupling terms, one for each mode as in Eq. (3.8c). Thus the sum over two modes in Eq. (3.8c) has to be replaced by a sum over $N + M$ modes.

Let us first consider the phenomenon of symmetry breaking in the multimode case. A simple calculation for the nontrivial stationary point $\mathbf{Q}^{(0)}$ of the lower adiabatic surface $V_1(\mathbf{Q})$ gives

$$Q_{gi}^{(0)} = \frac{\Delta - F}{L - D} \frac{\delta_i}{\omega_{gi}} - \frac{\sigma_i}{\omega_{gi}} \tag{3.25a}$$

$$Q_{uj}^{(0)} = \frac{\pm \lambda_j}{\omega_{uj}} \left[1 - \left(\frac{\Delta - F}{L - D} \right)^2 \right]^{1/2} \tag{3.25b}$$

where we have introduced in analogy to Eq. (3.12)

$$D = \sum_{i=1}^{N} \frac{\delta_i^2}{\omega_{gi}} \tag{3.26a}$$

$$F = \sum_{i=1}^{N} \frac{\delta_i \sigma_i}{\omega_{gi}} \tag{3.26b}$$

$$L = \sum_{j=1}^{M} \frac{\lambda_j^2}{\omega_{uj}} \tag{3.26c}$$

The stabilization energy [i.e., the lowering of $V_1(\mathbf{Q}^{(0)})$ relative to the minimum of $V_1(\mathbf{Q})$ in the absence of vibronic coupling ($\lambda_j = 0$)] is

$$E_s = \frac{(\Delta + D - L - F)^2}{2(L - D)} \tag{3.27}$$

The criteria for the existence of a stabilized ($E_s > 0$) minimum of a reduced symmetry ($Q_{uj}^{(0)} \neq 0$) are

$$L > D \tag{3.28a}$$

$$\left| \frac{\Delta - F}{L - D} \right| < 1 \tag{3.28b}$$

in complete analogy to the two-mode case.

Let us for the moment discard the tuning modes, namely, $\delta_i = (\kappa_i^{(2)} - \kappa_i^{(1)})/2 = 0$ for all i. Then Eq. (3.27) for the stabilization energy reduces to

$$\frac{E_s}{\Delta} = \begin{cases} \dfrac{(1-x)^2}{2x}, & x \geq 1 \\ 0, & x \leq 1 \end{cases} \tag{3.29a}$$

where

$$x = \frac{L}{\Delta} = \sum_{j=1}^{N} x_j \tag{3.29b}$$

$$x_j = \frac{\lambda_j^2}{\omega_{uj}\Delta} \tag{3.29c}$$

The functional dependence of E_s on the dimensionless parameter x given by Eq. (3.29a) has already been discussed in Section III.A.1 and is shown in Fig. 3. The interesting aspect here is that x is simply given by a sum of contributions of all coupling modes. The symmetry breaking is thus a *cumulative* effect of all coupling modes. In particular, consider a situation where all x_j values are smaller than unity. This means that the coupling of each single mode Q_{uj} is too weak to introduce a minimum of $V_1(\mathbf{Q})$ at $Q_{uj} \neq 0$. The sum of the x_j values may be larger than unity, however, and the symmetry will be broken. A startling example of such a multimode symmetry-breaking effect will be discussed in Section V.C.

Another interesting result follows from Eq. (3.25b). If the symmetry is broken, the symmetry lowering occurs with respect to *all* coupling modes Q_{uj}, except only modes for which λ_j is exactly zero. This implies that in large molecules with many coupling modes rather complicated distortions may occur. A closer examination of the stationary points (3.25b) shows that the lower adiabatic surface $V_1(\mathbf{Q})$ possesses two equivalent non-totally symmetric minima if the conditions (3.28) are fulfilled. The influence of the tuning modes is completely analogous to the two-mode situation discussed in Section III.A.1. Note that again the effect of the tuning modes is cumulative.

The second aspect of the adiabatic potential energy surfaces that deserves a more detailed discussion is the conical intersection in the multimode case. The conditions for the occurrence of a conical intersection of the adiabatic potential energy surfaces of the Hamiltonian (3.23) are simply

$$d = 0, \qquad c = 0 \tag{3.30}$$

These conditions define a hypersurface of dimension $N + M - 2$ in the $(N + M)$-dimensional coordinate space. In the case of one coupling mode and two tuning modes, for example, we obtain a line of conical intersections in three-dimensional space.

Let us consider the rather common case of a single coupling mode ($M = 1$) and N tuning modes in more detail. As we have seen in the preceding subsection, the energetic position of the conical intersection is of crucial importance for the vibronic fine structure of the absorption spectrum. We have seen that the adiabatic approximation fails completely above the point of intersection, whereas it becomes very accurate sufficiently below the energy of intersection. In the multidimensional case, an important quantity is thus the absolute minimum of the $(N + M - 2)$-dimensional hypersurface of conical intersections. As a rule, one can expect that the adiabatic approximation is applicable up to a few vibrational spacings below the minimum energy of intersection. Above the minimum energy of intersection, the adiabatic approximation will fail completely and one generally expects a dense and very complex vibronic line structure in the absorption spectrum (assuming that the interstate coupling constant is not too small). Starting from Eq. (3.24) for the adiabatic potentials, a simple calculation for the minimum energy of intersection gives

$$V_{min}^{(c)} = \Sigma + \frac{(F - \Delta)^2}{2D} - \frac{1}{2} \sum_{i=1}^{N} \frac{\sigma_i^2}{\omega_{gi}} \qquad (3.31)$$

The position of the minimum in the space of the tuning modes is

$$\left(Q_{gi}^{(c)} \right)_{min} = \frac{(\delta_i / \omega_{gi})(F - \Delta)}{D} - \frac{\sigma_i}{\omega_{gi}}, \qquad i = 1, \dots, N \qquad (3.32)$$

The shortest distance of the hypersurface of conical intersections from the origin is given by

$$Q_0^{(c)} = \Delta \left[\sum_{i=1}^{N} \delta_i^2 \right]^{-1/2} \qquad (3.33)$$

The last term in Eq. (3.31) describes a *monotonous* decrease of the minimum energy of intersection as more and more totally symmetric modes are included. Thus the onset of the breakdown of the adiabatic approximation will be shifted to lower and lower energy. An interesting quantity is also the minimum energy of intersection relative to the minimum of the upper adiabatic potential energy surface. This difference is given by

$$V_{min}^{(c)} - (V_2)_{min} = \frac{1}{2D} (\Delta - D - F)^2$$

$$= \frac{1}{2D} \left(\Delta - \sum_{i=1}^{N} \frac{\kappa_i^{(2)} \left(\kappa_i^{(2)} - \kappa_i^{(1)} \right)}{2\omega_{gi}} \right)^2 \qquad (3.34)$$

Assuming a relatively large vertical energy gap Δ, we see that all tuning modes, for which the two intrastate coupling constants $\kappa_i^{(1)}$ and $\kappa_i^{(2)}$ are of different sign, act to reduce $V_{min}^{(c)}$ relative to the minimum of $V_2(\mathbf{Q})$. When $V_{min}^{(c)}$ approaches $(V_2)_{min}$, the whole spectral band corresponding to the upper adiabatic surface will be affected by strong nonadiabatic coupling. A typical example of the lowering of the minimum energy of intersection by several totally symmetric modes is given in Section V.A.

Equation (3.33) shows that the minimum distance of the hypersurface of intersections from the center of the Franck–Condon zone decreases monotonously with the number of tuning modes included. This means that the hypersurface can closely approach the Franck–Condon zone even with a large vertical energy gap $E_2 - E_1$. These trends suggest that the optical spectra of large polyatomic molecules should be commonly affected by conical intersections.

It is clear that these conclusions pertain, strictly speaking, only to the simplified model Hamiltonian (3.22), since they depend on the choice of harmonic diabatic potentials and linear coupling. Nevertheless, the general trends following from the above analysis are probably representative of a wide class of vibronic coupling phenomena in molecular spectroscopy. Although only qualitative in nature, the present model Hamiltonian approach has several definite merits. It elucidates the mechanism that causes the symmetry lowering of molecules in excited or ionic states, and it provides, in particular, very simple rules for predicting probable symmetry-breaking effects. Once we know the symmetry species of the closely lying electronic states under consideration, the possible coupling modes follow from elementary symmetry considerations. The relevant coupling constants $\kappa_i^{(1,2)}$ and λ_j can be obtained from ab initio calculations[79,81] or estimated semiempirically.[140–144] In the case of ionization or electron attachment, simple molecular orbital considerations are often sufficient.[145] In principle, the adiabatic potential energy surface and its minima can be calculated directly using ab initio quantum chemical methods. For polyatomic molecules, however, it is hardly possible to explore the complete potential hypersurfaces numerically and additional minima at configurations of lower symmetry are easily overlooked. The qualitative considerations presented above might provide useful guidelines for the direct exploration of the potential hypersurfaces with ab initio methods.

To conclude this section, we should comment briefly on the dynamical aspects of multidimensional conical intersections. Explicit calculations of the nuclear dynamics for Hamiltonians of the type (3.22) including more than two modes are scarce in the literature. An interesting prototype is the vibronic interaction of the two lowest electronic states of the ethylene cation, which has been treated numerically as a nonseparable three-mode problem[31,79] (one coupling mode and two tuning modes) and is discussed below in Sections V.A, VI, and VII in some detail. The calculations demonstrate one important aspect of multidimensional conical intersections: with increasing number of tuning modes, the density of vibronic states increases dramatically. In the adiabatic picture, the energy levels of the upper surface are embedded in the dense manifold of levels of the lower surface and their spectral intensity is redistributed among all these very dense levels. Owing to the high level density, use of statistical methods to analyze the spectrum becomes more appropriate than does the assignment of individual levels (see

Section VI). As will be shown in Section VII, the time correlation function starts to exhibit irreversible behavior; that is, the dynamics can be characterized as a radiationless relaxation process. Since most polyatomic molecules possess several totally symmetric modes, conical intersections involving several tuning modes are probably a rather general phenomenon.

Another interesting regime of phenomena covered by the Hamiltonian (3.22) is the coupling of two electronic states via several non-totally symmetric coupling modes. Dynamical calculations using the Hamiltonian (3.22) with $N = 0$ (no tuning mode) and $M > 1$ (several coupling modes) have been performed in Refs. 146–149 (see Section V.E). These calculations illustrate the nonseparability of the coupling modes; we see that the vibronic spectrum is not simply a convolution of individual progressions in the coupling modes. The aspects of Herzberg–Teller intensity borrowing and the distortion of the normal coordinates (Dushinsky effect) induced by several coupling modes have been discussed in Refs. 150–152.

IV. VIBRONIC COUPLING INVOLVING DEGENERATE MODES AND DEGENERATE STATES

Degenerate electronic states are outstanding examples of the failure of the adiabatic approximation. In the case of linear molecules the vibronic coupling problem is known as Renner–Teller effect;[153] otherwise, as the Jahn–Teller effect.[27] The static and dynamic properties of these systems have been investigated by numerous workers, mostly for a single vibrational mode. Since comprehensive reviews exist both on the Jahn–Teller[28,29] and the Renner–Teller[154,155] effect we shall not enter a general discussion of the subject here. Rather, to stay in line with the main interests of the present work, we concentrate on the influence of several modes on the vibronic coupling problem. Starting with the Jahn–Teller effect, nearly all (nonlinear) molecules with degenerate electronic states possess *several* degenerate modes which can vibronically couple the components of these states. It is thus clear that we have to solve the multimode Jahn–Teller problem in order to arrive at an understanding of the interactions that occur in actual molecules. This is attempted next (in Section IV.A) for the case of twofold degenerate electronic states. We analyze in some detail the influence of the interaction between the modes on the vibronic ground state as well as on the band shape and the line structure arising upon an $A \rightarrow E$ type electronic transition. For multimode effects involving triply degenerate electronic states the reader is referred to the literature.[54,93,156–158]

In Jahn–Teller theory the degenerate electronic state under consideration is taken to be independent of other states. If other electronic states are energetically close to the degenerate state, this assumption is no longer justified and interstate vibronic coupling has to be considered in addition to the intrastate Jahn–Teller coupling terms. Of particular interest appears to be the case of a nondegenerate and a doubly degenerate electronic state, since for a wide class of molecules Jahn–Teller active vibrations can simultaneously lead to interstate vibronic coupling. This situation is discussed in Section IV.B. It will become clear that apart from the interaction between several Jahn–Teller active vibrations there are also multimode effects due to the totally symmetric modes.

Degenerate electronic states of linear molecules require special treatment, since the lowest order vibronic coupling terms are at least quadratic, rather than linear, in the nuclear displacements (the Renner–Teller effect[153]). Furthermore, the great majority of linear molecules studied to date are triatomics, having only a single bending mode. Thus, the influence of several Renner–Teller active (i.e., bending) modes[159] is less important in practice than that of several Jahn–Teller active modes. Being interested in those multimode effects which also occur in triatomics, we draw attention to the vibronic coupling between *different* electronic states of linear molecules. In Section IV.C we treat the case of $\Sigma - \Pi$ vibronic coupling and argue that this can lead to multimode effects in any linear molecule owing to the interplay between the bending and the totally symmetric stretching mode(s). Even the single-mode system has been discussed in the literature comparatively rarely. We shall therefore devote some space to the analysis of its properties and of its consequences on the excitation spectrum. The results will be seen also to be of relevance for the pseudo-Jahn–Teller problem treated in Section IV.B.

A.   The $E \times (\varepsilon + \varepsilon)$ Jahn–Teller Effect

1.   The Hamiltonian

Let us consider a twofold degenerate (E) electronic state of a nonlinear molecule. Jahn and Teller[27] have shown that there always exists a non-totally symmetric mode that can lift the degeneracy to first order and thus lead to vibronic coupling between the electronic component states (the Jahn–Teller effect). The symmetry of the vibrational mode has to be such that it is contained in the decomposition of the direct product $E \times E$. It is then found that in all but seven molecular point groups degenerate vibrations can be Jahn–Teller active, leading to the $E \times \varepsilon$ Jahn–Teller (JT) effect. In the point groups S_4, D_4, D_{2d}, D_{4h}, C_{4v}, C_{4h}, and C_4 only nondegenerate vibrations are JT active giving rise to the $E \times \beta$ Jahn–Teller effect.[132,160-163] In the latter case the single-mode problem can be solved trivially,[29] and the $E \times \beta$ Jahn–Teller effect has received only minor attention in the literature. It should be emphasized, however, that the nuclear dynamics becomes much more complicated in the presence of several modes. When the modes have different symmetry, they are not separable and the solution of the vibronic problem can no longer be given in closed form. The nuclear dynamics turns out to be very similar to that of the two-state–two-mode problem discussed in Section III. We therefore shall not discuss this problem separately here but refer to Section V.A, where the main features are illuminated by an example.

We now address ourselves to the case of degenerate JT active normal modes. Retaining for the moment only a single ε mode with Cartesian coordinates x_j, y_j and harmonic frequency ω_j the $E \times \varepsilon$ JT Hamiltonian can be written as follows:[78,164]

$$\mathcal{H}_j = h_j \mathbf{1} + \kappa_j \begin{pmatrix} x_j & y_j \\ y_j & -x_j \end{pmatrix} \tag{4.1}$$

Here h_j denotes the Hamiltonian of the two-dimensional harmonic oscillator with frequency ω_j; $\mathbf{1}$ stands for the 2×2 unit matrix, and the energy of the electronic E state at the symmetric conformation $x_j = y_j = 0$ has been arbitrarily put equal to zero. The Hamiltonian (4.1) has been frequently analyzed in the literature. Although its solutions cannot be obtained in closed form, their main properties are well understood.[28,29,78,164] To take advantage of the existence of a constant of motion, it is convenient to transform (4.1) to a complex electronic and vibrational basis

$$\mathbf{U}_c = \frac{1}{\sqrt{2}} \begin{pmatrix} 1 & 1 \\ i & -i \end{pmatrix} \tag{4.2}$$

$$x_j = r_j \cos \phi_j; \qquad y_j = r_j \sin \phi_j \tag{4.3}$$

in which the Hamiltonian reads

$$\mathbf{U}_c^\dagger \mathcal{H}_j \mathbf{U}_c = h_j \mathbf{1} + \kappa_j \begin{pmatrix} 0 & r_j \exp(-i\phi_j) \\ r_j \exp(i\phi_j) & 0 \end{pmatrix} \tag{4.4}$$

In this basis the vibronic angular momentum

$$\mathbf{J}_j = \frac{1}{i} \frac{\partial}{\partial \phi_j} \mathbf{1} + \frac{1}{2} \begin{pmatrix} 1 & 0 \\ 0 & -1 \end{pmatrix} \tag{4.5}$$

is diagonal and easily verified to commute with $\mathcal{H}_j$

$$[\mathbf{J}_j, \mathcal{H}_j] = 0 \tag{4.6}$$

The vibronic eigenstates are classified according to the quantum number $\pm 1/2, \pm 3/2, \pm 5/2, \ldots$ of the angular momentum $\mathbf{J}_j$. Since we shall consider in the following discussion the optical transition from an unperturbed ground state into the E electronic state according to Eq. (2.37), we have to deal here only with angular momentum $+1/2$ (or, equivalently, $-1/2$).

The vibronic problem is characterized by a single relevant parameter (κ_j/ω_j). A systematic survey of vibronic spectra as a function of κ_j/ω_j is, therefore, comparatively easy.[28,29,78,164] For weak coupling the vibronic structure of the band is characterized by uneven line spacings and intensities; for strong coupling, by a doubly peaked envelope of the intensity distribution.[78,164] In the strong coupling regime the adiabatic potential energy surfaces

$$V_{\pm}^{(j)} = \frac{\omega_j}{2} r_j^2 \pm \kappa_j r_j \qquad (4.7)$$

are helpful in interpretation of the spectrum. The surfaces (4.7) exhibit the characteristic "Mexican hat" shape[28,29] and so represent the first conical intersection that has been studied in detail in the literature. Their cylindrical symmetry reflects the conservation of angular momentum (4.5). Since near the bottom of the lower trough one degree of freedom is rotational and the other a harmonic vibration, the low-energy vibronic states are nearly equidistant for strong coupling and have lost half a vibrational quantum ω_j in zero point energy.[78] Up to the point of intersection the vibronic states are well approximated by the vibrational levels of the lower surface $V_{-}^{(j)}$. This is just the energy range where the first maximum of the band shape occurs. The second maximum, however, lies within the energy range of the upper potential energy surface $V_{+}^{(j)}$ and the vibronic states contributing to this maximum are strongly affected by nonadiabatic interactions.[165] This holds likewise for additional maxima that appear in the high-energy part of the band shape when the coupling strength is further increased.[87,166,167] The nonadiabatic interactions redistribute the spectral intensity of a hypothetical vibrational level of $V_{+}^{(j)}$ in such a way that the exact vibronic states form a regular sequence of lines with a bell-shaped intensity distribution.[87] It is centered at the position of the unperturbed vibrational level of $V_{+}^{(j)}$. Because of the high symmetries of the $E \times \varepsilon$ Jahn–Teller problem, the nonadiabatic interactions do not lead here to the demolition of any regular structure but instead to the formation of a single vibronic progression.[87]

Let us now turn to the multimode effects. Elementary symmetry considerations show that totally symmetric modes enter the JT effect only in a trivial way. Since their coupling constants for the two electronic component states are equal by symmetry, they are separable from the vibronic problem.

The full spectrum is thus obtained by convoluting the spectra of the individual modes, and the totally symmetric vibrations need not be considered here any further (see Section II.B.1). On the other hand, most molecules have at least two vibrational modes of the same ε-type symmetry, giving rise to the $E \times (\varepsilon + \varepsilon + \cdots)$ Jahn–Teller effect.[168-173] The nonseparability of the JT active modes makes it necessary to sum over all contributions $\mathscr{H}_j$ of the individual modes

$$\mathscr{H} = \sum_j^M \mathscr{H}_j \qquad (4.8)$$

and treat the total matrix Hamiltonian $\mathscr{H}$ as a whole rather than the individual terms separately. As a consequence, the vibronic symmetries are reduced considerably. The individual vibronic angular momenta (4.5) are no longer constants of the motion. It is only the total vibronic angular momentum

$$\mathbf{J} = \sum_j^M \frac{1}{i} \frac{\partial}{\partial \phi_j} \mathbf{1} + \frac{1}{2}\begin{pmatrix} 1 & 0 \\ 0 & -1 \end{pmatrix} \qquad (4.9)$$

that commutes with $\mathscr{H}$. In the adiabatic potential energy surfaces this manifests itself in a dependence of $V_\pm$ on the azimuthal angles ϕ_j of the individual modes. The potentials are invariant only under a common change of the angles of all vibrational modes and otherwise of a very complicated shape. Moreover, the locus of intersection is no longer a single point in coordinate space, but rather a subspace of dimension $2M - 2$. It must already be evident from these remarks that the multimode JT problem leads to much more complicated nuclear dynamics than does the single-mode problem. We shall indeed see that it is important to take these multimode effects into consideration in order to arrive at a realistic treatment of actual molecules.

The Hamiltonian (4.8) has been frequently studied in solid state physics when the modes $\{j\}$ represent the continuum of lattice vibrations.[168,170-173] The ground state energy, Ham's reduction factors, and the band shapes of optical spectra have been analyzed, often by introducing some type of cluster model.[172] In molecular physics usually comparatively few (especially two) vibrational modes are involved, which makes an exact numerical solution of the Hamiltonian (4.8) possible, though cumbersome. We will, therefore, present below a small selection of two-mode JT spectra that display typical multimode effects and enable us to test the quality of approximate solutions. Before doing this we will give some analytical results for the low-energy vibronic states and for the bandshape of an optical transition. Most of these results are easily extended to an arbitrary number of modes, but since the numerical results apply to two ε modes the other formulas are also quoted for two modes only.

2. The Low-Energy Vibronic States

The low-energy vibronic states are of special interest if the JT effect occurs in the electronic ground state or if the couplings are weak and the optical spectrum consists only of a small number of lines. We first consider the latter situation and treat the multimode effects by means of perturbation theory. As has been emphasized in previous sections, perturbation theory is generally inappropriate when we need to calculate vibronic spectra for the relevant range of parameters. Even if the couplings are weak and the excitation strengths of the various modes are well below unity, the applicability of a perturbation expansion is questionable. It has been argued that the radius of convergence of the perturbation series is always small[78] and decreases with the energy of the unperturbed vibrational state. Therefore, it is essential to confine the perturbation treatment to the JT ground state and to the first few excited vibronic states. Moreover, as will become clear below, we cite the results obtained by such perturbation treatment mainly for systematic reasons and not in order to reproduce actual spectra accurately.

In the light of these remarks, we apply Rayleigh–Schrödinger perturbation theory to the two-mode JT problem. The zeroth-order vibronic states are the harmonic oscillator wave functions (modes "r" and "s"), the JT coupling terms $\sim \kappa_r$ and $\sim \kappa_s$ are treated as perturbation. All odd orders of the perturbation series vanish, and in second order the modes are separable. The multimode or "mode-mixing" effects appear first in fourth order. The contribution $E_{00}^{(4)}$ to the ground state energy in this order reads

$$E_{00}^{(4)} = \frac{1}{2} \frac{\kappa_r^4}{\omega_r^3} + \frac{1}{2} \frac{\kappa_s^4}{\omega_s^3} + \frac{2\kappa_r^2\kappa_s^2}{\omega_r\omega_s(\omega_r + \omega_s)} \tag{4.10}$$

The last term in (4.10) describes the mode–mode interaction and is always positive. The independent-mode treatment therefore leads to too low a ground state energy. In the same order of perturbation theory, the energy difference $\Delta E_{10}^{(4)}$ between the singly excited state of mode "r" and the ground state is as follows:

$$\Delta E_{10}^{(4)} = -\frac{3}{2} \frac{\kappa_r^4}{\omega_r^3} + 2\kappa_r^2\kappa_s^2 \frac{3\omega_s^2 - \omega_r\omega_s - \omega_r^2}{\omega_r\omega_s^2(\omega_r^2 - \omega_s^2)} \tag{4.11}$$

An analogous formula, obtained by interchanging the indices r and s, applies to the other singly excited state. The divergence of the second term in (4.11) for $\omega_r = \omega_s$ reflects the inapplicability of perturbation theory in the case of near-degeneracies which inevitably occur for higher vibronic energies. The mode-mixing term in (4.11) is negative if $\omega_r < \omega_s$ or $\omega_r > 1.3\omega_s$, that is, for a wide range of parameters. The first spacing of the mode with smaller frequency will almost always be reduced by multimode effects; that of the mode with larger frequency will also be so reduced quite often.

These findings are in agreement with those of other approximation schemes and with numerical solutions of the two-mode problem.[169] The numerical calculations also reveal a substantial reduction of the intensities of the first excited states relative to the ground state, especially for the mode with smaller frequency.[169]

The numerical results to be presented below indicate that the exact ground state is invariably higher in energy than expected from the independent-mode treatment. This behavior, which has already been inferred from Eq. (4.10), is apparently not limited to weak coupling systems and can indeed be understood also in the strong coupling regime. For sufficiently strong coupling the lowest-energy vibronic states should be calculable using the harmonic approximation to the lower potential energy surface V_-. Therefore, we expand the lower of the surfaces

$$V_\pm = \frac{\omega_r}{2}\rho_r^2 + \frac{\omega_s}{2}\rho_s^2 \pm \sqrt{(\kappa_r x_r + \kappa_s x_s)^2 + (\kappa_r y_r + \kappa_s y_s)^2} \qquad (4.12)$$

in a Taylor series around its minimum and retain only the second-order terms. The energy of the minimum is

$$V_{\min} = -\frac{1}{2}\left(\frac{\kappa_r^2}{\omega_r} + \frac{\kappa_s^2}{\omega_s}\right) \qquad (4.13)$$

The four harmonic frequencies of V_- (corresponding to the four nuclear degrees of freedom) are obtained from the matrix of second derivatives by taking the kinetic energy properly into account:

$$\omega_1 = 0; \qquad \omega_2 = \omega_r; \qquad \omega_3 = \omega_s; \qquad \omega_4 = \sqrt{\frac{\kappa_s^2\omega_r^3 + \kappa_r^2\omega_s^3}{\kappa_s^2\omega_r + \kappa_r^2\omega_s}} \qquad (4.14)$$

The vanishing of the frequency ω_1 reflects the rotational invariance of the potential energy surfaces; ω_2 and ω_3 are the unperturbed frequencies ω_r and ω_s that also appear in the single-mode problem.[78] The multimode effects show up in the existence of another nonvanishing frequency, ω_4. In an independent-mode picture a twofold rotational invariance of V_- would exist and ω_4 would be zero. The static stabilization energy V_{min}, on the other hand, is additive and not affected by multimode effects. Therefore, the exact ground state energy is higher than that of an independent-mode treatment by the additional zero point energy $\omega_4/2$. Since this rise in energy has been derived both for the weak and strong coupling regime, it can be expected to hold also for intermediate coupling cases.

The above result also shows that in the low-energy range of the multimode JT spectrum there should appear additional lines not present in the convolution approximation. They can be interpreted as excitation of the additional mode with frequency ω_4. Since ω_4 is bounded according to $\min\{\omega_r, \omega_s\} \leq \omega_4 \leq \max\{\omega_r, \omega_s\}$, the first new line appears between the singly excited states of both modes. Although Eq. (4.14) will apply quantitatively only for strong coupling and low energies, the very appearance of new lines depends solely on the loss of vibronic symmetries and will, therefore, be observed quite generally and also for higher energies. This is indeed borne out by the numerical results shown below.

3. The Band Shape of the Multimode Jahn–Teller Spectrum

When the JT couplings are moderate or strong, the high-energy part of the $A \rightarrow E$ optical transition will often be unresolved. This holds the more when a great number of JT active modes exists (as in a solid) or when totally symmetric modes are also excited and mask the fine structure. In these cases a knowledge of the individual vibronic states is not of primary importance and it suffices to describe correctly the band shape of the excitation spectrum.

Several approximation methods have been suggested in the literature to achieve this goal. They are all based on the introduction of a suitable single JT mode, called the *cluster* or *effective single mode*, which replaces the multimode couplings of Eq. (4.8).[172] We describe here a version which has been shown to yield particularly accurate results.[57,174,175] It is based on the observation that for strong coupling the band shape should be dominated by the potential energy terms in $\mathscr{H}$ and that the kinetic energy should be less important. We therefore introduce new vibrational coordinates in such a way that all vibronic coupling terms are "absorbed" by a single mode.[57] Defining

$$\begin{pmatrix} X_1 \\ X_2 \end{pmatrix} = \frac{1}{\kappa} \begin{pmatrix} \kappa_r & \kappa_s \\ -\kappa_s & \kappa_r \end{pmatrix} \begin{pmatrix} x_r \\ x_s \end{pmatrix} \tag{4.15a}$$

$$\begin{pmatrix} Y_1 \\ Y_2 \end{pmatrix} = \frac{1}{\kappa} \begin{pmatrix} \kappa_r & \kappa_s \\ -\kappa_s & \kappa_r \end{pmatrix} \begin{pmatrix} y_r \\ y_s \end{pmatrix} \tag{4.15b}$$

$$\kappa = \sqrt{\kappa_r^2 + \kappa_s^2} \tag{4.15c}$$

we find the following expression for the Hamiltonian (4.8) in terms of the new vibrational coordinates (X_1, Y_1) and (X_2, Y_2):

$$\mathcal{H} = \mathcal{H}_1 + \mathcal{H}_2 + \mathbf{V}_{12} \tag{4.16a}$$

Here the first term $\mathcal{H}_1$ denotes a single-mode Jahn–Teller Hamiltonian

$$\mathcal{H}_1 = h_1 \mathbf{1} + \kappa \begin{pmatrix} X_1 & Y_1 \\ Y_1 & -X_1 \end{pmatrix} \tag{4.16b}$$

The second term $\mathcal{H}_2$ represents a harmonic oscillator

$$\mathcal{H}_2 = h_2 \mathbf{1} \tag{4.16c}$$

The third term $\mathbf{V}_{12}$ contains the interaction between the vibrational modes

$$\mathbf{V}_{12} = d\left(-\frac{\partial^2}{\partial X_1 \partial X_2} - \frac{\partial^2}{\partial Y_1 \partial Y_2} + X_1 X_2 + Y_1 Y_2 \right) \mathbf{1} \tag{4.16d}$$

Here we use the following abbreviations

$$h_i = \frac{\Omega_i}{2}\left(-\frac{\partial^2}{\partial X_i^2} - \frac{\partial^2}{\partial Y_i^2} + X_i^2 + Y_i^2 \right); \qquad i = 1, 2 \tag{4.16e}$$

$$\Omega_1 = \frac{\omega_r \kappa_r^2 + \omega_s \kappa_s^2}{\kappa^2} \tag{4.16f}$$

$$\Omega_2 = \frac{\omega_s \kappa_r^2 + \omega_r \kappa_s^2}{\kappa^2} \tag{4.16g}$$

$$d = \frac{\kappa_r \kappa_s |\omega_r - \omega_s|}{\kappa^2} \tag{4.16h}$$

In the new vibrational basis the JT coupling involves only a single mode at the expense of introducing the bilinear "hopping" term $\mathbf{V}_{12}$ into the Hamiltonian. Since the latter is of the order of the vibrational frequencies, we neglect it and thus reduce the two-mode JT problem to that of a single, "effective" mode with frequency and coupling constant:

$$\omega_{\text{eff}} \equiv \Omega_1 = \frac{\omega_r \kappa_r^2 + \omega_s \kappa_s^2}{\kappa^2} \tag{4.17}$$

$$\kappa_{\text{eff}} \equiv \kappa = \sqrt{\kappa_r^2 + \kappa_s^2}$$

The above treatment is easily extended to an arbitrary number of JT modes.[57] It should be mentioned that the same result can also be derived by transforming the Hamiltonian instead of introducing new vibrational modes.[175]

The effective single-mode approximation becomes exact if the frequencies ω_r and ω_s of the JT active modes coincide ($d = 0$). Therefore, the same holds trivially for the semiclassical limit $\omega_r \to 0, \omega_s \to 0$. The leading terms of *all* spectral moments for strong coupling are reproduced correctly by utilizing (4.17). In case of the even moments this is seen directly from the above treatment, since the leading terms involve no frequencies. It can be shown to hold likewise for the odd moments, although here the frequencies enter the leading terms.[174] As a consequence, the second and third moments, consisting only of a single term, are always exact within the present single-mode approach. One may, therefore, expect it to yield accurate band shape curves over a wide range of parameters.

Cluster models have also been considered with the goal of describing the vibronic ground state.[172] This is not the purpose of the above treatment. The static JT stabilization energy V_{min} is not given correctly by (4.17), and the band origin will in general not be reproduced accurately. For cluster approaches especially adapted to the vibronic ground state the reader is referred to the literature.[172]

The numerical results[57] show that multimode JT band shapes resemble very closely that of a single mode, the effective mode (4.17). Nevertheless, once the parameters of the multimode problem are known, pronounced multimode effects can be identified in the band shape curve. That is to say, the exact band shape curve usually differs considerably from the convolution of the single-mode band shape curves. For example, consider the semiclassical limit, where the band shape for a single mode j reads[28]

$$P_j(E) \sim \frac{|E|}{\kappa_j^2} \exp\left(\frac{-E^2}{\kappa_j^2} \right) \qquad (4.18)$$

It exhibits two symmetric maxima at $E = \pm \kappa_j / \sqrt{2}$ and vanishes at $E = 0$. Even without performing the convolution explicitly, it is obvious that it will lead to an agglomeration of intensity at the band center. When the coupling constants are approximately equal, even a maximum will be formed at $E = 0$.[57] The exact band shape, however, is still of the form (4.18) with the effective coupling constant (4.17). Therefore, the convolution will in general wash out the characteristic double peak of the JT band shape and often lead to singly peaked curves.

4. Numerical Results

We have performed numerical calculations of the vibronic structure of $A \to E$ type electronic bands for a wide range of parameters.[57] The secular matrix of the Hamiltonian (4.8) has been set up in a product basis of harmonic oscillator wave functions and diagonalized by means of the Lanczos algorithm. Calculations of this type have also been performed in Refs. 55, 56, and 176. If N is a typical maximum occupation number and M the number of ε modes considered, the rank of the secular matrix is of the order N^{2M-1}. This is more than the product of the single-mode numbers (N^M) and reflects the loss of vibronic symmetries which allows additional basis states to interact with each other. Since we are interested primarily in intermediate-to-strong coupling cases, we limit ourselves to $M = 2$. In the following we present three examples that serve to illustrate some interesting multimode effects. The parameters used in the calculation are collected in Table III. The mode ν_s is always taken to have the smaller frequency. Figures 16, 17, and 18 contain the exact two-mode JT spectrum in the uppermost panel (a), followed by the effective single-mode result (b) and the independent-mode treatment (c) (convolution of the single-mode JT spectra). Except for the first example (Fig. 16), we include at the bottom of these figures the corresponding band-shape curves (d). These are obtained by broadening the lines with Gaussians of sufficient width to smooth out the fine structure.

The first example (Fig. 16) belongs to the weak coupling regime. The spectrum is dominated by the 0-0 line, followed by a short, perturbed progression in ν_s, and ending with a number of weak, irregular lines. Compared with the convolution approximation, the fundamental line of the mode ν_r has dropped in absolute intensity by a factor of ~ 2.4; that of the mode ν_s has decreased by only $\sim 15\%$. (Note that the spectra are not drawn on the same intensity scale.) The marked decrease of the relative intensity of both fundamental lines with regard to the 0-0 line has been noticed previously.[169] This is a general phenomenon in the weak-to-intermediate coupling regime and primarily caused by the more energetic vibration. As far as the positions of the first two spectral lines are concerned, the couplings are too strong for the perturbation theory formulas (4.10) and (4.11) to apply quantitatively.

TABLE III

Coupling Constants and Frequencies Used in Calculating the Two-Mode JT Spectra of Figs. 16–18[a]

No.	Fig.	κ_r	κ_s	ω_r	ω_s	κ_{eff}	ω_{eff}
1	16	2.41	1.00	2.41	1.00	2.61	2.21
2	17	2.71	2.00	1.92	1.00	3.37	1.59
3	18	3.13	2.00	1.57	1.00	3.72	1.40

[a] The smaller frequency is denoted by ω_s and taken as energy unit. The effective frequency ω_{eff} and coupling constant κ_{eff} are defined in (4.17).

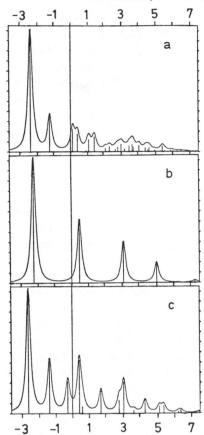

Fig. 16. Exact two-mode Jahn–Teller spectrum (*a*) in comparison with the spectra obtained via the effective single-mode approximation (*b*) and the convolution approximation (*c*). The input data are listed in Table III.

Qualitatively, however, there is agreement with Eqs. (4.10) and (4.11) in that the position of the first spectral line is higher in energy than in the independent-mode picture. This also causes a decrease in the first vibrational spacing by the multimode effects. Finally, although the line structure of the effective single-mode calculation is not of relevance in general, the 0-0 line is reproduced quite well in the present case.

In Fig. 17 we present a case of moderate to strong JT coupling. The maximum of the ν_s progression is shifted from the first to the second line, and also the ν_r progression is not very short. Since the frequency ratio ω_r/ω_s is close to an integer, the convoluted envelope (Fig. 17c) exhibits a series of peaks with spacings equal to ω_s which extends throughout the whole band. Perturbations of this structure occur only at high energies. Interestingly, these perturbations disappear in the exact spectrum where the lines are always

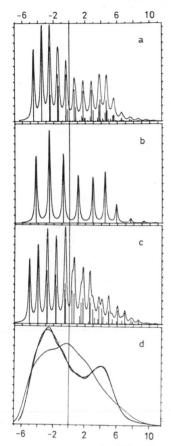

Fig. 17. Exact two-mode Jahn–Teller spectrum (*a*) in comparison with the spectra obtained via the effective single-mode approximation (*b*) and the convolution approximation (*c*). Panel *d* contains the band shapes corresponding to the spectra in panels *a*–*c*. The input data are listed in Table III.

grouped together in such a way that the envelope is very regular and resembles that of a single mode. However, this should not lead us to believe that we could reproduce this structure with a single-mode JT calculation and suitably chosen parameters. The effective single-mode spectrum exhibits the correct intensity distribution but too large a line spacing. In this respect it behaves in a manner opposite to that of the convoluted spectrum.

After smoothing out the vibronic structure, we can see that the band shapes of the exact and the effective single-mode spectra are very similar. The band shape of the convoluted spectrum, however, looks quite different and has a central maximum instead of a double peak. This confirms the idea that the overall intensity distribution of the spectrum depends primarily on the vibronic coupling term in $\mathscr{H}$. Remember that this term is treated exactly in the effective single-mode spectrum but not in the convolution approximation.

The band origin of Fig. 17 is amenable to a simple interpretation. The first few lines of the exact spectrum closely resemble the progression in ν_s but are shifted to a higher energy compared to the convoluted spectrum. According to the discussion in Section IV.A.2, this is attributable to the additional vibrational degree of freedom near the minimum of the lower potential energy surface. With (4.14) we find for the position of the 0-0 line a value of $-4.61\omega_s$, as compared to the exact result of $-4.56\omega_s$ (convolution: $-5.07\omega_s$). The agreement is very satisfactory in view of the moderate coupling strength, and the remaining deviations can be understood as a consequence of neglecting overall rotations.[78]

To give an idea of the wide range of possible vibronic structures, we present in Fig. 18 another two-mode JT spectrum with completely different behavior. The excitation strength of both modes is now the same as that of the low-frequency mode in Fig. 17. Since the frequency ratio is far from being an integer, even the convoluted spectrum looks quite complicated. This holds all the more for the exact result, where the spectral intensity at higher energies is dispersed over a quasi-continuum of vibronic lines that lack any regular pattern. Despite this difference from the previous example, the intensity distribution of the effective single-mode spectrum is essentially correct and the resulting band shape virtually coincides with the exact curve. The independent-mode treatment, however, accumulates too much intensity at the band center (see Fig. 18d).

In view of the larger coupling strength, the first few spectral lines should again be amenable to a description in terms of the harmonic frequencies (4.14). For the energy of the 0-0 line they yield $-5.79\omega_s$, which compares well with the exact result $-5.76\omega_s$ (convolution: $-6.27\omega_s$). The second and fourth spectral lines represent excitation of ν_s and ν_r, respectively. The fundamental line of the new mode ν_4 is the third spectral line. Its excitation energy of $1.33\omega_s$ is only slightly larger than the predicted value $\omega_4 = 1.25\omega_s$. (A vibronic eigenstate at an energy ω_4 above the JT ground state was also found in the case of the second example above (Fig. 17), but its spectral intensity was negligibly small.) The additional lines which occur in the exact spectrum at higher (but negative) energy can presumably be understood similarly by taking anharmonicity into account. At energies higher than the minimum of the multidmensional intersection of the potential energy surfaces, the nuclear motion is nonadiabatic. Nevertheless, one should be careful in attributing the *increase* in the line density of the exact two-mode spectrum relative to the convoluted spectrum only to the nonadiabatic effects. The single-mode JT line structure is also nonadiabatic at positive energies. What can be said safely is that the increase in line density may be ascribed to the loss of vibronic symmetry caused by the multimode interaction.

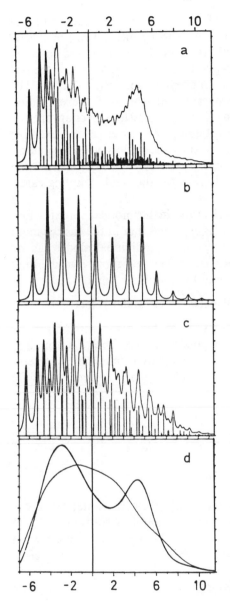

Fig. 18. Exact two-mode Jahn–Teller spectrum (*a*) in comparison with the spectra obtained via the effective single-mode approximation (*b*) and the convolution approximation (*c*). Panel *d* contains the band shapes corresponding to the spectra in panels *a*–*c*. The input data are listed in Table III.

A great number of other two-mode JT spectra have been calculated and found to exhibit interesting effects.[95] It proves to be difficult to predict the features of the vibronic structure directly from the frequencies and coupling constants. Generally the picture of independent modes works best if the frequencies are so different that the individual progressions do not overlap. The main multimode effect is then a (moderate) quenching of the excitation strength of the less energetic vibration. If the progressions overlap but the convoluted line structure exhibits line groupings as in our second example (Fig. 17), the regularities are often reinforced by the multimode interactions. For sufficiently strong coupling one usually observes a high-energy line structure as complicated as in Fig. 18. Except for the weak coupling cases, the band shape could always be very well reproduced by the effective single-mode calculation and the band origin by the harmonic approximation for the lower potential energy surface. It is expected that in the case of more than two JT modes even weaker couplings of the individual modes can be treated by these approximation schemes, since they rely mostly on the total coupling strength, which will increase with the number of modes.

B.   The $(E + A) \times (\varepsilon + \varepsilon)$ Pseudo–Jahn–Teller Effect

We now briefly discuss the interaction of a doubly degenerate (E) and a nondegenerate (A) electronic state through degenerate vibrational modes. The Hamiltonian for this problem can be set up by elementary symmetry considerations and reads as follows[141,177-179] (for a single degenerate mode j):

$$\mathcal{H}_j = h_j \mathbf{1} + \begin{pmatrix} E_E & \lambda_j r_j e^{i\phi_j} & \kappa_j r_j e^{-i\phi_j} \\ \lambda_j r_j e^{-i\phi_j} & E_A & \lambda_j r_j e^{i\phi_j} \\ \kappa_j r_j e^{i\phi_j} & \lambda_j r_j e^{-i\phi_j} & E_E \end{pmatrix} \quad (4.19)$$

Here $E_E(E_A)$ represents the energy of the $E(A)$ electronic state; $\mathbf{1}$ denotes the 3×3 unit matrix; λ_j is the interstate vibronic coupling constant, and the remaining symbols have been defined in Eqs. (4.1) and (4.3). The coupling terms $\sim \kappa_j$ describe the JT coupling within the E state, whereas the terms $\sim \lambda_j$ are responsible for the pseudo-Jahn–Teller (PJT) interaction between the E and the A electronic states (to avoid confusion, we mention explicitly that we reserve the name PJT coupling to the case where the vibronically active mode is degenerate). The above Hamiltonian extends the treatment of the previous Section IV.A, where the E state has been assumed to be isolated and not to interact with other electronic states. It can also be viewed as generalization of the two-state problem of Section III to non-Abelian point groups.

In Eq. (4.19) the same mode is assumed to be JT and PJT active. It should be pointed out that this situation does not pertain to all molecular point groups.[179] Rather, it applies to systems with threefold symmetry axes. In pentagonal molecules a given vibration can only be either JT or PJT active, and the interstate vibronic coupling problem for a single mode is thus characterized by $\kappa_j = 0$. In hexagonal systems the situation depends on the symmetry of the vibrational mode being considered. In the latter case, the symmetry specification A of the nondegenerate electronic state refers collectively to A- or B-type irreducible representations.

The solutions of the vibronic problem (4.19) have been examined several times in the literature, either by perturbation theory[141,178] or by numerical methods.[141,177,179] The first study appears to be that of Perrin and Gouterman[141] who pointed out the relation of (4.19) to the trimer problem and treated the special case $\kappa_j = \lambda_j$. Subsequently, model calculations over a wider range of the parameters were performed by van der Waals et al.[177] and especially by Zgierski and Pawlikowski,[179] who included also quadratic JT coupling terms. In the next section we shall present some results of such calculations for the case of vanishing JT coupling $\kappa_j = 0$. We shall pay special attention to the validity of the adiabatic approximation. Not unexpectedly, it will turn out that nonadiabatic interactions *between* the E and A states are important only if both states approach energetically up to 1 or 2 vibrational quanta. Even for a large $A - E$ energy gap there remain generally strong nonadiabatic effects *within* the E state. They originate from the instance that even the purely interstate vibronic coupling terms induce intrastate vibronic effects, the leading terms being of order r_j^2.

The various types of multimode effects can be classified in a way similar to that used in Section III:

First, there are always totally symmetric modes, and these modes will have different coupling constants in the A and E electronic states. Contrary to the pure JT case, the totally symmetric modes will, therefore, not decouple from the vibronic problem. The different slopes of the A and E potential energy surfaces as a function of the totally symmetric coordinates can lead to a strong decrease of the $A - E$ energy gap during the totally symmetric vibration and to a corresponding enhancement of the interstate nonadiabatic effects. In the extreme situation where the $A - E$ conical intersection is accessible energetically, the nonadiabatic effects will dominate the spectrum entirely and destroy any regular vibrational structure. Even if there are no such nonadiabatic effects, the modulation of the energy gap still modifies the unperturbed progressions in a notable way.[180]

The second type of multimode effects arises from the excitation of several ε vibrational modes. These are clearly nonseparable, either by virtue of their JT or their PJT activity. In view of the large number of independent parameters even for two modes (six), it is not possible here to analyze the various phenomena in a systematic way. Limiting ourselves to the case of a large $E - A$ energy gap, the multimode effects in the E term will be similar to those for the pure JT problem. The multimode effects in the A term depend on the relative ordering of the two electronic states. For later application, we consider the case where the A term is lower in energy. In this situation one encounters interesting multimode effects in the A term, which arises from the anharmonic shape of the adiabatic potential energy surfaces. For sufficiently strong coupling the absolute minimum will occur at nonzero values of the radial coordinate r_j. To obtain closed formulas, we put $\kappa_j = 0$ for all modes and find for the barrier height E_s by which the JT distorted geometry in the A term is stabilized energetically

$$E_s = \begin{cases} \dfrac{(e-\Delta)^2}{2e}, & e \geq \Delta \\ 0, & e \leq \Delta \end{cases} \tag{4.20a}$$

$$\Delta = \left| \frac{E_E - E_A}{2} \right|, \qquad e = 2\sum_j \frac{\lambda_j^2}{\omega_j} \tag{4.20b}$$

The coordinates of the minimum are

$$x_j = \sqrt{2}\, \frac{\lambda_j}{\omega_j} \cos\phi \sqrt{1 - \Delta^2/e^2} \tag{4.21a}$$

$$y_j = \sqrt{2}\, \frac{\lambda_j}{\omega_j} \sin\phi \sqrt{1 - \Delta^2/e^2} \tag{4.21b}$$

Apart from the arbitrary overall phase ϕ reflecting the cylindrical symmetry of the potential energy surfaces, these expressions are completely analogous to the previous Eqs. (3.25b) and (3.27) for the vibronic coupling problem involving nondegenerate modes. As before, it is only the sum of the individual coupling strengths $(2\lambda_j^2/\omega_j)$ that determines the stabilization energy E_s. Several ε modes therefore tend to mutually enhance the distortion and can lead to nonzero E_s even if each mode individually preserves the "original" molecular symmetry. An example of this effect will be given in Section V.C.

Finally, we draw attention to the nonseparability of ε_1 and ε_2 vibrations in pentagonal molecules. These modes can be only either JT or PJT active and thus occupy different positions in the matrix Hamiltonian (4.19). Although the problem is formally a special case of the one discussed above, it may nevertheless give rise to characteristic multimode effects.

C. Vibronic Coupling in Linear Molecules

The vibronic coupling within degenerate electronic states of linear molecules is termed the Renner–Teller (RT) effect.[154] In the original work of Renner on Π electronic states only the lowest order vibronic coupling terms were retained and treated by means of perturbation theory.[153] Later on the effects of electron spin,[181] of molecular rotation,[182] of electronic angular momentum $\Lambda = 2$,[183] and of higher-order coupling terms[184,185] have been included. The most recent treatments allow for a very general description of an isolated Π electronic state with arbitrary potential energy surfaces.[186–189] However, it is always assumed that other electronic states are sufficiently well separated energetically that their interaction with the Π (or Δ, ...) electronic state can be either neglected or included by means of perturbation theory.[190–192]

In the following subsections we adopt a different strategy and discard the higher-order coupling terms, but treat the interstate linear vibronic coupling problem exactly. As a prototype we consider the interaction of a Σ and a Π electronic state of a linear molecule through the bending mode. For the treatment of other cases—such as the $\Pi – \Delta$ vibronic coupling problem—we refer to previous work.[193] The interstate vibronic coupling problem is of special importance if both electronic states belong to the manifold of excited states. In this case they can be very close in energy and perturbation theory fails to describe their interaction. Examples of this type are known to occur in NCS[194] and HCN$^+$.[195] Even if the $\Sigma–\Pi$ energy gap amounts to several vibrational quanta and the nonadiabatic interaction between the Σ and Π states is weak, the $\Sigma–\Pi$ vibronic coupling approach will be seen to offer some distinct advantages over the concept of treating the Π state as isolated.

1. The Hamiltonian for $\Sigma – \Pi$ Vibronic Coupling

We now wish to describe vibronic coupling between Σ and Π electronic states ignoring spin-orbit coupling and rotation around axes perpendicular to the figure axis. For the moment, the influence of totally symmetric modes is omitted. We denote the electronic and vibrational azimuthal angles with respect to an arbitrary reference plane containing the figure axis by Θ and Φ, respectively. Here ρ represents the bending amplitude in dimensionless normal coordinate form, and Λ stands for the electronic angular momentum along the figure axis. The diabatic electronic basis functions that we use to represent the Hamiltonian correspond here to the electronic wave functions appropriate to the linear molecule limit. In a one-electron picture, the angular dependence of the basis functions is given by

$$\langle \Theta | \Lambda \rangle = e^{i\Lambda\Theta}, \qquad \Lambda = 0, \pm 1 \tag{4.22}$$

As in previous sections, we model the vibronic coupling problem by taking the unperturbed bending frequency ω to be the same in all diabatic states and by retaining only the leading terms of a Taylor series expansion of the vibronic coupling matrix elements in the bending amplitude ρ. Ordering the electronic basis states with increasing value of Λ from top left ($\Lambda = -1$) to bottom right ($\Lambda = +1$), the Hamiltonian for the bending motion reads as follows:[190-193]

$$\mathscr{H}_b = (T_N + V_0)\mathbf{1} + \begin{pmatrix} E_\pi & \lambda\rho e^{i\Phi} & \gamma\rho^2 e^{2i\Phi} \\ \lambda\rho e^{-i\Phi} & E_\sigma & \lambda\rho e^{i\Phi} \\ \gamma\rho^2 e^{-2i\Phi} & \lambda\rho e^{-i\Phi} & E_\pi \end{pmatrix} \qquad (4.23a)$$

$$T_N = -\frac{\omega}{2}\left(\frac{1}{\rho}\frac{\partial}{\partial\rho}\rho\frac{\partial}{\partial\rho} + \frac{1}{\rho^2}\frac{\partial^2}{\partial\Phi^2}\right) \qquad (4.23b)$$

$$V_0 = \frac{\omega}{2}\rho^2 \qquad (4.23c)$$

The first two terms of $\mathscr{H}_b$ represent the unperturbed bending motion, the last term causes the vibronic coupling; E_σ and E_π are the energies of the Σ and Π electronic states, respectively; the matrix elements $\sim \lambda\rho$ represent the (linear) $\Sigma - \Pi$ vibronic coupling and those $\sim \gamma\rho^2$ represent the (quadratic) RT coupling. Their form follows directly from the conservation of the projection of the total momentum on the molecular axis

$$\mathscr{L}_z = \frac{1}{i}\frac{\partial}{\partial\Phi}\mathbf{1} + \begin{pmatrix} -1 & 0 & 0 \\ 0 & 0 & 0 \\ 0 & 0 & 1 \end{pmatrix} \qquad (4.24)$$

The eigenvalues of $\mathscr{L}_z$ are integers and denoted by K.

Strictly speaking, the above labeling of the electronic states is not complete. In the case of $D_{\infty h}$ symmetry it is tacitly assumed that the transformation properties under the inversion operation are such that the vibronic coupling does not vanish. Furthermore, Eq. (4.23) applies to Σ^+ electronic states. By a simple change in sign of a nondiagonal element it can be used also for describing $\Sigma^- - \Pi$ vibronic coupling.[193]

When only the RT coupling is considered in (4.23), there is a well-known restriction on the magnitude of the RT coupling constant, namely, $|\gamma| < \omega/2$. This condition derives from the lower adiabatic potential energy surface which depends on ρ according to $(\omega - 2|\gamma|)\rho^2/2$ and which becomes unbounded from below if $|\gamma| > \omega/2$. Correspondingly, in the lowest order RT treatment both Π component states have a linear equilibrium geometry,[153] since the minima of both adiabatic potential energy surfaces occur at $\rho = 0$. To describe a non–linear equilibrium geometry of one Π component state, terms of at least fourth order in ρ have to be included in the Hamiltonian.[184,185]

We now demonstrate that the situation is quite different in the case of $\Sigma-\Pi$ vibronic coupling. Henceforth we neglect the direct RT coupling term ($\gamma = 0$). In this case the transformation $\mathbf{S}$, Eq. (2.43), to the adiabatic basis can easily be given in closed form:

$$\mathbf{S} = \frac{1}{\sqrt{2}} \begin{pmatrix} e^{i\Phi} & e^{i\Phi}w_- & e^{i\Phi}w_+ \\ 0 & \sqrt{2}\,w_+ & -\sqrt{2}\,w_- \\ -e^{-i\Phi} & e^{-i\Phi}w_- & e^{-i\Phi}w_+ \end{pmatrix} \qquad (4.25a)$$

where

$$w_\pm = \frac{1}{\sqrt{2}}\left[1 \pm \frac{\Delta}{w}\right]^{1/2} \qquad (4.25b)$$

$$\Delta = \frac{E_\sigma - E_\pi}{2}; \qquad w = \left[\Delta^2 + 2\lambda^2\rho^2\right]^{1/2} \qquad (4.25c)$$

The columns of the matrix $\mathbf{S}$ represent the expansion coefficients of the adiabatic basis states in terms of the diabatic electronic basis states. Utilizing Eq. (4.22), the coordinate dependence of the adiabatic electronic wave functions becomes

$$\langle \Theta | \Phi_\pi \rangle = \sqrt{2}\,\sin(\Phi - \Theta) \qquad (4.26a)$$

$$\langle \Theta | \Phi_\pm \rangle = \sqrt{2}\,w_\mp \cos(\Phi - \Theta) \pm w_\pm \qquad (4.26b)$$

Here Eq. (4.26a) refers to the first column of the matrix $\mathbf{S}$ and the upper (lower) sign in Eq. (4.26b) refers to the second (third) column of $\mathbf{S}$. The adiabatic basis state $|\Phi_\pi\rangle$ has pure Π electronic character and is antisymmetric under reflections σ_h in the instantaneous molecular plane. The basis states $|\Phi_\pm\rangle$ are symmetric under σ_h and represent a mixture of Σ and Π electronic parentage, the relative weight being determined by $|w_\pm|^2$. Written in the adiabatic electronic basis (also called "bent-molecule" basis[154]) the Hamiltonian takes the form

$$\tilde{\mathcal{H}}_b = \mathbf{S}^\dagger \mathcal{H}_b \mathbf{S} = T_N \mathbf{1} + \begin{pmatrix} V_\pi & 0 & 0 \\ 0 & V_+ & 0 \\ 0 & 0 & V_- \end{pmatrix} - \hat{\Lambda} \qquad (4.27a)$$

where

$$V_\pi = E_\pi + \frac{1}{2}\omega\rho^2 \tag{4.27b}$$

$$V_\pm = \frac{E_\sigma + E_\pi}{2} + \frac{1}{2}\omega\rho^2 \pm \sqrt{\Delta^2 + 2\lambda^2\rho^2} \tag{4.27c}$$

$$\hat{\Lambda} = -\mathbf{S}^\dagger[T_N, \mathbf{S}]$$

$$= -\omega\begin{bmatrix} \dfrac{1}{2\rho^2} & \dfrac{w_-}{\rho^2}\dfrac{1}{i}\dfrac{\partial}{\partial\Phi} & \dfrac{w_+}{\rho^2}\dfrac{1}{i}\dfrac{\partial}{\partial\Phi} \\[3mm] & \dfrac{w_-^2}{2\rho^2} + \dfrac{\lambda^2\Delta^2}{4w^4} & -\dfrac{\lambda\Delta}{\sqrt{2}\,w^2}\left[\dfrac{\Delta^2 - 2\lambda^2\rho^2}{2\rho w^2} + \dfrac{\partial}{\partial\rho}\right] + \dfrac{\lambda}{2\sqrt{2}\,\rho w} \\[3mm] \text{h.c.} & & \dfrac{w_+^2}{2\rho^2} + \dfrac{\lambda^2\Delta^2}{4w^4} \end{bmatrix}$$

$$\tag{4.27d}$$

The quantities V_π and $V_\pm$ are the adiabatic potential energy surfaces and are the eigenvalues of the potential energy matrix $\mathcal{H}_b - T_N\mathbf{1}$. The nonadiabatic coupling term arises from transforming the kinetic energy with $\mathbf{S}$ and is due to the coordinate dependence of the adiabatic electronic wave functions. The nondiagonal elements of $\hat{\Lambda}$ involving $|\Phi_\pi\rangle$ have the Coriolis-type form which is well known from the RT effect.[154] The nondiagonal element connecting $|\Phi_+\rangle$ and $|\Phi_-\rangle$ is of a different functional form and does, of course, not occur within the framework of RT theory; it represents the $\Sigma-\Pi$ nonadiabatic interaction. Note that all elements of $\hat{\Lambda}$ conserve the vibrational angular momentum. This is consistent with the observation that the total angular momentum (4.24) is purely vibrational in nature when written in the adiabatic basis

$$\tilde{\mathcal{L}}_z = \mathbf{S}^\dagger\mathcal{L}_z\mathbf{S} = \frac{1}{i}\frac{\partial}{\partial\Phi}\mathbf{1} \tag{4.28}$$

Although being equal to K, the eigenvalues of $\tilde{\mathcal{L}}_z$ are denoted by the different symbol l in order to distinguish the different nature of the angular momentum in the two electronic basis sets.

Let us now return to the adiabatic potential energy surfaces. From Eqs. (4.27b, c) it is seen that one of the Π-type surfaces, V_π, is not changed by the vibronic coupling and has the same shape as in the unperturbed electronic states. The other Π-type and the Σ-type surface repel each other in completely the same way as was found in the vibronic coupling problem involving nondegenerate modes (Section III.A). When the coupling is weak, the upper surface is slightly steeper and the lower surface is slightly flatter than for $\lambda = 0$, but both remain nearly harmonic. For sufficiently strong coupling, $2\lambda^2 > \omega|\Delta|$, the lower surface exhibits new minima at

$$\rho_0 = \sqrt{\frac{2\lambda^2}{\omega^2} - \frac{\Delta^2}{2\lambda^2}}, \qquad \Phi \text{ arbitrary} \qquad (4.29)$$

and a local maximum at the linear geometry $\rho = 0$. It depends only on the relative ordering of E_σ and E_π whether the nonlinearity occurs in the Σ or Π type potential energy surface. Thus $\Sigma-\Pi$ vibronic coupling is a mechanism which can lead to a non–linear equilibrium geometry of either Σ- or Π-type electronic states. It is interesting that such a situation can be de-

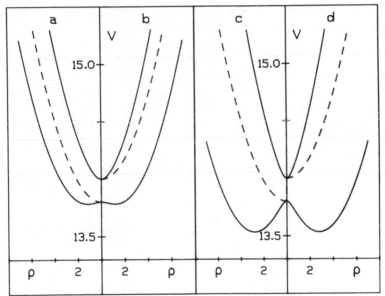

Fig. 19.   Potential energy curves $V_\pm$ (full lines) and V_π (dashed lines) for $\Sigma-\Pi$ vibronic coupling. The values of the parameters (all in electron volts) are as follows:

$$(a) \;\; E_\pi = 13.8, \qquad E_\sigma = 14.0, \qquad \lambda = \omega = 0.09;$$
$$(b) \;\; E_\pi = 14.0, \qquad E_\sigma = 13.8, \qquad \lambda = \omega = 0.09;$$
$$(c) \;\; E_\pi = 13.8, \qquad E_\sigma = 14.0, \qquad \lambda = 2\omega = 0.18;$$
$$(d) \;\; E_\pi = 14.0, \qquad E_\sigma = 13.8, \qquad \lambda = 2\omega = 0.18.$$

scribed even within the linear coupling approximation. This appears to be conceptually simpler than an RT treatment where terms of at least fourth order in ρ have to be considered to explain nonlinearity in the Π-type electronic state.

For illustrative purposes we have depicted in Fig. 19 the potential energy surfaces (4.27b, c) for realistic values of the parameters. Here the solid lines are the surfaces $V_{\pm}$; the dashed lines represent the unperturbed surface V_{π}. For comparison, both cases $E_{\sigma} > E_{\pi}$ and $E_{\sigma} < E_{\pi}$ have been combined in the figure. In Fig. 19a the values of the parameters represent the two lowest states of HCN^{+}. These will be discussed in detail in Section V.D. The energy gap of 0.2 eV is approximately twice the bending frequency ω, which in turn equals the vibronic coupling constant λ ($= 0.09$ eV). These parameters lead to a very shallow minimum at $\rho_0 \approx 1.2$, which is stabilized energetically by 0.018 eV [i.e., less than the zero point energy $\omega/2$ ($= 0.045$ eV)]. Upon doubling the value of λ (Fig. 19c, d) the minimum becomes much deeper and the nonlinear geometry $\rho_0 \approx 2.8$ is stabilized by 0.267 eV. In an electronic transition from a linear initial state to the interacting final states this leads to a substantial excitation of the bending mode.

One of the surfaces $V_{\pm}$ joins the unperturbed surface V_{π} at $\rho = 0$ and there becomes a member of a degenerate RT pair. The splitting of the Π-type potential energy surfaces for finite ρ reflects the RT effect within the Π state that is *induced* by the $\Sigma - \Pi$ vibronic coupling. If the $\Sigma - \Pi$ energy gap amounts to several vibrational quanta and the $\Sigma - \Pi$ nonadiabatic interaction is negligible, this is the main effect of the vibronic coupling (apart from anharmonicity). In this case the Hamiltonian for the induced RT effect takes the following form:[193]

$$\mathscr{H}_{\text{ind}} = T_N \mathbf{1} + \begin{pmatrix} \dfrac{V_{\alpha} + V_{\pi}}{2} & V_{\text{ind}} \\ V_{\text{ind}}^* & \dfrac{V_{\alpha} + V_{\pi}}{2} \end{pmatrix} \qquad (4.30a)$$

$$V_{\text{ind}} = \frac{1}{2}\left(\sqrt{\Delta^2 + 2\lambda^2 \rho^2} - |\Delta|\right) e^{2i\Phi} \qquad (4.30b)$$

where α is positive ($+$) or negative ($-$) depending on whether $E_{\pi} > E_{\sigma}$ or $E_{\pi} < E_{\sigma}$. Apart from the phase factor, the induced RT coupling term V_{ind} is one-half the difference of the Π-type potential energy surfaces; apparently, it contains arbitrary powers in ρ^2, although the original Hamiltonian consists only of linear coupling terms. This further underlines that the linear vibronic coupling approach describes a wider class of phenomena than a low-order RT treatment.

2. Model Spectra

In this section we present some model spectra for the $\Sigma - \Pi$ vibronic coupling problem. Even the single-mode problem has been studied comparatively infrequently. We therefore now digress from our general line and for pedagogical purposes consider only the bending mode. This does not mean, of course, that multimode effects are expected to be less important than in other vibronic coupling systems (see Section IV.C.3). We assume the initial electronic state to be linear and nondegenerate and not to interact with other electronic states. Vibronic coupling is operative only after the electronic transition. The problem is thus defined by the general transition probability, Eq. (2.37), and the Hamiltonian $\mathcal{H}_b$, Eq. (4.23), with $\gamma = 0$. When we compare the latter with the Hamiltonian (4.19) for the $(E + A) \times \varepsilon$ PJT effect, we readily see that they are identical if the intrastate (JT) coupling terms are put equal to zero. In this sense the following results also apply to the PJT problem. Within the framework of the PJT effect similar calculations have been performed in Refs. 141, 177, and 179.

To gain insight into the various dynamical effects of the vibronic coupling, we compare the exact result with the spectra obtained in the adiabatic and Franck–Condon approximations as was described in Section III.B. In the adiabatic approximation the Hamiltonian is transformed to the adiabatic basis and the operator $\hat{\Lambda}$ [Eq. (4.27d)] is put equal to zero. In the Franck–Condon (FC) approximation the coordinate dependence of the adiabatic electronic wave functions is neglected completely and the transformation matrix S [Eq. (4.25)] is put equal to unity (for details see Section II.B.2). It should be mentioned that both approximations behave pathologically in the present situation. The neglect of the divergent terms contained in $\hat{\Lambda}$ leads to a divergence of the adiabatic spectral moments.[193] This is similar to the conical intersection problem considered in Section III.B. In the FC spectrum the symmetries of the vibronic states are treated incorrectly. According to the FC principle only vibrational levels with angular momentum zero are excited. However, the form of the matrix S [Eq. (4.25)] shows that the actual transition involves vibrational levels of the Π electronic state with angular momentum one. Thus, both approximations must fail severely in case of the $\Sigma - \Pi$ vibronic coupling problem and can only be used to illustrate the influence of the nonadiabatic effects.

Figure 20a, b, c displays the $\Sigma - \Pi$ spectrum for parameters appropriate to the lowest states of HCN^+. The corresponding potential energy surfaces have been shown in Fig. 19, and for the values of the parameters the reader is referred to Table X. In the actual spectrum the totally symmetric modes play a crucial role (see Section V.D), but here we ignore them and thus depict the hypothetical photoelectron spectrum of HCN as if the bond lengths

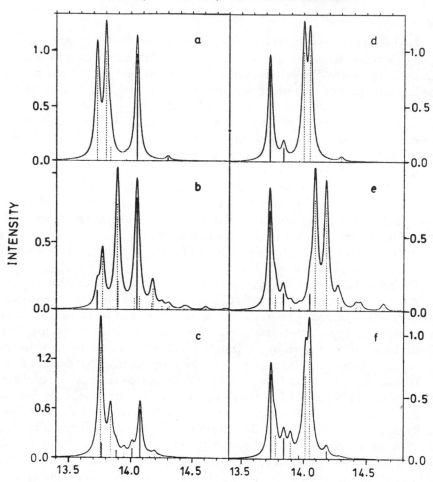

Fig. 20. Spectra for $\Sigma - \Pi$ vibronic coupling obtained in the Franck–Condon, adiabatic, and exact vibronic calculation, respectively, from top to bottom. The corresponding potential energy curves are shown in Fig. 19a,b. (a)–(c) The case $E_\pi < E_\sigma$. (d)–(f) The case $E_\pi > E_\sigma$.

were kept frozen at their values in the neutral ground state. The exact vibronic calculation (Fig. 20c) is compared with the adiabatic (Fig. 20b) and the FC spectrum (Fig. 20a). To distinguish the different sources of spectral intensity, those lines associated with an electronic factor $|\tau_\pi|^2$ are drawn dashed and those associated with $|\tau_\sigma|^2$ are drawn as full lines. In the exact spectrum the dashed lines represent Π vibronic levels ($K = 1$) and the full lines represent Σ vibronic levels ($K = 0$). In the adiabatic spectrum they analogously designate levels with vibrational angular momenta $l = 1$ and 0, respectively. In the case of the FC spectrum the full and dashed lines represent vibrational levels of the Σ- and Π-type potential energy surfaces, respectively, with $l = 0$. The intensity scale is an absolute one, assuming $\tau_\pi = \tau_\sigma = 1$.

Let us start with the discussion of the lower electronic state, the Π state. The vibronic spectrum consists of a rather short progression with unequal line spacings. It is reproduced neither by the adiabatic nor by the FC calculation. The latter yields a line with intensity unity at the energy E_π ($= 13.8$ eV) which represents the potential energy surface V_π and two other lines corresponding to the lowest $l = 0$ vibrational levels of the lower surface V_-. In the adiabatic spectrum the transition is to the $l = 1$ vibrational levels and the surface V_π is represented by a series of lines at energies $E_\pi + (2n + 1)\omega$, for $n = 0, 1, \ldots$. The lines corresponding to V_π are invariably the same in both the FC and adiabatic spectra for any values of the parameters. This is clearly an artefact which arises from neglecting the nonadiabatic interaction within the Π state. It is noteworthy that the excited Π vibronic states have clear counterparts in the FC spectrum. This has always been observed in weak coupling cases. The main deviation occurs in the Π vibronic ground state, which is not split in the exact treatment. This has been described in RT theory in terms of "unique" levels which are determined by the mean of both Π-type potential energy surfaces.[154]

We now turn to the Σ vibronic levels. These are not subject to RT nonadiabatic interaction, and for an isolated Π electronic state the nuclear motion corresponding to the Σ vibronic levels is confined to a single potential energy surface.[154] At low energy one observes the phenomenon of vibrational intensity borrowing within the Π electronic state.[190] This is accounted for qualitatively, but not quantitatively, by the adiabatic calculation. The remaining error is due to the neglect of the diagonal elements of $\hat{\Lambda}$. The Σ line structure beyond 13.85 eV is not reproduced by the adiabatic calculation. The appearance of several lines which share the intensity of the "adiabatic" line at 14.05 eV is caused by $\Sigma - \Pi$ nonadiabatic interaction. This is seen to have a marked influence on the line structure.

The spectra with $E_\pi > E_\sigma$ (Fig. 20d, e, f) seem to exhibit a close similarity between the FC and the exact Π vibronic line structure. In both cases there are two main lines at roughly the same energies. It must be emphasized that this similarity is purely accidental. In the FC spectrum the two lines in question represent the transitions to the ground vibrational levels of both Π-type potential energy surfaces. From RT theory, on the other hand, it is clear that an isolated Π electronic state should give rise to a single Π vibronic main line.[154] The split occuring in the exact calculation is, therefore, caused by interstate $(\Sigma - \Pi)$ nonadiabatic interactions. Both spectra of Fig. 20 are thus characterized by an induced RT effect of moderate strength and interstate nonadiabatic effects in the upper electronic state.

In Fig. 21 we present $\Sigma - \Pi$ spectra in which the value of λ has been doubled ($\lambda = 0.18$ eV) compared to Fig. 20. The corresponding potential energy surfaces have been shown in Fig. 19c, d. The increase of the vibronic coupling constant leads to a substantial stabilization of the bent equilibrium geometry in the lowest electronic (component) state and to a strong excitation of the bending mode in the first spectral band. The present situation is, therefore, complementary to the weak-to-intermediate coupling case of Fig. 20. Let us start with the energetic ordering $E_\pi < E_\sigma$ (left-hand part of Fig. 21). The adiabatic and FC spectra exhibit a long progression of the bending mode corresponding to the lowest potential energy surface V_- and the artificial structure due to the other Π-type surface V_π. The adiabatic calculation in addition shows the effect of vibrational intensity borrowing in the lower as well as in the upper electronic state. The exact $K = 1$ spectrum (dashed lines) exhibits in its low energy band an even longer progression in the bending mode which peaks at higher energies than in the adiabatic treatment. This is caused by the RT nonadiabatic interactions within the π electronic state. For strong coupling these lead to a dilution of the vibrational levels of V_π among several vibronic eigenstates in such a way that a single vibronic progression is formed. Below the minimum of V_π this requires a downward shift in energy of the vibronic compared to the adiabatic levels of V_-. It leads to a reordering of the $K = 1$ vibronic states and their $K = 0$ counterparts near the barrier to linearity at 13.8 eV (note that $K = 0$ vibronic states are not changed by the RT nonadiabatic interaction). This *negative* value of the effective rotational constant $A = E_{v, K=1} - E_{v, K=0}$ has been discussed previously in the case of the pure RT effect.[154] Here we see that it occurs likewise within the framework of the linear vibronic coupling treatment.

At higher energies, the exact $K = 0$ vibronic progression is also quite different from the adiabatic one and the $\Sigma - \Pi$ nonadiabatic interaction is seen to fill the gap between the vertical transition energies E_σ and E_π with spectral intensity. The formation of a regular vibronic progression is typical for strong coupling cases. If the energy gap $|E_\sigma - E_\pi|$ is still smaller than in Fig. 21, the intensity distribution may even be singly peaked and thus bear no reminscence to the existence of two electronic states.

When the ordering of E_σ and E_π is reversed (Fig. 21d, e, f), the Π electronic state is higher in energy than the Σ state and the induced RT effect belongs to the weak coupling regime, since both Π component states remain linear. The $\Sigma - \Pi$ nonadiabatic effects now affect the Π vibronic states in very much the same way as they affected the Σ vibronic states. Their in-

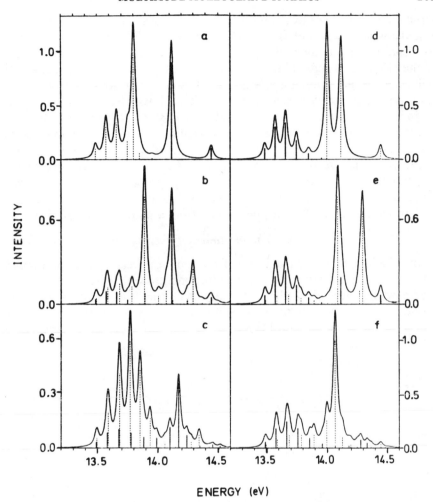

Fig. 21.   Spectra for $\Sigma - \Pi$ vibronic coupling obtained in the Franck–Condon, adiabatic, and exact vibronic calculation, respectively, from top to bottom. The corresponding vibrational energy curves are shown in Fig. 19c,d. (a)–(c) The case $E_\pi < E_\sigma$. (d)–(f) The case $E_\pi > E_\sigma$.

fluence on the line structure is thus largely independent of the vibronic symmetries. The low-energy electronic band, on the other hand, is well described adiabatically as in the corresponding example of Fig. 20d, e, f, although the vibronic coupling is stronger now.

With the above analysis we have achieved an overview over the most important dynamic effects of $\Sigma - \Pi$ vibronic coupling. These trends have been confirmed by a great number of additional calculations.[193] When the energy gap amounts to one or two vibrational quanta, there are always strong $\Sigma-\Pi$ nonadiabatic effects and the $\Sigma-\Pi$ vibronic coupling approach is a prerequisite to correct calculation and interpretation of the vibronic structure of the electronic bands. If the electronic states are well separated energetically, the vibronic coupling effects consist in an anharmonicity of the adiabatic potential energy surfaces, in vibrational intensity borrowing, and in an induced RT effect within the Π electronic state. In principle, the electronic states could then be treated as isolated by employing potential energy surfaces and RT coupling terms of high order in ρ and utilizing a Herzberg–Teller treatment to account for the intensity borrowing effects. Nevertheless, the interstate vibronic coupling approach appears to be conceptually advantageous also for a large $\Sigma-\Pi$ energy gap, since it allows us to describe all these different phenomena from a unified point of view.

3. The Multimode Effects

In principle, there are two types of multimode effects in linear molecules. The first is the excitation of two bending modes, which is of relevance in four-atomic molecules.[159] Here we shall not venture into a discussion of this problem. Rather, we turn to the impact which totally symmetric modes exert on the vibronic coupling problem. These modes are included by adding the following term:

$$\mathscr{H}_t = (T_N + V_0)\mathbf{1} + \sum_{i=1}^{N} \begin{pmatrix} \kappa_i^{(\pi)}Q_i & 0 & 0 \\ 0 & \kappa_i^{(\sigma)}Q_i & 0 \\ 0 & 0 & \kappa_i^{(\pi)}Q_i \end{pmatrix} \tag{4.31a}$$

$$T_N = -\sum_{i=1}^{N} \frac{\omega_i}{2} \frac{\partial^2}{\partial Q_i^2} \tag{4.31b}$$

$$V_0 = \sum_{i=1}^{N} \frac{\omega_i}{2} Q_i^2 \tag{4.31c}$$

to the Hamiltonian $\mathscr{H}_b$ of Eq. (4.23), that is,

$$\mathscr{H} = \mathscr{H}_b + \mathscr{H}_t \tag{4.31d}$$

The nonseparability of the modes is expressed by

$$[\mathscr{H}_b, \mathscr{H}_t] \neq 0 \tag{4.32}$$

The summation in Eq. (4.31) is over all totally symmetric modes.

For any totally symmetric mode the coupling constants in both Π component states coincide by symmetry. Therefore, if the Π electronic state were isolated, the Hamiltonians $\mathscr{H}_b$ and $\mathscr{H}_t$ would commute and the stretching modes would not interfere with the bending motion. Indeed, even in the most recent general treatments of the (direct) RT effect, the stretching modes are usually taken to be negligible.[186–189] The situation is quite different for $\Sigma - \Pi$ vibronic coupling, and the totally symmetric modes are nonseparable from the bending motion even in the present lowest-order treatment. This derives from the difference of the coupling constants in the Σ and Π electronic states. As a consequence, the energy gap between the interacting states becomes a function of the bond lengths and can strongly decrease during the totally symmetric motion. Therefore these modes will enhance the strength of the $\Sigma - \Pi$ nonadiabatic effects. This behavior is completely analogous to the vibronic coupling problem involving nondegenerate modes, which has already been analyzed in Section III.B. In particular, the crossing of the Σ and Π potential energy surfaces for $\rho = 0$ gives rise to a conical intersection of these surfaces.[116] If this intersection is accessible energetically, the vibronic states are a complete mixture of many vibrational levels of different electronic states. An example of this type occurs in the photoelectron spectrum of HCN and will be investigated in Section V.D.

V. MULTIMODE VIBRONIC COUPLING EFFECTS IN SPECTROSCOPY

In the preceding sections we have presented various classes of vibronic coupling systems and analyzed in some detail the nature of the nonadiabatic and multimode effects. These calculations represent model studies in the sense that simple model Hamiltonians have been adopted and that the energies and coupling constants entering the Hamiltonian have been considered as free parameters. We now want to present an illustrative set of applications wherein vibronic coupling effects have been identified in electronic spectra. This presentation shall serve to demonstrate that the general features of the various coupling mechanisms elucidated above are indeed operative in real polyatomic molecules. This pertains, in particular, to the multimode effects. To our knowledge, there are only two cases where the interactions in polyatomic molecules may be successfully described in terms of a vibronic coupling model involving only a single nondegenerate mode. These are the dimer[35,36,46,197] and the core hole[198] problems. In the case of core electron ionization in symmetric triatomic molecules, the vibronic coupling between the two resulting core hole states is a single-mode problem to a very good approximation.[198] The interaction between excited states of molecular dimers is usually described in the literature in terms of a single-mode vibronic coupling model.[35,36] Here, the applicability of the model in practice is less certain than in the case of the core hole problem.[199] In our opinion these systems constitute the exception rather than the rule, and we will, therefore, not consider them in the following. Rather, we hope to demonstrate that in general the inclusion of the multimode effects—with the ensuing enhancement of the nonadiabatic effects—is crucial to reproduction of the observations. As a by-product, we will see that the models discussed so far can reliably describe the gross features of the nuclear dynamics and that the corrections neglected modify only the details of the spectrum.

A. Conical Intersections in the Cumulene Cations

The family of the cumulenes C_nH_4 ($n = 2,\ldots,5$) presents a series of compounds that well illustrate the nature of the multimode vibronic coupling effects and their impact on electronic spectra. All members $n > 2$ of the series contain cumulated double bonds. Those molecules with an even number of carbon atoms are planar (symmetry group D_{2h}); in those with an odd number of carbon atoms the terminal CH_2 groups are mutually perpendicular (symmetry group D_{2d}). In all cases there exists a remarkably strong vibronic coupling between the lowest electronic states of the cation. We shall see that despite the different symmetry groups of the even- and odd-carbon members the vibronic coupling mechanism is very similar.

1. Butatriene

The vibronic coupling mechanism is relatively simple in case of the butatriene radical cation, $C_4H_4^+$, involving only two vibrational modes: a single nondegenerate coupling mode and a single tuning mode. This constitutes precisely the conical intersection model analyzed in detail in Section III. To make contact with this analysis we start therefore with the butatriene radical cation.

The low-energy photoelectron (PE) spectrum of butatriene exhibits seemingly three bands with a peculiar vibronic structure.[200] They are reproduced in Fig. 22a. Ab initio calculations have shown that there are only two electronic states of the ion in this energy range.[200,201] They arise from ionization out of the high-lying b_{3u} and b_{3g} molecular orbitals. The shape of these orbitals and their occupancy in the leading electron configuration of the corresponding $^2B_{3u}$ and $^2B_{3g}$ states is schematically indicated in Fig. 23b. These two electronic states have been tentatively associated with the low- and high-energy peaks of the experimental PE spectrum. The central structure remained unexplained and was termed the "mystery band".[200]

Since the two ionic states are separated by only ~ 0.5 eV, vibronic coupling effects can be expected to be important. To investigate this possibility, we first have to consider the symmetries of the various vibrational modes.

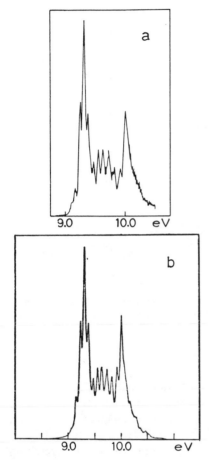

Fig. 22. The photoelectron spectrum of butatriene between 9 and 11 eV. (a) The experimental recording.[200] (b) The result of the vibronic coupling calculation (linewidth fwhm = 0.04 eV). For the values of the parameters see Table IV.

Butatriene, C_4H_4, belongs to the symmetry group D_{2h}, and its 18 normal modes of vibration are of the following species:[202]

$$\Gamma = 4A_g + 3B_{2g} + 2B_{3g} + A_u + 3B_{1u} + 2B_{2u} + 3B_{3u} \qquad (5.1)$$

According to the general discussion in Section II, there can be a linear vibronic coupling between the ionic states through a vibrational mode that transforms as the direct product of their symmetry species (i.e., as $B_{3g} \times B_{3u}$ = A_u). From Eq. (5.1) we see that there is indeed such a mode, namely, the torsional mode, consisting in the torsion of the CH_2 terminal groups. In addition, all four totally symmetric modes should be taken into account. The vibronic problem can thus be described by the Hamiltonian (3.22) with $N = 4$ and $M = 1$.

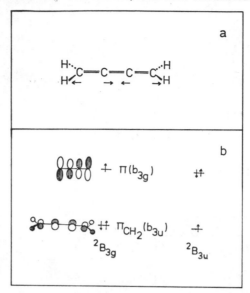

Fig. 23. (a) Schematic drawing of the C—C stretching mode which acts as tuning mode in the vibronic coupling in $C_4H_4^+$. (b) The two high-lying orbitals of butatriene and their occupancy in the lowest states of $C_4H_4^+$.

To determine the energies and coupling constants entering the Hamiltonian, we have performed ab initio Green's function calculations of these quantities as described in Section II.A.4. The ab initio calculations reveal a strong vibronic coupling between the two ionic states through the torsional mode.[81] Among the four totally symmetric modes only the C—C stretching mode ν_2, sketched in Fig. 23a, is predicted to be excited significantly. The other totally symmetric modes are excited very weakly and can therefore be neglected.[81] This reduces the problem to $N = 1$ in Eq. (3.22). Calculating the spectrum with the general formula (2.37) and using the ab initio data for the relevant parameters, we obtain qualitative agreement with experiment. We have subsequently readjusted the parameters and tried to reproduce the observations as accurately as possible. The changes in the numerical values are rather small in all cases and remain within the error limits of the ab initio calculation. The best result has been achieved with the set of parameters collected in Table IV. It yields the theoretical spectrum which has already been presented and discussed in Section III.B.2 and which is reproduced on a more suitable scale in Fig. 22b.

The agreement between theory and experiment is very satisfactory. Almost every detail of the observed band envelope is reproduced by the calcu-

TABLE IV
Energies, Vibrational Frequencies, and Coupling
Constants Used in the Calculation of the
PE Spectrum of Butatriene (Fig. 22)[a]

	$^2B_{3g}$	$^2B_{3u}$
E	9.450	9.850
κ/ω_g	$-0.212/0.258$	$0.255/0.258$
λ/ω_u	0.318/0.091	

[a] The frequencies are those of the neutral ground
state.[202] All quantities are in electron volts.

lation. We can thus be sure that we have identified the mechanism that leads
to the peculiar shape of the PE band system of butatriene. Further evidence
comes from the close similarity between the adjusted and the ab initio
calculated values of the parameters. The coupling constants of Table IV differ from the ab initio data[81] by at most 0.04 eV. The energy difference
$E_2 - E_1$ is given less accurately by the Green's function calculation, since it
results from subtracting two numbers of roughly equal magnitude. The
vibronic structure, on the other hand, depends very sensitively on the values
of the parameters. As a consequence, ab initio calculations of sufficient accuracy for vibronic calculations of this type are hardly possible at present.
This is a general experience of our work in the past years and not limited to
butatriene. Nevertheless, ab initio calculations are of great value, since they
show which modes are of importance and fix the relevant range of the
parameters. This greatly facilitates the search for a "best fit" and practically
excludes the possibility of fortuitous agreement between theory and experiment.

The theoretical spectrum has been analyzed in detail in Section III.B.2,
and we limit ourselves here to a brief summary of the results. It has been
shown in Section III.B.2 that only the low-energy part up to 9.5 eV in Fig.
22b represents a simple vibrational progression in the torsional mode. For
energies beyond 9.5 eV, regular vibrational progressions are broken up into
numerous lines and the nuclear motion behaves completely nonadiabati-
cally. Due to the limited experimental resolution, the peaks of the observed
envelope in this energy range contain several lines. They do not represent
single vibronic states of the molecular ion, and their appearance will change
when the spectrum is recorded with different resolution. The accumulation
of several lines under a single peak of the envelope also gives rise to the formation of the mystery band, which represents a nonadiabatic effect (see Section III.B.2).

The strength of the nonadiabatic and non-Condon effects has been analyzed in Section III in terms of the adiabatic potential energy surfaces and the nonadiabatic coupling terms. These have been depicted for the present example of $C_4H_4^+$ in Figs. 4–6 and 7, respectively. The lower potential energy surface is characterized by a double-minimum along the torsional coordinate that accounts for the torsional progression in the low-energy part of Fig. 22b. Searching for the minimum of the lower surface one obtains a torsional angle of $\sim 46°$ as the equilibrium geometry of $C_4H_4^+$. The upper potential energy surface touches the lower surface at $Q_g = -0.86$ and $Q_u = 0.0$ and an energy of 9.73 eV. This results in a conical intersection of the potential energy surfaces (see Fig. 6), where the vibrational motion on the upper surface is coupled with particular strength (see Fig. 7) to that on the lower surface. This explains the strong nonadiabatic effects in the spectrum of Fig. 22b (or, equivalently, Fig. 8c) at energies beyond 9.5 eV. We see from the good agreement with the experimental spectrum that the mechanisms analyzed in Section III are indeed operative in a real molecule.

The low-energetic position of the conical intersection (it is at the minimum of the upper potential energy surface) rests on the opposite signs and the large absolute values of the coupling constants κ_1 and κ_2 (see Fig. 4 and Table IV). This behavior of the coupling constants has a simple chemical origin in the nodal pattern of the molecular orbitals out of which the ionization takes place.[145] The totally symmetric mode in question and the b_{3u} and b_{3g} molecular orbitals are depicted in Fig. 23. Although the drawing is only schematic, this does not affect the qualitative conclusion to be drawn here, namely, that both orbitals have opposite bonding properties with respect to the C—C stretching coordinate Q_g. Whereas the b_{3g} orbital is bonding (antibonding) with regard to the outer (inner) C—C bonds, the b_{3u} orbital is antibonding (bonding). Thus, the orbital energies—and hence the energies of the ionic states—should exhibit a marked dependence on Q_g with different slopes. As discussed above, this is precisely the behavior that leads to the strong multimode effects in the PE spectrum.

2. Ethylene

The vibronic coupling between the two lowest electronic states of the ethylene radical cation is qualitatively very similar to that just discussed in butatriene. Quantitatively, there are differences concerning the number of vibrational modes and the energy gap between the interacting states. Consider the PE spectrum of ethylene, reproduced from the work of Mintz and Kuppermann[203] in Fig. 24. In the energy range below 18 eV it exhibits four electronic bands that arise from ionization out of the four high-lying molecular orbitals. The first two bands correspond to the B_{2u} ionic ground state and the B_{2g} first excited ionic state.[204,205] These are described by a single

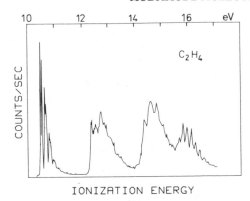

COUNTS/SEC

IONIZATION ENERGY

Fig. 24. Experimental photoelectron spectrum of ethylene between 10 and 17 eV as obtained by Mintz and Kuppermann.[203] The first two bands are involved in the vibronic coupling mechanism.

electronic configuration in Fig. 25b, and the relevant molecular orbitals, which are included in the figure, are seen to have analogous bonding properties as in butatriene (Fig. 23b). In contrast to C_4H_4, the corresponding first two spectral bands are separated by almost 2 eV and do not overlap.

To investigate the possibility of vibronic coupling between the ionic states, we start by considering the symmetries of the vibrational modes. Ethylene, C_2H_4, belongs to the symmetry group D_{2h}, and its 12 normal modes of

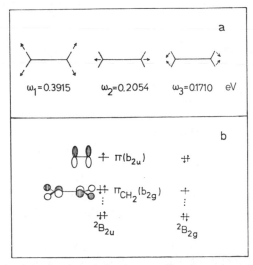

Fig. 25. (a) Schematic drawing of the three totally symmetric modes of ethylene. The numbers below each graph are the harmonic frequencies in the ground state of the neutral molecule.[79] (b) The two high-lying molecular orbitals of ethylene and their occupancy in the lowest states of $C_2H_4^+$.

vibration are of the following species:[206]

$$\Gamma = 3A_g + A_u + 2B_{1g} + B_{1u} + B_{2g} + 2B_{2u} + 2B_{3u} \qquad (5.2)$$

The direct product of the symmetry species of the lowest ionic states is B_{2u} $\times B_{2g} = A_u$. According to Eq. (5.2), there is a single vibrational mode which transforms as A_u and can thus linearly couple the two ionic states, and this is again the torsional mode. Including the three totally symmetric normal modes, the vibronic problem can be described by the Hamiltonian (3.22) with $N = 3$, $M = 1$, and the expression (2.37) for the PE spectrum.

We have performed ab initio Green's function calculations in order to determine the relevant parameters as described in Section II.A.4. The ab initio data reveal a substantial interaction between the two ionic states.[79] In addition, all three totally symmetric modes of ethylene are found to partake in the vibronic coupling mechanism as tuning modes. They are sketched in Fig. 25a. Two of these, the C—H stretching mode ν_1 and the C—C stretching mode ν_2, are nonseparable from the torsional mode and lead to genuine multimode effects. The totally symmetric C—H bending mode ν_3 has roughly equal coupling constants for both ionic states and can therefore be treated in the convolution approximation[79] (see Section II.B.1).

Let us start with the first PE band. The pure ab initio result is in partial agreement with experiment and shows that the vibronic coupling mechanism can, in principle, account for the observations.[79] There is, however, an additional effect present in this band which precludes a quantitative description of its fine structure within a strictly linear coupling scheme: the change of the harmonic vibrational frequencies in the diabatic states that occurs upon ionization. In the ground state of $C_2H_4^+$ this frequency change is particularly pronounced, leading to a reordering of the harmonic frequencies ω_2 and ω_3.[207,208,213] In searching for the best set of parameters, we have therefore also adjusted the vibrational frequencies suitably. The ratios $\kappa_i^{(1)}/\omega_{gi}$ and $\kappa_i^{(2)}/\omega_{gi}$ have been kept fixed at their ab initio values to avoid the variation of too many parameters. These ratios determine the excitation strength of the totally symmetric modes in the absence of vibronic coupling (see Section II.B.1 and Ref. 82). The best result has been achieved with the data collected in Table V.[79] The corresponding theoretical first PE band is reproduced in Fig. 26 and compared there with the recent high-resolution recording of Pollard et al.[209]

The calculated and experimental PE spectra agree in all essential details. The same holds for the PE band of the deuterated species, which has been calculated without further readjustments.[79] We furthermore emphasize that the numerical values of the energies and the vibronic coupling constants in Table V are rather close to and within the error limits of the ab initio data.[79]

TABLE V
Energies, Vibrational Frequencies, and Coupling
Constants Used in Calculating the PE Spectrum
of Ethylene (Figs. 26–29)[a]

	$^2B_{2u}$	$^2B_{2g}$
E	10.750	12.650
κ_1/ω_1	0.027/0.392	−0.362/0.360
κ_2/ω_2	−0.178/0.155	0.330/0.205
κ_3/ω_3	0.155/0.180	0.102/0.171
λ/ω_4	0.402/0.110	

[a] The frequency value given in the left-hand column has been used for the first band, that in the right hand column for the second band. All quantities are in electron volts.

This is conclusive evidence that the vibronic coupling mechanism is also operative in the lowest ionic states of ethylene and affects the nuclear motion in the ionic ground state.

To obtain an insight into the nature of the nuclear dynamics, we draw in Fig. 27a−d four cuts through the adiabatic potential energy surfaces along the four normal coordinate axes. In the order a−d these are the coordinates of the totally symmetric modes $\nu_1 − \nu_3$ and of the torsional mode ν_4, respectively. In Fig. 27d the diabatic potential energy curves (defined by $\lambda = 0$) are included as dashed lines. First and foremost, the figure gives evidence that the vibrational energy of the levels populated in the first PE band (lower curves) is always well below the minimum energy of the first excited ionic state (upper curves). From the general analysis of Section III.B.2 it may therefore be anticipated—and has been confirmed by numerical computations—that the adiabatic and Franck–Condon approximations remain applicable for this band. The lines in Fig. 26b can therefore be assigned in the usual way. The four lines following the 0-0 line represent (in the order of increasing energy) excitation of double and quadruple quanta of the torsional mode and of a single quantum of the C—C stretching and the C—H bending mode, respectively.

The excitation of the C—C stretching mode reflects the elongation of the C—C bond in the ionic compared to the neutral ground state (1.41 Å versus 1.33 Å; see Fig. 27b). Similarly, the torsional mode is excited due to the lowering of the molecular symmetry that occurs upon ionization. This lowering of the symmetry is the main effect of the vibronic coupling in the case of the first PE band. Using Eq. (3.25) and the data of Table V, we get a torsional angle of 25° for the equilibrium geometry of $C_2H_4^+$. The stabilization

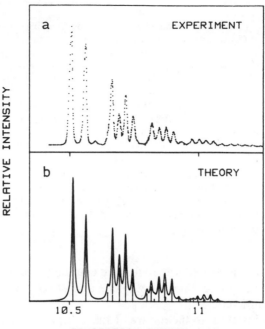

Fig. 26. The first band in the photoelectron spectrum of ethylene. (*a*) The experimental recording according to Pollard et al.[209] (*b*) The result of the vibronic coupling calculation (linewidth fwhm = 0.01 eV). For the values of the parameters see Table V.

energy of the nonplanar conformation is [using Eq. (3.27)] found to be 0.03 eV. The nonplanarity of $C_2H_4^+$ is also displayed by the lower adiabatic potential energy curves in Fig. 27*d*. It should, however, be mentioned that the actual stabilization energy (0.03 eV) is considerably smaller than that in Fig. 27*d*, where the totally symmetric coordinates are fixed at zero. In this latter, hypothetical, case the nonplanar conformation would be stabilized by 0.09 eV, that is, three times as much as it actually is. The quenching of the distortion by the totally symmetric modes results from the anharmonicity of the adiabatic potential energy surfaces, as has been discussed in Section III.A.1.

Although the barrier to planarity is small, its existence is considered certain in view of the good agreement between calculated and experimental spectra. Also the Green's function data predict $C_2H_4^+$ to be nonplanar, but the torsional angle comes out too small.[79] The most extensive previous ab initio investigations obtained a planar cation,[210] even when taking electron correlation into account.[211] We believe that one reason for this failure has to

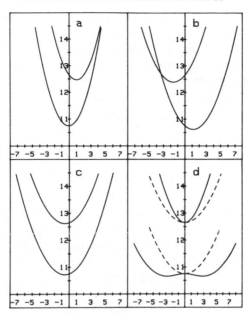

Fig. 27.   Four cuts of the multidimensional potential energy surfaces of the ground and first excited states of $C_2H_4^+$ along the four normal coordinate axes. (a) The C—H stretching mode Q_1. (b) The C—C stretching mode Q_2. (c) The C—H bending mode Q_3. (d) The torsional mode Q_4. The dashed lines in panel d are the diabatic potentials, obtained by putting $\lambda = 0$.

do with the variational treatment, during which the energy of the lower cationic state $^2B_{2u}$ is minimized and, simultaneously, the energy of the upper cationic state $^2B_{2g}$ becomes much too high. Since the geometry of the cation depends on the coupling of these two states, the ab initio treatment tends to underestimate their mixing, leading to a planar cation. This reasoning indicates that great care is necessary when calculating vibronically coupled states and that a balanced treatment of both states is required to obtain reliable results.[79,80]

Let us return to Fig. 27 and consider also the upper potential energy curves, corresponding to the first excited ionic state. Although for $Q = 0$ the two states are separated by 1.9 eV, their energy gap diminishes quickly for positive values of the C—H stretching coordinate Q_1 or negative values of the C—C stretching coordinate Q_2. In Fig. 27b this leads to a crossing of the curves at an energy which almost equals the vertical ionization potential of the first excited ionic state. In (Q_2, Q_4) coordinate space this crossing constitutes a conical intersection of the potential energy surfaces. Following the analysis of Section III.B.2, and expecially the discussion of Fig. 11, we therefore can expect strong nonadiabatic effects in the high-energy band.

The second PE band of ethylene looks indeed quite different from the first band. It is reproduced from the book of Turner et al.[212] on an enlarged scale in Fig. 28a and consists of three broad humps that are much broader than the experimental resolution function. Even under higher resolution no detailed line structure could be resolved.

Can one explain the observations in terms of the present vibronic coupling mechanism? In principle, all parameters of the model Hamiltonian are fixed from the first PE band. However, in the former calculation we have readjusted the vibrational frequencies, which is an ad hoc procedure. In the case of the second PE band we used for ν_2 and ν_3 the harmonic frequencies of the ground state of the neutral molecule, but a smaller frequency value was needed for the mode ν_1.[31] The values are included in Table V. All the other parameters are the same as used for the calculation of the first PE band.

To calculate the vibronic structure we use again the general formula (2.37) for the transition probability and treat the mode ν_3 in the convolution approximation (see Section II.B.1). In view of the high energy relative to the minimum of the ionic ground state, one must employ large maximum occupation numbers of the modes to eliminate truncation errors: 6, 20, and 30 quanta of the vibrational modes ν_1, ν_2, and ν_4, respectively, have been included in setting up the vibronic secular matrix (see Section II.B.3). This leads to a secular matrix of dimension 3600 (see Section II.B.3). The resulting spectrum is drawn in Fig. 28b together with its envelope, which has been obtained by convoluting the line spectrum with Lorentzians of suitable width (fwhm = 0.04 eV). For comparison we have also computed the corresponding Franck–Condon spectrum as described in Section II.B.2. It is included in Fig. 28c.

The striking feature of the theoretical spectrum is the very high line density and the irregular intensity distribution of the individual vibronic lines. The envelope is in rather good agreement with experiment. It consists of three broad humps and has different slopes of the low- and high-energy wings. Although the details of the calculated line structure certainly depend on the adopted model, the gross features of the nuclear dynamics are described correctly by the present treatment. As expected, there are many vibronic lines under each of the humps. The latter do not even approximately represent vibronic states, and an assignment of vibrational quantum numbers to the observed peaks is inappropriate.

The Franck–Condon spectrum in Fig. 28c shows that the FC principle fails completely to reproduce the observations. From the analysis of Section III.B it is clear that this failure pertains likewise for the adiabatic spectrum. The strong nonadiabatic effects extend throughout the entire energy range of the band and not even the lowest vibrational level is unaffected by them. The total number of vibronic lines exceeds those of the FC spectrum by two

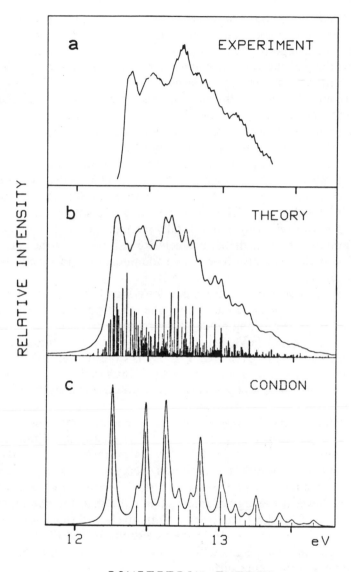

Fig. 28. The second band in the photoelectron spectrum of ethylene. (*a*) The experimental recording of Turner et al.[212] (*b*) The result of the vibronic coupling calculation. For the values of the parameters, see Table V (linewidth fwhm = 0.04 eV). (*c*) The Franck–Condon spectrum, obtained with the same values of the parameters as in panel *b*.

orders of magnitude and their density roughly equals that of the vibrational levels of the ionic ground state in the energy range of the second PE band. Even without computing vibronic eigenvectors explicitly, this observation makes it clear that the vibronic states are a complete mixture of vibrational levels of both potential energy surfaces.

The second PE band of ethylene represents a convincing confirmation of the general conclusions established in Section III.B, namely, that in multimode vibronic coupling systems the nonadiabatic effects depend mainly on the energetic position of the conical intersection. Its location in coordinate space or the energetic separation of the electronic states within the FC zone are almost immaterial. This underlines that nonadiabatic effects can be strong even in isolated electronic bands. The conclusions in Section III.B were derived with the aid of a model system consisting of a single (totally symmetric) tuning mode and a single (non-totally symmetric) coupling mode. In the present example the nonadiabatic effects are reinforced by the existence of two tuning modes. This gives rise to two additional multimode effects, which have already been discussed in Section III.C.

The first of these effects consists in a lowering of the minimum energy of intersection relative to the minimum of the upper potential energy surface. If only the C—C stretching mode Q_2 were operative, the conical intersection would occur at 12.7 eV, that is, 0.3 eV above the minimum of the upper (Q_2) potential energy curve (see Fig. 27b). Considering only the C—H stretching mode Q_1 (Fig. 27a), the energy of intersection is 2 eV above the accessible energy range. In (Q_1, Q_2) space, however, the conditions (3.30) for intersection define a line rather than isolated points and the resulting locus of intersection is a parabola within our model Hamiltonian. This parabola is depicted in Fig. 29 for $C_2H_4^+$, where it can be seen that the minimum energy of intersection (12.3 eV) is lower by 0.4 eV than in case of the C—C stretching mode Q_2 alone. This minimum energy of intersection is only 0.07 eV above the minimum of the upper surface. Apparently, several tuning modes here lower the energy of intersection more than the minimum of the upper surface. This effect has been quantitatively described for the multimode case in Eqs. (3.30)–(3.34). In the case of the second PE band of ethylene it causes the nonadiabatic effects to extend down to the lowest vibrational energies of the upper surface.

There is a further characteristic by which the three-mode problem differs from the two-mode vibronic coupling problem. It arises from the high density of vibronic lines. The calculated spectrum of Fig. 28b consists of ~1000 vibronic lines, and hence the complete determination of all the vibronic eigenvectors is a rather hopeless numerical task. But the question arises whether in such systems a complete knowledge of every quantum state is of main interest at all or whether another type of description is more ap-

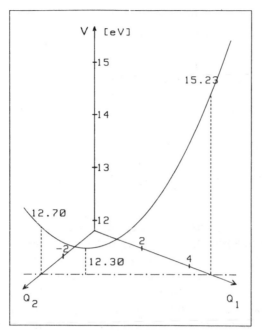

Fig. 29.   A portion of the curve of intersection between the $^2B_{2u}$ and the $^2B_{2g}$ ionic states in $Q_1 - Q_2$ normal coordinate space (full line). Its projection on the $Q_1 - Q_2$ plane (dashed–dotted line) is given by $0.57Q_2 - 0.39Q_1 + 1.9 = 0$.

propriate. For example, one could consider *distributions* of physical quantities rather than their values in every single vibronic state.[214] We believe that such an approach can yield valuable insight into multimode vibronic coupling systems and present first attempts in that direction in Section VI, below, using $C_2H_4^+$ as an example. Another phenomenon related to the high line density is the occurrence of irreversible motion of the optically excited initial state.[127] It is caused by the large number of vibrational levels of the lower surface with which the levels of the upper surface can interact and leads to the process of nonradiative decay.[103,104] This will be explored for the ethylene radical cation in Section VII.

To summarize, the PE spectrum of ethylene exhibits in a transparent way the various phenomena associated with strong vibronic coupling. The low-energy band is characterized by excitation of double quanta of the non-totally symmetric coupling mode, thus reflecting the symmetry lowering in the ionic ground state. The high-energy band exhibits strong nonadiabatic effects caused by the multidimensional conical intersection. It is therefore considered as the example of central importance in this work. Due to the larger energy gap and the participation of several totally symmetric tuning modes, the nuclear dynamics in the ethylene cation is somewhat different from that in the butatriene cation. However, the *structural* features of the coupling mechanism are very similar in both cases,[145] as becomes evident by comparing Figs. 23 and 25. The coupling mode is always the torsional mode, and in ethylene as in butatriene it is the C—C stretching mode that plays a central role among the tuning modes. The different slopes of the potential energy curves along this coordinate (see Fig. 27b) can be rationalized in terms of the different bonding properties of the b_{2u} and b_{2g} molecular orbitals across the C—C bond (see Fig. 25b). Generally, whenever the bonding properties of the orbitals partaking in the vibronic coupling are opposite, one expects the tuning and hence the multimode effects to be strong.[145]

3. Allene

The even-carbon members of the cumulene series, being of symmetry D_{2h}, possess only nondegenerate electronic states and vibrational modes. In the symmetry group D_{2d} of the odd-carbon members there are also (twofold) degenerate electronic states and Jahn–Teller (JT) effects may occur. This is the case for the lowest ionic states of allene, C_3H_4, and pentatetraene, C_5H_4, since the highest occupied molecular orbital is of symmetry E.

The JT effect in these compounds differs from the $E \times \varepsilon$ Jahn–Teller effect discussed in Section IV.A in that it involves only nondegenerate vibrational modes. This follows generally from the decomposition of the symmetrized product of the irreducible representation E with itself. For the following seven symmetry groups it contains no representation of degenerate modes[61] (the indices g and u are omitted):

$$[E^2] = A_1 + B_1 + B_2 \quad \text{for} \quad C_{4v}, D_4, D_{2d}, D_{4h}$$
$$= A_1 + B \quad \text{for} \quad C_4, C_{4h}, S_4 \tag{5.3}$$

The coupling of the components of an E electronic state through vibrational modes of type B is termed the $E \times \beta$ Jahn–Teller effect.[160–163] Retaining for simplicity only two modes, Q_1 and Q_2, of symmetry B_1 and B_2, respectively, one obtains the following form of the $E \times \beta$ Hamiltonian:

$$\mathscr{H} = (T_N + V_0 + E_0)\mathbf{1} + \begin{pmatrix} \kappa Q_2 & \lambda Q_1 \\ \lambda Q_1 & -\kappa Q_2 \end{pmatrix} \tag{5.4}$$

Here $T_N + V_0$ represents the Hamiltonian for the unperturbed vibrational motion as in Eq. (3.1), and E_0 is the vertical transition energy. Since the totally symmetric modes decouple from the vibronic problem, they are not included in $\mathscr{H}$. Equation (5.4) is formally a special case of the two-state–

two-mode problem discussed in Sections III.A and III.B. It emerges from the latter by putting the energy gap and the arithmetic mean of the coupling constants of the "tuning mode" Q_2 equal to zero. In this case the conical intersection is situated right at the center of the Franck–Condon zone, which always leads to strong nonadiabatic effects. Furthermore, at the minima of the lower adiabatic potential energy surface, only either of the B_1 and B_2 coordinates can be different from zero. Depending on the relative magnitude of the stabilization energies $E_s^{(1)} = \lambda^2/\omega_1$ and $E_s^{(2)} = \kappa^2/\omega_2$, the minima of the lower surface occur either at

$$Q_1^{(0)} = \pm \frac{\lambda}{\omega_1}; \qquad Q_2^{(0)} = 0 \qquad (5.5a)$$

(when $E_s^{(1)} > E_s^{(2)}$) or at

$$Q_2^{(0)} = \pm \frac{\kappa}{\omega_2}; \qquad Q_1^{(0)} = 0 \qquad (5.5b)$$

(when $E_s^{(1)} < E_s^{(2)}$). Those two of the four geometries (5.5) which are not minima represent saddle points of the potential energy surface. Thus, there is a total quenching of the distortion along one coordinate by the other mode. This is a special example of the general effect discussed in Section III.A.

Allene, C_3H_4, belongs to the symmetry group D_{2d}, and its 15 normal modes of vibration are of the following species:

$$\Gamma = 3A_1 + B_1 + 3B_2 + 4E \qquad (5.6)$$

There are four modes of symmetry B_1 or B_2 which can be operative in the $E \times \beta$ Jahn–Teller effect. The mode of B_1 symmetry is the torsional mode. In addition, all three totally symmetric modes can be excited in any electronic transition.

The first PE band of allene, depicted in Fig. 30a, exhibits in the low-energy part a line spacing similar to the torsional frequency.[212] The torsional distortion has also been found to lead to a marked lowering of the ionic energy.[215] Therefore, the coupling to this mode should be substantial as in the even-carbon cumulenes. The complexity of the high-energy part of the first PE band suggests that at least one of the B_2 modes must also be important. For simplicity, we try to reproduce the observations by retaining only a single B_2 mode and neglecting the totally symmetric modes.[216] Guided by the previous results for butatriene, we choose the (asymmetric) C—C stretching mode. Upon varying the parameters within a range found realis-

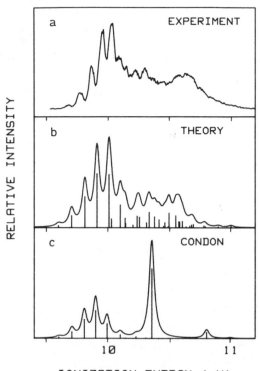

Fig. 30.   The first band in the photoelectron spectrum of allene. (*a*) The experimental re-cording of Turner et al.[212] (*b*) The result of the vibronic coupling (Jahn–Teller) calculation. For the values of the parameters see Table VI (linewidth fwhm = 0.05 eV). (*c*) The Franck–Condon spectrum, obtained with the same values of the parameters as in panel *b*.

tic for $C_4H_4^+$, we arrive[216] at the spectrum depicted in Fig. 30*b* and the cor-responding Franck–Condon spectrum of Fig. 30*c*. The values of the param-eters are collected in Table VI. A similar result has been obtained also for pentatetraene,[216] and the corresponding parameters are included in the same table.

The agreement between theory and experiment is quite satisfactory in view of the simple analysis using only two vibrational modes. The sequence of regular lines up to ~10 eV represents excitation of the torsional mode. The fading of the vibronic structure beyond ~10.1 eV is attributed to the de-molition of vibrational progressions that occurs as a consequence of multi-mode effects. If only the torsional mode would be involved in the $E \times \beta$ JT effect, this gives rise to a simple Poisson distribution.[29]

The energy range of the irregular vibronic lines in Fig. 30*b* is precisely the same where the Franck–Condon spectrum of Fig. 30*c* fails. The low-

TABLE VI

Energies, Vibrational Frequencies, and
Coupling Constants Pertinent to the JT
Effect in the Ground States of the Allene
and Pentatetraene Radical Cations[a]

	$C_3H_4^+$	$C_5H_4^+$
E	10.150	9.250
κ/ω_2	0.216/0.246	0.311/0.258
λ/ω_1	0.339/0.107	0.297/0.080

[a]The vibration of B_1 (or B_2) symmetry is the torsional (or asymmetric C—C stretching) mode. All quantities are in electron volts.

energy torsional progression is well approximated by the Franck–Condon calculation. Note that the C—C stretching mode is practically not discernible in the low-energy range, although the "Poisson parameter" $\frac{1}{2}(\kappa/\omega_2)^2 =$ 0.38 (see Section II.B.1) should lead to a marked excitation of this mode. This is a consequence of the mutual quenching of the modes, discussed above after Eq. (5.5) and in Section III. Since in $C_3H_4^+$ the stabilization energy of the torsional mode (λ^2/ω_1) exceeds that of the C—C stretching mode (κ^2/ω_2), the latter coordinate is zero at the ionic equilibrium geometry according to (5.5a). Consequently, the excitation of the C—C stretching mode is strongly quenched by the torsional mode.

The reader should note the similarity between the PE spectra of allene and butatriene (see Fig. 22). In either case the low-energy part is dominated by the torsional progression. At energies beyond the band center, the conical intersection leads to an erratic line structure that is not resolved experimentally. Indeed, the coupling terms are comparable in magnitude in both cases. Although butatriene has a finite energy gap of 0.4 eV, it is "washed out" within the zero point amplitude $|Q_g| \leq \sqrt{2}$. Also, the coupling constants of ν_g in both ionic states have different signs and are at least approximately equal in magnitude as are those of the B_2 mode in allene. The nonadiabatic behavior of the nuclear motion in the case of the JT effect is well known in the literature.[29] The analogy of $C_4H_4^+$ and $C_3H_4^+$ allows us to understand the nuclear dynamics near a conical intersection as a generalized JT effect. Therefore JT-type phenomena are not restricted to highly symmetrical molecules with degenerate electronic states but will also occur in many others that are less symmetric and possess only nondegenerate states.

Finally, we would like to draw attention to the remarkably uniform behavior of the coupling mechanisms in all cumulenes studied.[145] In all cases the different electronic states (or component states) are mixed by the torsional mode and their energy gap is modulated by a C—C stretching mode. The torsional frequency and the vibronic coupling constant decrease monotonously in the series $C_2H_4, \ldots, C_5H_4$, and so does the energy gap in the subset of the even-carbon members. Concerning the distortion of the molecular framework, also in allene and pentatetraene the equilibrium torsional angle changes upon ionization and the ionic equilibrium values are 52° and 38°, respectively, compared to 90° in the neutral ground state. Thus, the change of the torsional angle upon ionization is 25°, 38°, 46°, and 52° in the order C_2H_4 to C_5H_4. Although these numbers are not claimed to be of high accuracy because of the use of simple model Hamiltonians, they should reliably describe the trend that the geometry change increases monotonously when going from smaller to larger members in the cumulene series. A similarly uniform picture emerges when one considers the effect of chemical substitution in the individual cumulene molecules.[145]

B.   The Visible Absorption Spectrum of NO_2

The visible absorption spectrum of NO_2 represents a spectroscopic problem of outstanding complexity. Since the first published account of Brewster of this "phenomenon so extraordinary in its aspect",[217] numerous workers have been engaged in the attempt to unravel the principles behind this spectrum and to arrive at least at a partial assignment of quantum numbers to the spectral lines. For a comprehensive list of references as well as a high-resolution absorption spectrum, the reader is referred to the book of Zare and coworkers.[59] Briefly, the zero-temperature absorption starts at ~ 8900 Å,[218] extending throughout the visible range. Two different electronic transitions can be distinguished. The perpendicular bands form a regular vibrational progression[219] with a spacing resembling the bending frequency of NO_2. The upper electronic state of the transition is of $^2B_1$ symmetry. The parallel bands, resulting from a $^2B_2-^2A_1$ transition,[220] are completely irregular and, except for the lowest energies,[221] no vibrational quantum numbers could be assigned. The complexity is due to both an unusual rotational structure and the high density of band origins, which leads to an overlapping of the different rotational subbands.[59] The number of vibronic bands exceeds that expected from the Franck–Condon principle by an order of magnitude, and their intensities do not follow any systematic pattern.[222] Further bits of anomalous behavior of NO_2 are irregular g factors[223] and radiative lifetimes which are 10–100 times longer than expected from the integrated absorption coefficient.[123,224]

A source of explanation of the observations is provided by the ab initio calculations of Jackels and Davidson[225] and of Gillispie et al.[226] These authors find three low-lying doublet excited states, which are listed here in order of increasing energy together with the $^2A_1$ ground state and their dominant electron configurations:

$$\cdots (1a_2)^2(4b_2)^2(6a_1)^1 \quad {}^2A_1$$

$$\cdots (1a_2)^2(4b_2)^1(6a_1)^2 \quad {}^2B_2$$

$$\cdots (1a_2)^2(4b_2)^2(2b_1)^1 \quad {}^2B_1 \qquad\qquad (5.7)$$

$$\cdots (1a_2)^1(4b_2)^2(6a_1)^2 \quad {}^2A_2$$

The transition to the $^2A_2$ state is electric dipole forbidden. Whereas the $^2B_1$ state has a linear equilibrium geometry and correlates there with the same $^2\Pi_u$ state as the $^2A_1$ ground state, the $^2B_2$ and $^2A_2$ states are more strongly bent than the ground state of NO_2. This is illustrated by Fig. 31a, which shows the ab initio calculated bending potential energy curves of Gillispie et al.[226] for these four states. In principle there are three different vibronic coupling mechanisms operative among these four states. The first is the Renner–Teller effect in the $(^2A_1, {}^2B_1)$ electronic manifold at the linear conformation. The locus of degeneracy is, however, well outside the Franck–Condon zone with respect to energy and nuclear geometry. Therefore, the Renner–Teller effect may lead to perturbations of the rotational structure but will not drastically influence the vibrational motion. Furthermore, the $^2B_2$ and $^2A_1$ states as well as the $^2B_1$ and $^2A_2$ states cross each other for decreasing bond angle. In both cases linear vibronic coupling is possible through the asymmetric stretching mode, which has the proper symmetry B_2 and converts the crossing to a conical intersection. Whereas the $^2B_1$–$^2A_2$ intersection occurs at rather high energy and is therefore less important for the $^2B_1$ absorption system, the $^2A_1$–$^2B_2$ intersection is situated near the minimum of the $^2B_2$ state and should have a profound impact on the nuclear motion in this state. The anomalous behavior of NO_2 is likely to be caused by the strong nonadiabatic interactions associated with this conical intersection.

Although the above interpretation is plausible, it has not yet been shown to what extent a dynamical calculation can reproduce the qualitative features of the experimental findings. This will substantiate (or dismiss) the validity of the concepts currently used for the explanation of the spectrum. It may contribute toward a better understanding of the interactions and of the relative importance of the above three mechanisms. Furthermore, by using ab initio data in the calculation, one may obtain direct evidence as to whether the ab initio coupling constants are compatible with the observed strength of the nonadiabatic effects.

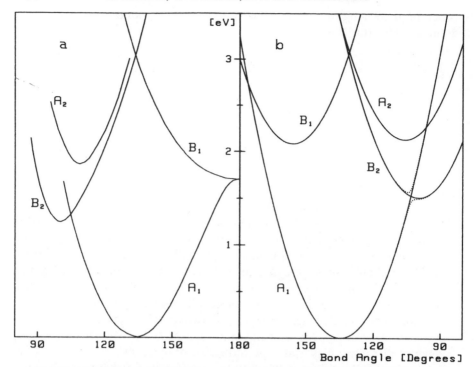

Fig. 31.   Bending potential energy curves of the four lowest electronic states of NO_2. (*a*) The result of the ab initio calculation of Gillispie et al.[226] (*b*) These curves are obtained when fitting the potential energy surfaces (3.24) of the present vibronic coupling model to the data of Gillispie et al.[226] The dots represent the two lowest A' potential energy curves for unequal NO bond lengths ($r_{NO(1)} - r_{NO(2)} = 0.05$ bohr).

Guided by these ideas we have performed[89] a theoretical calculation of the absorption spectrum of NO_2 based on the ab initio data of Gillispie et al.[226] and of Jackels and Davidson.[225] We consider only the vibronic structure for excitation from the $^2A_1$ vibrational ground state. As usual, we employ harmonic diabatic potential curves and the linear coupling scheme. Moreover, all three vibrational modes are treated as one dimensional. As will become clear below, these approximations are more severe in NO_2 than in other cases of vibronic coupling studied in this work. On the other hand, both totally symmetric modes—the symmetric stretching mode ν_1 and the bending mode ν_2—are retained in the calculation (tuning modes). The coupling mode is the asymmetric stretching mode ν_3. Both the $^2A_1 - ^2B_2$ and the $^2B_1 - ^2A_2$ vibronic problems are thus described by the Hamiltonian (3.22) with $N = 2$ and $M = 1$, and the Renner–Teller effect is neglected in the vibronic treatment. The latter approximation renders the two vibronic problems independent.

The energies and coupling constants entering the Hamiltonian are obtained by fitting the adiabatic potential energy surfaces (3.24) of the vibronic coupling model to the ab initio calculated energies of the electronic states. The general procedure has been described in Section II.A.4. The data for the totally symmetric modes are taken from the work of Gillispie et al.[226] The vibronic coupling constant is computed from the repulsion of the potential energy surfaces caused by an asymmetric bond length change. Ab initio calculations for such geometries have been performed by Jackels and Davidson in the case of the $^2A_1$ and $^2B_2$ electronic states.[225] To our knowledge no data are available for the high-energy $^2A_2-^2B_1$ conical intersection. Here, the dominant electronic configurations of the interacting states differ by the quantum numbers of two electrons and the vibronic coupling is expected to be small. We rather arbitrarily take λ in this case to be one-half of the $^2A_1-^2B_2$ vibronic coupling constant and consider this as an upper limit to the correct value.

In Table VII we present the numerical results for the vertical energies and coupling constants deduced from the above analysis. The resulting bending potential energy curves are depicted in Fig. 31b for all four states. Near the crossing between the $^2A_1$ and $^2B_2$ electronic states we include the analogous curves for unequal N—O bond lengths ($\Delta R = 0.05$ bohr) as dashed lines. The magnitude of their splitting determines the vibronic coupling constant. The comparison between Figs. 31a and 31b shows that the most severe error of the present treatment occurs at the linear geometry. Both the $^2A_1$ and the $^2B_1$ electronic energies are much too high in this region and do not properly join at the linear conformation. It has, however, already been argued above that this region is well outside the Franck–Condon zone and is therefore of minor importance for the vibronic structure. It should also be mentioned that the nonadiabatic interactions of the Renner–Teller effect are energetically rather localized around the point of degeneracy and do not extend far above it[154] (for zero angular momentum they even vanish exactly;[154] see Section IV.C).

TABLE VII

Energies, Vibrational Frequencies, and Coupling Constants Used in Calculation of the Visible Absorption Spectrum of $NO_2{}^a$

	$^2A_1$	$^2B_2$	$^2B_1$	$^2A_2$	ω
E	0	3.40	2.80	3.40	—
κ_1	0	-0.58	-0.01_4	-0.41	0.170
κ_2	0	0.51	-0.38	0.43	0.094
λ		0.28		0.14	0.210

^{a}All quantities are in electron volts.

The potential energy curves within the Franck–Condon zone and especially the occurrence of the $^2A_1-^2B_2$ conical intersection near the minimum of the $^2B_2$ state are reproduced correctly by the vibronic coupling model. We shall see below that this conical intersection dominates the vibronic structure of the spectrum entirely. We expect, therefore, that the present crudely simplified description of the vibronic coupling in NO_2 will give a qualitative explanation of the experimental findings.

To calculate the vibronic structure, one should use the general formula (2.34), which includes the $^2A_1$ ground state within the subset of vibronically coupled states. It has recently been argued by Innes that the $^2A_1-^2B_2$ vibronic coupling leads to an unusually strong anharmonicity of the asymmetric stretching mode in the electronic ground state.[227] Nevertheless, considering only excitation out of the $^2A_1$ vibrational ground state, we will ignore this effect and use the simpler formula (2.37) in which the electronic ground state is taken to be vibronically uncoupled. This will not introduce large errors in the calculation. In setting up the vibronic secular matrices, one has to use large maximum occupation numbers to eliminate truncation errors. They originate from the large values of the coupling constants relative to the frequencies (see Table VII). For the $^2B_2-^2A_1$ system the required occupation numbers are 23, 45, and 18 for the modes ν_1, ν_2, and ν_3, respectively, leading to a secular matrix of the order of 18,630. Also for the $^2A_2-^2B_1$ system the dimension is large (24,000). The total CPU time for generating both spectra with the aid of the Lanczos algorithm, performing 10,000 iterations, amounts to about 6 hours on an IBM 370/168. In Fig. 32 we show the B_1 and the B_2 spectra separately and compare them with the experimental low-resolution absorption spectrum quoted by Hsu et al.[59]

Let us start with the B_1 spectrum. At low energies, it consists of a regular progression of the bending mode as expected from the FC principle and in accord with the observations.[219,228] At high energies, when the $^2B_1-^2A_2$ conical intersection is accessible to the nuclear motion, the lines are split into several closely spaced components as a result of nonadiabatic interactions. The splittings are rather small, however, and the unperturbed progression is still discernible as underlying the calculated line structure. If we recall that the value of λ employed in the calculation is rather an upper limit, it becomes clear that the $^2B_1$ absorption spectrum is essentially well behaved and the $^2B_1-^2A_2$ conical intersection is only of minor importance for the observed spectrum.

Precisely the opposite observation applies to the $^2B_2-^2A_1$ electronic transition. The calculated vibronic line structure is highly erratic throughout the whole energy range and contains no unperturbed vibrational progressions. To be compared properly with experimental results, both theoretical spectra should be suitably superimposed and rotations be taken into account. This

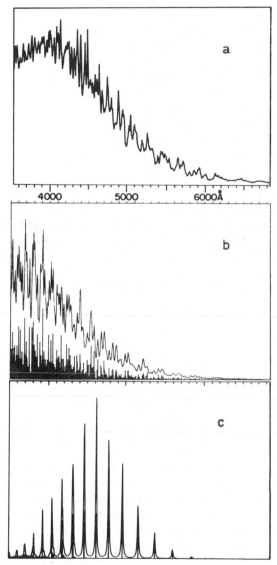

Fig. 32. The visible absorption spectrum of NO_2. (a) The low-resolution recording of Hsu et al.[59] (b) The result of the vibronic coupling calculation for the $^2B_2 \leftarrow {}^2A_1$ transition. (c) The result of the vibronic coupling calculation for the $^2B_1 \leftarrow {}^2A_1$ transition. For the values of the parameters used in panels b and c, see Table VII (linewidth fwhm ~ 50 cm^{-1}).

is not attempted here, since we cannot expect quantitative agreement from the present simplified calculation. The deviation in the overall band shape between theory and experiment indicates that the parameters of the model (which have been taken from ab initio calculations) are not yet optimally chosen. However, the qualitative features of the observed spectrum such as the irregular intensity fluctuations are reproduced authentically by theory. Also, the high density of vibronic lines agrees well with the experimentally deduced distribution of vibronic band origins of Smalley et al.[222] This distribution has been obtained from the fluorescence excitation spectrum of rotationally cooled NO_2. The remaining rotational excitation in the $^2A_1$ ground state is weak enough to permit a separation of the numerous vibronic bands but allows a determination of vibronic intensities[222] only up to a factor of ~ 5. The experimental and theoretical results are compared in Fig. 33. The number of calculated vibronic lines exceeds that expected from the Franck–Condon principle by an order of magnitude, and their density is essentially that of the ground state vibrational levels. These findings establish beyond doubt that the complexity of the $^2B_2$ absorption spectrum of NO_2 is caused by the conical intersection between the $^2B_2$ and $^2A_1$ electronic states and the associated strong nonadiabatic interactions. This conclusion is considered the more unambiguous as it derives solely from ab initio values for the energies and coupling constants entering the vibronic calculation.

The quantitative interpretation of the observed vibronic structure in absorption of NO_2 might be even more complicated than believed so far. It has been attempted, for example, to "deconvolute" the experimental vibronic structure and thus determine directly from experiment the energies and coupling matrix elements of the vibrational levels of the diabatic potential energy surfaces.[229] To simplify the analysis, each vibronic eigenstate has been assumed to derive its spectral intensity from a single vibrational level of the upper potential energy surface. However, it has already been shown in Section III.B.2 for butatriene that in the presence of conical intersections and strong coupling the vibronic states are a mixture of many vibrational levels of the lower and the *upper* surface, both in the adiabatic and in the diabatic representation. This phenomenon has been termed "vibronic interference."[139] In the case of NO_2 we have two totally symmetric tuning modes, compared to one such mode in the example of Section III.B.2. The nonadiabatic effects and therefore also the "vibronic interference" are thus even stronger than found before. Any attempt to neglect it in a conical intersection situation may lead to erroneous conclusions and should therefore be considered with great care.

In order to reproduce more quantitatively the observations, one should augment the present Hamiltonian suitably by including higher-order terms and allowing for the Renner–Teller effect. In the general case this leads to a

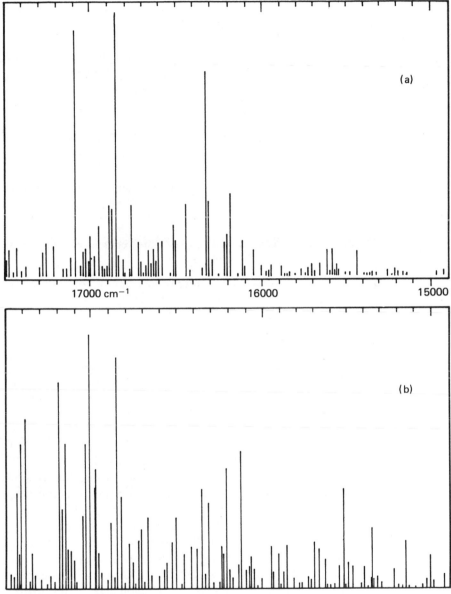

Fig. 33. Vibronic line structure of the NO_2 excitation spectrum between 15000 and 17500 cm^{-1}. (*a*) The vibronic band origins deduced by Smalley et al.[222] from the jet-cooled fluorescence excitation spectrum. (*b*) The result of the vibronic coupling calculation for the $^2B_2 \leftarrow ^2A_1$ transition.

secular matrix that is of similar dimension to the present one but is no longer a sparse matrix. Clearly, an adjustment of the parameters by trial and error becomes prohibitively expensive. Therefore, such a procedure requires the ab initio determination of the relevant parameters to a high degree of accuracy. Even if this were possible, the result would hardly lead to an understanding of the individual vibronic lines in the traditional spectroscopic sense. As has already been pointed out for ethylene, we believe that an interpretation in terms of vibrational quantum numbers is neither appropriate nor useful in cases like the $^2B_2$ absorption system of NO_2. The question arises whether spectra as erratic as the one being considered here are not more adequately analyzed by focusing on the *distribution* of the various properties rather than on every single quantum state.[214] Some attempts toward such a statistical description of the $^2B_2 - {}^2A_1$ system of NO_2 will be presented in Section VI.

C. Multimode Jahn–Teller and Pseudo-Jahn–Teller Effects in $C_6F_6^+$ and BF_3^+

In this section we want to survey some interesting examples of multimode vibronic coupling involving electronic states of E symmetry. We start with the pure intrastate problem (Jahn–Teller effect). As was discussed in Section IV.A., one has to distinguish between the two cases of degenerate and nondegenerate Jahn–Teller active modes. If the vibrations are nondegenerate, the Jahn–Teller problem is very similar to the more general two-state–two-mode problem discussed in Section III.B. An example of this type has already been presented in Section V.A.3. Therefore, we limit ourselves in the following to the case of degenerate Jahn–Teller active modes.

1. Four-Mode Jahn–Teller Effect in $C_6F_6^+$

High resolution spectroscopic investigations of the $E \times (\varepsilon + \varepsilon + \cdots)$ JT effect have been performed recently in a series of chloro- and fluoro-substituted benzene cations.[230-237] If the substitution is sufficiently symmetric, the ionic ground state is of E symmetry and exhibits an irregular, anharmonic level structure characteristic of the JT effect. In the *sym*trifluorobenzene ion, for example, three of the seven vibrations of proper ε symmetry are excited with noticeable strength in the emission spectrum from the $\tilde{B}^2A_2''$ excited to the $\tilde{X}^2E''$ ground state.[232-235] Intermode interactions have been made responsible by Sears et al.[56] for anomalies in the distribution of spectral intensities. Here we select another example—the $C_6F_6^+$ radical cation.[231,236,237] All four ε_{2g} modes show up in the $\tilde{B}^2A_{2u} \rightarrow \tilde{X}^2E_{1g}$ emission spectrum, giving rise to a complicated spectral pattern that could be resolved experimentally only by advanced laser techniques. In Fig. 34a we display the emission line spectrum for vibronic angular momentum $j =$

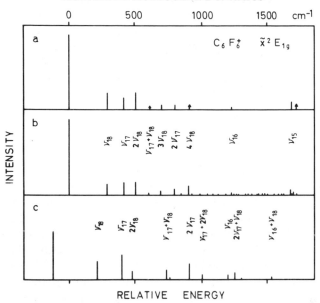

Fig. 34. Vibronic line structure of the $\tilde{B}^2A_{2u} \to \tilde{X}^2E_{1g}$ emission spectrum of $C_6F_6^+$. (a) The experimental result.[176] (b) The result of the four-mode Jahn–Teller calculation.[176] For the values of the parameters see Table VIII. (c) The result of the independent-mode treatment, using the same values of the parameters as in panel b.

1/2, which has been extracted from the experimental data.[176] For clarity, the totally symmetric mode ν_2, which is also active, has been dropped in the drawing. Since the initial state of the emission process is the $\tilde{B}^2A_{2u}$ vibrational ground state, the spectrum is described by the same formulas as were used in previous sections. The line structure is dominated by the 0-0 transition, but a number of additional lines occurring at higher energy are clearly seen and exhibit an irregular intensity distribution. The observations were well reproduced by Sears et al.[176] with the aid of a JT calculation involving all four ε_{2g} modes of C_6F_6, namely, ν_{15} (C—C stretch), ν_{16} (C—F stretch), ν_{17} (C—C—C bend) and ν_{18} (C—F bend). No ab initio calculation has been performed, but the relevant parameters have been adjusted by trial and error to fit the observations as accurately as possible. The linear coupling scheme has been employed to render the calculation tractable, although minor effects of quadratic coupling term have also been ascertained for $j = 3/2$ and $5/2$. We reproduce Sears and associates' result[176] for $j = 1/2$ in Fig. 34b and observe that a very satisfactory agreement with experiment has been achieved. This holds also for a number of other data that are not reproduced here for the sake of brevity, such as the $\tilde{B}^2A_{2u} \leftarrow \tilde{X}^2E_{1g}$ absorption

TABLE VIII

Energies, Vibrational Frequencies, and Coupling
Constants Pertinent to the JT Effect in the Ground
State of $C_6F_6^+$ as Determined by Sears et al.[a,b]

Mode	ν_{15}	ν_{16}	ν_{17}	ν_{18}
κ_i	1092	384	496	231
ω_i	1610	1215	425	265

[a]See ref. 176.
[b]All quantities are in cm^{-1}.

spectrum or $\tilde{X}$ levels of higher vibronic angular momentum $j = 3/2$ and $5/2$. Therefore, the parameters of Table VIII, which have been used in the calculation, represent an accurate description of the Jahn–Teller effect in $C_6F_6^+$. To get evidence on the nature and strength of the multimode effects, the Jahn–Teller problem has also been solved by Sears et al.[176] for each mode separately. In their Fig. 1 they normalize the results of their single-mode calculation in such a way that the 0-0 line is given the same position and intensity as the 0-0 line of the four-mode calculation. Strictly speaking, however, such a procedure is incorrect. The picture of independent modes does not conserve position and intensity of the 0-0 line, but rather the zeroth and first spectral moments. This follows from the analysis of Sections II.B and IV.A. Also, the combination lines have been omitted previously when giving the single-mode results. We have therefore included the latter and rescaled the energies and intensities in such a way that the zeroth and first spectral moments of the single-mode calculation coincide with the exact ones. The convoluted spectrum thus obtained is included in Fig. 34c. For ease of interpretation the mode assignments are indicated above the respective lines and can be correlated with those of the full spectrum. (Of course, in the latter case the assignments are only approximate; they refer to the dominant contribution to vibronic eigenstates which usually exists for low energies.)

Clearly, the independent-mode picture is unable to account for the observations. Even at the lowest energies the predicted line spacings are considerably in error. The properly normalized spectrum allows us to identify the origin of this error. Apparently, the absolute positions of the first few excited vibronic levels are reproduced quite well in the convolution approximation. It is rather the wrong position of the 0-0 line which causes the excitation energies to be invariably too high in the independent-modes picture. This contradicts the usual idea that the nonseparability of the various modes is minor for the ground state and increases with vibrational energy content. The multimode effects for the JT ground state have been explained in Section IV.A.2 in terms of the artificially high symmetries of the independent-mode treatment. For each mode the vibronic angular momentum, Eq. (4.5), is a conserved quantity leading to an N-fold rotational invariance in case of N independent JT modes. The full Hamiltonian (4.8), however, has only a single constant of motion and the reduction in zero point energy that occurs upon increasing the coupling strength is correspondingly smaller than in the convolution approximation.

At higher energies the interaction between the modes leads to a reordering of the lines and a complete redistribution of the intensities. The full spectrum consists here of a great number of lines which are too weak to be detected experimentally.[176] As before, the high line density is attributed to the reduction of the vibronic symmetries in the multimode problem. Most of the states contributing to the vibronic structure are forbidden in the independent-modes picture due to the additional artificial symmetries it possesses.

In the case of $C_6F_6^+$ the analysis of the vibronic interactions has proven possible using solely experimental information. Although the couplings are rather weak and only few vibronic levels are excited with noteable strength, considerable experimental effort has been necessary to observe the line structure unambiguously.[231,236,237] For larger coupling strengths the detection of the line structure will be the more difficult and, correspondingly, electronic bands exhibiting the JT effect are in general only poorly resolved. To gain more information on the JT effect, ab initio calculations are therefore of considerable value.

2. Two-mode Jahn–Teller Effect in BF_3^+

The outer-valence photoelectron (PE) spectrum of BF_3 has been recorded by several authors.[238-242] Between 15 and 22 eV it consists of six electronic bands. In order of increasing energy, the ionic final states corresponding to these bands are the $1A_2'$, $1E''$, $3E'$, $1A_2''$, $2E'$, and $2A_1'$ states.[239,241,242] There is thus a variety of JT and pseudo-JT coupling mechanisms which can be operative between these ionic states.[242] The bands exhibit a substantial excitation of the vibrational modes but do not permit an analysis of the vibronic coupling effects from experiment alone.

We have performed ab initio Green's function calculations for the ionization energies and the various JT and vibronic coupling constants in the way described in Section II.A.4. All four vibrational modes, transforming according to

$$\Gamma = A_1' + A_2'' + 2E' \tag{5.8}$$

have been considered. The coupling to the out-of-plane bending mode ν_2 turns out to be negligible. The other vibrations—the symmetric stretching mode ν_1 and the two degenerate modes ν_3 (asymmetric stretching) and ν_4 (asymmetric bending)—are all excited in the PE spectrum. The energies and coupling constants for three ionic states are listed in Table IX. The ab initio data have been used without further readjustment to calculate the vibronic structure of the PE bands, and satisfactory agreement with experiment has been obtained in all cases.[242] As an example, we present in Fig. 35a the calculated line structure for the second PE band of BF_3, corresponding to the $1E''$ ionic state.[242] The Hamiltonian (4.8) with $N = 2$ and the expression (2.37) for the transition probability have been used in the calculation, and the very weakly coupling totally symmetric mode ν_1 has been neglected. The exact result, obtained from the parameters of Table IX, is contrasted to that of the independent-mode treatment. Both degenerate modes ν_3 and ν_4 are excited with considerable strength, leading to a complicated line structure even in the convolution approximation. Comparing the two parts of the figure, we again observe a rise in energy of the JT ground state due to mode-mixing effects. Since, however, the next two lines of the convoluted spectrum are similarly shifted to higher energy, their line spacings are influenced only slightly by multimode effects. The similar pattern of the intensities in both spectra allows to assign the first three intense lines of the exact spectrum mainly as a ν_4 progression. On the other hand, the strong line at ~ 16.95 eV in Fig. 35b that represents the singly excited ν_3 mode is "dissolved" severely by multimode effects and can no longer be uniquely identified in the full spectrum. For energies beyond 17 eV the exact line structure is quite erratic, indicating that an assignment in terms of quantum numbers of the individual modes is generally impossible.

TABLE IX

Ab Initio Calculated Ionization Energies and Coupling
Constants of the Individual Vibrational Modes
in the Photoelectron Spectrum of BF_3[a]

	$1A_2'$		$3E'$	$1E''$	ω
E	16.26		17.48	17.14	—
κ_1	0.03		-0.06	-0.04	0.11
κ_3	—		0.11	0.29	0.18
λ_3		0.28			
κ_4	—		0.09	0.08	0.06
λ_4		0.13			

[a] The frequencies ω are the harmonic frequencies of the neutral ground state.[242] Here ν_1 refers to the totally symmetric mode; ν_3 and ν_4, to the JT modes. All quantities are in electron volts.

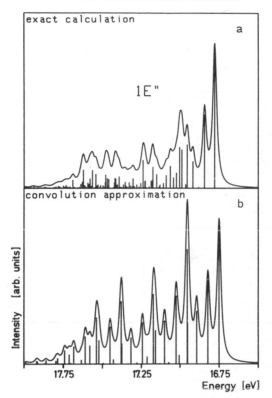

Fig. 35. The theoretically calculated $1E''$ band in the photoelectron spectrum of BF_3. (*a*) The result of the two-mode Jahn–Teller calculation. For the values of the parameters see Table IX. (*b*) The result of the independent-mode treatment, using the same values of the parameters as in panel *a*.

The importance of the mode–mode interaction is further illustrated by the perspective drawing of the $1E''$ potential energy surfaces in Fig. 36. Here we show the potential energy surfaces (4.12) in ν_3 coordinate space for three different values of the ν_4 radial coordinate R_4 as parameter.[242] The value $R_4 = 0$ corresponds to the conical intersection, the value $R_4 = \kappa_4/\omega_4$ to the minimum, and $R_4 = 2\kappa_4/\omega_4$ to the second zero of the potential energy surface of the one-mode JT problem of the ν_4 mode, respectively. The geometry $R_3 = 0$ is indicated by an open circle in each case. The potential energy surface at the top of Fig. 36 is identical to the "Mexican hat" of the one-mode JT problem of the ν_3 mode. An increasing distortion of the surface is observed toward the bottom of the figure, giving evidence of the complexity of the two-mode JT problem. The potential energy surfaces in the indepen-

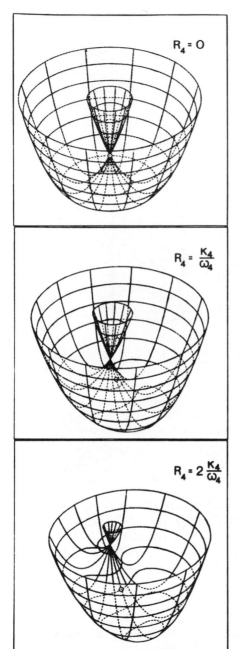

Fig. 36. Potential energy surface of the two-mode Jahn–Teller problem of the $1E''$ state of BF_3^+. The coordinates of the ν_3 mode are treated as variables, whereas the radial coordinate R_4 of the mode ν_4 is taken as the parameter. The geometry $R_3 = 0$ is marked by an open circle in each case.

dent-mode picture, on the other hand, retain their shape for arbitrary values of R_4.

As has been mentioned above, the individual lines are not yet resolved experimentally. The observed band shape, however, is in good accord with that derived from the theoretical spectrum. The multimode effects show up in the asymmetry of the band profile (which is given incorrectly by the convolution approximation). It was argued in Section IV.A.3 that the band shape is in general rather sensitive to the nonseparability of the modes. This allows the identification of multimode effects also in strong coupling cases, when the spectra are only poorly resolved. Another example of this type occurs in the PE spectrum of NH_3.[243]

3. Two-mode Pseudo-Jahn–Teller Effects in C_6H_6 and BF_3^+

Let us now turn to the $(A + E) \times (\varepsilon + \varepsilon)$ pseudo-Jahn–Teller (PJT) effect. A well-known spectroscopic example of this type of vibronic interaction are the low-energy $^3B_{1u}$ and $^3E_{1u}$ triplet states of benzene.[141,177,244] Their mutual coupling through the ε_{2g} modes ν_8 and ν_9 has been invoked to explain three different observations: the existence of the $^3B_{1u} \leftrightarrow {}^1A_{1g}$ absorption and phosphorescence spectra,[245] the nonhexagonality of the electron spin distribution in the phosphorescent triplet state,[246] and the remarkable low-frequency doublet in the $^3B_{1u} \leftarrow {}^1A_{1g}$ absorption spectrum.[247] The first fact is attributed to the borrowing of intensity from the $^3E_{1u}$ state.[248] The latter two observations are explained in terms of the symmetry lowering[249] and the associated anomalously small vibrational spacings which occur as a consequence of the repulsion of the $^3B_{1u}$ and $^3E_{1u}$ potential energy surfaces.[177] The first vibrational spacing in the $^3B_{1u}$ state is smaller by a factor of 6 than the unperturbed vibrational frequency ω_8 with which it correlates diabatically.[247] The extent of the reduction has been ascribed by Egmond and van der Waals[250] to mode mixing between the PJT active modes ν_8 and ν_9.

Another interesting example of two-mode PJT coupling occurs in the first PE band of BF_3.[242] In BF_3^+, the ground state (symmetry A_2') interacts with the second excited ionic state (symmetry E') through both ε vibrational modes, leading to a pronounced vibronic structure in the first PE band. In Fig. 37 we depict the result of the theoretical calculation,[242] using the Green's function data of Table IX. The interstate and intrastate coupling terms for the ε modes ν_3 and ν_4 have been treated on an equal footing. Simultaneous inclusion of both ε modes leads to a rather large secular matrix (on the order of $\sim 40,000$), which has been diagonalized using the Lanczos algorithm.[242] The characteristic features of the fine structure, namely, the long progression with a spacing resembling the bending frequency ω_4 and the irregular background on the high-energy side of the band, are numerically authentic. Unfortunately, no high-resolution recording of this band is available at

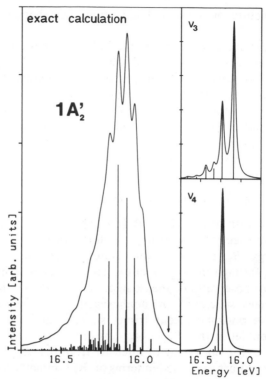

Fig. 37. The theoretically calculated $1A'_2$ band in the photoelectron spectrum of BF_3. The parameters are listed in Table IX (linewidth fwhm = 0.05 eV). For comparison, the one-mode spectra of the modes ν_3 and ν_4 are shown on the right-hand side. The arrow indicates the adiabatic ionization potential.

present and the theoretical spectrum is therefore to be considered as a prediction that must still await experimental verification. The calculation, however, is consistent with the observed bandwidth of ~ 0.25 eV, which is much larger than the spectrometer resolution.[242]

The theoretical line spacing might suggest that essentially the bending mode ν_4 is excited in the band. The single-mode calculations displayed on the right-hand side of Fig. 37 make clear that this is not at all the case. It may be seen that both vibrations are excited rather weakly, the mode ν_4 even weaker than ν_3. Recall that in an independent-mode picture the line structure is given by convoluting these single-mode spectra. Their convolution bears no resemblance to the full result, which is thus characterized by a strong mutual enhancement of the excitation of both vibrations.

The remarkable multimode effects in the first PE band of BF_3 should be adiabatic in nature, since the ground state of BF_3^+ is nondegenerate and the energy range is below the minimum of the $3E'$ ionic state (17.4 eV). These effects can indeed be rationalized by considering the potential energy surface of the ionic ground state and the energetic stabilization due to the JT modes. For the moment, we ignore the purely intrastate (JT) coupling terms and follow the analysis of Sections III.C and IV.B. It has been argued there that in the multimode PJT problem the stabilization energies are not additive. Rather, the quantity λ^2/ω must be summed over all modes and then inserted into the expression (4.20) for the energy lowering due to JT distortion. In BF_3^+ we encounter the situation where either mode alone is very near the threshold for producing a JT distortion—the bending mode ν_4 being slightly below it, the stretching mode ν_3 being slightly above it. Their combined effect, however, is far beyond threshold, producing a substantial energetic stabilization (cf. Fig. 3). Since in the present range of the parameters the stabilization energy changes in a very nonlinear way, this makes the mutual enhancement of the excitation strength not altogether surprising.

The distorted nuclear conformation is further stabilized energetically when taking the JT coupling within the $3E'$ state into account. In Fig. 38 we show the potential energy surface of the ground state of BF_3^+ that results from the combined JT and PJT coupling of the ν_3 and ν_4 vibrational modes. The surface is represented by equipotential lines in (R_3, R_4) radial coordinate space. Along the R_4 axis the potential energy increases monotonously and

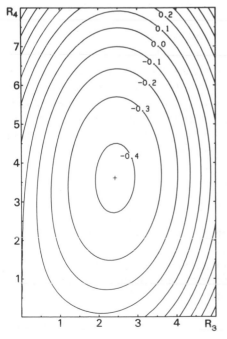

Fig. 38. Equipotential lines of the lowest potential energy surface of the $(1A_2' + 3E') \times (\nu_3 + \nu_4)$ pseudo-Jahn–Teller problem in BF_3^+. The polar angles of the two modes are both zero. The potential energy is measured in electron volts relative to the vertical ionization potential of the $1A_2'$ state.

no energetic stabilization of a JT distorted geometry occurs. Along the R_3 axis the minimum of the potential energy is at some $R_3 \neq 0$, but the energy lowering relative to $R_3 = 0$ is quite small (0.084 eV). The absolute minimum of the surface, however, occurs at completely different values of the radial coordinates, the value of R_4 being even larger than that of R_3. Also, the full stabilization energy (0.42 eV) exceeds the sum of the single-mode stabilization energies by a factor of ~ 5 and amounts to several vibrational quanta.

Upon searching for the minimum of the full potential energy surface and transforming back to usual internal coordinates, we find that one B—F bond is elongated by ~ 0.11 Å and the other two are shortened by ~ 0.055 Å compared to the equilibrium distance in the neutral ground state of ~ 1.30 Å. The angle between the two shortened bonds is widened by $\sim 24°$, and each of the other two angles is reduced by $\sim 12°$. The lowest triplet state of benzene and the ground state of BF_3^+ are thus examples of a JT-type distortion in a nondegenerate electronic state.

D. Vibronic Coupling and Conical Intersections in Linear Molecules: HCN $^+$

The Renner–Teller (RT) effect in Π electronic states of linear molecules represents probably the vibronic coupling system which has been best documented experimentally in the literature.[154,155] For a long time the interest focused exclusively on the purely intrastate coupling mechanism. The g-factor anomaly in NCO represented an early hint of perturbations of Π states through interstate vibronic coupling with Σ or Δ electronic states.[251] Subsequently, two other phenomena have been attributed to these effects. The first is a reordering of the Σ and Δ vibronic sublevels of the $\nu_2 = 1$ ($\nu_2 =$ quantum number of the bending mode) state as compared to the predictions of standard RT theory.[252] This is now commonly ascribed to the so-called g_K correction first derived by Brown by treating the $\Sigma - \Pi$ vibronic coupling in perturbation theory.[191] The other is the appearance of forbidden parallel bands in the $\tilde{A}^2\Sigma \leftarrow \tilde{X}^2\Pi$ absorption systems of many nonsymmetric triatomic radicals.[194,253] This has also been explained in terms of a (weak) interaction between the $\tilde{X}$ and $\tilde{A}$ states exerted by the bending motion.[190]

The identification of the g_K correction and of the forbidden bands in many triatomics gives evidence of the importance of the $\Sigma - \Pi$ vibronic coupling mechanism.[190,191] In most of the previous work the electronic ground state was considered. Owing to the large energy separation (2–3 eV) between the ground and excited electronic states the $\Sigma - \Pi$ vibronic coupling is rather weak and can be treated in perturbation theory. If both the Σ and Π states belong to the manifold of excited states, the vibronic coupling effects will often be stronger. Strong perturbations of this type have been reported to occur in the absorption spectrum of NCS[194] and in the PE spectra of HCN^{195} and $C_2N_2.^{254}$ In the case of NCS the origins of the $\tilde{A}(^2\Pi) \leftarrow$ $\tilde{X}(^2\Pi)$ and $\tilde{B}(^2\Sigma) \leftarrow \tilde{X}(^2\Pi)$ transitions are separated by only ~ 833 cm^{-1} and both totally symmetric modes are excited in the spectrum.[194] The vibronic structure is complicated by RT coupling within the $\tilde{X}$ and $\tilde{A}$ states as well as by $\tilde{B} - \tilde{A}$ ($\Sigma - \Pi$) vibronic interactions. The situation appears to be slightly simpler in the case of the first PE band system of HCN. Here the $\tilde{X}^2\Pi$ and $\tilde{A}^2\Sigma$ ground and first excited states of HCN^+ are also very close in energy,[195] but only a single totally symmetric mode is excited and the initial state (ground state of HCN) is of Σ symmetry and does not suffer from RT coupling. Since the PE spectrum of HCN could almost quantitatively be explained by theory,[255] we shall consider it here in some more detail.

The experimental recording of Fridh and Åsbrink[256] is reproduced in Fig. 39a. The width of the individual lines of the spectrum is smaller than the nominal spectrometer resolution. This has been achieved by numerically deconvoluting the instrumental function from the raw data.[256] The two intense lines below 13.9 eV may be associated with excitation of the C—N stretch-

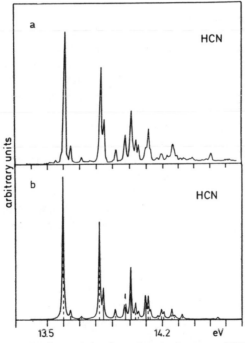

Fig. 39. The first band system of the photoelectron spectrum of HCN. (a) The envelope as deduced by Fridh and Åsbrink from the experimental recording.[256] (b) The result of the theoretical calculation. The values of the parameters are given in Table X (linewidth fwhm = 0.008 eV). The Π and Σ vibronic states are drawn as dashed and full lines, respectively.

ing mode and ionization out of the outermost occupied π molecular orbital.[257] Their "fine structure" has been interpreted as excitation of the bending mode,[195] that is, as involving the RT effect within the $^2\Pi$ ground state of HCN^+. Of further interest are the irregularities at energies beyond 13.95 eV. Since the first excited ionic state is of Σ symmetry and has been calculated to lie in this energy range,[257] these perturbations have been tentatively attributed to $\Sigma-\Pi$ vibronic coupling.[195]

It has proven possible to reproduce the observations almost quantitatively in terms of the vibronic coupling mechanism and to arrive at a detailed assignment of the individual lines.[255] We again work in the linear coupling approximation, that is, neglect the direct RT coupling within the Π state. The energies and coupling constants entering the Hamiltonian are obtained by ab initio Green's function[82] calculations, and it is confirmed that the C—H stretching mode ν_1 can be ignored in the vibronic treatment. The C—N stretching mode ν_3, on the other hand, acts as a tuning mode in the vibronic coupling. The Hamiltonian used in the calculation is thus given by Eqs. (4.23) and (4.31), with $\gamma = 0$ and $N = 1$. The parameters ΔE, λ, and ω_3 are subsequently slightly readjusted[255] to improve the agreement with the observations. As in other cases of vibronic coupling, the final values thus determined are quite close to and clearly within the error limits of the ab initio data. The need to perform the readjustment rests on the sensitivity with which the fine structure depends on the parameters. The best result has been obtained[255] with the values quoted in Table X and is reproduced in Fig. 39b. The general expression (2.37) for the transition probability has been used with $\tau_\pi = \tau_\sigma$. The vibronic secular matrices included 30 basis functions of the bending mode ν_2 and 15 basis functions of the C—N stretching mode ν_3, respectively. This leads to a dimension of 450 of the secular matrix. To facilitate interpretation of the spectrum, we have distinguished vibronic Σ and Π states by drawing them as full and dashed lines, respectively. Except for the highest energies, the agreement between theory and experiment is almost quantitative. The remaining smaller discrepancies may be due to the linear coupling approximation or to errors in the coupling constants for the C—N stretching mode ν_3 that have not been adjusted. It should be mentioned that the spectrum has also been calculated for DCN without further fitting but rather by applying the isotope rules to the parameters of Table X. Equally good agreement has been obtained as for HCN.[255] This is considered very satisfactory in view of the simplicity of the Hamiltonian, which should therefore represent a reliable description of the vibronic coupling between the lowest states of HCN^+.

The relatively small number of vibronic lines in the spectrum makes it possible to decompose each vibronic eigenstate into a few dominating zeroth-order states. With this generalization, the usual spectroscopic inter-

TABLE X
Energies, Vibrational Frequencies, and Coupling
Constants Used in Calculating the
PE Spectrum of HCN (Fig. 39)[a]

	$^2\Pi$	$^2\Sigma$	ω
E	13.800	14.000	—
κ	0.283	-0.069	0.225
λ		0.090	0.090

[a]All quantities are in electron volts.

pretation of the vibronic structure proves to be feasible in HCN^+. By zeroth-order states we mean here the vibrational levels of the potential energy surfaces for $\lambda = 0$ (diabatic surfaces). They are defined by the set of quantum numbers $\{\Lambda; v_2^{l_2}v_3\}$, where Λ and l_2 denote the electronic and vibrational[258] angular momenta, respectively; v_2 stands for the quantum number of the bending mode and v_3 for that of the C—N stretching mode. In Table XI we list the expansion coefficients of the eigenstates up to 14.11 eV, as far as they appear in the spectrum with noticeable strength. Only those zeroth-order states are included which are of interest for answering the following two questions: (i) From which zeroth-order states do the vibronic eigenstates mainly derive their spectral intensity? (These are the vibrational levels with quantum numbers $v_2 = l_2 = 0$.) (ii) Which zeroth-order states dominate in a given eigenstate? According to the table, the first two intense lines at 13.61 and 13.83 eV correspond to transitions to the $v_3 = 0$ and $v_3 = 1$ vibrational levels of the electronic ground state of HCN^+, as expected. The two weak "satellites" at 13.65 and 13.72 eV represent excitation of one and two quanta of the bending mode, respectively ($v_3 = 0$). The latter state is an almost equal mixture of vibrational angular momenta $l_2 = 0$ and $l_2 = 1$, which reflects the induced RT effect within the Π electronic state (recall that the direct RT coupling has been neglected in the calculation). Note the considerable reduction of the line spacing compared to the unperturbed bending frequency $\omega_2 = 0.09$ eV. It is also interesting that the $v_2 = 1$ line at 13.65 eV derives its intensity from a (small) admixture of the $|0; 0^0 0\rangle$ zeroth-order state, that is, by intensity borrowing from the first excited ionic state. Nevertheless, the $v_3 = 0$ bending progression can in principle also be described by treating the Π electronic state as independent from the Σ state, if one allows for an arbitrary coordinate dependence of the potential energy surfaces and transition dipole moments in the RT treatment. The next progression of lines at 13.83, 13.85, and 13.92 eV can in a similar way be assigned to levels with $v_3 = 1$ and $v_2 = 0$, 1, and 2. However, the mixing of zeroth-order states with $\Lambda = 0$ is now considerably larger than for the $v_3 = 0$ progression since they are much closer in energy than before. In the energy range beyond 13.95 eV, new structures appear which can no longer be interpreted as progressions even in the above generalized sense. Restricting ourselves to 13.95 eV $\leq E \leq 14.12$ eV, the unperturbed spectrum ($\lambda = 0$) consists here of two lines corresponding to $|\Pi; 0°2\rangle$ (at 14.11 eV) and $|\Sigma; 0°0\rangle$ (at 13.98 eV).[255] The vibronic coupling redistributes each of these zeroth-order intensities over mainly three vibronic eigenstates. The latter contain several zeroth-order states, and all are a severe mixture of Σ and Π electronic parentage. This is the effect of genuine $\Sigma - \Pi$ nonadiabatic interactions, and the line structure in this energy range cannot be understood by treating the Σ and Π ionic states as independent.

It is evident from the above analysis that the spectral features arise from an intricate interplay of both the bending and C—N stretching vibrational modes. The spectrum without excitation of ν_3 but with the same values for ω_2, λ, E_1, and E_2 as in Table X has already been presented in Fig. 20 (see Section IV.C.2). The $v_3 = 0$ and $v_3 = 1$ bending "progressions" of the full spectrum are qualitatively similar to the single-mode result, but quantitatively substantial deviations occur; in particular the line spacings between the different members of either the $v_3 = 0$ or $v_3 = 1$ bending "progression" are much larger than those of the single-mode calculation. Moreover, the strong $\Sigma - \Pi$ nonadiabatic interactions in the high-energy part of the PE band vanish almost completely if the totally symmetric mode is neglected.

The nature of the multimode effects can be visualized by suitable cuts through the adiabatic potential energy surfaces, shown in Fig. 40. For fixed bond lengths (Fig. 40a, c) the vibronic coupling leads to a repulsion of the Σ- and Π-type potential energy surfaces upon bending the molecular ion. Moreover, the Π-type surface splits into two components (induced static RT effect). The additional (totally symmetric) mode leads to a modulation of the $\Sigma - \Pi$ energy gap as shown in Fig. 40b. Therefore, the bending potential energy surfaces depend on the C—N stretching coordinate Q_g. Near the minimum of the lower surface (Fig. 40a) their energetic separation is larger and their repulsion is smaller than for $Q_g = 0$ (Fig. 40c). Correspondingly, the line spacings of the $v_3 = 0$ bending "progression" in Fig. 39 deviate less from the harmonic ones than in the single-mode calculation of Fig. 20. The other effect of the modulation is to bring the surfaces close together, which leads to a crossing at $Q_3 = 0.57$ and an energy of 14.0 eV. Apart from the additional surface V_π [see Eq. (4.27)], the topology of this crossing is the same as for nondegenerate modes: it is of the conical type.[116] The main features of the potential energy surfaces are therefore the same as for vibronic coupling systems involving nondegenerate vibrational modes. By analogy with

TABLE XI
Eigenvector coefficients for the $\tilde{X}(^2\Pi)-\tilde{A}(^2\Sigma)$ Vibronic Coupling System of HCN^{+} [a]

A. Vibronic angular momentum $K = 0$

E (eV)	13.65	13.85	13.98	14.01	14.10
$\lvert 0; 0^0 0\rangle$	-0.21	0.43	0.40	0.61	0.36
$\lvert 0; 0^0 1\rangle$	0.12	$\cdots$	$\cdots$	0.22	-0.17
$\lvert 0; 2^0 0\rangle$	-0.18	0.21	-0.35	$\cdots$	-0.42
$\lvert \pm1; 1^{\mp1}0\rangle$	0.65	0.11	$\cdots$	$\cdots$	$\cdots$
$\lvert \pm1; 1^{\mp1}1\rangle$	$\cdots$	-0.56	$\cdots$	0.36	$\cdots$
$\lvert \pm1; 1^{\mp1}2\rangle$	$\cdots$	-0.10	$\cdots$	-0.31	0.33
$\lvert \pm1; 3^{\mp1}0\rangle$	0.14	0.11	-0.12	$\cdots$	$\cdots$
$\lvert \pm1; 3^{\mp1}1\rangle$	$\cdots$	-0.11	0.47	$\cdots$	-0.13
$\lvert \pm1; 3^{\mp1}2\rangle$	$\cdots$	$\cdots$	0.12	$\cdots$	$\cdots$
$\lvert 0; 2^0 1\rangle$	$\cdots$	$\cdots$	$\cdots$	0.14	-0.17
$\lvert 0; 4^0 0\rangle$	$\cdots$	$\cdots$	$\cdots$	$\cdots$	0.20
$\lvert \pm1; 5^{\mp1}1\rangle$	$\cdots$	$\cdots$	$\cdots$	$\cdots$	-0.29

B. Vibronic angular momentum $K = 1$

E (eV)	13.61	13.72	13.83	13.92	14.04	14.06	14.11
$\lvert 1; 0^0 0\rangle$	0.97	-0.21	$\cdots$	$\cdots$	$\cdots$	$\cdots$	$\cdots$
$\lvert 1; 0^0 1\rangle$	$\cdots$	$\cdots$	0.92	-0.33	$\cdots$	$\cdots$	$\cdots$
$\lvert 1; 0^0 2\rangle$	$\cdots$	$\cdots$	$\cdots$	$\cdots$	0.66	0.23	0.61
$\lvert 1; 2^0 0\rangle$	$\cdots$	0.61	$\cdots$	$\cdots$	$\cdots$	$\cdots$	$\cdots$
$\lvert 1; 2^0 1\rangle$	$\cdots$	$\cdots$	0.12	0.52	0.21	-0.46	0.14
$\lvert 1; 2^0 2\rangle$	$\cdots$	$\cdots$	$\cdots$	$\cdots$	$\cdots$	$\cdots$	-0.31
$\lvert 1; 4^0 1\rangle$	$\cdots$	$\cdots$	$\cdots$	$\cdots$	$\cdots$	0.40	$\cdots$
$\lvert -1; 2^2 0\rangle$	0.13	0.60	-0.17	$\cdots$	$\cdots$	$\cdots$	$\cdots$
$\lvert -1; 2^2 1\rangle$	$\cdots$	$\cdots$	0.16	0.51	-0.54	$\cdots$	0.23
$\lvert -1; 2^2 2\rangle$	$\cdots$	$\cdots$	$\cdots$	$\cdots$	0.16	$\cdots$	-0.34
$\lvert -1; 4^2 1\rangle$	$\cdots$	$\cdots$	$\cdots$	$\cdots$	$\cdots$	0.39	$\cdots$
$\lvert 0; 1^1 1\rangle$	$\cdots$	$\cdots$	$\cdots$	$\cdots$	-0.19	$\cdots$	0.22
$\lvert 0; 1^1 0\rangle$	-0.14	-0.21	-0.24	-0.35	-0.29	$\cdots$	0.40
$\lvert 0; 3^1 0\rangle$	$\cdots$	$\cdots$	$\cdots$	-0.24	$\cdots$	-0.27	$[0.17]^{b}$
$\lvert 0; 5^1 0\rangle$	$\cdots$	$\cdots$	$\cdots$	$\cdots$	$\cdots$	-0.25	$\cdots$

[a] The expansion is in terms of the vibrational levels $\lvert \Lambda; v_2^{\ell_2} v_3\rangle$ of the diabatic surfaces, defined by $\lambda = 0$. The states with vibronic angular momentum $K = \Lambda + \ell_2 = 1$ are degenerate with the states with $K = -1$. Only those zeroth-order states are included which have largest coefficients in a given eigenvector; otherwise the entry is marked by dots.

[b] For $\lvert 0; 3^1 1\rangle$.

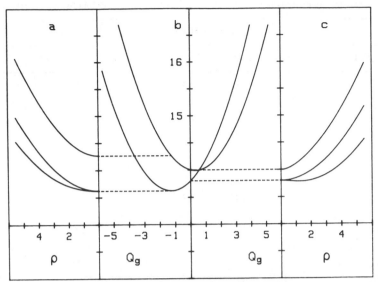

Fig. 40.   Potential energy curves of the ground and first excited states of HCN$^+$ as obtained from the parameters of Table X. (*a*) The bending potential energy curves for the ionic optimum bond length ($Q_g = -1.26$). (*b*) The C—N stretching potential energy curves for linear geometry ($\rho = 0$). (*c*) The bending potential energy curves for bond lengths appropriate to the ground state of the neutral molecule ($Q_g = 0$).

the analysis of Section III.B, one expects a strong enhancement of the interstate nonadiabatic effects in the energy range at and above the intersection. This has indeed been found in the analysis of the spectrum.

The nonseparability of the modes is also of importance for the structure of the molecular ion. If the bond lengths were kept frozen at their values in the neutral molecule, the vibronic coupling would lead to a non-linear-equilibrium geometry of the ion (see Fig. 40c). The additional mode of relaxation leads to a weakening of the vibronic coupling strength near the optimum ionic bond lengths and thus to a linear equilibrium geometry of HCN$^+$. The interatomic distances are found to be $r(CH) = 1.09$ Å and $r(CN) = 1.21$ Å, which are 0.025 Å and 0.07 Å, respectively, larger than in the neutral molecule.[255] The case of HCN$^+$ represents a good example of the general rule, established in Section III, that the additional totally symmetric modes tend to enhance the nonadiabatic effects but quench the static distortion due to the vibronic coupling. We would like to recall here the example of BF$_3^+$, where the excitation of two coupling modes acts precisely in the opposite way and greatly enhances the distortion. This comparison illustrates the profound difference between both types of multimode effects.

E. Other Vibronic Coupling Systems

Apart from the multimode Jahn–Teller effect, the above applications were dealing mainly with the impact of totally symmetric vibrational modes on vibronic coupling systems. This derives from our primary interest in strong nonadiabatic effects. It has been argued in Section III that the totally symmetric modes are usually more effective in enhancing the nonadiabatic effects than are additional coupling modes: it is the former vibrations which exert a tuning effect and bring the adiabatic potential energy surfaces close together, whereas the latter ones mainly pull them further apart.

It should be mentioned that this tuning also leads to an interesting type of borrowing effect. Progressions of totally symmetric modes in vibronically induced spectra do not reflect the geometry of the electronic state that borrows the intensity, but rather the geometry of the electronic state that lends the spectral intensity.[133] An example of this type has been reported[135] to occur in the $\tilde{A} \leftarrow \tilde{X}$ absorption transition of SO_2. This transition is dipole forbidden and gains its intensity from vibronic coupling with the $\tilde{B}$ electronic state via the antisymmetric stretching mode. It exhibits a long progression in the bending mode, which is much longer than expected from the Franck–Condon principle and the difference in OSO angle between the $\tilde{X}$ and $\tilde{A}$ electronic states. This Hennecker et al.[135] explained by noting that the $\tilde{B}$ state, which lends the spectral intensity to the $\tilde{A}$ state, has quite a different equilibrium OSO angle than the $\tilde{X}$ and $\tilde{A}$ electronic states. A similar but less pronounced effect has also been identified[180] in the case of the $2A'_1$ state of BF_3^+.

1. Several Nondegenerate Coupling Modes

A considerable amount of work in the literature has also been devoted to the case of several nondegenerate vibronically active (i.e., coupling) modes. We therefore shall mention here a few examples wherein dynamical calculations have been carried out.

The first electronic absorption transition of the benzyl radical, the $^2A_2 \leftarrow {}^2B_2$ transition at 2.73 eV, has been analyzed by Cossart–Magos and Leach.[148] They attributed the unusual vibronic structure and the lack of the mirror-image relationship between absorption and emission to vibronic coupling between $^2A_2$ and the lowest excited $^2B_2$ electronic states. Two modes of B_1 symmetry, ν_{6b} and ν_{18b}, have been assigned in the band. Including both in the vibronic coupling calculation, Cossart–Magos and Leach were able to reproduce the observed vibronic origins accurately and give a fair account of the spectral intensities.[148] Since the interacting electronic states are separated by only 462 cm^{-1}, the nonadiabatic interactions should be quite strong. This expectation is in accord with the numerically computed eigenvectors, which also display a substantial mixing between both vibronically active modes.[148]

Along similar lines, Ross and coworkers (see Lacey et al.[146] and Chappell and Ross[147]) have analyzed electronic transitions in azulene and quinoxaline. In the $n \to \pi^*$ spectrum of quinoxaline, three modes of B_1 symmetry are weakly excited. They can couple the $B_1(n, \pi^*)$ to the $A_1(\pi, \pi^*)$ state, which is about 4000 cm^{-1} higher in energy. All three modes have been included in the vibronic calculation by Chappell and Ross.[147] Besides a reduction in vibrational spacing they obtain—as the dominant multimode effect—a substantial transfer of spectral intensity from the more strongly coupled mode $2b_1$ to the less strongly coupled mode $4b_1$. This is a regular feature of multimode coupling.[150] In the case at hand it explains why the intensity ratio $I(4b_1)/I(2b_1)$ of the corresponding $\Delta v = 1$ transitions is much larger in absorption than in emission.[147]

The second excited singlet $\tilde{B}(A_1)$ of azulene is severely perturbed by vibronic coupling with the fourth excited $\tilde{D}(A_1)$ state, which is about 8000 cm^{-1} higher in energy.[146] This leads to a complicated $\tilde{B} \leftarrow \tilde{X}$ absorption spectrum with an unusually strong dependence of the vibronic structure on the crystalline host into which the molecule is embedded. Since the coupling modes are now totally symmetric, they can simultaneously act as tuning modes (i.e., exhibit "Franck–Condon" and "vibronic" activity). This complication was, however, ignored by Lacey et al., who treated the problem only in terms of two coupling modes.[146] The calculation accounts again for different intensity ratios of thé $\Delta v = 1$ transitions in absorption and in emission and for part of the additional complicated structure. Moreover, the medium dependence of the spectral intensities is rationalized by noting that the host influences the $\tilde{B} - \tilde{D}$ energy gap of the interacting states.[146]

As the last example in this list we mention the coupling of the S_0 and S_1 states of s-tetrazine[149] by the two B_{3u} vibrations ν_{16b} and ν_{17b}. In addition to the phenomena quoted above, the coupling gives rise to an interesting intensity borrowing effect.[149] Since S_0 is the electronic ground state, the borrowing of intensity from S_1 can lead to extraordinarily strong infrared transitions of the active mode. This has indeed been observed[259] for ν_{16b}.

We now turn to the case where the coupling modes simultaneously modulate the energy gap between the uncoupled (i.e., diabatic) electronic states. Within the framework of our model Hamiltonian (2.25, 2.30), this occurs when the interacting electronic states have the same spatial symmetries and are coupled by totally symmetric vibrations. This leads to a loss of the "internal" vibronic symmetries and to several complications compared to the coupling by non-totally symmetric modes.[42,45] For example, the vibronic secular matrix no longer factorizes into two submatrices, which leads to an interference of the transition moments of the interacting electronic states (see Section II.B.3). The potential energy surfaces and the possible double minimum are asymmetric. If two or more vibrations are involved, there is a conical intersection of the two potential energy surfaces. Its location in coordinate space is, however, not fixed by symmetry and is in this sense accidental. It may be located anywhere in coordinate space.

Vibronic coupling by two totally symmetric modes has been identified[260] in the valence ionization spectrum of N_2O. The interaction occurs between the configuration with a vacancy in the 1π molecular orbital and a linear combination of configurations where the 2π molecular orbital is doubly and the 3π (virtual) molecular orbital is singly occupied.[260] In a shorthand notation these are referred to as the $(1\pi)^{-1}$ and the $(2\pi)^{-2}(3\pi)$ electronic configurations. They interact statically (configuration mixing) and dynamically (vibronic coupling). The $(2\pi)^{-2}(3\pi)$ configuration acquires spectral intensity by both mechanisms and thus leads to severe perturbations of the third photoelectron band of N_2O. The vibronic coupling is caused mainly by the asymmetric stretching mode ν_3, but the symmetric stretching mode ν_1 also influences the calculated line structure.[260] The numerical data indicate that the nonseparability of two totally symmetric coupling modes is a very pronounced effect that is expected to show up also in other examples. Generally, this type of vibronic coupling should often occur in molecules of low symmetry, since this increases the probability that neighboring electronic states transform according to the same irreducible representation of the molecular point group.

2. Multistate Vibronic Coupling

Throughout this work we have been concerned with the influence of several vibrational modes on the interaction of two electronic states. To conclude this section, we briefly consider the opposite situation and call attention to the vibronic coupling of several electronic states. The two-state problem is clearly of relevance for lower excitation energies since here the density of electronic states is small and a close approach of different potential energy surfaces is a relatively rare phenomenon. For higher excitation energies the density of electronic states grows rapidly and many surfaces will undergo several curve crossings; this will couple several electronic states simultaneously. Another specific situation is the crossing of a valence state with a series of Rydberg states.[261] Since the latter are practically parallel, the crossing will usually involve several members of the series.

Multistate vibronic coupling between two valence and three Rydberg states of $^2\Pi$ symmetry in NO has been treated by Gallusser and Dressler and deperturbed potential energy curves and interaction constants have been deduced from the analysis.[262] Another well-known example of interaction between a valence state and a series of Rydberg states occurs in H_2.[263] The valence excited state is the $2\,^1\Sigma_g^+$ state with the dominating electron configuration $(2p\sigma_u)^2$. The series of Rydberg states can be characterized by the electron configurations $(1s\sigma_g)(ns\sigma_g)$, for $n \geq 2$, and $(1s\sigma_g)(md\sigma_g)$, for $m \geq 3$. Although at low energies the vibronic coupling is a two-state problem,[264] at sufficiently high energies it will involve several Rydberg states simultaneously.

The inner-valence ionization spectra of molecules provide another example of densely lying electronic states.[288,289] In the inner-valence ionization process of N_2, for example, the single-hole configuration with a vacancy in the $2\sigma_g$ orbital interacts with many higher excited ionic configurations.[265,290] Their potential energy curves differ considerably in shape, which gives rise to pronounced multistate vibronic coupling effects. The resulting peaks in the spectrum can no longer be identified with the electronic states of the molecular ion nor with vibrational excitation. Rather, they represent the joint action of both degrees of freedom.[265] Model studies have shown that in multistate vibronic coupling systems one often gets agglomerations of lines far from the vertical excitation energies, which tend to fill in the gaps in the fixed-nuclei (vertical) spectrum with spectral intensity.[266] Another typical feature is the formation of regular peaks of the envelope with a spacing different from the vibrational frequency. Each of the peaks accommodates several or many vibronic lines, usually because of strong nonadiabatic effects.[266] Considerable progress both in experimental techniques and in accuracy of ab initio calculations will be required to arrive at a more detailed understanding of these phenomena in actual molecules.

VI. STATISTICAL BEHAVIOR OF VIBRONIC ENERGY LEVELS

In diatomic molecules, where the density of vibronic energy levels is relatively low, we are usually in the position to analyze in detail properties of individual energy levels. Such a detailed analysis requires a much greater effort for polyatomic molecules and may become prohibitively difficult in some cases of even the smallest polyatomic molecules, for example, triatomic molecules. The experimental $^2B_2 \leftarrow\,^2A_1$ excitation spectrum of NO_2, for instance, and its analysis in terms of nonadiabatic effects presented in the preceding sections make it clear beyond doubt that the observed individual energy levels are not amenable to a simple assignment and interpretation. This also holds for many other systems that exhibit strong multimode vibronic coupling effects. The excitation spectra of these systems are rather complicated and exhibit a vast number of irregularly spaced vibronic lines. Here, the question suggests itself as to whether it is useful at all to investigate detailed properties of the observed individual energy levels. An alternative approach is to investigate statistical properties of the levels and their intensities. One resorts to statistical studies because detailed properties of the spectrum are not really amenable to calculation. However, even in systems where the individual levels and their quantum descriptions are available, statistical methods are often essential for a more complete understanding.[214]

Wigner[267,268] has introduced an ensemble of real symmetric matrices whose matrix elements obey certain statistical laws. Wigner's ensemble is called the Gaussian orthogonal ensemble (GOE). This and other *random-matrix ensembles* supply models for the behavior of complicated systems.[214] Instead of averaging a quantity over the spectrum (or a part of it) of a single large matrix that describes the system under consideration, averaging over the ensemble is used. Some theoretical justifications for the use of random-matrix ensembles are given by Pandey[269] and Brody et al.[214] An important advantage of the GOE is that the ensemble-averaged value of certain measures of the energy-level fluctuations (i.e., departures from uniformity in the spectrum) and of the fluctuations of the transition strengths (intensities) are calculable analytically. These measures as well as other related quantities determine the statistical behavior of the eigenvalues and eigenvector components of the matrices associated with the random-matrix ensemble and should be compared with their counterparts deduced from experiment. Random-matrix ensemble methods have been extensively applied to nuclei.[214,270] It has been found that the statistical properties of the energy levels and transition strengths are well described by the GOE. This applies to experimentally measured levels as well as to results of extended shell-model calculations. Very recently it has been demonstrated that the statistical behavior of the energy levels of neutral and ionized rare-earth atoms is also in agreement with the predictions of the GOE.[271]

Most useful for investigation of the statistical behavior of energy levels is a long sequence of states, all with the same symmetry (symmetry quantum numbers) and spanning a narrow energy range. When studying atoms one is thus confined to heavy atoms and/or to highly excited electronic states. Owing to their additional degrees of freedom, molecules provide a class of systems naturally suitable for statistical investigations. As mentioned above, even the smallest polyatomic molecules at low energy (e.g., the *first* electronic excitation of NO_2) may exhibit extremely complex spectra with a relatively high density of energy levels characterized by the same quantum numbers. The energy levels of coupled Morse oscillators in the strongly coupled regime have recently been shown to be amenable to statistical studies.[272]

Let us start with the nearest-neighbor spacing distribution $P(S)$; then $P(S)\,dS$ is the number of nearest-neighbor energy spacings in the small interval $(S, S + dS)$. We may also consider $P(S)\,dS$ as the probability that, given a level at E, the *next* level be in the interval $(E + S, E + S + dS)$. Then we have

$$P(S)\,dS = P_1(S)P_2(S)\,dS \qquad (6.1)$$

where $P_1(S)$ is the probability that there are no levels between E and $E + S$; that is,

$$P_1(S) = 1 - \int_0^S P(x)\,dx \qquad (6.2)$$

and $P_2(S)\,dS$ is the conditional probability that the interval of length dS contains one level, when no levels are contained in $(E, E + S)$. For arbitrary functions $P_2(S)$ one immediately obtains[268]

$$P(S) = P_2(S)\exp\left\{ -\int_0^S P_2(x)\,dx \right\} \qquad (6.3)$$

The nearest-neighbor distribution (6.3) is normalized to unity and the mean spacing D is given by

$$D = \int_0^\infty SP(S)\,dS \qquad (6.4)$$

For a random sequence of energy levels the conditional probability will be independent of whether or not there is a level in the interval $(E, E + S)$; that is, $P_2(S)$ is a constant. From (6.3) one finds the Poisson distribution

$$P(S) = \left(\frac{1}{D}\right)e^{-S/D} \qquad (6.5)$$

Assuming, on the other hand, a linear repulsion of the energy levels defined by $P_2(S) \sim S$, one obtains the Wigner distribution

$$P(S) = \left(\frac{\pi S}{2D^2}\right)e^{-\pi S^2/(4D^2)} \qquad (6.6)$$

It now becomes clear why the sequence of levels to be analyzed should contain levels of one symmetry only. Levels of different quantum numbers can be thought of as emerging from different matrices and will thus be uncorrelated; that is, in the limit of many different quantum numbers we always encounter the Poisson distribution.

To make contact with the GOE we first define it more precisely. The d-dimensional GOE is an ensemble of d-dimensional real symmetric matrices

M with distinct matrix elements M_{ij} independently distributed as Gaussian random variables. The distributions of the diagonal elements M_{ii} and off-diagonal elements M_{ij}, for $i \neq j$, are $\rho(x, \sqrt{2} v)$ and $\rho(x, v)$, respectively, where $\rho(x, v) = (2\pi v^2)^{-1/2} \exp[-x^2/(2v^2)]$. It is illustrative to calculate the nearest-neighbor spacing distribution for the two-dimensional GOE.[214] The eigenvalue spacing S of a two-dimensional matrix **M** is given by

$$S = \left[(M_{11} - M_{22})^2 + 4M_{12}^2 \right]^{1/2} \tag{6.7}$$

The nearest-neighbor distribution follows from

$$\int P(S) \, dS = \int \int \int dM_{11} \, dM_{22} \, dM_{12} \rho \left(M_{11}, \sqrt{2} \, v \right) \rho \left(M_{22}, \sqrt{2} \, v \right) \rho \left(M_{12}, v \right)$$
$$\tag{6.8a}$$

Introducing new variables $(M_{11} + M_{22})/2$, $\rho = \text{arctg}[(M_{11} - M_{22})/(2M_{12})]$ and S, and integrating over the former two yields

$$P(S) = \left(\frac{S}{4v^2} \right) e^{-S^2/(8v^2)} \tag{6.8b}$$

From Eq. (6.4) it follows that $(2\pi v^2)^{1/2} = D$ and we recover the Wigner distribution (6.6). This distribution, which has been derived above assuming a linear repulsion of the eigenvalues of a single matrix, has been rederived here as a distribution of the level spacings of the ensemble members. In general $(d > 2)$, the Wigner spacing distribution is an accurate approximation to the spacing distribution predicted by the GOE.[214]

Figure 41 shows two nearest-neighbor spacing histograms for the vibronic energy levels of B_2 vibronic symmetry of NO_2. The first histogram (Fig. 41a) has been drawn using the 140 levels given by Smalley et al.[222] that lie in the energy range from 14900 to 17500 cm^{-1} in the excitation spectrum of NO_2 (see Fig. 33 in Section V.B). These levels represent vibronic band origins and were extracted from the fluorescence excitation spectrum of rotationally cooled NO_2. Although they have not been measured directly, we refer to them as "experimental" levels. It should be noted, however, that some of the levels might be subject to considerable error since the rotational structure could not be assigned unambiguously.[222] Our calculation on the spectrum of NO_2 described in Section V.B predicts 201 energy levels in the same energy range. These levels have been used to construct the second histogram (Fig. 41b). In view of the relatively small number of levels, the data are in agreement with the Wigner distribution (6.6) and clearly display a level repulsion.

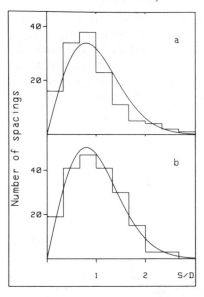

Fig. 41. Nearest-neighbor spacing histogram for energy levels of NO_2 with vibronic symmetry B_2. (a) Experimental energy levels of Smalley et al. in the energy range between 14900 and 17500 cm^{-1} above the ground state (140 levels). (b) Calculated levels for the same energy range (201 levels). In both cases the Wigner distribution is also shown.

To discuss fluctuations we have to compare the spectrum with a fluctuation-free spectrum (smoothed spectrum) that derives from it. In Fig. 42 a staircase plot of the cumulative number of energy levels N as a function of energy E is shown. The plot contains the 505 energy levels of the ethylene cation (vibronic symmetry B_{2g}) calculated to lie in the energy range between 12.1 and 13.6 eV. For the details of the calculation see Section V.A. This staircase function can be fitted by the smooth function

$$N(E) = \sum_{i=0}^{3} a_i E^i \qquad (6.9)$$

which appropriately describes the secular behavior of the original eigenvalue density dN/dE of the three harmonic vibrational modes considered in the calculation. We have obtained $a_0 = 3.31 \times 10^4$, $a_1 = -7.26 \times 10^3$, $a_2 = 5.06 \times 10^2$, and $a_3 = -10.9$. Most fluctuation measures are defined with respect to a *constant* density of states of the smoothed spectrum, and we therefore transform the increasing level density to a constant level density. The smooth function in Eq. (6.9) enables us to carry out this transformation $N(E) \rightarrow N(\tilde{E}) = b_0 + b_1 \tilde{E}$, where the energy $\tilde{E} = b_1^{-1}[N(E) - b_0]$ defines the new scale. The straight line $N(\tilde{E})$ is also plotted in Fig. 42 and the staircase plot fluctuates about it.

The nearest-neighbor spacing histogram as well as the rth-nearest-neighbor spacing histograms should be evaluated using the data of the

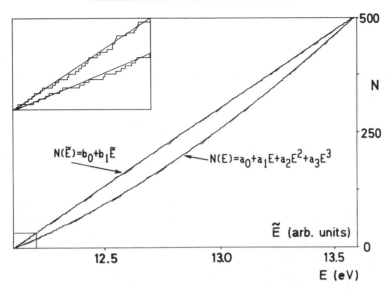

Fig. 42. Two staircase plots of the cumulative number N of levels as a function of energy for the calculated levels of $C_2H_4^+$ with vibronic symmetry B_{2g} (505 levels are included). To each plot there corresponds a smooth (fluctuation-free) curve. The plot corresponding to the nonlinear smooth curve represents the original data. The second staircase plot fluctuating about the straight line $N(\tilde{E})$ is the cumulative number of levels with the secular variation in level density unfolded. The inset shows the lower left corner of the figure on an enlarged scale.

staircase plot corresponding to $N(\tilde{E})$. In our first example (excitation spectrum of NO_2) the calculated energy levels belong to a very small energy interval and the level density dN/dE is constant offhand. Figure 43 shows the nearest-, third-nearest-, and fifth-nearest-neighbor spacing histograms of the calculated ethylene cation data. The nearest-neighbor spacing histogram compares satisfactorily with the Wigner distribution, but it compares better with a Brody distribution discussed below. The fifth-nearest-neighbor spacing histogram resembles a Gaussian distribution. It is well-known that rth-nearest-neighbor distributions become Gaussian for large r (and large dimensionality d of the members of the random matrix ensemble).[214] Since the centroid of the rth-nearest-neighbor spacing distribution is rD, the only significant measure connected with the distribution for large r is the width $\sigma(r)$. The "speed" by which the rth-nearest-neighbor distribution approaches (with increasing r) a Gaussian distribution is a measure for long-range correlations between energy levels.

To analyze energy levels, one usually calculates characteristic parameters of the underlying distributions (called fluctuation measures or statistics). In

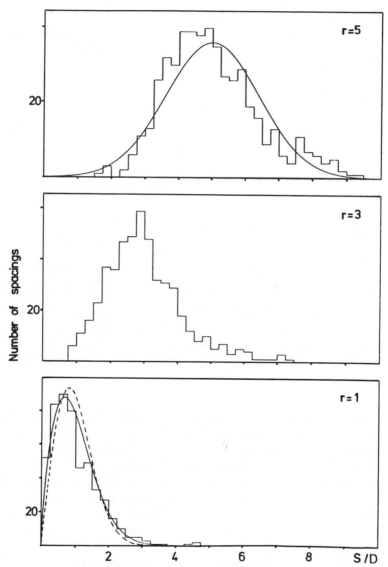

Fig. 43.   Three rth nearest-neighbor spacing histograms ($r = 1$, 3, and 5) for the 505 calcu-lated energy levels of $C_2H_4^+$ corresponding to Fig. 42. For $r = 1$ (nearest-neighbor spacing histogram) the histogram is compared with the Brody distribution $\alpha x^\omega \exp(-\beta x^{1+\omega})$, for $x = S/D$, with $\omega = 0.71$ (full line) and with the Wigner distribution (dashed line). For $r = 5$ the histogram is compared with a Gaussian distribution.

the literature, several fluctuation measures have been discussed in detail and computed according to random-matrix theory.[214,273,274] Of these we mention here only a few. The Δ_3 statistic, introduced by Dyson and Mehta,[273] is a measure of the fluctuations of the cumulative number of levels $N(\tilde{E})$ about a best-fitting straight line in an interval of length $2L$:

$$\Delta_3 = \frac{1}{2L} \min_{b_0, b_1} \int_{\tilde{E}-L}^{\tilde{E}+L} \left[N(\tilde{E}) - (b_0 + b_1\tilde{E}) \right]^2 d\tilde{E} \qquad (6.10)$$

The parameters b_0 and b_1 are chosen to minimize the integral. Obviously Δ_3 depends on the number n of levels contained in the interval ($2L = nD$). For the GOE one obtains[273]

$$\Delta_3(n) = \pi^{-2}(\ln n - 0.0687) \qquad (6.11)$$

with a standard deviation of 0.11. The Δ_3 statistic reflects both long- and short-range correlations between the levels. It should be mentioned that evaluation of Δ_3 from Eq. (6.10) may be sensitive to the procedure used to unfold the spectrum,[275] that is, to map to a constant the secular (fluctuation-free) level density.

Correlation coefficients between spacings yield important information about a spectrum.[276,277] Out of this class of fluctuation measures we briefly discuss here the correlation coefficients for adjacent rth-order spacings. The deviation of the rth-order spacings from their mean value is

$$S_i(r) = (E_{i+r} - E_i) - rD \qquad (6.12a)$$

For a given spectrum the variance of these quantities is defined as usual by a spectral averaging

$$\sigma^2(r) = \frac{1}{n} \sum_{i=1}^{n} [S_i(r)]^2 \qquad (6.12b)$$

When one is studying matrix ensembles, the variance is, of course, not calculated by (6.12b) but via an ensemble averaging for a fixed i and is assumed to be independent of i (stationarity). The correlation coefficients for adjacent rth-order spacings are defined by ($r \geq 1$)

$$C(r) = \frac{\sum_i S_i(r) S_{i+r}(r)}{\sum_i [S_i(r)]^2} \qquad (6.13a)$$

They reflect the correlations between levels E_i and E_{i+r}. Via Eq. (6.12) it is easily verified that these coefficients can be expressed by variances σ^2 alone:

$$C(r) = \frac{\sigma^2(2r)}{2\sigma^2(r)} - 1 \qquad (6.13b)$$

For the GOE the correlation coefficients are simply given by the numbers[277] -0.271, -0.333, and -0.359 for $r = 1$, 2, and 3, respectively, and $C(r)$ approaches -0.5 for large r.

In Table XII we have collected Δ_3 and $C(1)$ for the experimental and calculated data on NO_2 corresponding to Figs. 41a and 41b, respectively, as well as for the computed $C_2H_4^+$ data (Figs. 42 and 43). Since the spectrum corresponding to Fig. 42 contains as many as 505 levels, we cannot expect to have unfolded it to a high accuracy by the simple *ansatz* (6.9). We, therefore, also consider in Table XII a smaller set of levels along which the local density is nearly constant offhand. One such set lies, for example, in the energy range between 12.95 and 13.1 eV and contains 55 levels. For this set the optimum parameters (in the sense of a least-square fit) of the smooth cumulative number of levels are $a_0 = -1.76 \times 10^6$, $a_1 = 4.07 \times 10^5$, $a_2 = -3.14 \times 10^4$, and $a_3 = 8.07 \times 10^2$. To compute Δ_3 and $C(1)$ for the larger set of 505 levels, we have used the unfolding procedure suggested by Wong and French[294] where the density of states is locally averaged making use of several levels nearest to each individual level. For both sets of levels of $C_2H_4^+$ the calculated Δ_3 values compare well with the GOE results and the values of the correlation coefficients $C(1)$ are somewhat smaller than predicted by GOE. It is interesting to note that for a random sequence of n levels one finds[273] $\Delta_3 = n/15$ and $C(1) = 0$. For the data of NO_2 both Δ_3 and $C(1)$ agree very nicely with the predictions of GOE. This agreement should, however, be considered with care. As mentioned above these data might be subject to errors and the statistical tests are sensitive to the quality of the data. Indeed, the corresponding cumulative number of experimental levels is not smooth enough as a function of energy to allow for an unambiguous unfolding procedure as is possible for the data on $C_2H_4^+$. We, therefore, concentrated here on the smaller energy range 16600–17300 cm^{-1}, which is amenable to an unobjectionable unfolding procedure via the *ansatz* (6.9). Unfortunately, the calculated energy levels of NO_2 in this energy range are also subject to errors: having used the Lanczos diagonalization procedure (see Sections II.B.3, V.B, and VII.A) with only 10,000 iterations, the corresponding eigenvalues of the 18,630 dimensional matrix cannot be considered to be fully converged. Only the first eigenvalues (about 200) are fully converged. No such

TABLE XII

Statistics of Pure Sequence Experimental and Theoretical Data[a]

	n	Energy (eV)	Δ_3	Δ_3^{GOE}	C(1)
NO$_2$ (experiment[b])	46	2.06–2.14	0.38	0.38±0.11	−0.32
NO$_2$ (theory[b])	201	1.85–2.17	0.44	0.53±0.11	−0.23
C$_2$H$_4^+$ (theory)	55	12.95–13.1	0.49	0.40±0.11	−0.17
C$_2$H$_4^+$ (theory)	505	12.1–13.6	0.76	0.62±0.11	−0.14

[a] The results are compared with the predictions of GOE: Δ_3 is from Eq. (6.11) and C(1) = −0.27; n is the number of levels considered.

[b] The underlying energy levels are subject to errors (see the text).

problems arise for the data on C$_2$H$_4^+$, where the corresponding matrix is of much lower dimension and has been diagonalized accurately (see Section II.B.3).

An important criterion for the applicability of statistical methods is that the energy levels should represent complicated states. As we have demonstrated in preceding sections, the molecular vibronic states in conical intersection situations can be regarded as being complicated states and are thus amenable to statistical analysis. In this section we have presented evidence that statistics of the calculated and measured data compare well with the predictions of random-matrix theory (GOE). These GOE predictions also compare well with nuclear and atomic data and seem to have some "universal" character. To gain more insight into the specific statistical properties of molecular states, we should also analyze the data in relation to other types of random-matrix ensembles and models. For instance, the nearest-neighbor spacing histogram in Fig. 43 compares better with the Brody distribution[278]

$$P(S) = \alpha S^\omega \exp\left(-\beta S^{1+\omega}\right)$$

$$\alpha = (1+\omega)\beta; \qquad \beta = \left[D^{-1}\Gamma\left(\frac{2+\omega}{1+\omega}\right)\right]^{1+\omega}$$

(6.14)

than with the Wigner distribution, the optimum value of ω being 0.71. The distribution (6.14) includes the Poisson and Wigner distributions as special cases and is easily obtained from the general expression (6.3) by assuming a non-linear-level repulsion $P_2(S) = \alpha S^\omega$. Although very preliminary, the present results are promising and call for more work and suitable molecular data on energy levels as well as on their intensities and lifetimes.

VII. CONICAL INTERSECTIONS AND TIME EVOLUTION OF FLUORESCENCE

The major body of this work has been devoted to spectroscopic investigations of vibronic coupling. Of course, there are many other interesting processes where vibronic coupling effects are of importance, two of them being radiative and nonradiative decay. When only a bound molecular level structure is involved, these decays should be amenable to a description in terms of our model Hamiltonian. We therefore complement our above studies by analyzing in the following discussion the decay processes in two vibronic coupling systems for which the spectra have already been discussed in Section V: $C_2H_4^+$ and NO_2. Besides explaining experimental findings, we will in this way also gain further insight into the nature of the nonadiabatic interaction and the nuclear dynamics.

In all vibronic coupling systems treated in this work, the density of vibronic states is much smaller than radiative decay times. Therefore we are not concerned here with the "statistical limit" but rather with small to intermediate-sized molecules, in the usual nomenclature.[103,292] When the initial state in the decay process is a single vibronic level, the decay is radiative and its characteristic feature is the quenching of the radiative rate (the Douglas effect[123]). This situation applies, for example, to NO_2, where the bandwidth of the exciting radiation is smaller than typical energy spacings between adjacent vibronic levels. When the initial state in the decay process is a coherent superposition of several or many vibronic levels, one gets, in addition, for short times the phenomenon of nonradiative decay.[291] We shall argue below that this situation is realized in the photoionization experiment and analyze the nonradiative decay process for the example of $C_2H_4^+$.

A. Quenching of the Radiative Rate: NO_2 and $C_2H_4^+$

One of the numerous anomalies observed in NO_2 is the quenching of the radiative rate. It has been found by many workers[224,279-286] that the radiative decay rate for emission in electronically excited NO_2 is some one to two orders of magnitude smaller than expected from the integrated absorption coefficient. For example, exciting NO_2 with radiation near $\lambda = 600$ nm, the observed radiative decay times range from ~ 20 to ~ 200 μsec.[224] The value deduced from the integrated absorption coefficient, on the other hand, is only ~ 1.5 μsec.[224] The argument linking these two quantities together is based on the assumption that the initial and final electronic states are well separated energetically and that the upper electronic state can be described within the adiabatic approximation.[287] This has lead Douglas to the suggestion that the quenching of the radiative rate, observed in NO_2 and other molecules, is caused by nonadiabatic interactions.[123]

The analysis of the visible absorption spectrum of NO_2 in Section V.B established beyond doubt that there are indeed strong nonadiabatic interactions between the $^2A_1$ ground and the $^2B_2$ first excited electronic states. They are responsible for the complicated vibronic structure of the $^2B_2 \leftarrow ^2A_1$ absorption spectrum. We have therefore performed a theoretical calculation in order to clarify to what extent these interactions can also explain the quenching of the radiative rates of the B_2 vibronic states.[89]

The calculation is based on the concept of the radiative damping matrix as described in Section II.C. For the present example this leads to the following form of the effective Hamiltonian (2.64):

$$\mathcal{H}_{\text{eff}} = \mathcal{H} - \frac{i}{2} \begin{pmatrix} 0 & 0 \\ 0 & \Gamma \end{pmatrix} \qquad (7.1)$$

The first term, $\mathcal{H}$, is the vibronic coupling Hamiltonian used for the calculation of the $^2B_2 \leftarrow ^2A_1$ absorption spectrum [Eq. (3.22) with $N = 2$ and $M = 1$]. The second term in (7.1) is the radiative damping matrix; Γ is the golden rule radiative decay rate for the upper diabatic electronic state and is taken to be a constant in the following analysis. Then, in the absence of vibronic coupling ($\lambda = 0$), the radiative width of the vibrational levels of the upper diabatic electronic state (briefly called upper vibrational levels) is also given by Γ. This quantity is the unquenched radiative decay rate. The vibrational levels of the lower diabatic electronic state (the ground state of NO_2) decay only by infrared emission and their width can be neglected. The elements of the radiative damping matrix are thus (up to a constant) identity operators in vibrational space.

The radiative damping matrix can be represented as a supermatrix within the basis of the vibrational wave functions of the diabatic electronic states. If the vibronic coupling is weak, each vibronic state involves a single upper vibrational level and the supermatrix can be decomposed into several submatrices, one for each upper vibrational level. Consequently, the radiative and nonradiative decays can then be treated for each upper vibrational level separately. For strong vibronic coupling, however, the vibronic states have been shown (Section III.B) to contain several upper vibrational levels. Therefore, in cases like NO_2 the effective Hamiltonian (7.1) must be treated as a whole.

It should be mentioned that the approximations used in writing (7.1) are not strictly valid in NO_2. As discussed at the end of Section II.C, they require that the final electronic state be well below the emitting state and does not interact with it vibronically (external emission). In the case of NO_2 the emission takes place within the manifold of vibronically coupled states and therefore the expression (2.88) for the radiative damping matrix should be used in order to obtain quantitative results. However, even the diagonalization of the simplified Hamiltonian (7.1) amounts to a formidable numerical task in the cases of NO_2 and $C_2H_4^+$, and this task becomes virtually unsolvable on present day computers when the form (2.88) of the damping matrix is used. Therefore, we shall work with the simplified expression (7.1), bearing in mind that this will place an upper limit on the actual quenching constants in NO_2 and $C_2H_4^+$ (see Section VII.B.1). Furthermore, we may consider the results discussed in the following as representing a hypothetical molecule having the same parameters, but where the emission is external.

In small molecules like NO_2 the rate Γ ($\approx 10^{-8}$ eV) is much smaller than the average next-neighbor distance of vibronic states ($\approx 10^{-3}$ eV). Therefore, the eigenstates $|\chi_\nu\rangle$ and the real part of the eigenenergies of $\mathcal{H}_{eff}$ and $\mathcal{H}$ can be taken to coincide, and the imaginary parts $-\Gamma_\nu/2$ of the eigenenergies of $\mathcal{H}_{eff}$ can be approximated by

$$q_\nu = \frac{\Gamma_\nu}{\Gamma} = \langle \chi_\nu | \begin{pmatrix} 0 & 0 \\ 0 & 1 \end{pmatrix} | \chi_\nu \rangle \tag{7.2}$$

The ratios Γ_ν/Γ are smaller than unity, and the radiative rates Γ_ν of the vibronic eigenstates are thus quenched by the vibronic coupling (for a more detailed discussion, see Section II.C). According to Eq. (7.2), the "quenching factors" Γ_ν/Γ can be calculated from the solutions of the vibronic coupling Hamiltonian $\mathcal{H}$ provided that the vibronic eigenstates are completely known. This appears to be impossible for NO_2, in view of the large dimension of the secular matrix (see Section V.B). It proves to be easier to solve the Schrödinger equation for the effective Hamiltonian (7.1) itself and then calculate the quenching from the imaginary parts of the resulting eigenvalues.

The diagonalization of the nonhermitian effective Hamiltonian has been achieved with the aid of the Lanczos algorithm (see Section II.B.3).[89] A larger value for Γ (10^{-4} eV) has been used than the actual radiative width to speed up numerical convergence. This is immaterial for the quenching factors as long as it does not influence the real parts of the eigenvalues. Furthermore, the Lanczos procedure is known to lead to spurious eigenvalues when an insufficient number of iteration steps is performed. Since in our calculation the number of these steps is much smaller than the dimension of the secular matrix, we obtain many eigenvalues which are not fully converged. Nevertheless, by a careful search we have been able to extract from our data a total of 175 numbers for which the quenching factors are converged to at least two digits.[89] The energies of these states are up to 0.7 eV above the minimum of the potential energy surface of the $^2B_2$ state. Moreover, the real parts of these eigenvalues coincide with those obtained in the calculation of the spectrum (using $\mathcal{H}$ instead of $\mathcal{H}_{eff}$) to within 10^{-6} eV. Therefore, these numbers represent a partial solution of the problem of diagonalizing (7.1). When plotted as a function of energy, the quenching factors scatter irregularly between 0.02 and 0.6. We therefore do not give this graph here. Rather, we show in Fig. 44 the resulting *distribution* of quenching factors, that is, the number of vibronic states versus the quenching factor. The theoretical data should be compared to experimental results obtained with a wavelength of the exciting radiation near 600 nm. Such measurements have been reported several times in the literature.[224,281,283,286]

The quenching factors of Fig. 44 are markedly smaller than unity, typical values ranging from 0.04 to 0.4. More specifically, the theoretical distribution exhibits two major peaks at ~ 0.05 and ~ 0.22. This is interesting, since the fluorescence decay times have been observed to cluster around two quite different values, often referred to as long-lived and short-lived species.[224] The associated decay times are ~ 200 and ~ 30 μsec, respectively. Assuming that the unquenched radiative decay rate corresponds to a decay time of 1.5 μsec,[224] this leads to quenching factors of 0.01 and 0.05. The calculation is thus able to reproduce the major experimental findings, although the calculated absolute values of the quenching factors appear to be too large by a factor of 4-5. This discrepancy is partly due to deficiencies of the theoretical model discussed above and partly to uncertainties in the experimental determination of the unquenched radiative rate. We conclude that the anomalously small fluorescence decay rates of the excited B_2 vibronic states of NO_2 are indeed caused by the strong nonadiabatic interaction between the $^2A_1$ and $^2B_2$ electronic states. The intuitive suggestion put forward by Douglas[123] and other workers[229] is thus confirmed by our theoretical calcu-

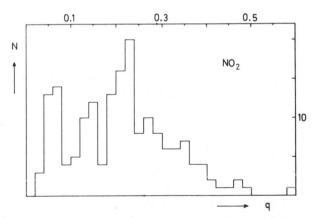

Fig. 44.   Quenching factor histogram for the B_2 vibronic states of NO_2. Here N = number of levels.

lation. (It should be remembered that the parameters entering the Hamiltonian are taken from ab initio data.)

The form of the theoretical distribution of quenching factors, shown in Fig. 44, deserves some further comment. Can the formation of essentially two peaks be explained or at least interpreted in physical terms? The meaning of the quenching factors Γ_ν / Γ is already apparent from (7.2): they represent the weight of the upper ($^2B_2$) electronic state in the vibronic eigenstates. Consider now the hypothetical case of vanishing vibronic coupling ($\lambda = 0$). For this case the vibronic states are just the vibrational levels of either the upper or the lower (diabatic) potential energy surfaces. For our model Hamiltonian and in the appropriate energy interval, the number of these levels is 24 and 120, respectively. The hypothetical distribution of quenching factors for $\lambda = 0$ is thus composed of 120 entries for $q = 0$ and 24 entries for $q = 1$, yielding an average value of 0.17. Clearly, the effect of the vibronic coupling is to bring these two disjoint sets together and form a comparatively narrow distribution. The mean of the distribution of Fig. 44 (0.21) is rather close to the average relative level density of 0.17, but its variance is much smaller than that for $\lambda = 0$ (0.012 versus 0.139). This indicates that the mixing of the vibrational levels of different electronic states is close to being statistical. On the other hand, the two-peaked shape of the distribution is still reminiscent of the two isolated peaks found for $\lambda = 0$, suggesting that the vibronic states can approximately be grouped into two sets with almost purely $^2A_1$ electronic character or mixed electronic character. Despite the complexity of the absorption spectrum, the vibronic interaction between the two electronic states is not strong enough to ensure that the quenching factors are more or less equal for a close-lying group of levels.

It is instructive to compare these findings with those obtained for another example given in Section V, the ethylene radical cation. In Section V.A.2 we have seen that the interaction between the two lowest ionic states can be described with the present vibronic coupling model, including two tuning modes and a single nondegenerate coupling mode. The type of the vibronic coupling mechanism and the number of vibrational modes is thus the same as were used for NO_2. We have calculated quenching factors for $C_2H_4^+$ along the lines described above and obtained 147 converged numbers in the energy range between 12.1 and 13.0 eV.[89] The resulting distribution is drawn in Fig. 45 in the same way as for NO_2. Again, we observe that all quenching factors are far below unity. Moreover, the distribution of Fig. 45 exhibits only a single peak in contrast to the distribution of Fig. 44. This indicates that the electronic mixing in the vibronic states is more uniform in the case of $C_2H_4^+$ than in the case of NO_2 and that it is virtually complete; that is, no "preferential" electronic character can any longer be distinguished. Note that the mean of the distribution in Fig. 45 deviates markedly from the relative

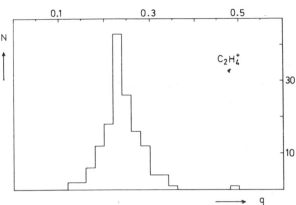

Fig. 45. Quenching factor histogram for the B_{2g} vibronic states of $C_2H_4^+$. Here $N =$ number of levels.

density of vibrational levels (0.1). This is evidence of the strength of the vibronic coupling, as is the observation that the number of vibronic states in the present energy range exceeds that of the zeroth-order vibrational levels by a factor of ~ 2.

The difference in the quenching factor distribution between $C_2H_4^+$ and NO_2 might seem unexpected when the two spectra in Figs. 28 and 32 are compared. In the latter respect, NO_2 looks more complicated than $C_2H_4^+$. It has already been noted in Section V.B that the coupling constants of the totally symmetric modes are very large for NO_2. This causes the appearance of many lines in the spectrum and is a major reason why the spectrum looks so complicated. In $C_2H_4^+$ the excitation strengths of the totally symmetric modes [see Eq. (2.41)] are much smaller than in NO_2. The vibronic coupling strength, on the other hand, shows the opposite behavior: comparing Table V with Table VII, we can see that the interstate coupling constant λ is larger and the frequency of the coupling mode is smaller in $C_2H_4^+$ than in NO_2. Furthermore, the values of the vertical energy gaps are different (1.9 versus 3.4 eV). According to the analysis of Section III these factors favor a stronger vibronic interaction between the electronic states in $C_2H_4^+$ and thus rationalize the different behavior of the quenching factors in the two molecules.

B. Ultrafast Nonradiative Decay in $C_2H_4^+$

1. Short-Time Behavior

When treating the decay processes in molecular ions, we should first consider the nature of the optically excited initial state. In the case of neutrals, the transition that generates the initial state is direct; that is, the energy of the exciting photon matches that of the vibronic state(s) to be examined. Therefore the energy spread of the initial state is fixed by that of the exciting radiation (typically a few cm^{-1}). In the case of ions the situation is different since in most experimental arrangements an initial electron or photon is used to *ionize* the originally neutral species.[110,111] The energy of the ionizing electron or photon is usually considerably higher than that of the vibronic states of the ion under study. The excess energy is carried off as kinetic energy by the photoelectron. As long as its kinetic energy is not measured, the initial state in the decay process is therefore a coherent superposition of all ionic states accessible energetically to the ionizing photon. To be sure, in actual experiments often an energy selection is performed by measuring the kinetic energy of the photoelectron.[110,111] However, the resolving power is comparatively poor, resulting in an energy spread of the initial state of typically ~ 500 cm^{-1}. In complex spectral bands such as the second photoelectron band of $C_2H_4^+$ the spread will therefore comprise several or many vibronic levels. To simplify the discussion, we take the initial state of the decay process to be a coherent superposition of all vibronic states of a given electronic band. Neglecting the effect of intensity borrowing, this initial state $|\psi(0)\rangle$ is just the vibrationless level $|0\rangle$ of the neutral ground state, transferred vertically up to the respective diabatic potential energy surface.

Let us follow the time evolution of this initial state for the second photoelectron band of C_2H_4, studied in Section V.A.2. In this ion, we have identified strong vibronic coupling effects between the ground and first excited electronic states. Now the question arises: How do these effects influence the time evolution of the initial state $|\psi(0)\rangle$, Eq. (2.80), on the coupled potential energy surfaces? Following the discussion in Section II.C.3, we start with the autocorrelation function (2.81). For sufficiently short times the effective Hamiltonian $\mathcal{H}_{eff}$ in (2.81) can be replaced by $\mathcal{H}$ and the autocorrelation function becomes

$$C(t) = \langle \psi(0)|\psi(t)\rangle = \langle \psi(0)|e^{-i\mathcal{H}t}|\psi(0)\rangle \qquad (7.3)$$

Here $\mathcal{H}$ is the Hamiltonian used for calculating the spectrum. Dropping the separable C—H bending mode[127] (see also Section V.A.2), it is given by Eq. (3.22) with $N=2$ and $M=1$. Physically, $C(t)$ represents the probability amplitude that the wavepacket $|\psi(t)\rangle$ at time t is still the same as at starting time $t=0$. Mathematically, $C(t)$ is seen from Eq. (2.37) to be the Fourier transform of the respective electronic band (putting $\tau_2 = 1$ and $\tau_1 = 0$). The modulus $|C(t)|$ is drawn for the second photoelectron band of ethylene in Fig. 46a. For comparison we also show in Fig. 46b the diabatic $|C(t)|$, defined by $\lambda = 0$.

Both curves in Fig. 46 decay on a femtosecond time scale. However, while the diabatic $|C(t)|$ exhibits pronounced recurrences and a large average value of ~ 0.4, the exact curve is quite irregular and fluctuates around a constant value of only ~ 0.08. This difference reflects the different appearance of the corresponding spectra. The exact spectrum (see Fig. 28b) exhibits many irregularly spaced lines. The diabatic spectrum ($\lambda = 0$) looks very similar to the Franck–Condon spectrum of Fig. 28c and is composed of a small number of lines.[127] We have also computed $|C(t)|$ up to and beyond the typical recurrence time[293] $2\pi\rho$, where ρ is the average density of vibronic levels in the second photoelectron band of ethylene ($\sim 130 \times 10^{-14}$ sec). No indication of a recurrence in the exact $|C(t)|$ has been observed; that is, the recurrence is completely washed out[293] by the fluctuations of the vibronic energy levels (see Section VI). This absence of recurrences even for longer times demonstrates the irreversible motion of the initial wavepacket $|\psi(0)\rangle$ on the coupled potential energy surfaces. The irreversibility and the low average value of the exact, as compared to the diabatic, $|C(t)|$ are caused

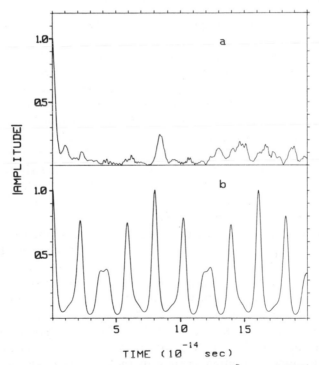

Fig. 46. Moduli of the autocorrelation functions for the $\tilde{A}$ state of $C_2H_4^+$. (a) The full autocorrelation function according to (7.3). (b) The diabatic autocorrelation function, defined by $\lambda = 0$.

by the nonradiative transition from the upper to the lower surface. Care should be taken, however, when inquiring about the nonradiative decay time from the autocorrelation function. The rapid initial decrease of $|C(t)|$ reflects the motion of the wave packet on the upper surface rather than the radiationless decay. It is governed by the overall width of the spectral band, which is almost the same in the exact and the diabatic ($\lambda = 0$) calculation. A nonradiative decay time can be estimated only indirectly by noting that the first recurrence in the diabatic $|C(t)|$ after $\sim 2 \times 10^{-14}$ sec is already absent in the full calculation. Therefore, the decay time in question will be about 10^{-14} sec.

To obtain more direct information on the nonradiative decay we calculate the time-dependent expectation value of the radiative damping matrix. As discussed in Section II.C, this is proportional to the total time-resolved intensity of fluorescence and constitutes the appropriate measure of the nonradiative decay in cases of strong vibronic coupling. Limiting ourselves to sufficiently short times, $\mathcal{H}_{\text{eff}}$ can again be replaced by $\mathcal{H}$ and the general Eq. (2.91) becomes

$$I(t) = \langle \psi(0)|e^{i\mathcal{H}t}\Gamma e^{-i\mathcal{H}t}|\psi(0)\rangle \tag{7.4}$$

Two expressions for the radiative damping matrix Γ have been discussed in Section II.C. When the fluorescence is to a third, vibronically uncoupled electronic state (external emission), the form (2.92) is appropriate, while one should use Eq. (2.88) when the fluorescence takes place within the vibronically coupled electronic manifold as in $C_2H_4^+$ (internal emission). In the case of the time-dependent calculations we are in the position to determine both quantities numerically for the parameter values of interest. This is accomplished by numerically integrating the time-dependent Schrödinger equation (see Section II.C) to obtain $|\psi(t)\rangle$.[127] The results for the ethylene cation are displayed in Fig. 47 which shows the time-dependent probability of photon emission, normalized to unity at time $t = 0$.

Let us start with the curve with crosses. It represents the solution of the external-emission problem when all three modes are included. Remember that from Eq. (2.86) this quantity equals the probability $P_2(t)$ of being on the second diabatic potential energy surface. It is seen from Fig. 47 that $P_2(t)$ exhibits an initial decrease with a decay time of $\sim 10^{-14}$ sec and subsequent oscillations around an average value of ~ 0.3. The overall behaviour of $P_2(t)$ makes clear that the notion of a "decay" is indeed appropriate, as has been anticipated from Fig. 46. For comparison we have also included the result of the single-mode calculation where only the coupling mode ν_4 is retained and the totally symmetric modes ν_1 and ν_2 are dropped (full line). This latter curve exhibits virtually no decay, as is expected from the large vertical energy gap of the interacting states. This illustrates that the decay of $P_2(t)$ occurs by virtue of the combined action of the coupling and the totally symmetric tuning modes.

The dashed line in Fig. 47 denotes the result for internal emission and is thus the proper solution of the nonradiative-decay problem in $C_2H_4^+$. Apparently, the initial decay of the fluorescence is even faster and the average value for longer times considerably smaller than that of $P_2(t)$. The corresponding numbers are 0.3×10^{-14} sec for the decay time and 0.02 for the average value. We have thus achieved a direct calculation of the non-radiative decay time in $C_2H_4^+$ which is based on ab initio data (see Section V.A.2) and represents an exact quantum mechanical solution of the model Hamiltonian adopted.[302]

In most examples of internal conversion studied theoretically, the nonradiative decay times range from $\sim 10^{-6}$ sec to $\sim 10^{-11}$ sec.[103,104] These times are much longer than typical vibrational periods of the molecule (10^{-13}–10^{-14} sec), and the molecule thus has ample time to vibrate before it undergoes the radiationless transition. In this sense the latter constitutes only a small perturbation of the nuclear motion. In our example, however, the nonradiative decay time is of the same order as a typical vibrational period

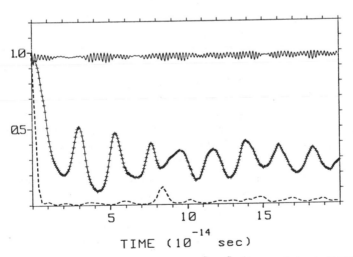

$$TIME \ (10^{-14} \ sec)$$

Fig. 47.   Normalized probability $I(t)/I(0)$ of $\tilde{A} \to \tilde{X}$ photon emission in $C_2H_4^+$, treated at three different levels of sophistication. Note that for external emission this quantity equals the probability $P_2(t)$ of being in the second diabatic state. *Full line:* result for external emission, Eq. (2.92), retaining only the coupling mode. *Full line with crosses:* result for external emission, Eq. (2.92), including all three modes. *Dashed line:* result for internal emission, Eq. (2.88), including all three modes. This represents the proper solution of the nonradiative-decay problem in $C_2H_4^+$.

and the decay takes place during the first few vibrations. This "ultrafast" decay can no longer be considered a small perturbation. It just represents the strong nonadiabatic effects found earlier in the spectrum (Section V.A.2) in a different manner. The strong nonadiabatic effects have been attributed in Section V.A to the existence of a conical intersection of the ionic potential energy surfaces. Now we may say that conical intersections are a mechanism that can lead to ultrafast nonradiative decay, that is, to decay times in the subpicosecond time range.

We conclude this subsection with two remarks. First, there is an apparent similarity between the autocorrelation function, Fig. 46, and the probability of photon emission, Fig. 47. We emphasize that this similarity is purely accidental. It is easy to conceive situations where the time dependence of the two quantities is quite different.[302] Second, Fig. 47 gives evidence that treating the internal emission process as external will overestimate the intensity of fluorescence. Therefore, the quenching factors as calculated in Section VII.A will tend to be too large.

2. Long-Time Behavior

In the preceding subsection we have been concerned with the time evolution of $C_2H_4^+$ on a subpicosecond time scale. Decay times of this order are not directly observable at present. In the experimental search for $\tilde{A} \rightarrow \tilde{X}$ emission in $C_2H_4^+$ the time evolution was followed from $\sim 10^{-9}$ to $\sim 10^{-5}$ sec.[295] No emission has been observed, even if the internal energy of the ion is below the threshold for fragmentation (at 13.1 eV[296]). This places an upper limit of $\sim 10^{-4}$ on the fluorescence quantum yield,[295] to be explained in terms of purely intramolecular radiationless decay.

To what extent can the present calculation explain the absence of detectable emission in $C_2H_4^+$? First of all, we have to consider the other vibrational modes of $C_2H_4^+$ which have been neglected so far. We have to estimate the total density ρ of vibrational levels of the $\tilde{X}$ state in the energy range of the second band. Using the present frequencies ω_1, ω_2 and ω_4 and those of the neutral ground state for the other modes,[297] we obtain at the band center $1/\rho \approx 0.001$ cm^{-1}. This is clearly much larger than the rate k_{ir} of infrared emission ($k_{ir} \lesssim 10^4$ sec^{-1}) and therefore $C_2H_4^+$ represents an intermediate size molecule where proper nonradiative relaxation cannot occur.[103,292] In other words, even when all vibrational modes are included we expect near-to-unity quantum yield for infinitely long observation time.

We suggest that the absence of detectable emission is explained in terms of an anomalously small radiative decay rate. More precisely, the radiative decay rate is expected to be so small that the quantum yield $Y_0 = Y(t_0)$ within the observation time t_0 is below the detection limit[295] of $\sim 10^{-4}$. Let us see to what extent the present calculations support this conjecture. On the experimentally relevant time scale of $10^{-9}-10^{-6}$ sec the oscillations of Fig. 47 are washed out and the curves are to be replaced by their average values I_∞. Therefore, one has simply

$$Y_0 = I_\infty t_0 \tag{7.5}$$

Without a knowledge of the electronic matrix element g_0 entering Eqs. (2.88, 89) we have to estimate the average radiative decay rate I_∞ by referring to a related molecule, such as the cis-difluoroethylene radical cation. Here the $\tilde{A} \to \tilde{X}$ radiative decay rate has been found[303] to be 2×10^5 sec^{-1}, the mean energy separation between the $\tilde{A}$ and $\tilde{X}$ electronic states being[295] ~ 3.5 eV. Since the intensity $I(t=0)$ is roughly proportional to the third power of this energetic separation it will be smaller in $C_2H_4^+$ than in cis-$C_2H_2F_2^+$ by a factor $\approx (3.5/1.9)^3 \approx 6$. From the time-dependent calculations reported above we have inferred $I(t=0)/I_\infty \approx 55$ and thus obtain $I_\infty \approx 2 \times 10^5/(6 \times 55)$ sec$^{-1} \approx 6 \times 10^2$ sec^{-1}. Furthermore, on a microsecond time scale also the other vibrational modes of $C_2H_4^+$, neglected so far, will influence the emission. Classical trajectory calculations[302] show that these other modes may reduce I_∞ by a factor ~ 3. Including this reduction factor and taking the time of flight of the ions in the beam ($\leq 10^{-5}$ sec)[295] as an upper limit to the observation time t_0 we arrive at $Y_0 \approx 10^{-3}$. This is considered a rather cautious estimate because the influence of the other modes is presumably larger than has been assumed and because the emission is dipole allowed in cis-$C_2H_2F_2^+$ but not in $C_2H_4^+$ (D_{2h} symmetry). In view of these considerations it appears very likely that the anomalously small emission rates associated with the conical intersection prevent the detection of fluorescence in $C_2H_4^+$. We mention that the strong nonadiabatic interactions between the $\tilde{A}$ and $\tilde{X}$ ionic states also enhance the efficiency of collisional quenching processes[298] which will further reduce the quantum yield.

To illustrate the impact of conical intersections on fluorescence, we can cite two other examples. In HCN$^+$ there is no emission $\tilde{B} \to \tilde{X}$ or $\tilde{B} \to \tilde{A}$ even if the internal energy of the $\tilde{B}$ state is below the relevant dissociation asymptote.[299] We have performed calculations of the ionic energies for various bond lengths and found a crossing of the $\tilde{B}$ and $\tilde{X}$ states upon shortening of the C—N bond. Since there is a linear vibronic coupling through the bending mode, this crossing constitutes a conical intersection of the potential energy surfaces. In NH$_3^+$, although the $\tilde{A}^2E$ state is dissociated only to 83%,[301] no emission $\tilde{A} \to \tilde{X}$ could be detected. Our data show that also in this case there exists a crossing of the $\tilde{X}$ and $\tilde{A}$ potential energy surfaces, namely, for increasing pyramidalization. A linear vibronic coupling is possible through the Jahn–Teller active normal modes ν_3 and ν_4. In both examples the crossing occurs near the minimum of the upper potential energy surface, although the two electronic states are separated vertically by almost 6 eV. This can rationalize in a natural way the lack of emission down to the lowest vibrational energies. It also underlines that electronic states may strongly interact vibronically, even when the corresponding spectral bands do not overlap.

VIII. CONCLUSIONS

The study of multimode molecular dynamics on coupled electronic surfaces clearly presents a challenging theoretical and computational problem. Here we have reviewed a particular approach to this problem, based on the adoption of simple model Hamiltonians. The essential simplifications are the assumption of harmonic diabatic potentials and linear (in the nuclear coordinates) coupling of the diabatic electronic states. We have concentrated on systems where the interplay of the (generally non–totally symmetric) coupling mode(s) with the totally symmetric tuning mode(s) leads to a conical intersection of the adiabatic potential energy surfaces. The simplifying model assumptions allow the exact numerical solution of the Schrödinger equation via the diagonalization of large sparse matrices. These techniques enabled us to delve into a new regime of non-Born–Oppenheimer phenomena: strong vibronic coupling involving several nonseparable vibrational modes.

The results reveal a wealth of interesting physical phenomena. The vibronic structure of electronic spectra has been analyzed for a typical conical intersection situation involving a single nondegenerate coupling mode and a single totally symmetric tuning mode. The calculated spectra have been interpreted in terms of the associated intersecting adiabatic potential energy surfaces. The lower of these surfaces exhibits, in general, multiple minima at configurations that correspond to a reduced symmetry of the nuclear framework. Associated with this breaking of the molecular symmetry is a pronounced anharmonicity of the adiabatic potential energy surfaces. Vibronic energy levels that lie well below the point of intersection are generally accurately described in the adiabatic Born–Oppenheimer approximation, whereas the adiabatic approximation fails completely in the whole spectral range above the point of intersection. The diabatic limit is approached for energy levels high above the conical intersection, though only very slowly. In a wide energy range near and above the point of intersection the vibronic spectrum is very irregular and complex in the strong coupling cases considered here. The important effect of additional tuning modes is that they generally lower the minimum of the locus of intersection and thus enhance the nonadiabatic effects. The symmetry-breaking effect, on the other hand, which is purely adiabatic in nature, is generally quenched by the tuning modes.

The typical spectroscopic effects of conical intersections have been identified and more or less quantitatively reproduced in the photoelectron spectra of the cumulene series, in particular ethylene and butatriene, the photoelectron spectrum of HCN, and the absorption spectrum of NO_2. The conceptually closely related multimode Jahn–Teller effect has been discussed for the examples of $C_6F_6^+$ and BF_3^+. The ab initio calculation of the vertical energy gap and the electron-vibrational coupling constants entering the model Hamiltonians is an essential ingredient of the present approach. It opens the way for the analysis of very complex spectra and excludes the possibility of accidental agreement when fitting low-resolution spectra. The existence of a conical intersection dominates completely the calculated spectra, rendering the approximations introduced through the assumption of harmonic diabatic potentials and linear coupling rather irrelevant.

Above the minimum of the locus of intersection the calculated vibronic spectra exhibit an enormous complexity, especially when more than two modes are involved. A level-by-level description of the spectrum is then neither possible nor desirable. We have set out to apply statistical analysis to the (calculated or spectroscopically observed) data and have presented first results for the vibronic spectra of $C_2H_4^+$ and NO_2. The statistical aspects of the energy level pattern associated with conical intersections deserve further experimental and theoretical investigation.

The high density of vibronic levels and their statistical distribution is associated with the onset of irreversible behavior. The models considered in this work are suitable for describing quantitatively the radiationless decay within the small-molecule limit. Particular aspects of this phenomenon are the quenching of radiative decay rates of individual vibronic levels (the Douglas effect) and the ultrafast decay of the occupation probability of the upper electronic state. We have performed a qualitative calculation of the former phenomenon in NO_2, while the latter has been analyzed in detail for $C_2H_4^+$. When the additional modes not involved in the vibronic coupling mechanism and the effect of collisions are considered, the calculated ultrafast decay explains the absence of emission from the $\tilde{A}^2B_{2g}$ state of $C_2H_4^+$. More generally, one expects the absence of detectable fluorescence whenever there exists a conical intersection close to the minimum of the corresponding adiabatic potential energy surface. We hope we have shown that the exact (numerical) solution of multimode vibronic coupling models yields a unified description of the vibronic structure of electronic spectra, the radiative rate of individual levels as well as the nonradiative decay of electronic states.

The examples presented in the foregoing sections are typical strong coupling cases; that is, the relevant nonadiabatic coupling matrix elements are large and the density of vibronic states is comparatively low. It is probably not accidental that all systems considered are either open-shell molecules

(NO_2) or cations of closed-shell molecules ($C_nH_4^+$, HCN^+, BF_3^+, $C_6F_6^+$). In open-shell systems, adjacent electronic states usually differ in the quantum number of a single electron. The vibronic coupling constant λ is essentially given by the matrix element of $\partial U / \partial Q$, where U is the electron–nucleus interaction term in the Hamiltonian and can be quite large. In closed-shell molecules, on the other hand, neighboring electronic states often differ in the quantum number of two or more electrons. Then the matrix element of $\partial U / \partial Q$, which is a single-particle operator, is zero and vibronic coupling occurs via the residual electron–electron interaction. Consequently the vibronic coupling constant λ will often be rather small for low-lying excited states of closed-shell molecules. Strong vibronic coupling and conical intersections are thus typical and probably ubiquituous phenomena in open-shell systems. Experimental progress in high-resolution photoelectron spectroscopy and optical spectroscopy and future ab initio calculations of multidimensional potential energy surfaces will certainly reveal many additional examples of conical intersections.

The types of intersections we have considered are the simplest possible. They are characterized by linear interstate coupling involving either nondegenerate or doubly degenerate modes. Extensions of this concept may involve, for example, quadratic intrastate or bilinear interstate coupling terms, leading to new topological features of the adiabatic potential energy surfaces such as closed hypersurfaces of conical intersections or glancing intersections. The matrix representation of the corresponding Hamiltonian will still be a sparse matrix, rendering the exact numerical solution of the dynamical problem possible along the lines adopted in the present work.

Spectroscopy is the most powerful and accurate tool used to investigate the nuclear dynamics on multidimensional intersecting potential energy surfaces, and we have concentrated on this aspect. There are, however, many other phenomena where conical intersections are expected to play an important role. For a recent monograph on the Jahn–Teller effect and vibronic interactions with particular emphasis on chemical applications, see Bersuker.[300] The unimolecular decay of ions or excited species, for example, will often proceed through conical intersections. The adiabatic potential energy surfaces of even the simplest atom–diatom collision systems are known to exhibit conical intersections. The accurate calculation of unimolecular decay rates and reactive collision cross sections is an important problem but is beyond the capabilities of the present theoretical approach, which is confined to the calculation of bound level structures.

Acknowledgments

The authors are indebted to W. von Niessen for the ab initio Green's function results and for a very close collaboration. Special thanks are due to

E. Haller, who implemented the Lanczos algorithm and has actively taken part in our research on the vibronic coupling problem. We also thank D. Meyer for many useful discussions and E. Zanders for a careful typing of the manuscript. Financial support by the Deutsche Forschungsgemeinschaft and by the Fonds der Chemischen Industrie is gratefully acknowledged. Computer time has generously been provided by the Universitätsrechenzentrum Heidelberg.

REFERENCES

1. M. Born and R. Oppenheimer, *Ann. Physik* (Leipzig) **84**, 457 (1927).

2. M. Born, *Nachr. Akad. Wiss. Göttingen, Math.-Physik. Kl. IIa* p. 1 (1951); M. Born and K. Huang, *Dynamical Theory of Crystal Lattices*, Oxford University Press, New York, 1954.

3. For a careful discussion of the sometimes confusing nomenclature, see C. J. Ballhausen and A. E. Hansen, *Ann. Rev. Phys. Chem.* **23**, 15 (1972).

4. R. G. Bray and M. J. Berry, *J. Chem. Phys.* **71**, 4909 (1979); P. R. Stannard, M. L. Elert, and W. M. Gelbart, ibid. **74**, 6050 (1981); R. T. Lawton and M. S. Child, *Mol. Phys.* **44**, 709 (1981).

5. J. Ford, *Advan. Chem. Phys.* **24**, 155 (1973); I. C. Percival, ibid. **36**, 1 (1977); D. W. Noid, M. L. Koszykowski, and R. A. Marcus, *J. Chem. Phys.* **71**, 2864 (1979) and references therein.

6. D. F. Heller and S. Mukamel, *J. Chem. Phys.* **70**, 463 (1979); K. F. Freed and A. Nitzan, ibid. **73**, 4765 (1980).

7. W. H. Miller, ed., Dynamics of Molecular Collisions, Pt. B, Plenum Press, New York, 1976.

8. M. S. Child, in *Atom–Molecular Collision Theory*, R. B. Bernstein, ed., Plenum Press, New York, 1979, Chap. 13.

9. B. DiBartolo, ed., *Radiationless Processes*, Plenum Press, New York, 1980.

10. L. D. Landau, *Physik. Z. Sowjetunion* **2**, 46 (1932); C. Zener, *Proc. Roy. Soc.* (London) **A137**, 696 (1932); E. C. G. Stueckelberg, *Helv. Phys. Acta* **5**, 369 (1932).

11. W. Lichten, *Phys. Rev.* **131**, 229 (1963); **139**, A27 (1965); **164**, 131 (1967).

12. H. C. Longuet-Higgins, *Advan. Spectrosc.* **2**, 429 (1961).

13. T. F. O'Malley, *Phys. Rev.* **162**, 98 (1967).

14. F. T. Smith, *Phys. Rev.* **179**, 111 (1969).

15. W. Kolos and L. Wolniewicz, *Rev. Mod. Phys.* **35**, 473 (1963); W. Kolos, *Advan. Quantum Chem.* **5**, 99 (1970).

16. See, for example, A. Lagerqvist and E. Miescher, *Helv. Phys. Acta* **31**, 221 (1958); K. Dressler, *Can. J. Phys.* **47**, 547 (1969).

17. G. Herzberg, *Molecular Spectra and Molecular Structure*, Vol. I: *Spectra of Diatomic Molecules*, 2nd ed., Van Nostrand, New York, 1950, Chap. VII.

18. E. E. Nikitin, *Advan. Chem. Phys.* **28**, 317 (1975).

19. J. Von Neumann and E. Wigner, *Physik. Z.* **30**, 467 (1929).

20. H. C. Longuet-Higgins, *Proc. Roy. Soc.* (London) **A344**, 147 (1975).

21. A. J. Stone, *Proc. Roy. Soc.* (London) **A351**, 141 (1976).

238   H. KÖPPEL, W. DOMCKE, AND L. S. CEDERBAUM

22. J. Katriel and E. R. Davidson, *Chem. Phys. Lett.* **76**, 259 (1980).

23. E. Teller, *J. Phys. Chem.* **41**, 109 (1937).

24. G. Herzberg and H. C. Longuet-Higgins, *Discuss. Faraday Soc.* **35**, 77 (1963); G. Herzberg, *Molecular Spectra and Molecular Structure*, Vol. III: *Electronic Spectra and Electronic Structure of Polyatomic Molecules*, Van Nostrand, New York, 1966, p. 422.

25. T. Carrington, *Discuss. Faraday Soc.* **53**, 27 (1972); *Accts. Chem. Res.* **7**, 20 (1974).

26. E. R. Davidson, *J. Amer. Chem. Soc.* **99**, 397 (1977).

27. H. A. Jahn and E. Teller, *Proc. Roy. Soc.* (London) **A161**, 220 (1937).

28. M. D. Sturge, *Solid State Phys.* **20**, 91 (1967).

29. R. Englman, *The Jahn–Teller Effect*, Wiley, New York, 1972.

30. W. Domcke, H. Köppel, and L. S. Cederbaum, *Mol. Phys.* **43**, 851 (1981).

31. H. Köppel, L. S. Cederbaum, and W. Domcke, *J. Chem. Phys.* **77**, 2014 (1982).

32. E. E. Nikitin, in *Chemische Elementarprozesse*, H. Hartmann, ed., Springer, New York, 1968, p. 43. M. Vaz Pires, C. Galloy, and J. C. Lorquet, *J. Chem. Phys.* **69**, 3242 (1978); M. Desouter-Lecomte, C. Galloy, J. C. Lorquet, and M. Vaz Pires, ibid. **71**, 3661 (1979).

33. J. C. Tully and R. K. Preston, *J. Chem. Phys.* **55**, 562 (1971).

34. W. H. Miller and T. F. George, *J. Chem. Phys.* **56**, 5637 (1972).

35. R. L. Fulton and M. Gouterman, *J. Chem. Phys.* **35**, 1059 (1961).

36. R. L. Fulton and M. Gouterman, *J. Chem. Phys.* **41**, 2280 (1964).

37. M. Gouterman, *J. Chem. Phys.* **42**, 351 (1965).

38. R. Lefebvre and M. Garcia Sucre, *Int. J. Quant. Chem.* **1S**, 339 (1967).

39. M. Garcia Sucre, F. Gény, and R. Lefebvre, *J. Chem. Phys.* **49**, 458 (1968).

40. R. L. Fulton, *J. Chem. Phys.* **56**, 1210 (1972).

41. A. R. Gregory, W. H. Henneker, W. Siebrand, and M. Z. Zgierski, *J. Chem. Phys.* **65**, 2071 (1976).

42. A. R. Gregory, W. H. Henneker, W. Siebrand, and M. Z. Zgierski, *J. Chem. Phys.* **67**, 3175 (1977).

43. W. H. Henneker, A. P. Penner, W. Siebrand, and M. Z. Zgierski, *J. Chem. Phys.* **69**, 1884 (1978).

44. J. Brickmann, *Mol. Phys.* **35**, 155 (1978).

45. R. A. Friesner and R. Silbey, *J. Chem. Phys.* **75**, 3925 (1981).

46. E. A. Chandross, J. Ferguson, and E. G. McRae, *J. Chem. Phys.* **45**, 3546 (1966).

47. G. C. Nieman, *J. Chem. Phys.* **50**, 1660 (1969).

48. G. Fisher and G. J. Small, *J. Chem. Phys.* **56**, 5934 (1972).

49. S. B. Piepho, E. R. Krausz, and P. N. Schatz, *J. Amer. Chem. Soc.* **100**, 2996 (1978).

50. A. E. W. Knight and C. S. Parmenter, *Chem. Phys.* **43**, 257 (1979).

51. Y. Udagawa, M. Ito, I. Susuka, W. Siebrand, and M. Z. Zgierski, *Chem. Phys. Lett.* **68**, 258 (1979).

52. M. Z. Zgierski, *Chem. Phys. Lett.* **69**, 608 (1980).

53. D. L. Narva and D. S. McClure, *Chem. Phys.* **56**, 167 (1981).

54. M. C. M. O'Brien and S. N. Evangelou, *J. Phys.* **C13**, 611 (1980).

55. C. Cossart-Magos and S. Leach, *Chem. Phys.* **48**, 329 (1980).

56. T. Sears, T. A. Miller, and V. E. Bondybey, *J. Chem. Phys.* **72**, 6070 (1980).

57. E. Haller, L. S. Cederbaum, and W. Domcke, *Mol. Phys.* **41**, 1291 (1980).

58. P. M. Champion and A. C. Albrecht, *Ann. Rev. Phys. Chem.* **33**, 353 (1982).

59. D. K. Hsu, D. L. Monts, and R. N. Zare, *Spectral Atlas of Nitrogen Dioxide*, Academic Press, New York, 1978.

60. See, for example, R. Englman and J. Jortner, *Mol. Phys.* **18**, 145 (1970).

61. E. B. Wilson, Jr., J. C. Decius, and P. C. Cross, *Molecular Vibrations*, McGraw-Hill, New York, 1955.

62. O. Atabek, A. Hardisson, and R. Lefebvre, *Chem. Phys. Lett.* **20**, 40 (1973).

63. T. F. O'Malley, *Advan. Atomic Mol. Phys.* **7**, 223 (1971).

64. M. Baer, *Chem. Phys. Lett.* **35**, 112 (1975).

65. M. Desouter-Lecomte, C. Galloy, J. C. Lorquet, and M. Vaz Pires, *J. Chem. Phys.* **71**, 3661 (1979).

66. C. A. Mead and D. G. Truhlar, *J. Chem. Phys.* **77**, 6090 (1982); C. A. Mead, ibid. **78**, 808 (1983).

67. M. Baer, *Mol. Phys.* **40**, 1011 (1980); *Chem. Phys.* **15**, 49 (1976).

68. Z. H. Top and M. Baer, *Chem. Phys.* **25**, 1 (1977); M. Baer and J. A. Beswick, *Phys. Rev.* **A19**, 1559 (1979).

69. I. H. Zimmerman and T. F. George, *J. Chem. Phys.* **63**, 2109 (1975).

70. R. K. Preston and J. C. Tully, *J. Chem. Phys.* **54**, 4297 (1971); J. C. Tully, ibid. **60**, 3042 (1974).

71. M. Desouter-Lecomte and J. C. Lorquet, *J. Chem. Phys.* **66**, 4006 (1977).

72. G. Hirsch, P. J. Bruna, R. J. Buenker, and S. D. Peyerimhoff, *Chem. Phys.* **45**, 335 (1980).

73. H.-J. Werner and W. Meyer, *J. Chem. Phys.* **74**, 5802 (1981).

74. J. Hendekovic, *Chem. Phys. Lett.* **90**, 193 (1982).

75. I. Özkan and L. Goodman, *Chem. Rev.* **79**, 275 (1979).

76. J. des Cloizeaux, *Phys. Rev.* **135**, A685 (1964).

77. B. H. Brandow, *Rev. Mod. Phys.* **39**, 771 (1967); V. Kvasnicka, *Advan. Chem. Phys.* **36**, 345 (1977).

78. H. C. Longuet-Higgins, U. Öpik, M. H. L. Pryce and R. A. Sack, *Proc. Roy. Soc.* (London) **A244**, 1 (1958).

79. H. Köppel, W. Domcke, L. S. Cederbaum, and W. von Niessen, *J. Chem. Phys.* **69**, 4252 (1978).

80. H.-J. Werner and W. Meyer, *J. Chem. Phys.* **74**, 5794 (1981).

81. L. S. Cederbaum, W. Domcke, H. Köppel, and W. von Niessen, *Chem. Phys.* **26**, 169 (1977).

82. L. S. Cederbaum and W. Domcke, *Advan. Chem. Phys.* **36**, 205 (1977).

83. W. Domcke, L. S. Cederbaum, H. Köppel, and W. von Niessen, *Mol. Phys.* **34**, 1759 (1977).

84. G. Herzberg and E. Teller, *Z. Physik. Chem.* (Leipzig) **B21**, 410 (1933).

85. H. Cramér, *Mathematical Methods of Statistics*, Princeton Univ. Press, Princeton, N.J., 1964.

86. C. A. Mead and D. G. Truhlar, *J. Chem. Phys.* **70**, 2284 (1979); C. A. Mead, *Chem. Phys.* **49**, 23 (1980).

87. H. Köppel, E. Haller, L. S. Cederbaum, and W. Domcke, *Mol. Phys.* **41**, 669 (1980).

88. B. N. Parlett, *The Symmetric Eigenvalue Problem*, Prentice-Hall, Englewood Cliffs, N.J., 1980.

89. E. Haller, Thesis, University of Heidelberg (1983); E. Haller, H. Köppel, and L. S. Cederbaum, to be published.

90. C. Lanczos, *J. Res. Nat. Bur. Stand.* **45**, 255 (1950).

91. C. C. Paige, *J. Inst. Math. Appl.* **10**, 373 (1972).

92. R. R. Whitehead, A. Watt, B. J. Cole, and I. Morrison, *Advan. Nucl. Phys.* **9**, 123 (1977).

93. N. Sakamoto and S. Muramatsu, *Phys. Rev.* **B17**, 868 (1978).

94. M. C. M. O'Brien and S. N. Evangelou, *J. Phys.* **C13**, 611 (1980).

95. E. Haller, Dipl. Thesis, University Freiburg (1980); E. Haller, L. S. Cederbaum, and W. Domcke, *Mol. Phys.* **41**, 1291 (1980).

96. G. Moro and J. H. Freed, *J. Chem. Phys.* **74**, 3757 (1981).

97. R. G. Gordon and T. Messenger, in *Electron-Spin Relaxation in Liquids*, L. T. Muus and P. W. Atkins, eds., Plenum Press, New York, 1972.

98. Y. V. Vorobyev, *Method of Moments in Applied Mathematics*, Gordon & Breach, New York, 1965.

99. R. R. Whitehead, in *Moment Methods in Many-Fermion Systems*, B. J. Dalton, S. M. Grimes, J. P. Vary, and S. A. Williams, eds., Plenum Press, New York, 1980.

100. D. O. Harris, G. G. Engerholm, and W. D. Gwinn, *J. Chem. Phys.* **43**, 1515 (1965).

101. A. S. Dickinson and P. R. Certain, *J. Chem. Phys.* **49**, 4209 (1968).

102. C. S. Lin and G. W. F. Drake, *Chem. Phys. Lett.* **16**, 35 (1972).

103. K. F. Freed, *Topics Appl. Phys.* **15**, 23 (1976) and references therein.

104. P. Avouris, W. M. Gelbart, and M. A. El-Sayed, *Chem. Rev.* **77**, 793 (1977) and references therein.

105. N. Shimakura, Y. Fujimura, and T. Nakajima, *Chem. Phys.* **19**, 155 (1977).

106. J. Chaiken, Th. Benson, M. Gurnick, and J. D. McDonald, *Chem. Phys. Lett.* **61**, 195 (1979).

107. W. A. Wassam and E. C. Lim, *Chem. Phys.* **38**, 217 (1979).

108. W. Siebrand and M. Z. Zgierski, *J. Chem. Phys.* **75**, 1230 (1981).

109. H. Hornburger and J. Brand, *Chem. Phys. Lett.* **88**, 153 (1982).

110. G. Dujardin, S. Leach, G. Taieb, J. P. Maier, and W. M. Gelbart, *J. Chem. Phys.* **73**, 4987 (1980).

111. J. P. Maier, *Angew. Chem.* **93**, 649 (1981).

112. M. Kasha, *Discuss. Faraday Soc.* **9**, 14 (1950).

113. M. Allan, E. Kloster-Jensen, and J. P. Maier, *J. Chem. Soc. Faraday Trans. II* **73**, 1406 (1977).

114. G. G. Balint-Kurti and R. N. Yardley, *Chem. Phys. Lett.* **36**, 342 (1975).

115. C. F. Jackels, *J. Chem. Phys.* **72**, 4873 (1980).

116. L. S. Cederbaum, H. Köppel, and W. Domcke, *Int. J. Quant. Chem.* **S15**, 251 (1981).

117. R. M. Hochstrasser and C. Marzzacco, *J. Chem. Phys.* **49**, 971 (1968).

118. M. Bixon, J. Jortner, and Y. Dothan, *Mol. Phys.* **17**, 109 (1969).

119. C. Tric, *J. Chem. Phys.* **55**, 4303 (1971).

120. F. K. Gel'mukhanov, L. N. Mazalov, A. V. Nikolaev, A. V. Kondratenko, V. G. Smirnii, P. I. Wadash, and A. P. Sadovskii, *Dokl. Akad. Nauk SSSR* **225**, 597 (1975).

121. F. K. Gel'mukhanov, L. N. Mazalov, and A. V. Kondratenko, *Chem. Phys. Lett.* **46**, 133 (1977); L. N. Mazalov, A. V. Kondratenko, F. K. Gel'mukhanov, V. V. Murakhtanov, and T. I. Guzhavina, *Theor. Chim. Acta* **44**, 257 (1977).

122. F. Kaspar, W. Domcke, and L. S. Cederbaum, *Chem. Phys.* **44**, 33 (1979).

123. A. E. Douglas, *J. Chem. Phys.* **45**, 1007 (1966).

124. R. D. Levine, *Quantum Mechanics of Molecular Rate Processes*, Oxford Univ. Press (Clarendon), New York, 1969.

125. E. J. Heller, *J. Chem. Phys.* **68**, 2066 (1978); ibid. p. 3891 (1978).

126. A. J. Lorquet, J. C. Lorquet, J. Delwiche, and M. J. Hubin-Franskin, *J. Chem. Phys.* **76**, 4692 (1982).

127. H. Köppel, *Chem. Phys.* **77**, 359 (1983).

128. F. B. Hildebrand, *Introduction to Numerical Analysis*, McGraw-Hill, New York, 1974.

129. W. Henneker, W. Siebrand, and M. Z. Zgierski, *J. Chem. Phys.* **74**, 6560 (1981).

130. I. Özkan and L. Goodman, *J. Chem. Phys.* **72**, 6777 (1980).

131. M. S. Child, *Mol. Phys.* **3**, 601 (1960).

132. J. T. Hougen, *J. Mol. Spectrosc.* **13**, 149 (1964).

133. A. R. Gregory, W. Siebrand, and M. Z. Zgierski, *J. Chem. Phys.* **64**, 3145 (1976).

134. J. Brickmann, *Ber. Bunsenges.* **80**, 917 (1976).

135. W. H. Henneker, W. Siebrand, and M. Z. Zgierski, *Chem. Phys. Lett.* **68**, 5 (1979).

136. W. Siebrand and M. Z. Zgierski, *Chem. Phys.* **52**, 321 (1980).

137. G. W. Robinson and C. A. Langhoff, *Chem. Phys.* **5**, 1 (1974).

138. W. Rhodes, in *Radiationless Transitions*, S. H. Lin, ed., Academic Press, New York, 1980, p. 219.

139. J. Wessel and D. S. McClure, *Mol. Cryst. Liquid Cryst.* **58**, 121 (1980).

140. W. D. Hobey, *J. Chem. Phys.* **43**, 2187 (1965).

141. M. H. Perrin and M. Gouterman, *J. Chem. Phys.* **46**, 1019 (1967).

142. M. F. Merienne-Lafore and H. P. Trommsdorff, *J. Chem. Phys.* **64**, 3791 (1976).

143. G. Orlandi, *Chem. Phys. Lett.* **44**, 277 (1976).

144. G. Orlandi and G. Marconi, *Chem. Phys. Lett.* **53**, 61 (1978); G. Marconi and G. Orlandi, *J. Chem. Soc. Faraday Trans. II* **78**, 565 (1982).

145. H. Köppel, L. S. Cederbaum, W. Domcke, and S. S. Shaik, *Angew. Chem., Int. Ed. Engl.* **22**, 210 (1983).

146. A. R. Lacey, E. F. McCoy, and I. G. Ross, *Chem. Phys. Lett.* **21**, 233 (1973).

147. P. J. Chappell and I. G. Ross, *Chem. Phys. Lett.* **43**, 440 (1976).

148. C. Cossart-Magos and S. Leach, *J. Chem. Phys.* **64**, 4006 (1976).

149. K. K. Innes, *J. Chem. Phys.* **76**, 2100 (1982).

150. B. Scharf and B. Honig, *Chem. Phys. Lett.* **7**, 132 (1977).

151. G. J. Small, *J. Chem. Phys.* **54**, 3300 (1971).

152. A. R. Gregory and R. Silbey, *J. Chem. Phys.* **65**, 4141 (1976).

153. R. Renner, *Z. Physik* **92**, 172 (1934).

154. Ch. Jungen and A. J. Merer, in *Molecular Spectroscopy: Modern Research*, K. N. Rao, ed., Vol. 2, Academic Press, New York, 1976, p. 127.

155. G. Duxbury, in *Molecular Spectroscopy*, Specialist Periodical Report, Vol. 3, Chemical Society, London, 1975, p. 497.

156. M. C. M. O'Brien, *J. Phys.* **C9**, 3153 (1976).

157. C. H. Leung and W. H. Kleiner, *Phys. Rev.* **B10**, 4434 (1974).

158. V. Z. Polinger, *Sov. Phys. JETP* **50**, 754 (1979).

159. A. N. Petelin and A. A. Kiselev, *Int. J. Quantum Chem.* **6**, 701 (1972); R. Colin, M. Herman, and I. Kopp, *Mol. Phys.* **37**, 1397 (1979).

160. M. S. Child, *Mol. Phys.* **3**, 601 (1960).

161. C. J. Ballhausen, *Theor. Chim. Acta* **3**, 368 (1965).

162. B. M. Hoffmann and M. A. Ratner, *Mol. Phys.* **35**, 901 (1978).

163. S. Muramatsu and N. Sakamoto, *J. Phys. Soc. Japan* **44**, 1640 (1978).

164. W. Moffit and W. Thorson, *Phys. Rev.* **108**, 1251 (1957); also in R. A. Daudel, ed., *Calcul des fonctions d'onde moléculaires*, CNRS, Paris, 1958, p. 141; W. Thorson and W. Moffit, *Phys. Rev.* **168**, 362 (1968).

165. J. C. Slonczewski and V. L. Moruzzi, *Physics* **3**, 237 (1967).

166. P. Habitz and W. H. E. Schwarz, *Theor. Chim. Acta* **28**, 267 (1973).

167. M. C. M. O'Brien and S. N. Evangelou, *Solid State Commun.* **36**, 29 (1980).

168. J. C. Slonczewski, *Phys. Rev.* **131**, 1596 (1963).

169. C. S. Sloane and R. Silbey, *J. Chem. Phys.* **56**, 6031 (1972).

170. J. R. Fletcher, *J. Phys.* **C5**, 852 (1972).

171. M. C. M. O'Brien, *J. Phys.* **C5**, 2045 (1972); S. N. Evangelou, M. C. M. O'Brien, and R. S. Perkins, ibid. **13**, 4175 (1980).

172. R. Englman and B. Halperin, *Ann. Phys.* **3**, 453 (1978).

173. V. Z. Polinger and I. G. Bersuker, *Phys. Status Solidi* **95**, 403 (1979); **96**, 153 (1979).

174. J. R. Fletcher, M. C. M. O'Brien, and S. N. Evangelou, *J. Phys.* **A13**, 2035 (1980).

175. L. S. Cederbaum, E. Haller, and W. Domcke, *Solid State Commun.* **35**, 879 (1980).

176. T. J. Sears, T. A. Miller, and V. E. Bondybey, *J. Chem. Phys.* **74**, 3240 (1981).

177. J. H. van der Waals, A. M. D. Berghuis, and M. S. de Groot, *Mol. Phys.* **13**, 301 (1967); **21**, 497 (1971).

178. P. J. Stephens, *J. Chem. Phys.* **51**, 1995 (1969).

179. M. Z. Zgierski and M. Pawlikowski, *J. Chem. Phys.* **70**, 3444 (1979).

180. E. Haller, H. Köppel, L. S. Cederbaum, G. Bieri, and W. von Niessen, *Chem. Phys. Lett.* **85**, 12 (1982).

181. J. A. Pople, *Mol. Phys* **3**, 16 (1960).

182. J. T. Hougen, *J. Chem. Phys.* **36**, 519, 1874 (1962).

183. A. J. Merer and D. N. Travis, *Can. J. Phys.* **43**, 1795 (1965).

184. J. A. Pople and H. C. Louguet-Higgins, *Mol. Phys.* **1**, 372 (1958).

185. R. N. Dixon, *Mol. Phys.* **9**, 357 (1965).

186. T. Barrow, R. N. Dixon, and G. Duxbury, *Mol. Phys.* **27**, 1217 (1974).

187. Ch. Jungen and A. J. Merer, *Mol. Phys.* **40**, 1, 95 (1980); Ch. Jungen, K.-E. J. Hallin, and A. J. Merer, ibid. **40**, 25, 65 (1980); D. Gauyacq and Ch. Jungen, ibid. **41**, 383 (1980).

188. M. Peric and J. Radic-Peric, *Chem. Phys. Lett.* **67**, 138 (1979); M. Peric, ibid. **76**, 573 (1980); R. J. Buenker, M. Peric, S. D. Peyerimhoff, and R. Marian, *Mol. Phys.* **43**, 987 (1981).

189. J. M. Brown and F. Jørgensen, *Mol. Phys.* **47**, 1065 (1982).
190. P. S. H. Bolman and J. M. Brown, *Chem. Phys. Lett.* **21**, 213 (1973).
191. J. M. Brown, *J. Mol. Spectrosc.* **68**, 412 (1977).
192. J. F. M. Aarts, *Mol. Phys.* **35**, 1785 (1978).
193. H. Köppel, W. Domcke, and L. S. Cederbaum, *J. Chem. Phys.* **74**, 2945 (1981).
194. R. N. Dixon and D. A. Ramsay, *Can. J. Phys.* **46**, 2619 (1968).
195. D. C. Frost, S. T. Lee, and C. A. McDowell, *Chem. Phys. Lett.* **23**, 472 (1973).
196. S. F. Fischer, *Chem. Phys. Lett.* **91**, 367 (1982).
197. A. Witkowski and W. Moffit, J. Chem. Phys. **33**, 872 (1960).
198. W. Domcke and L. S. Cederbaum, *Chem. Phys.* **25**, 189 (1977).
199. M. Pawlikowski and M. Z. Zgierski, *J. Chem. Phys.* **76**, 4789 (1982).
200. F. Brogli, E. Heilbronner, E. Kloster-Jensen, A. Schmelzer, A. S. Manocha, J. A. Pople, and L. Radom, *Chem. Phys.* **4**, 107 (1974); G. Bieri, J. D. Dill, E. Heilbronner, J. P. Maier, and J. L. Ripoll, *Helv. Chim. Acta* **60**, 629 (1977).
201. W. von Niessen, G. H. F. Diercksen, L. S. Cederbaum, and W. Domcke, *Chem. Phys.* **18**, 469 (1976).
202. F. A. Miller and I. Matsubara, *Spectrochim. Acta* **22**, 173 (1966).
203. D. M. Mintz and A. Kuppermann, *J. Chem. Phys.* **71**, 3499 (1979).
204. I. Fischer-Hjalmars and P. Siegbahn, *Theor. Chim. Acta* **31**, 1 (1973).
205. M. B. Robin, R. R. Hart, and N. A. Kuebler, *J. Chem. Phys.* **44**, 1803 (1966).
206. G. Herzberg, *Infrared and Raman Spectra of Polyatomic Molecules*, Van Nostrand, New York, 1945.
207. A. D. Baker, C. Baker, C. R. Brundle, and D. W. Turner, *Int. J. Mass Spectrom. Ion Phys.* **1**, 285 (1968).
208. R. Stockbauer and M. G. Inghram, *J. Electron Spectrosc.* **7**, 492 (1975).
209. J. E. Pollard, D. J. Trevor, Y. T. Lee, and D. A. Shirley, *Rev. Sci. Instrum.* **52**, 1837 (1981).
210. W. A. Lathan, W. J. Hehre, and J. A. Pople, *J. Amer. Chem. Soc.* **93**, 808 (1971).
211. W. R. Rodwell, M. F. Guest, D. T. Clark, and D. Shuttleworth, *Chem. Phys. Lett.* **45**, 50 (1977).
212. D. W. Turner, C. Baker, A. D. Baker, and C. R. Brundle, *Molecular Photoelectron Spectroscopy*, Wiley (Interscience), New York, 1970.
213. T. Cvitas, H. Güsten, and L. Klasinc, *J. Chem. Phys.* **70**, 57 (1979).
214. T. A. Brody, J. Flores, J. B. French, P. A. Mello, A. Pandey, and S. M. S. Wong, *Rev. Mod. Phys.* **53**, 385 (1981).
215. E. Haselbach, *Chem. Phys. Lett.* **7**, 428 (1970).
216. L. S. Cederbaum, W. Domcke, and H. Köppel, *Chem. Phys.* **33**, 319 (1978).
217. D. Brewster, *Trans. Roy. Soc. Edinburgh* **12**, 519 (1834).
218. A. J. Merer and K.-E. J. Hallin, *Can. J. Phys.* **56**, 838 (1978).
219. A. E. Douglas and K. P. Huber, *Can. J. Phys.* **43**, 74 (1965).
220. C. G. Stevens and R. N. Zare, *J. Mol. Spectrosc.* **56**, 167 (1975); T. C. Lee and E. R. Peck, ibid. **65**, 249 (1977); R. Schmiedl, I. R. Bonilla, F. Paech, and W. Demtröder, ibid. **68**, 236 (1977).
221. J. C. D. Brand, W. H. Chan, and J. L. Hardwick, *J. Mol. Spectrosc.* **56**, 309 (1975).

222. R. E. Smalley, L. Wharton, and D. H. Levy, *J. Chem. Phys.* **63**, 4977 (1975).

223. F. Bylicki and H. G. Weber, *Chem. Phys. Lett.* **79**, 517 (1981); P. J. Brucat and R. N. Zare, *J. Chem. Phys.* **78**, 100 (1983).

224. V. M. Donnelly and F. Kaufman, *J. Chem. Phys.* **66**, 4100 (1977).

225. C. F. Jackels and E. R. Davidson, *J. Chem. Phys.* **64**, 2908 (1976).

226. G. D. Gillispie, A. U. Khan, A. C. Wahl, R. P. Hosteny, and M. Krauss, *J. Chem. Phys.* **63**, 3425 (1975).

227. K. K. Innes, *J. Mol. Spectrosc.* **96**, 331 (1982).

228. J. L. Hardwick and J. C. D. Brand, *Chem. Phys. Lett.* **21**, 458 (1973).

229. J. C. D. Brand and A. R. Hoy, *J. Mol. Spectrosc.* **65**, 75 (1977).

230. M. Allen, J. P. Maier, and O. Marthaler, *Chem. Phys.* **26**, 136 (1977); J. P. Maier and O. Marthaler, ibid. **32**, 419 (1978).

231. C. Cossart-Magos, D. Cossart, and S. Leach, *J. Chem. Phys.* **69**, 4313 (1978); *Mol. Phys.* **37**, 793 (1979).

232. C. Cossart-Magos, D. Cossart, and S. Leach, *Chem. Phys.* **41**, 345, 363 (1979).

233. T. A. Miller, V. E. Bondybey, and J. H. English, *J. Chem. Phys.* **70**, 2919 (1979); **71**, 1088 (1979); *J. Mol. Spectrosc.* **81**, 455 (1980).

234. T. J. Sears, T. A. Miller, and V. E. Bondybey, *J. Chem. Phys.* **72**, 6749 (1980).

235. V. E. Bondybey, T. J. Sears, T. A. Miller, and J. H. English, *J. Chem. Phys.* **73**, 2063 (1980).

236. V. E. Bondybey and T. A. Miller, *J. Chem. Phys.* **70**, 138 (1979); **73**, 3053 (1980).

237. T. J. Sears, T. A. Miller, and V. E. Bondybey, *J. Amer. Chem. Soc.* **103**, 326 (1981).

238. A. W. Potts, H. J. Lempka, D. G. Streets, and W. C. Price, *Phil. Trans. Roy. Soc. London* **A268**, 59 (1970).

239. P. J. Bassett and D. R. Lloyd, *Chem. Commun.* p. 36 (1970); *J. Chem. Soc., A* p. 1955 (1971).

240. G. H. King, S. S. Krishnamurthy, M. F. Lappert, and J. B. Pedley, *Faraday Discuss. Chem. Soc.* **54**, 70 (1972).

241. C. F. Batten, J. A. Taylor, B. P. Tsai, and G. G. Meisels, *J. Chem. Phys.* **69**, 2547 (1978).

242. E. Haller, H. Köppel, L. S. Cederbaum, W. von Niessen, and G. Bieri, *J. Chem. Phys.* **78**, 1359 (1983).

243. E. Haller, L. S. Cederbaum, W. Domcke, and H. Köppel, *Chem. Phys. Lett.* **72**, 427 (1980).

244. A. D. Liehr, *Z. Naturforsch.* **A16**, 641 (1961). ·

245. M. Mizushima and S. Koide, *J. Chem. Phys.* **20**, 765 (1952).

246. A. M. Ponte-Goncalves and C. A. Hutchison, *J. Chem. Phys.* **49** 4235 (1968), M. S. de Groot, I. A. M. Hesselmann, and J. H. van der Waals, *Mol. Phys.* **16**, 45 (1969).

247. D. M. Burland, G. Castro, and G. W. Robinson, *J. Chem. Phys.* **52**, 4100 (1970).

248. A. C. Albrecht, *J. Chem. Phys.* **33**, 169 (1960).

249. G. C. Nieman and D. Tinti, *J. Chem. Phys.* **46**, 1432 (1967); G. W. King and E. H. Pinnington, *J. Mol. Spectrosc.* **15**, 394 (1965).

250. J. V. Egmond and J. H. van der Waals, *Mol. Phys.* **28**, 457 (1974).

251. A. Carrington, A. R. Fabris, B. J. Howard, and N. J. D. Lucas, *Mol. Phys.* **20**, 961 (1971).

252. P. S. H. Bolman, J. M. Brown, A. Carrington, I. Kopp, and D. A. Ramsay, *Proc. Roy. Soc.* (London) **A343**, 17 (1975).

253. R. N. Dixon, *Phil. Trans. Roy. Soc. London* **A252**, 165 (1960); C. Devillers and D. A. Ramsay, *Can. J. Phys.* **49**, 2839 (1971).

254. L. S. Cederbaum, W. Domcke, J. Schirmer, and H. Köppel, *J. Chem. Phys.* **72**, 1348 (1980).

255. H. Köppel, L. S. Cederbaum, W. Domcke, and W. von Niessen, *Chem. Phys.* **37**, 303 (1979).

256. C. Fridh and L. Åsbrink, *J. Electron Spectrosc.* **7**, 119 (1975).

257. S. P. So and W. G. Richards, *J. Chem. Soc. Faraday Trans. II* **71**, 62 (1975); W. von Niessen, L. S. Cederbaum, W. Domcke, and G. H. F. Diercksen, *Mol. Phys.* **32**, 1057 (1976).

258. The definition of l_2 in Section V.D differs from that of l in Section IV.C. Whereas previously l denoted the vibrational angular momentum in the adiabatic basis, l_2 refers to the vibrational angular momentum in the diabatic electronic basis.

259. L. A. Franks, A. J. Merer, and K. K. Innes, *J. Mol. Spectrosc.* **26**, 458 (1968).

260. H. Köppel, L. S. Cederbaum, and W. Domcke, *Chem. Phys.* **69**, 175 (1982).

261. R. S. Mulliken, *Chem. Phys. Lett.* **46**, 197 (1977).

262. R. Gallusser and K. Dressler, *J. Chem. Phys.* **76**, 4311 (1982).

263. L. Wolniewicz and K. Dressler, *J. Mol. Spectrosc.* **67**, 416 (1977).

264. K. Dressler, R. Gallusser, P. Quadrelli, and L. Wolniewicz, *J. Mol. Spectrosc.* **75**, 205 (1979).

265. L. S. Cederbaum and H. Köppel, *Chem. Phys. Lett.* **87**, 14 (1982).

266. L. S. Cederbaum, *J. Chem. Phys.* **78**, 5714 (1983).

267. E. P. Wigner, *Ann. Math.* **53**, 36 (1951).

268. E. P. Wigner, *SIAM Rev.* **9**, 1 (1967).

269. A. Pandey, *Ann. Phys.* (N.Y.) **119**, 170 (1979).

270. R. U. Hag, A. Pandey, and O. Bohigas, *Phys. Rev. Lett.* **48**, 1086 (1982).

271. H. S. Camarda and P. D. Georgopulos, *Phys. Rev. Lett.* **50**, 492 (1983).

272. V. Buch, R. B. Gerber, and M. A. Ratner, *J. Chem. Phys.* **76**, 5397 (1982).

273. F. J. Dyson and M. L. Mehta, *J. Math. Phys.* **4**, 701 (1963).

274. F. J. Dyson, *J. Math. Phys.* **3**, 166 (1962).

275. R. Venkatamaran, *J. Phys.* **B15**, 4293 (1982).

276. J. D. Garrison, *Ann. Phys.* (N.Y.) **30**, 269 (1964).

277. O. Bohigas and M. J. Giannoni, *Ann. Phys.* (N.Y.) **89**, 393 (1975).

278. T. A. Brody, *Lett. Nuovo Cimento* **7**, 482 (1973).

279. D. Neuberger and A. B. F. Duncan, *J. Chem. Phys.* **22**, 1693 (1954).

280. L. F. Keyser, F. Kaufman, and E. C. Zipf, *Chem. Phys. Lett.* **2**, 523 (1968).

281. S. E. Schwartz and H. S. Johnston, *J. Chem. Phys.* **51**, 1286 (1969).

282. P. B. Sackett and J. T. Yardley, *Chem. Phys. Lett.* **6**, 323 (1970); **9**, 612 (1971); *J. Chem. Phys.* **57**, 152 (1972).

283. C. G. Stevens, M. W. Swagle, R. Wallace, and R. N. Zare, *Chem. Phys. Lett.* **18**, 465 (1973).

284. R. Solarz and D. H. Levy, *J. Chem. Phys.* **60**, 842 (1974).

285. F. Paech, R. Schmiedl, and W. Demtröder, *J. Chem. Phys.* **63**, 4369 (1975).

286. C. H. Chen, S. D. Kramer, D. W. Clark, and M. G. Payne, *Chem. Phys. Lett.* **65**, 419 (1979).

287. M. Jeunehomme and A. B. F. Duncan, *J. Chem. Phys.* **41**, 1692 (1964).

288. L. S. Cederbaum, W. Domcke, J. Schirmer, and W. von Niessen, *Phys. Scripta* **21**, 481 (1980).

289. G. Wendin, *Struct. Bonding* **45** (1981).

290. J. Schirmer, L. S. Cederbaum, W. Domcke, and W. von Niessen, *Chem. Phys.* **26**, 149 (1977); M. F. Herman, K. F. Freed, and D. F. Yeager, ibid. **32**, 437 (1978).

291. F. Lahmami, A. Tramer, and C. Tric, *J. Chem. Phys.* **60**, 4431 (1974).

292. M. Bixon and J. Jortner, *J. Chem. Phys.* **50**, 3284 (1969).

293. W. M. Gelbart, D. F. Heller, and M. L. Elert, *Chem. Phys.* **7**, 116 (1975).

294. S. S. M. Wong and J. B. French, *Nucl. Phys.* **A198**, 188 (1972).

295. J. P. Maier, O. Marthaler, and G. Bieri, *Chem. Phys.* **44**, 131 (1979).

296. R. Stockbauer and M. G. Inghram, *J. Chem. Phys.* **62**, 4862 (1975).

297. J. L. Duncan, M. C. McKean, and P. D. Mallinson, *J. Mol. Spectrosc.* **45**, 221 (1973).

298. W. M. Gelbart and K. F. Freed, *Chem. Phys. Lett.* **18**, 470 (1973).

299. J. P. Maier, private communication.

300. I. B. Bersuker, *The Jahn-Teller Effect and Vibronic Interactions in Modern Chemistry*, Plenum Press, New York, 1984.

301. I. Powis, *J. Chem. Soc. Faraday Trans. II* **77**, 1433 (1981).

302. H.-D. Meyer and H. Köppel, *J. Chem. Phys.*, in the press.

303. J. P. Maier and F. Thommen, *J. Chem. Soc. Faraday Trans. II* **77**, 845 (1981).

JAHN–TELLER TRAJECTORIES

B. R. JUDD

Physics Department
The Johns Hopkins University
Baltimore, Maryland

CONTENTS

I. INTRODUCTION

The Jahn–Teller (JT) effect provides, on an atomic scale, an example that broken symmetry is more of a commonplace in nature than we might have anticipated, given the simplicity and symmetry of the basic laws of physics. The first statement of the effect was made at the Washington meeting of the American Physical Society in the spring of 1936.[1] The abstract runs:

As a general rule the electronic state of a polyatomic molecule can be degenerate only if the atomic configuration has a sufficiently high degree of symmetry. If the atomic nuclei are displaced the degenerate state may split up and if the splitting is a linear function of the displacement the original symmetrical configuration, and with it the original degenerate state, does not correspond to an equilibrium state of the molecule. A group-theoretical investigation shows that except for molecules in which all atoms lie on a straight line only undegenerate states or the doubly degenerate states of molecules with an odd number of electrons can correspond to stable configurations. . . .

In other words, nonlinear molecules with electronic degeneracy of the non-Kramers type are expected to spontaneously distort to remove the degeneracy.

The origin of the JT effect comes from the fact that a displacement of atoms (or *ligands*, to use a more suggestive word) from a symmetric configuration usually produces splittings of a degenerate electronic energy level that are linear with respect to the displacement, while all elastic restoring forces are derived from potentials that are quadratic. Equilibrium is therefore achieved for a nonzero value of the displacement.

The above remarks refer to what is often called the static JT effect. Once the kinetic energy of the constituent atoms is included, motions about their displaced sites take place. It is the purpose of the present work to explore the trajectories of these oscillating ligands. It might be thought that such a task would be of only moderate interest. For example, we might think that the usual normal-mode analysis should present few difficulties and the resulting description would completely define the physical situation. However, a solution in these terms is inadequate in two respects: it leaves the actual motions of the ligands at one remove from the normal-mode description; and, what is more significant, it provides very little physical insight into those JT systems where, in the face of what is ostensibly a symmetry-breaking interaction, the final energy-level structure can be labeled by the irreducible representations of groups that are sometimes higher (that is, they comprise more elements or more generators) than the point group for the JT molecule in its undistorted form. As an example of this, we mention the octahedral system $\Gamma_8 \otimes \tau_2$, which was discovered some years ago by Moffitt and Thorson[2] to possess the symmetry of the rotation group in three dimensions, R_3 [or, in the more modern notation, the special orthogonal group in three dimensions, SO(3)].

In the present work the current method of labeling JT systems is used. That is, the irreducible representation of the degenerate electronic state (Γ_8 in the example above) is separated by the symbol $\otimes$ from the irreducible representation(s) of the mode or modes that describe the type of displacement under study. The usual representation labels are converted to their greek forms for the modes (as in the use of τ_2 for the three-dimensional irreducible representation T_2 of the octahedral group O).

It would not be practicable to work through the enormous number of JT systems that have been studied over the years. All that can be done is to illustrate the main features with examples. To give as much coherence to this review article as possible, these examples are all drawn from octahedral JT systems.

Classical Oscillations

The notion of a trajectory is classical by its very nature. Well-defined positions on the trajectories are connected by equally well-defined momentum vectors. This apparent conflict with the tenets of quantum mechanics is less of a difficulty than might be imagined. We are mainly concerned with understanding how the motions of the atoms are consistent with the symmetry groups of the various Hamiltonians, rather than using the trajectories to calculate in full detail the properties of the JT systems. We can always justify the significance of the trajectories by taking the quantum mechanical solution in the limit of large quantum numbers, of course, though this is a procedure we prefer to think of as being carried out in principle rather than in fact. The import of quantum mechanics is felt in such topics as tunneling, where classical physics fails.

The simplest case to treat is that of a single atom, of mass m, displaced from its original site to a new equilibrium position E, which we take to be the origin of coordinates. For a conservative field, we know that we can write $\mathbf{f} = -\nabla U$ for the force on the atom, where U is a potential function. At E we have $\mathbf{f} = 0$. A new small displacement $\delta\mathbf{r}$ from E provokes a restoring force

$$\delta\mathbf{f} = -(\delta\mathbf{r}\cdot\nabla)\nabla U \equiv -\mathbf{A}\,\delta\mathbf{r}$$

where $\mathbf{A}$ is a symmetric matrix. Thus $\delta\mathbf{f}$ and $\delta\mathbf{r}$ are not in general parallel. Writing $\mathbf{r}$ for $\delta\mathbf{r}$, we find that the equation of motion for the atom is $\ddot{\mathbf{r}} = \mathbf{A}\mathbf{r}/m$, which suffers from the defect that $\ddot{x}$ involves not only x but y and z as well. To remedy this, we rotate the coordinates by setting $\mathbf{r} = \mathbf{D}\mathbf{r}'$, where $\mathbf{D}$ is an orthogonal matrix with three rows and three columns. The equation of motion now becomes

$$\mathbf{D}\ddot{\mathbf{r}}' = -\frac{\mathbf{A}\mathbf{D}\mathbf{r}'}{m}$$

whence

$$\ddot{\mathbf{r}}' = -\frac{(\mathbf{D}^{-1}\mathbf{A}\mathbf{D})\mathbf{r}'}{m} \tag{1}$$

Since $\mathbf{A}$ is a symmetric matrix, we can choose $\mathbf{D}$ so that $\mathbf{D}^{-1}\mathbf{A}\mathbf{D}$ takes the diagonal form

$$\begin{pmatrix} m\omega_x^2 & 0 & 0 \\ 0 & m\omega_y^2 & 0 \\ 0 & 0 & m\omega_z^2 \end{pmatrix}$$

The form of the elements has been chosen so that the solutions to (1) can be written

$$\begin{aligned} x' &= x_0'\cos(\omega_x t - \sigma_x) \\ y' &= y_0'\cos(\omega_y t - \sigma_y) \\ z' &= z_0'\cos(\omega_z t - \sigma_z) \end{aligned} \tag{2}$$

The trajectory represented by Eqs. (2) is not, in general, simple. A twisting path is threaded out as t increases. It lies within the rectangular parallelepiped bounded by the planes

$$x' = \pm x_0', \qquad y' = \pm y_0', \qquad z' = \pm z_0'$$

If ω_x, ω_y, and ω_z are relatively rational (so that we can write $\omega_x = n_x\omega$, $\omega_y = n_y\omega$, and $\omega_z = n_z\omega$, where n_x, n_y, and n_z are integers with no common factor other than unity), the path closes on itself after a period $2\pi/\omega$.

If $\omega_x = \omega_y = \omega_z$, the trajectory represented by (2) becomes an ellipse the plane of whose orbit possesses a normal with direction cosines proportional to

$$\frac{\sin(\sigma_y - \sigma_z)}{x_0'}, \quad \frac{\sin(\sigma_z - \sigma_x)}{y_0'}, \quad \frac{\sin(\sigma_x - \sigma_y)}{z_0'} \tag{3}$$

Multiplication of the right-hand sides of Eqs. (2) by these quantities and addition yields zero, showing that the trajectory lies in a plane passing through the origin E. It is more difficult to confirm the elliptical nature of the trajectory from (2); in fact, it is simplest to return to the equation of motion $\ddot{\mathbf{r}} = -\omega^2\mathbf{r}$ and appeal to the well-known result for an isotropic three-dimensional harmonic oscillator.[3]

For isotropic media we have $\omega_x = \omega_y = \omega_z$. Elliptical trajectories are thus to be expected and, indeed, have been recognized for many years in octahedral and other systems of high symmetry. The reader is referred to the remarkable article of Liehr[4] and especially to Figures 30a–c therein for some examples.

II. THE OCTAHEDRAL GROUP

Since all our examples are drawn from octahedral complexes, some background to these systems and to the octahedral group O seems necessary. We shall only touch on those aspects of group theory that are of direct use to us either in the construction of the normal modes of vibration of an octahedron or in the elucidation of the properties of the electronic state that is coupled to these modes.

A. Classes

Two things are needed to bring group theory into play: the character table and sample operations from each of the classes. Character tables are readily available for all the groups of interest to molecular spectroscopists (see, for example, Koster et al.[5]). If we augment O with the operations of inversion and reflection, we obtain the group O_h. Its characters are set out in Table I. The ten classes C_i each correspond to a different kind of rotation or, for classes C_6 through C_{10}, rotation and inversion. The angles of rotation are 0, π, π, $\frac{2}{3}\pi$, and $\frac{1}{2}\pi$ for C_1, C_2, C_3, C_4, and C_5, respectively. There are two classes characterized by the same angle π because there are two essentially distinct classes of rotations by π that send an octahedron into itself: for C_2 the axis joins the midpoints of opposite edges; for C_3 the axis passès through opposite vertices of the octahedron. The angle of rotation for C_{5+i} is the same as that for C_i, but an inversion operation is included.

Let us set up an orthogonal reference frame as indicated in Fig. 1. Each operation of O_h can be represented by a transformation matrix acting on the column matrix having successive elements x, y, and z. For example, when the axis of an operation belonging to C_2 lies in the xy plane and makes an angle of $\frac{1}{4}\pi$ with the positive x and y axes, the transformation matrix is

$$\begin{pmatrix} 0 & 1 & 0 \\ 1 & 0 & 0 \\ 0 & 0 & -1 \end{pmatrix}$$

This corresponds to the simple substitution $(x, y, z) \rightarrow (y, x, -z)$. In fact, our sample operations from all 10 classes can be chosen so that the symbols x, y, and z are merely permuted with a possible change of sign. It is not difficult to confirm that the following substitutions constitute an acceptable set for each of the 10 classes $C_1, C_2, \ldots, C_{10}$:

$$\begin{aligned} (x, y, z) \rightarrow & (x, y, z), (y, x, -z), (x, -y, -z), (z, x, y), \\ & (x, -z, y), (-x, -y, -z), (-y, -x, z), \\ & (-x, y, z), (-z, -x, -y), (-x, z, -y) \end{aligned} \tag{4}$$

TABLE I

Characters for O_h

Irreducible Representation	Class									
	C_1	C_2	C_3	C_4	C_5	C_6	C_7	C_8	C_9	C_{10}
A_{1g}	1	1	1	1	1	1	1	1	1	1
A_{2g}	1	-1	1	1	-1	1	-1	1	1	-1
E_g	2	0	2	-1	0	2	0	2	-1	0
T_{1g}	3	-1	-1	0	1	3	-1	-1	0	1
T_{2g}	3	1	-1	0	-1	3	1	-1	0	-1
A_{1u}	1	1	1	1	1	-1	-1	-1	-1	-1
A_{2u}	1	-1	1	1	-1	-1	1	-1	-1	1
E_u	2	0	2	-1	0	-2	0	-2	1	0
T_{1u}	3	-1	-1	0	1	-3	1	1	0	-1
T_{2u}	3	1	-1	0	-1	-3	-1	1	0	1

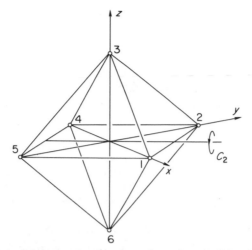

Fig. 1.   The octahedron of ligands. The six vertices $1, 2, \ldots, 6$ of the regular octahedron define the directions of the positive and negative x, y, and z axes. A rotation by π about the symmetry axis labeled C_2 interchanges the pairs of ligands $(1,2)$, $(4,5)$, and $(3,6)$.

The traces, or characters, of the corresponding 10 matrices are at once seen to be $3, -1, -1, 0, 1, -3, 1, 1, 0, -1$. This sequence of numbers appears in the character table for O_h labeled by T_{1u}. We can conclude that the three functions x, y, and z form a basis for the irreducible representation T_{1u} of O_h.

The substitutions (4) can be used to find how more complicated functions of x, y, and z transform. The triple product xyz, for example, consistently transforms into itself (with a possible sign reversal); and, from the sequence obtained by taking each substitution of (4) in succession, we may rapidly confirm that xyz forms by itself a basis for the irreducible representation A_{2u} of O_h.

B.   Normal Coordinates

To describe the normal modes of oscillation of an octahedron, we set up similarly oriented coordinate frames at the equilibrium positions for each site, as shown in Fig. 2. We have now to form linear combinations of the 18 displacements $x_1, y_1, \ldots, z_6$ that transform according to the irreducible representations of O_h. The first step is to find which irreducible representations occur. This is not as difficult a task as might at first appear. Under the action of the identity element every displacement goes into itself, so the transformation matrix is diagonal with a trace of 18. If we take the operation of C_2 described in Section II.A, the sites 1 and 2 interchange, as do 4 and 5 as well as 3 and 6. Therefore no displacement contributes to the diagonal of

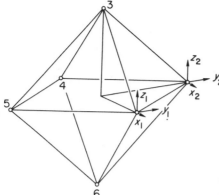

Fig. 2. Coordinate axes (x_i, y_i, z_i) to describe the displacements of ligands i. At each vertex of the regular octahedron, a similarly oriented right-handed set of axes is set up. This is illustrated here for sites 1 and 2.

the transformation matrix and so the character is 0. Proceeding in this way, we arrive at the sequence

$$18, 0, -2, 0, 2, 0, 2, 4, 0, 0 \qquad (5)$$

for the characters of the 10 classes of O_h. The decomposition of this sequence into the rows appearing in Table I can be carried out by using the orthogonality properties of the rows of the character table (when each entry is weighted by the number of elements in the corresponding class). However, for our purposes we can simply state that (5) corresponds to the superposition

$$A_{1g} + E_g + T_{1g} + 2T_{1u} + T_{2g} + T_{2u} \qquad (6)$$

a result that can be confirmed by adding the appropriate entries in Table I.

The next step is to construct the linear combinations of $x_1, y_1, \ldots, z_6$ that form bases for the irreducible representations appearing in (6). As an example, take T_{2g}. In the context of the simple Cartesian forms of (4), we could construct a basis by taking the three functions yz, zx, and xy. There is no way that we can directly carry these results over to the problem in hand; but we can at least try to find transformation matrices that match those combing from (yz, zx, xy). In other words, we seek normal coordinates Q_{yz}, Q_{zx}, and Q_{xy} that transform like yz, zx, and xy, and that are linear combinations of $x_1, y_1, \ldots, z_6$. For example, the sample operation C_2 listed in (4) yields

$$(yz, zx, xy) \rightarrow (-zx, -yz, xy) \qquad (7)$$

On the other hand, the corresponding rotation of the octahedron (with its vertices displaced) yields

$$(x_1, y_1, z_1, x_2, y_2, z_2, x_3, y_3, z_3, x_4, y_4, z_4, \ldots, z_6)$$
$$\rightarrow (y_2, x_2, -z_2, y_1, x_1, -z_1, y_6, x_6, -z_6, \ldots, -z_3) \qquad (8)$$

as can be seen from Figs. 1 and 2. Suppose we try $Q_{yz} = (x_1 + \cdots)$. A cyclic permutation of the coordinate axes yields $Q_{zx} = (y_2 + \cdots)$. Now (7) indicates that we must have $Q_{yz} \rightarrow -Q_{zx}$: however, (8) contains $x_1 \rightarrow y_2$, which implies $Q_{yz} \rightarrow Q_{zx}$. This conflict can be resolved only by dropping x_1 from Q_{yz} and y_2 from Q_{zx}.

Suppose, on the other hand, we try $Q_{yz} = (z_2 + \cdots)$, which implies $Q_{zx} = (x_3 + \cdots)$ and $Q_{xy} = (y_1 + \cdots)$ by cyclic permutation. (In the last parenthesis y_1 rather than y_4 appears because $x \rightarrow y \rightarrow z \rightarrow x \rightarrow \cdots$ corresponds to $1 \rightarrow 2 \rightarrow 3 \rightarrow 1 \rightarrow \cdots$) As before, (7) indicates $Q_{yz} \rightarrow -Q_{zx}$; but now (8) leads to $Q_{zx} = (z_1 + \cdots)$, $Q_{yz} = (-y_6 + \cdots)$, and $Q_{xy} = (x_2 + \cdots)$. The conflict has disappeared; in fact, we have gained considerable information because, for consistency, we must have

$$Q_{yz} = (z_2 - y_6 + \cdots), \quad Q_{zx} = (x_3 + z_1 + \cdots), \quad Q_{xy} = (y_1 + x_2 + \cdots)$$

We can now use the fact that T_{2g} is even with respect to inversion, as the suffix g (for *gerade*) indicates. Thus z_2 must be combined with $-z_5$, and y_6 with $-y_3$, etc. Thus we get

$$Q_{yz} = (z_2 - y_6 - z_5 + y_3 + \cdots)$$
$$Q_{zx} = (x_3 + z_1 - x_6 - z_4 + \cdots) \qquad (9)$$
$$Q_{xy} = (y_1 + x_2 - y_4 - x_5 + \cdots)$$

By considering other sample operations, we find, remarkably, that the coordinates in Eqs. (9) are complete and each ellipsis can be deleted. Including the (arbitrary) normalizing factors $\frac{1}{2}$ and rearranging the entries to bring out the symmetry better, we conclude that

$$Q_{yz} = \tfrac{1}{2}(z_2 - z_5 + y_3 - y_6)$$
$$Q_{zx} = \tfrac{1}{2}(x_3 - x_6 + z_1 - z_4) \qquad (10)$$
$$Q_{xy} = \tfrac{1}{2}(y_1 - y_4 + x_2 - x_5)$$

Other normal coordinates can be constructed in a similar way. In particular, it can be shown that the two normal coordinates Q_θ and Q_ε belonging to E_g can be written, to within a constant factor, as

$$Q_\theta = \left(\tfrac{1}{12}\right)^{1/2}(2z_3 - 2z_6 - x_1 + x_4 - y_2 + y_5)$$
$$Q_\varepsilon = \tfrac{1}{2}(x_1 - x_4 - y_2 + y_5) \tag{11}$$

This pair transforms in a parallel way to $\left(\tfrac{1}{12}\right)^{1/2}(3z^2 - r^2)$ and $\tfrac{1}{2}(x^2 - y^2)$.

C. Normal Modes

Equations (10) and (11) give linear combinations of the displacements that transform according to the irreducible representations T_{2g} and E_g of O_h. In keeping with the notation described in Section I.A, the roman symbols T and E are replaced by τ and ε when labeling the normal modes themselves. To visualize these modes, it is convenient to make use of a property of the normal coordinates: the matrix transforming the 18 displacements $x_1, y_1, \ldots, z_6$ into the 18 normal coordinates is orthogonal.[6] This has a very remarkable consequence: if we pick any normal coordinate and set the displacements $x_1, y_1, \ldots, z_6$ equal to the coefficients that preceed them, only that particular normal coordinate is nonzero. For example, if we take Q_{yz} and set $z_2 = 1$, $z_5 = -1$, $y_3 = 1$, $y_6 = -1$, and all other displacements equal to zero, then it turns out that all 18 of the Q's vanish with the sole exception of Q_{yz}. Evidently the displacements $(1, -1, 1, -1)$ for (z_2, z_5, y_3, y_6) represent Q_{yz} in a direct and transparent way. The normal modes corresponding to τ_{2g} and ε_g are sketched in Figs. 3 and 4.

A classically vibrating octahedron would not exhibit normal-mode oscillations unless it were specially set up to do so. A decomposition of the typically complex classical motion into normal modes would be very much of a mathematical exercise. Quantum mechanics gives a much greater significance to the normal modes because, in breaking the Hamiltonian up into uncoupled parts, each normal mode becomes separately quantized. Quantum mechanics forces us to pay attention to JT systems labeled by specific normal modes; but our present interest in the corresponding trajectories im-

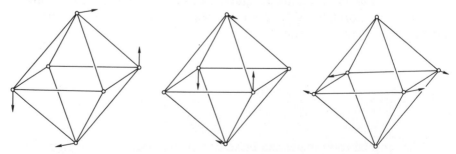

Fig. 3. The three components of the octahedral mode τ_{2g}. From left to right the corresponding normal coordinates are Q_{yz}, Q_{zx}, and Q_{xy}.

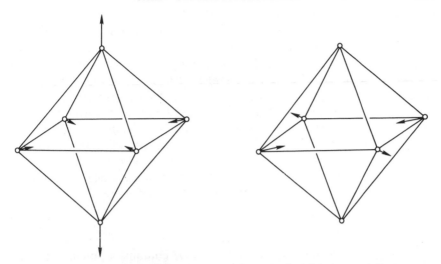

Fig. 4.   The two components of the octahedral mode ε_g. On the left is the set of displacements corresponding to Q_θ; that for Q_ε is shown on the right.

plies going to the limit of large quantum numbers for one mode while somewhat artificially excluding all others. This is rather paradoxical, of course, from a physical point of view. However, we gain insight into the structure of particular JT systems that might well stand us in good stead were more complex systems to come under study.

III. $T_1 \otimes \tau_{2g}$

As a first example of a trajectory analysis, the octahedral system $T_1 \otimes \tau_{2g}$ is selected. Since the three states $|x\rangle$, $|y\rangle$ and $|z\rangle$ of a p electron belong to the irreducible representation T_1 of O, the system can be thought of as a p electron trapped at a site of octahedral symmetry in a crystal and interacting with a localized vibrational mode of τ_{2g} symmetry. Early analyses have been made by Van Vleck,[7] Öpik and Pryce,[8] Moffitt and Thorson,[2] and, in summary form, by Englman.[9] Our principal interest in this JT system lies in the fact that only a finite number (namely, four) static distortions are energetically favored, and so the trajectories correspond to atoms with well-defined equilibrium positions.

A.   Static Distortions

The simplest coupling between the τ_2 mode and the electronic state T_1 is linear in the normal coordinates and necessarily takes the form

$$V = V_T \left(Q_{yz} T_{yz} + Q_{zx} T_{zx} + Q_{xy} T_{xy} \right)$$

where V_T is a coupling parameter and the three operators T_{yz}, T_{zx}, and T_{xy} act between the states of the p electron. In order that V be an octahedral scalar, the T_{ij} operators must belong to T_{2g} of O_h; in fact, they must transform as their suffixes indicate. This makes it easy to relate their matrix elements to one another. The only nonvanishing ones are of the type $\langle i|T_{ij}|j\rangle$, since the product of $\langle i|$ and $|j\rangle$ transforms like the Cartesian product ij, and this is precisely the function that is needed to be combined with T_{ij} to produce a resultant that contains a scalar component—the only component that survives the integration implied in the matrix element. If this argument seems less than totally convincing, the sample operations of O_h can be brought into play to show that other types of matrix elements are the negatives of themselves and thus vanish.[10] We follow convention and write

$$\langle i|T_{ij}|j\rangle = \langle j|T_{ij}|i\rangle = -1$$

In addition to V, the static Hamiltonian H contains a potential function which we write in the form

$$U = \tfrac{1}{2}\mu\omega^2\left(Q_{yz}^2 + Q_{zx}^2 + Q_{xy}^2\right) \tag{12}$$

where μ is an effective mass and ω an angular frequency. We now set up the equation $H|\psi\rangle = E|\psi\rangle$, where $|\psi\rangle$ is some linear combination of $|x\rangle$, $|y\rangle$, and $|z\rangle$. This leads to the secular equation

$$\begin{vmatrix} U-E & -V_TQ_{xy} & -V_TQ_{zx} \\ -V_TQ_{xy} & U-E & -V_TQ_{yz} \\ -V_TQ_{zx} & -V_TQ_{yz} & U-E \end{vmatrix} = 0 \tag{13}$$

that is, to

$$(U-E)^3 - (U-E)V_T^2\left(Q_{yz}^2 + Q_{zx}^2 + Q_{xy}^2\right) - 2V_T^3 Q_{yz}Q_{zx}Q_{xy} = 0 \tag{14}$$

There are no specifically quantum features to this equation, which could be used to determine the potential surface for the ligand trajectories. This will be done shortly: for the moment, however, we seek the new equilibrium positions of the ligands. The displacements of the ligands from their original sites at the vertices of a regular octahedron can be determined by insisting that

$$\frac{\partial E}{\partial Q_{yz}} = \frac{\partial E}{\partial Q_{zx}} = \frac{\partial E}{\partial Q_{xy}} = 0$$

Differentiating (14) with respect to Q_{yz} and using (12) leads to

$$3(U-E)^2\mu\omega^2 - \mu\omega^2 V_T^2\left(Q_{yz}^2 + Q_{zx}^2 + Q_{xy}^2\right)$$

$$-2(U-E)V_T^2 - \frac{2V_T^3 Q_{zx}Q_{xy}}{Q_{yz}} = 0$$

Of all the terms in this equation, only the last term on the right-hand side changes under cyclic permutations. We can therefore deduce that

$$\frac{Q_{zx}Q_{xy}}{Q_{yz}} = \frac{Q_{xy}Q_{yz}}{Q_{zx}} = \frac{Q_{yz}Q_{zx}}{Q_{xy}} = Q_e \qquad (15)$$

say, where Q_e is an equilibrium coordinate. These equations yield

$$Q_{yz}^2 = Q_{zx}^2 = Q_{xy}^2 = Q_e^2 \qquad (16)$$

For real coordinates, just four solutions are possible. The obvious one is

$$Q_{yz} = Q_{zx} = Q_{xy} = Q_e \qquad (17)$$

but we can immediately see that minus signs can be inserted in front of a pair of Q_{ij}'s without violating Eqs. (15) or (16). These four solutions cover all possible ways in which an octahedron can distort while preserving the trigonal symmetry axes. The one corresponding to (17) is sketched in Fig. 5. The analysis above leads to $Q_e = 2V_T/3\mu\omega^2$ and to an eigenvalue E equal to $-Q_e V_T$.

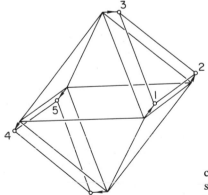

Fig. 5. The trigonal distortion of the octahedral complex described by Eqs. (17). There are three similar distortions corresponding to the different ways of picking a trigonal axis of the octahedron.

B. Dynamics

We have now to face the first example of a calculation of trajectories. In short, we have to work out how the ligands will move if they are assigned some kinetic energy. For the moment it will be supposed that this kinetic energy is small compared to the eigenvalues $|E|$ of Eq. (13). The atoms execute oscillations about their displaced equilibrium positions. If we work with the first of the four possible distorted configurations [i.e., the one specified by (17) and drawn in Fig. 5], we should more appropriately use the coordinates D_{ij} given by

$$D_{yz} = Q_{yz} - Q_e, \qquad D_{zx} = Q_{zx} - Q_e, \qquad D_{xy} = Q_{xy} - Q_e$$

The natural approach is to express $V + U$ as a function of the D_{ij} rather than the Q_{ij}. Doing this, we find

$$
\begin{aligned}
V + U = {} & V_T Q_e \left(T_{yz} + T_{zx} + T_{xy} \right) + D_{yz} \left(V_T T_{yz} + \mu\omega^2 Q_e \right) \\
& + D_{zx} \left(V_T T_{zx} + \mu\omega^2 Q_e \right) + D_{xy} \left(V_T T_{xy} + \mu\omega^2 Q_e \right) \\
& + \tfrac{1}{2}\mu\omega^2 \left(D_{yz}^2 + D_{zx}^2 + D_{xy}^2 \right) + Q_e V_T
\end{aligned}
\tag{18}
$$

The ground electronic state, which follows from the secular matrix from which equation (13) is constructed, is $\left(\tfrac{1}{3}\right)^{1/2}(|x\rangle + |y\rangle + |z\rangle)$. Within this state each operator T_{ij} possesses the expectation value $-\tfrac{2}{3}$ and the three terms in (18) linear in D_{ij} vanish. This is only to be expected, since the whole point of the analysis of the preceding section is to eliminate linear distorting terms by choosing new equilibrium sites. We also see that

$$V_T Q_e \langle T_{yz} + T_{zx} + T_{xy} \rangle = -2 V_T Q_e$$

so $V + U$ corresponds to the potential-energy term $\tfrac{1}{2}\mu\omega^2(D_{yz}^2 + D_{zx}^2 + D_{xy}^2)$ together with the constant $-Q_e V_T$, which is just the expected eigenvalue E.

Having manipulated $V + U$ into such a simple form, we might easily jump to the conclusion that we have an isotropic three-dimensional oscillator and that the trajectories are ellipses corresponding to an angular frequency ω. This would be incorrect. As Moffitt and Thorson have pointed out,[2] $V + U$ is *not* diagonal with respect to the electronic states. Although, in solving the determinantal equation (13), we have effectively diagonalized $V + U$, the resulting eigenstate $\left(\tfrac{1}{3}\right)^{1/2}(|x\rangle + |y\rangle + |z\rangle)$ is valid only for $Q_{yz} = Q_{zx} = Q_{xy}$ and not for the nonequilibrium case where they are not equal.

In setting up the secular matrix again, it is more convenient to take the following linear combinations of states and operators:

$$|\alpha\rangle = \left(\tfrac{1}{3}\right)^{1/2}\left(|x\rangle + |y\rangle + |z\rangle\right)$$

$$|\theta\rangle = \left(\tfrac{1}{6}\right)^{1/2}\left(2|z\rangle - |x\rangle - |y\rangle\right)$$

$$|\varepsilon\rangle = \left(\tfrac{1}{2}\right)^{1/2}\left(|x\rangle - |y\rangle\right)$$

$$T_\alpha = \left(\tfrac{1}{3}\right)^{1/2}\left(T_{yz} + T_{zx} + T_{xy}\right)$$

$$T_\theta = \left(\tfrac{1}{6}\right)^{1/2}\left(2T_{xy} - T_{yz} - T_{zx}\right) \tag{19}$$

$$T_\varepsilon = \left(\tfrac{1}{2}\right)^{1/2}\left(T_{yz} - T_{zx}\right)$$

$$D_\alpha = \left(\tfrac{1}{3}\right)^{1/2}\left(D_{yz} + D_{zx} + D_{xy}\right)$$

$$D_\theta = \left(\tfrac{1}{6}\right)^{1/2}\left(2D_{xy} - D_{yz} - D_{zx}\right)$$

$$D_\varepsilon = \left(\tfrac{1}{2}\right)^{1/2}\left(D_{yz} - D_{zx}\right)$$

Equation (18) now becomes

$$V + U = V_T Q_e\left(1 + \sqrt{3}\,T_\alpha\right) + V_T\left(D_\alpha T_\alpha + D_\varepsilon T_\varepsilon + D_\theta T_\theta\right)$$
$$+ \mu\omega^2\left[Q\sqrt{3}\,D_\alpha + \tfrac{1}{2}\left(D_\alpha^2 + D_\varepsilon^2 + D_\theta^2\right)\right] \tag{20}$$

and the matrix of $V + U$, taken in the basis $|\alpha\rangle$, $|\theta\rangle$, and $|\varepsilon\rangle$, turns out to be

$$
\begin{pmatrix}
-Q_e V_T & V_T\sqrt{\tfrac{1}{3}}\,D_\theta & V_T\sqrt{\tfrac{1}{3}}\,D_\varepsilon \\
+\tfrac{1}{2}\mu\omega^2\left(D_\alpha^2 + D_\varepsilon^2 + D_\theta^2\right) & & \\
& & \\
V_T\sqrt{\tfrac{1}{3}}\,D_\theta & 2Q_e V_T + V_T\left(\sqrt{3}\,D_\alpha - \sqrt{\tfrac{2}{3}}\,D_\theta\right) & V_T\sqrt{\tfrac{2}{3}}\,D_\varepsilon \\
& +\tfrac{1}{2}\mu\omega^2\left(D_\alpha^2 + D_\varepsilon^2 + D_\theta^2\right) & \\
& & \\
V_T\sqrt{\tfrac{1}{3}}\,D_\varepsilon & V_T\sqrt{\tfrac{2}{3}}\,D_\varepsilon & 2Q_e V_T + V_T\left(\sqrt{3}\,D_\alpha + \sqrt{\tfrac{2}{3}}\,D_\theta\right) \\
& & +\tfrac{1}{2}\mu\omega^2\left(D_\alpha^2 + D_\varepsilon^2 + D_\theta^2\right)
\end{pmatrix}
$$

We need not diagonalize this matrix because, in the limit $Q_e \gg |D_{ij}|$, corresponding to amplitudes of oscillation about the new equilibrium sites that are small compared to the displacements from the original sites, perturbation theory is adequate. The lowest eigenvalue is given by

$$E = -Q_e V_T + \tfrac{1}{2}\mu\omega^2\big(D_\alpha^2 + D_\varepsilon^2 + D_\theta^2\big) - \frac{V_{T\frac{2}{3}}^{\frac{1}{1}} D_\theta^2}{3Q_e V_T} - \frac{V_{T\frac{2}{3}}^{\frac{1}{1}} D_\varepsilon^2}{3Q_e V_T}$$

$$= -Q_e V_T + \tfrac{1}{2}\mu\omega^2 D_\alpha^2 + \tfrac{1}{3}\mu\omega^2\big(D_\theta^2 + D_\varepsilon^2\big) \tag{21}$$

The energy levels are given by the superposition of those of a one-dimensional oscillator with angular frequency ω with those of a two-dimensional oscillator with angular frequency ω', where $\omega' = \big(\tfrac{2}{3}\big)^{1/2}\omega$.

C. Trajectories

We have now to unravel the coordinates. At first sight we might think that the three equations (10) are scarcely adequate to yield values for the 12 displacements $z_2, z_5, \ldots, x_5$ in terms of the three normal coordinates Q_{yz}, Q_{zx}, and Q_{xy}. It would appear that we must impose the conditions that all other normal coordinates vanish. However, we can avoid doing this explicitly by appealing to the orthogonality of the matrix that converts displacements to normal coordinates. If $\mathbf{A}$ is an orthogonal matrix with real elements a_{ij}, then the elements of $\mathbf{A}^{-1}$ are a_{ji}. Thus we can at once write down

$$y_3 = -y_6 = z_2 = -z_5 = \tfrac{1}{2}Q_{yz} = \tfrac{1}{2}Q_e + \tfrac{1}{2}D_{yz}$$

$$= \tfrac{1}{2}Q_e + \big(\tfrac{1}{12}\big)^{1/2}D_\alpha + \big(\tfrac{1}{8}\big)^{1/2}D_\varepsilon - \big(\tfrac{1}{24}\big)^{1/2}D_\theta$$

$$z_1 = -z_4 = x_3 = -x_6 = \tfrac{1}{2}Q_{zx} = \tfrac{1}{2}Q_e + \tfrac{1}{2}D_{zx}$$

$$= \tfrac{1}{2}Q_e + \big(\tfrac{1}{12}\big)^{1/2}D_\alpha - \big(\tfrac{1}{8}\big)^{1/2}D_\varepsilon - \big(\tfrac{1}{24}\big)^{1/2}D_\theta \tag{22}$$

$$x_2 = -x_5 = y_1 = -y_4 = \tfrac{1}{2}Q_{xy} = \tfrac{1}{2}Q_e + \tfrac{1}{2}D_{xy}$$

$$= \tfrac{1}{2}Q_e + \big(\tfrac{1}{12}\big)^{1/2}D_\alpha + \big(\tfrac{1}{6}\big)^{1/2}D_\theta$$

All other displacements (that is, x_1, y_2, z_3, x_4, y_5, and z_6) are zero.
Take ligand 1. Setting

$$D_\alpha = a(12)^{1/2}\cos(\omega t - \sigma_\alpha)$$

$$D_\varepsilon = b(8)^{1/2}\cos(\omega' t - \sigma_\varepsilon)$$

$$D_\theta = c(6)^{1/2}\cos(\omega' t - \sigma_\theta)$$

corresponding to classical simple-harmonic motion consistent with the potential form (21), we get

$$x_1 = 0$$

$$y_1 = \tfrac{1}{2}Q_e + a\cos(\omega t - \sigma_\alpha) + c\cos(\omega' t - \sigma_\theta)$$

$$z_1 = \tfrac{1}{2}Q_e + a\cos(\omega t - \sigma_\alpha) - b\cos(\omega' t - \sigma_\varepsilon) - \tfrac{1}{2}c\cos(\omega' t - \sigma_\theta)$$

$$= \tfrac{1}{2}Q_e + a\cos(\omega t - \sigma_\alpha) + d\cos(\omega' t - \sigma')$$

(23)

where d and σ' are related to b, c, σ_ε, and σ_θ in a way that need not concern

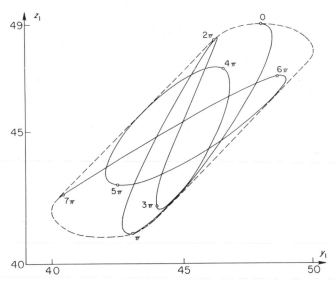

Fig. 6.   The trajectory for an atom in the JT system $T_1 \otimes \tau_{2g}$ in the limit of strong coupling. The special case chosen here corresponds to

$$y_1 = 45 + 3\cos\theta + 2\sin\sqrt{\tfrac{2}{3}}\,\theta$$

$$z_1 = 45 + 3\cos\theta + \cos\sqrt{\tfrac{2}{3}}\,\theta$$

where $\theta = \omega t$. Points defined by $\theta = n\pi$, where $n = 0, 1, \ldots, 7$ are indicated by the small open circles. The trajectory never closes, but sweeps out an area bounded by the dotted lines. The envelope comprises the two ellipses

$$(y_1 - 45 \pm 3)^2 + 4(z_1 - 45 \pm 3)^2 = 4$$

and the pair of common tangents to them.

us. We see that the trajectory of atom 1 lies in a plane and is the superposition of two harmonic motions of angular frequencies ω and ω'. Since the latter are relatively irrational, the trajectory never closes on itself. An example is shown in Fig. 6.

D. Tunneling

In the picture presented above, there is no possibility of the distorted configuration of Fig. 5 converting itself into one of the other three equivalently distorted configurations of trigonal symmetry. The trajectories of Fig. 6 assume a rigid equilibrium site at $y_1 = z_1 = Q_e/2$, $x_1 = 0$. However, the quantum mechanical eigenfunction for the ground state must necessarily show no preference for any one of the four distorted structures. In fact, it must consist of a linear superposition of functions corresponding to each of the four distortions. This paradox is resolved by noting that the four distinct configurations can interconvert by means of quantum mechanical tunneling. The fact that an eternity may be necessary to give a high probability of this happening is immaterial. But because this phenomenon is in essence quantum mechanical, there is no way in which we can represent the trajectories of the constituent atoms when the conversion occurs. The mechanism cannot be recast in the classical framework we have adopted to draw Fig. 6.

IV. $E \otimes \varepsilon_g$

The octahedral JT system $E \otimes \varepsilon_g$, in which a doubly degenerate electronic state E is coupled to a vibrational mode ε_g with normal coordinates Q_θ and Q_ε given in Eq. (11), has received much attention over the years. It is probably the simplest system that exhibits an infinite spatial degeneracy: the static problem possesses an infinity of solutions instead of the mere four of $T_1 \otimes \tau_{2g}$. Although this extraordinary degeneracy can be removed by including coupling terms that go beyond the purely linear ones in the normal coordinates, the degeneracy is of enormous interest in its own right.

The first detailed analysis was carried out by Longuet-Higgins et al.,[11] who showed that the dynamical problem involves the diagonalization of an infinite tridiagonal matrix. An approximate analytic solution has recently been provided by Barentzen et al.[12] Some isolated exact solutions have been found,[13] and they have been set into a wider context by Reik et al.[14] O'Brien has obtained approximate eigenfunctions for the excited states by means of the WKB approximation,[15] while Ballhausen has introduced variational techniques into the problem.[16] Accurate eigenfunctions near the strong-coupling limit have been expressed as a perturbation series by O'Brien and Pooler.[17] Apart from such theoretical articles as the ones cited here, there is an enormous literature on observations of $E \otimes \varepsilon$ systems. The reader is referred to the book by Englman[9] and to the somewhat more recent review article by Bates.[18]

A. The Static Problem

In analogy to the JT system $T_1 \otimes \tau_{2g}$, we introduce a coupling term V and a potential function U given by

$$V = V_E(Q_\theta T_\theta + Q_\varepsilon T_\varepsilon), \qquad U = \tfrac{1}{2}\mu\omega^2\big(Q_\theta^2 + Q_\varepsilon^2\big) \qquad (24)$$

The matrix elements of T_θ and T_ε, taken between the electronic kets $|\theta\rangle$ and $|\varepsilon\rangle$ and the bras $\langle\theta|$ and $\langle\varepsilon|$, can probably be obtained most readily by making use of the correspondences (as far as their transformation properties go)

$$T_\theta \leftrightarrow \big(\tfrac{1}{12}\big)^{1/2}(3z^2 - r^2), \qquad T_\varepsilon \leftrightarrow \tfrac{1}{2}(x^2 - y^2) \qquad (25)$$

$$|\theta\rangle \leftrightarrow |d,0\rangle, \qquad |\varepsilon\rangle \leftrightarrow \sqrt{\tfrac{1}{2}}|d,2\rangle + \sqrt{\tfrac{1}{2}}|d,-2\rangle \qquad (26)$$

where the kets $|d,m\rangle$ are angular-momentum states with an azimuthal quantum number l of 2 and a magnetic quantum number m. Angular-momentum theory can now be used to evaluate all the matrix elements (see, for example, Edmonds[19]). With a suitable normalization of T_θ and T_ε, we find

$$T_\theta|\theta\rangle = -|\theta\rangle, \qquad T_\theta|\varepsilon\rangle = |\varepsilon\rangle$$
$$T_\varepsilon|\theta\rangle = |\varepsilon\rangle, \qquad T_\varepsilon|\varepsilon\rangle = |\theta\rangle$$

The secular equation, the analog of (13), is

$$\begin{vmatrix} U - V_E Q_\theta - E & V_E Q_\varepsilon \\ V_E Q_\varepsilon & U + V_E Q_\theta - E \end{vmatrix} = 0 \qquad (27)$$

namely

$$(U - E)^2 - V_E^2\big(Q_\theta^2 + Q_\varepsilon^2\big) = 0 \qquad (28)$$

The conditions

$$\frac{\partial E}{\partial Q_\theta} = \frac{\partial E}{\partial Q_\varepsilon} = 0$$

yield

$$\begin{aligned} 2(U - E)\mu\omega^2 Q_\theta - 2V_E^2 Q_\theta &= 0 \\ 2(U - E)\mu\omega^2 Q_\varepsilon - 2V_E^2 Q_\varepsilon &= 0 \end{aligned} \qquad (29)$$

with the result that the displacements Q_θ and Q_ε cannot be uniquely determined. Instead, we write

$$Q_\theta = Q\cos\phi, \qquad Q_\varepsilon = Q\sin\phi \tag{30}$$

For the value Q_e of Q satisfying (28) and (29), we have $Q_e = V_E/\mu\omega^2$ and $E = -\frac{1}{2}\mu\omega^2 Q_e^2$. The electronic eigenfunction corresponding to this value of E turns out to be

$$|\phi_-\rangle = \cos\tfrac{1}{2}\phi|\theta\rangle - \sin\tfrac{1}{2}\phi|\varepsilon\rangle \tag{31}$$

The locus of acceptable solutions for the minimum potential energy is a circle in the space of the coordinates Q_θ, Q_ε. To see what this means in terms of the positions of the ligands in real space, we need to invert the Eqs. (11), just as we did earlier for (10). The results are

$$x_1 = -x_4 = \tfrac{1}{2}Q_\varepsilon - \left(\tfrac{1}{12}\right)^{1/2}Q_\theta = \left(\tfrac{1}{3}\right)^{1/2}Q\cos\left(\phi - \tfrac{2}{3}\pi\right)$$
$$y_2 = -y_5 = -\tfrac{1}{2}Q_\varepsilon - \left(\tfrac{1}{12}\right)^{1/2}Q_\theta = \left(\tfrac{1}{3}\right)^{1/2}Q\cos\left(\phi + \tfrac{2}{3}\pi\right) \tag{32}$$
$$z_3 = -z_6 = \left(\tfrac{1}{3}\right)^{1/2}Q_\theta = \left(\tfrac{1}{3}\right)^{1/2}Q\cos\phi$$

Unlike the τ_2 modes, the ε modes are purely radial. A way to represent the actual displacements by means of trilinear coordinates[20] is shown in Fig. 7. If $Q_\theta/\sqrt{3}$ and $Q_\varepsilon/\sqrt{3}$ are represented by the two sides of the triangle GHI, then the lengths of the perpendiculars to the three sides of an equilateral triangle whose center coincides with G can be taken as the actual distances from the center of the octahedral complex under study to the six displaced atoms. Further details are given in the caption to Fig. 7.

B. Dynamics

We now relax the constraint that the normal coordinates correspond to the minimum of $\langle U + V\rangle$. In other words, we retain (30) but no longer insist that $Q = Q_e$. This prepares the ground for the inclusion of the kinetic energy of the ligands. Equation (28) can be written as

$$E = \tfrac{1}{2}\mu\omega^2 Q^2 \pm V_E Q \tag{33}$$

which is independent of ϕ. The two surfaces of revolution obtained by plotting E against Q for the two possible signs preceding $V_E Q$ in (33) is often referred to as a "Mexican hat." It is in this potential that the motion takes place.

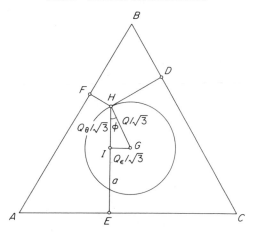

Fig. 7. Conversion of normal coordinates $(Q_\theta, Q_\varepsilon)$ for $E \otimes \varepsilon_g$ to atomic displacements x_1, y_2, z_3, x_4, y_5, and z_6 of a regular octahedral complex. The two sides HI and IG of the right-angle triangle HIG are drawn equal to $Q_\theta/\sqrt{3}$ and $Q_\varepsilon/\sqrt{3}$. If the undisplaced atoms of the regular octahedron are all a distance a from its center, an equilateral triangle ABC of side $2a\sqrt{3}$, with its center at G, is drawn such that HI is perpendicular to AC. Then the lengths of the perpendiculars HD, HE, and HF dropped from H to the sides of the triangle give the new distances of atoms 1 (or 4), 3 (or 6) and 2 (or 5) respectively from the center of the octahedron. Subtracting a (the length IE) from these distances gives the actual displacements (such as HI for atom 3).

The experience we have gained with the JT system $T_1 \otimes \tau_{2g}$ alerts us to the necessity of considering the effects of the electronic state

$$|\phi_+\rangle = \sin\tfrac{1}{2}\phi|\theta\rangle + \cos\tfrac{1}{2}\phi|\varepsilon\rangle \tag{34}$$

the orthogonal companion to $|\phi_-\rangle$ of (31). However, it is straightforward to show that the matrix of $U + V$ is diagonal in the basis provided by $|\phi_\pm\rangle$, so the complications of Section III.B that lead to the two frequencies ω and ω' do not recur.

It is clear from the nature of the potential that the motion in the space defined by the coordinates $(Q_\theta, Q_\varepsilon)$ comprises a rotation (described by the angular coordinate ϕ) and a vibration (described by Q). The rotation is in two dimensions and the vibration in one. A pure rotation corresponds to motion along the circle of Fig. 7. An alternative way of visualizing the rotation is to replace the displacements x_1, y_2, and z_3 of the three separate atoms of mass μ by the displacement in three dimensions with components x_1, y_2, and z_3 of a single atom of mass μ. A similar coalescence can be imagined for atoms 4, 5, and 6. Since

$$x_1^2 + y_2^2 + z_3^2 = x_4^2 + y_5^2 + z_6^2 = \tfrac{1}{2}Q^2$$

from (32), the rotation now corresponds to that of a dumb-bell possessing a moment of inertia μQ^2 and turning in the plane $x + y + z = 0$.

The kinetic energy T is given by

$$T = \tfrac{1}{2}\mu\left(\dot{x}_1^2 + \dot{y}_2^2 + \dot{z}_3^2 + \dot{x}_4^2 + \dot{y}_5^2 + \dot{z}_6^2\right) = \tfrac{1}{2}\mu\dot{Q}^2 + \tfrac{1}{2}\mu Q^2\dot{\phi}^2$$

from Eqs. (32). The canonical momenta are defined by

$$p_Q = \frac{\partial T}{\partial \dot{Q}} = \mu\dot{Q} \to -\frac{i\hbar\,\partial}{\partial Q}$$

$$p_\phi = \frac{\partial T}{\partial \dot{\phi}} = \mu Q^2\dot{\phi} \to -\frac{i\hbar\,\partial}{\partial \phi} \tag{35}$$

Ignoring for the moment the quantum mechanical substitutions on the right of these equations, we see that we can write the classical Hamiltonian for $E \otimes \varepsilon_g$ as

$$H = \frac{1}{2\mu}\left(p_Q^2 + \frac{p_\phi^2}{Q^2}\right) + \frac{1}{2}\mu\omega^2 Q^2 \pm V_E Q \tag{36}$$

where the sign option matches that of $|\phi_\pm\rangle$. Since $\partial H/\partial\phi = 0$, the angular momentum p_ϕ can be set equal to a constant, say k. We can now write (36) in the form

$$H = \left(\frac{1}{2\mu}\right)p_Q^2 + U'(Q) \tag{37}$$

where

$$U'(Q) = \frac{k^2}{2\mu Q^2} + \frac{1}{2}\mu\omega^2 Q^2 \pm V_E Q$$

An example of a pair of potentials $U'(Q)$, corresponding to the two choices of sign for the term linear in Q, is shown in Fig. 8.

The Hamiltonian (37) corresponds to one-dimensional motion in the potential U'. For a total energy E, we find, setting $p_Q = \mu\,dQ/dt$, that

$$dt = (\tfrac{1}{2}\mu)^{1/2}(E - U')^{-1/2}\,dQ \tag{38}$$

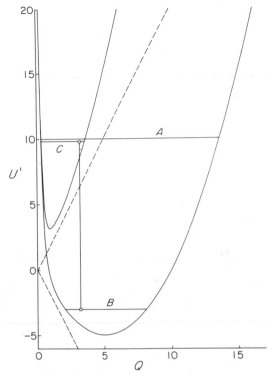

Fig. 8. A plot of the potential energy U' versus Q for the two special cases

$$U' = Q^{-2} + \frac{Q^2}{5} \pm 2Q$$

corresponding to identical angular momenta. Motion in the ranges of the radial coordinate Q defined by the lines A, B, and C correspond to total energies E_A, E_B, and E_C of 10.0, -3.0, and 9.8, respectively. The dotted lines represent the linear terms $U' = \pm 2Q$. The vertical line represents a transition from C to B at the most likely value of Q.

while

$$dt = \left(\frac{\mu Q^2}{k} \right) d\phi \tag{39}$$

In this way $d\phi$ can be related to dQ and the trajectories plotted by numerical integration. Some examples are shown in Fig. 9. To find the actual motions of the ligands we can superpose the equilateral triangle of Fig. 7 and use trilinear coordinates; or we can use the alternative scheme of the dumbbell.

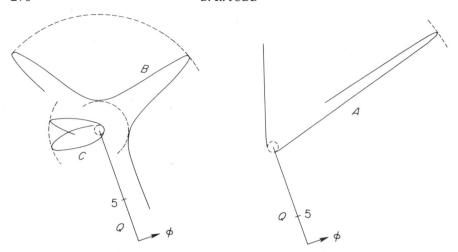

Fig. 9. Trajectories in (Q, ϕ) space for the JT system $E \otimes \varepsilon_g$. The labels A, B, and C correspond to the ranges of Q and associated energies E represented in Fig. 8. Thus A and B correspond to motion of atoms constituting the octahedral complex for which the coupled electronic state is $|\phi_-\rangle$; for C the orthogonal companion $|\phi_+\rangle$ must be chosen. One way to find the actual motions of the atoms is to centrally superpose an equilateral triangle on the diagrams, as shown in Fig. 7.

C. Coupling to the Upper Surface

A point that we have not considered is that the electronic states $|\phi_\pm\rangle$ involve the angle ϕ in terms of which the normal coordinates are described. Thus the term $p_\phi^2/2\mu Q^2$ in the kinetic energy T can produce contributions to the expectation values of T from the electronic states as well as from the motion of the ligands. What is rather more surprising, non-vanishing terms connecting $\langle\phi_\pm|$ to $|\phi_\mp\rangle$ also occur even though

$$\langle\phi_\pm|p_\phi^2|\phi_\mp\rangle \rightarrow -\hbar^2\langle\phi_\pm|\left(\frac{\partial^2}{\partial\phi^2}\right)|\phi_\mp\rangle = 0$$

The coupling arises because, in quantum mechanics, the total eigenfunction involves a product of $|\phi_\pm\rangle$ with an oscillator state $|\psi\rangle$ which also is a function of ϕ. Thus

$$p_\phi^2|\phi_\mp\rangle|\psi\rangle = |\psi\rangle p_\phi^2|\phi_\mp\rangle + 2\left(p_\phi|\phi_\mp\rangle\right)\left(p_\phi|\psi\rangle\right) + |\phi_\mp\rangle p_\phi^2|\psi\rangle$$

and since

$$p_\phi|\phi_\mp\rangle \rightarrow \left(\frac{\hbar}{i}\right)\left(\frac{\partial}{\partial\phi}\right)|\phi_\mp\rangle = \mp\frac{1}{2}\left(\frac{\hbar}{i}\right)|\phi_\mp\rangle$$

we see that

$$\langle \phi_{\pm} | p_{\phi}^2 | \phi_{\mp} \rangle \equiv \mp \left(\frac{\hbar}{i} \right) p_{\phi} \qquad (40)$$

when the left-hand side of this equivalence is regarded as an operator, as it properly should be. However, the origin of (40) is essentially quantum mechanical in nature and, for the classical trajectories that are our main concern here, we set aside the coupling terms deriving from this source.

D. Slonczewski Resonances

It was Slonczewski,[21] who first treated the quasi-stability of the upper electronic state $|\phi_+\rangle$ and, with it, the associated atomic motions such as that represented in Figs. 8 and 9 by the lines labeled C. His original analysis has been extended by Slonczewski and Moruzzi.[22] The most likely mode of decay can be discussed in the context of the Franck–Condon principle. However, for the classical orbits that we are considering, the analysis can be cast in a slightly different form. The paths in phase space corresponding to the three trajectories A, B, and C are drawn in Fig. 10. We note first that the curves A and C do not intersect (though they have an extended region of close approach for small Q). Classically, the orbits are distinct. From the point of view of quantum mechanics, however, the existence of coupling terms [such as that occurring on the right-hand side of (40)] mixes the states corresponding to orbits of the same energy (i.e., the orbits A and C). The approximate forms of these eigenfunctions have been given by O'Brien.[15] From our viewpoint, the corresponding blending of the trajectories A and C is most likely to occur through a transition from one trajectory to the other when Q is small, for it is there that the quantum mechanical overlap is greatest. This is the mechanism analyzed by Slonczewski.

Like the tunneling phenomenon discussed in Section III.D, the mixing of the trajectories A and C is a quantum phenomenon that may have a low probability. However, the decay of A into B, with the emission of the corresponding difference in energy, can be described very well in terms of our classical trajectories. There are two points of intersection in Fig. 10 for which no change in either momentum or position is required for a transition to take place. An example of such a transition is shown in Fig. 11. Quantum mechanically, the eigenfunctions (of the oscillating octahedral complex) that correspond to the trajectories A and C have a particularly large overlap at each of the intersections in phase space. For the transition actually to occur, there must be a perturbing term H' in the total Hamiltonian for which $\langle \phi_+ | H' | \phi_- \rangle \neq 0$. Possible candidates for H' include the electric-quadrupole component of the radiation field or the electrostatic interaction with a

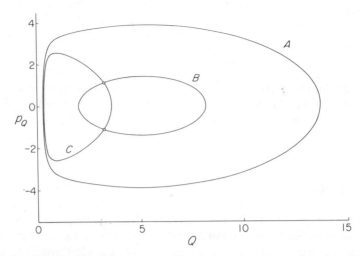

Fig. 10. Paths in phase space corresponding to the trajectories A, B, and C of Fig. 9. The curves are plotted for

$$p_\phi^2 = E - \left(Q^{-2} + \frac{Q^2}{5} \pm 2Q \right)$$

where the signs match those of the associated electronic states $|\phi_\pm\rangle$. In the absence of the centrifugal term Q^{-2}, the paths would be elliptical, corresponding to simple-harmonic motion. The fact that B approximates to an ellipse indicates that centrifugal terms do not play an important role in determining the corresponding trajectory of Fig. 9. The two intersections of B with C correspond to points where both p_ϕ and Q are the same for B as for C. The intersections should thus be associated with relatively high probabilities for the transitions $|\phi_+\rangle \to |\phi_-\rangle$.

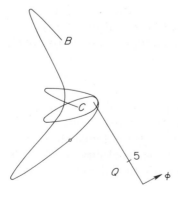

Fig. 11. A trajectory exhibiting a transition from C to B at the point indicated by the small open circle. At that point there is no change in position or momentum.

272

neighboring complex that could absorb the energy difference $E_C - E_B$ through some excitation process. In actual atomic systems, such mechanisms would have to compete with more likely transitions such as $E \rightarrow A$.

V. $\Gamma_8 \otimes \tau_{2g}$

The SO(3) symmetry of $\Gamma_8 \otimes \tau_{2g}$ has already been mentioned in Section I. Our trajectory analysis of this system is therefore directed toward understanding how a complex of six atoms vibrating in the mode τ_{2g} can exhibit rotational symmetry in three dimensions. In short, we have to find out what is rotating and how it can be described.

A. Group Theory

The symbol Γ_8 refers to an irreducible representation of the *double* octahedral group. This is the group that is an extended form of O, devised so that the transformation properties of spinors can be properly described and labeled. The need for such an extension becomes apparent when it is noticed that the spin-up and spin-down eigenfunctions α and β for an electron, when rotated by 2π, become $-\alpha$ and $-\beta$. There is no way in which functions with properties as bizarre as this can be accommodated within the framework of O or O_h as they stand, since rotations by 0 and 2π have been assumed to be equivalent. The details of how the situation is remedied need not concern us here: the result is that we have to introduce three new irreducible representations Γ_6, Γ_7, and Γ_8 to the five already existing in O. Both Γ_6 and Γ_7 are two-dimensional; Γ_8 is four-dimensional. The letter notation $E_{1/2}$, $E_{3/2}$, and $G_{3/2}$ would be more in keeping with that already being used (and, in fact, is preferred by some writers such as Englman[9]), but we shall continue to use the traditional forms.

The four kets $|\frac{3}{2}\rangle, |\frac{1}{2}\rangle, |-\frac{1}{2}\rangle$, and $|-\frac{3}{2}\rangle$ that define the components $|M\rangle$ of an angular-momentum state for which $S = 3/2$ belong to Γ_8. Of course, it is more common to meet Γ_8 states that are linear combinations of orbital and spin states; however, we are concerned here only with general properties, and for these a knowledge of the transformation properties of the states suffices. We can always choose the components $|\Gamma_8 M\rangle$ of a particular Γ_8 state to transform like the states $|S, M\rangle$ with $S = 3/2$.

One complication arises because the reduction of the triple Kronecker product $\Gamma_8 \times T_2 \times \Gamma_8$ yields two identity representations A_1 in its decomposition. This means that electronic matrix elements of the type $\langle \Gamma_8 M | T_{ij} | \Gamma_8 M' \rangle$ cannot be assumed to be proportional to $\langle S, M | T_{ij} | S, M' \rangle$ for all M and M'. Two independent sets of matrix elements are, in general, required. Now, it is highly convenient for us to use the bases $|S, M\rangle$ (with, of course, $S = 3/2$), which means that we would prefer the T_{ij} to be functions of S. Two independent sets of three operators, both of which transform according to T_2, are given by the equations

$$W_{yz} = S_y S_z + S_z S_y$$
$$W_{zx} = S_z S_x + S_x S_z \quad (41)$$
$$W_{xy} = S_x S_y + S_y S_x$$

and

$$W'_{yz} = S_x\left(S_y^2 - S_z^2\right) + \left(S_y^2 - S_z^2\right)S_x$$
$$W'_{zx} = S_y\left(S_z^2 - S_x^2\right) + \left(S_z^2 - S_x^2\right)S_y \quad (42)$$
$$W'_{xy} = S_z\left(S_x^2 - S_y^2\right) + \left(S_x^2 - S_y^2\right)S_z$$

The interaction term in the Hamiltonian can be written as

$$V = V_T\left(Q_{yz}W_{yz} + Q_{zx}W_{zx} + Q_{xy}W_{xy}\right) + V'_T\left(Q_{yz}W'_{yz} + Q_{zx}W'_{zx} + Q_{xy}W'_{xy}\right)$$
$$(43)$$

in analogy to previous analyses. In contrast to what has gone before, we seem to need two coupling constants, V_T and V'_T, rather than just one.

However, as Child has pointed out,[23] the Hamiltonian should not contain terms that are not time-reversal invariant. Since, under time reversal, we have $\mathbf{S} \to -\mathbf{S}$, it is clear that $W'_{ij} \to -W'_{ij}$, and hence we can set V'_T equal to zero. If we were to include such terms we would allow the Kramers degeneracy associated with Γ_8 to be broken. The reader may ask why, instead of the expressions for W'_{ij} given in (42), we do not construct another set in which even products of S_x, S_y, and S_z appear. However, there is no way in which we could obtain operators whose matrix elements are not multiples of those for W_{ij}, since we know that, at most, two linearly independent sets of matrix elements can occur and one of them is provided by the operators (42).

B. Trajectories

The construction of the secular matrix is simplified if the orthonormal linear combinations

$$|1\rangle = \sqrt{\tfrac{1}{2}}\left(|-\tfrac{3}{2}\rangle - i|\tfrac{1}{2}\rangle\right)$$
$$|2\rangle = \sqrt{\tfrac{1}{2}}\left(-|\tfrac{3}{2}\rangle + i|-\tfrac{1}{2}\rangle\right)$$
$$|3\rangle = \sqrt{\tfrac{1}{2}}\left(-|-\tfrac{3}{2}\rangle - i|\tfrac{1}{2}\rangle\right) \quad (44)$$
$$|4\rangle = \sqrt{\tfrac{1}{2}}\left(|\tfrac{3}{2}\rangle + i|-\tfrac{1}{2}\rangle\right)$$

are chosen. With the same U as in (12), we get, for the static problem,

$$\begin{vmatrix} V'Q_{xy} + U - E & V'(Q_{yz} - iQ_{zx}) & 0 & 0 \\ V'(Q_{yz} + iQ_{zx}) & -V'Q_{xy} + U - E & 0 & 0 \\ 0 & 0 & -V'Q_{xy} + U - E & -V'(Q_{yz} + iQ_{zx}) \\ 0 & 0 & -V'(Q_{yz} - iQ_{zx}) & V'Q_{xy} + U - E \end{vmatrix} = 0 \tag{45}$$

whence

$$E = U \pm V'\left(Q_{yz}^2 + Q_{zx}^2 + Q_{xy}^2\right)^{1/2} \tag{46}$$

where $V' = \sqrt{3}\, V_T$. Each energy E, given by (46), is doubly degenerate as expected.

Equation (46) exhibits the expected rotational symmetry. Spherical polar coordinates expose it in an obvious way. We put

$$Q_{yz} = Q \sin\theta \cos\phi, \qquad Q_{zx} = Q \sin\theta \sin\phi, \qquad Q_{xy} = Q \cos\theta \tag{47}$$

and the motion takes place in the potential

$$\tfrac{1}{2}\mu\omega^2 Q^2 \pm V'Q \tag{48}$$

The addition of the kinetic energy term $(1/2\mu)(p_{yz}^2 + p_{zx}^2 + p_{xy}^2)$ does nothing to lower the spherical symmetry of the problem. In fact, we have free rotations in three dimensions, with the possibility of a one-dimensional vibration described by the coordinate Q. If we limit our attention to the lower of the two potentials given in (48), the inclusion of a small amount of rotational kinetic energy activates rotations for which $Q = Q_e = V'/\mu\omega^2$. The dynamics of the problem is the same as that of a point mass free to move on the surface of a sphere. The motion takes place in a great circle: the projection of this circle on the yz, zx, and xy planes yields three ellipses whose coordinates, from (47), are given by

$$\begin{aligned} X_1 &= 0, & Y_1 &= Q \sin\theta \sin\phi, & Z_1 &= Q \cos\theta \\ X_2 &= Q \sin\theta \cos\phi, & Y_2 &= 0, & Z_2 &= Q \cos\theta \\ X_3 &= Q \sin\theta \cos\phi, & Y_3 &= Q \sin\theta \sin\phi, & Z_3 &= 0 \end{aligned} \tag{49}$$

From the first four equalities on the left-hand side of equations (22), the actual displacements of the atoms at sites 1, 2, and 3 are given by

$$
\begin{aligned}
x_1 &= 0 & y_1 &= \tfrac{1}{2}Q\cos\theta, & z_1 &= \tfrac{1}{2}Q\sin\theta\sin\phi \\
x_2 &= \tfrac{1}{2}Q\cos\theta, & y_2 &= 0, & z_2 &= \tfrac{1}{2}Q\sin\theta\cos\phi \\
x_3 &= \tfrac{1}{2}Q\sin\theta\sin\phi, & y_3 &= \tfrac{1}{2}Q\sin\theta\cos\phi, & z_3 &= 0
\end{aligned}
\tag{50}
$$

Apart from the factors of $\tfrac{1}{2}$, there is a perfect match between Eqs. (49) and (50) if, in each of the nonzero pairs on the right-hand sides of (49), a lateral exchange is made. Since the trajectory of each point (X_i, Y_i, Z_i) is an ellipse, that of ligand i must be so too. The planes of the ellipses for ligands $1, 2, \ldots, 6$ are perpendicular to the radial vectors leading out to the respective sites from the center of the undistorted octahedron.

To specify these ellipses in greater detail, we note that a unit circle for which the normal to its plane possesses direction cosines l, m, and n and for which the origin coincides with its center is projected into the ellipse

$$
x^2(l^2 + n^2) + y^2(m^2 + n^2) + 2xylm = n^2
\tag{51}
$$

in the xy plane. The ellipses obtained by projecting the circle on to the yz and zx planes can be obtained by cyclic permutation. The exchange $x \leftrightarrow y$ corresponding to the differences between (49) and (50) leads to the ellipse

$$
x^2(m^2 + n^2) + y^2(l^2 + n^2) + 2xylm = n^2
\tag{52}
$$

This possesses a semimajor axis of length 1 and a semiminor axis of length n. The eccentricity e is $(l^2 + m^2)^{1/2}$. There appears to be no special relation between the directions of the major axes of the ellipses for the various sites (except, of course, that the axis for the ellipse at site 1 is parallel to that at site 4, etc.) However, the eccentricities are constrained by

$$
\sum_{i=1}^{6} e_i^2 = 2(l^2 + m^2) + 2(m^2 + n^2) + 2(n^2 + l^2) = 4
\tag{53}
$$

The uniform motion on the great circle becomes projected into nonuniform motion on an ellipse. If we pick ξ and η axes so that only the ξ coordinate is reduced in a particular projection, then the uniform motion, which we may, without loss of generality, write as

$$
\xi = a\cos(\omega t - \sigma), \qquad \eta = a\sin(\omega t - \sigma)
$$

becomes, upon projection,

$$
\xi = b\cos(\omega t - \sigma), \qquad \eta = a\sin(\omega t - \sigma)
\tag{54}
$$

Equations (54) are of the same form as (2), showing that the motion of ligand *i* around the ellipse centered at vertex *i* is consistent with the analysis of Section I.A.

Since the dimensions of the ellipses are proportional to Q, any change in this coordinate, such as that associated with oscillations in the potential (48), immediately leads to a ripple being superposed on the smooth trajectories of the ellipses. As in the case of $E \otimes \varepsilon_g$, the effective one-dimensional oscillation of Q can only be properly considered when centrifugal terms are added to $U + V$. Because of this similarity, there seems little need to carry out a parallel analysis for the $\Gamma_8 \otimes \tau_{2g}$ system.

VI. $\Gamma_8 \otimes \varepsilon_g$

Having treated a system involving a Γ_8 electronic state and purely angular motions of the ligands, it is of interest to turn to a JT system such as $\Gamma_8 \otimes \varepsilon_g$, where the motions are radial. The first point to check is that the triple Kronecker product $\Gamma_8 \times E \times \Gamma_8$ contains the identity representation A_1 only once in its decomposition. This indeed turns out to be the case, so we can immediately pick the two functions

$$T_\theta = S_z^2 - \frac{S(S+1)}{3}, \qquad T_\varepsilon = \frac{\left(S_x^2 - S_y^2\right)}{\sqrt{3}} \tag{55}$$

which match the correspondences (25), and substitute them into Eqs. (24) to get the part $V + U$ of the total Hamiltonian. Instead of the linear combinations (44), it is simpler to take the four states $|1/2\rangle$, $|-3/2\rangle$, $|3/2\rangle$, and $|-1/2\rangle$ as the Γ_8 basis. The matrix of $V + U$ is

$$\begin{pmatrix} U - V_E Q_\theta & V_E Q_\varepsilon & 0 & 0 \\ V_E Q_\varepsilon & U + V_E Q_\theta & 0 & 0 \\ 0 & 0 & U + V_E Q_\theta & V_E Q_\varepsilon \\ 0 & 0 & V_E Q_\varepsilon & U - V_E Q_\theta \end{pmatrix} \tag{56}$$

As can be seen from (27), this corresponds to a pair of $E \otimes \varepsilon_g$ systems. The trajectories are identical to those for $E \otimes \varepsilon_g$. The only difference is that the electronic states become doubled.

VII. $\Gamma_8 \otimes (\varepsilon_g \oplus \tau_{2g})$

A. Coupling of Modes via the Electronic State

We are now ready to study a system in which two modes are involved, ε_g and τ_{2g}. From the analyses of Sections V and VI, we can write down the

Hamiltonian for $\Gamma_8 \otimes (\varepsilon_g \oplus \tau_{2g})$ as

$$
\begin{aligned}
H = (1/2\mu)\big(p_\theta^2 + p_\varepsilon^2 + p_{yz}^2 + p_{zx}^2 + p_{xy}^2 \big) + V_E(Q_\theta T_\theta + Q_\varepsilon T_\varepsilon) \\
+ V_T(Q_{yz}W_{yz} + Q_{zx}W_{zx} + Q_{xy}W_{xy}) + \tfrac{1}{2}\mu\omega_E^2(Q_\theta^2 + Q_\varepsilon^2) \\
+ \tfrac{1}{2}\mu\omega_T^2\big(Q_{yz}^2 + Q_{zx}^2 + Q_{xy}^2 \big)
\end{aligned} \tag{57}
$$

At a first glance the Hamiltonian might seem easily tractable, since it appears to break up into two parts, H_E for $\Gamma_8 \otimes \varepsilon_g$ and H_T for $\Gamma_8 \otimes \tau_{2g}$, which have already been analyzed. This overlooks the coupling that exists between the ε_g and τ_{2g} modes in virtue of their separate coupling to the electronic state Γ_8. If we represent this state through its four components $|1\rangle$, $|2\rangle$, $|3\rangle$, and $|4\rangle$, defined in (44), the terms in V_E and V_T can be represented by a matrix. Making the replacements (47), and, in analogy with (30), introducing Q' and χ through the equations

$$
Q_\theta = Q' \cos\chi, \qquad Q_\varepsilon = Q' \sin\chi
$$

we find that the matrix for the potential-energy terms in (55) becomes

$$
\begin{pmatrix}
U + V'Q\cos\theta & V'Qe^{-i\phi}\sin\theta & -V_E Q'e^{i\chi} & 0 \\
V'Qe^{i\phi}\sin\theta & U - V'Q\cos\theta & 0 & -V_E Q'e^{i\chi} \\
-V_E Q'e^{-i\chi} & 0 & U - V'Q\cos\theta & -V'Qe^{-i\phi}\sin\theta \\
0 & -V_E Q'e^{-i\chi} & -V'Qe^{i\phi}\sin\theta & U + V'Q\cos\theta
\end{pmatrix} \tag{58}
$$

where $V' = \sqrt{3}\, V_T$ as before, and

$$
U = \tfrac{1}{2}\mu\omega_E^2 Q^2 + \tfrac{1}{2}\mu\omega_T^2 Q^2 \tag{59}
$$

From (58) we can write the equation

$$
\big(\textstyle\sum p_j^2/2\mu - E + U + V'Q\cos\theta\big)|1\rangle + V'Qe^{-i\phi}\sin\theta|2\rangle - V_E Q'e^{i\chi}|3\rangle = 0
$$

together with three similar ones. We can guarantee the consistency of these equations if we set the determinant of the matrix (58), to which $\sum p_j^2/2\mu - E$ has been added on the diagonal, equal to zero. When this determinant is worked out, it is found that θ, ϕ, and χ do not explicitly appear. We get, in fact,

$$
\Big[\big(\textstyle\sum p_j^2/2\mu - E + U\big)^2 - V'^2 Q^2 - V_E^2 Q'^2\Big]^2 = 0
$$

whence

$$E = \sum p_j^2/2\mu + U \pm \left(V'^2Q^2 + V_E^2 Q'^2\right)^{1/2} \qquad (60)$$

This equation is the generalization of (33) and (46). It differs from them in two ways: the kinetic energy is explicitly included, and the potential-energy terms reflect the presence of two modes instead of one.

B. Local Oscillations

Since the angular variables do not appear in (60), the rotational motions for $\Gamma_8 \otimes \varepsilon_g$ and $\Gamma_8 \otimes \tau_{2g}$ are unaffected by bringing these systems together. The symmetry group is the direct product $SO(2) \times SO(3)$. In contrast to this simplicity, the radial coordinates Q and Q' cannot be treated independently. From a purely static point of view, the regular octahedron may distort in either the ε_g mode or the τ_{2g} mode. The corresponding energies of the minima are $-V_E^2/2\mu\omega_E^2$ and $-V'^2/2\mu\omega_T^2$. Suppose $(V'/\omega_T)^2 > (V_E/\omega_E)^2$. The τ_{2g} mode is preferred, and the trough of the potential-energy surface permits rotations (characterized by θ and ϕ) of the type described in Section V. As soon as vibrations are considered, allowance must be made for variations in Q'. Near the static solution, defined by $Q_e = V'/\mu\omega_T^2$ and $Q'_e = 0$, we can write $Q = Q_e + q$. We find

$$U - \left(V'^2Q^2 + V_E^2 Q'^2\right)^{1/2} \simeq -\frac{V'^2}{2\mu\omega_T^2} + \frac{1}{2}\mu\omega_T^2 q^2 + \frac{1}{2}\mu\omega_{\text{eff}}^2 Q'^2 \qquad (61)$$

to quadratic terms in q and Q'. The effective angular frequency ω_{eff} is defined by

$$\omega_{\text{eff}} = \left(\omega_E^2 - \frac{\omega_T^2 V_E^2}{V'^2}\right)^{1/2} \qquad (62)$$

Thus the ripple described in Section V.B as being superposed on the purely elliptical motion of a ligand (and lying in the plane of the ellipse) has a new component perpendicular to the plane of the ellipse and characterized by an angular frequency ω_{eff}. The motions out of the planes of the ellipses for the various ligands are 120° out of phase with those of their neighbors, as indicated in Eqs. (32)—with, of course, the Q there being replaced by Q'.

C. Rotations in Five Dimensions

If we set $\omega_E = \omega_T = \omega$ and $V_E = V'$, a remarkable simplification occurs in the JT system $\Gamma_8 \otimes (\varepsilon_g \oplus \tau_{2g})$. The coordinates Q_θ, Q_ε, Q_{yz}, Q_{zx}, and Q_{xy} appear on an equal footing, and the quadratic functions of S_x, S_y, and S_z [which appear when T_θ, T_ε, W_{yz}, W_{zx}, and W_{xy} are given their explicit forms (55) and (41)] are equally weighted. The result is that the Hamiltonian H now possesses the rotation group in five dimensions, SO(5), as a symmetry group. The existence of this group was first noticed by O'Brien.[24] Some discussion from a group-theoretical standpoint has been given,[25] and a detailed analysis has been provided by Pooler and O'Brien.[26]

The approach to the high-symmetry limit can be studied with Eqs. (60) and (62). In addition to the angles θ, ϕ, and χ, we now introduce a fourth, say β. We make the replacements

$$Q \rightarrow Q \cos \beta, \qquad Q' \rightarrow Q \sin \beta$$

and (60) becomes

$$E = \sum p_j^2/2\mu + \tfrac{1}{2}\mu\omega^2 Q^2 \pm V'Q \tag{63}$$

The angular frequency ω_{eff} becomes zero, and only one oscillatory frequency remains. Instead of the two oscillations associated with Q and Q', we have one oscillation (associated with the new Q) and a rotation corresponding to the motion of a point on the surface of a sphere in five dimensions. The complete description of the normal coordinates runs

$$Q_\theta = Q \cos \chi \sin \beta = Q_1 \text{ (say)}$$
$$Q_\varepsilon = Q \sin \chi \sin \beta = Q_2$$
$$Q_{yz} = Q \sin \theta \cos \phi \cos \beta = Q_3$$
$$Q_{zx} = Q \sin \theta \sin \phi \cos \beta = Q_4$$
$$Q_{xy} = Q \cos \theta \cos \beta = Q_5$$

and the momenta are given by $p_j = \mu\dot{Q}_j$. The Cartesian forms (p_j, Q_j) are rather more convenient for giving a formal solution to the problem in hand. The Hamiltonian

$$\sum p_j^2/2 + \tfrac{1}{2}\mu\omega^2 \sum Q_j^2 - V'\left(\sum Q_j^2\right)^{1/2}$$

at once leads to the equations of motion

$$-\mu\ddot{Q}_j = \left(\mu\omega^2 - \frac{V'}{Q}\right)Q_j \tag{64}$$

So

$$\frac{\ddot{Q}_1}{Q_1} = \frac{\ddot{Q}_2}{Q_2} = \cdots = \frac{\ddot{Q}_5}{Q_5} = -\omega_r^2 \qquad (65)$$

where

$$\omega_r = \left(\omega^2 - \frac{V'}{\mu Q}\right)^{1/2} \qquad (66)$$

For a constant Q, corresponding to rotations in the five-dimensional space, all five normal coordinates can be cast in the same form:

$$Q_j = Q_{j0}\cos(\omega_r t - \sigma_j) \qquad (67)$$

It can be readily verified from (63) that the static equilibrium value Q_e of Q is $V'/\mu\omega^2$. Thus the rotational frequency ω_r is small as long as Q is near Q_e. The faster the rotation, the greater Q becomes. This is what would be expected when centrifugal forces come into play.

D. Trajectories

If we restrict our attention to rotations in five dimensions, all displacements must have the form of (67) since they themselves are linear combinations of the normal coordinates. For example,

$$x_1 = x_{10}\cos(\omega_r t - \sigma_{x1})$$
$$y_1 = y_{10}\cos(\omega_r t - \sigma_{y1})$$
$$z_1 = z_{10}\cos(\omega_r t - \sigma_{z1})$$

This shows that the trajectory of every ligand closes on itself and forms a planar ellipse. Moreover, each ellipse is traversed in the period $2\pi/\omega_r$.

We might wonder what constraints are imposed on the six ellipses. As already mentioned, the major axes of the ellipses for $\Gamma_8 \otimes \tau_{2g}$ are not related in any special way, so it is unlikely that a further relaxation of the possible motions of the ligands would produce anything of interest in that regard. The ellipses of $\Gamma_8 \otimes \tau_{2g}$ for atoms 1, 2, and 3 lie in three mutually perpendicular planes, while radial motions of $\Gamma_8 \otimes \varepsilon_g$ also take place in orthogonal directions. It is thus natural to speculate that the ellipses for atoms 1, 2, and 3 of $\Gamma_8 \otimes (\varepsilon_g \oplus \tau_{2g})$ might lie in mutually orthogonal planes. It takes only a little work to convince oneself that the freedom inherent in solutions of the type (67), even though they are constrained by $\Sigma Q_j^2 = Q^2$, is sufficiently great to permit ellipses to lie in nonorthogonal planes. In fact, it is in the lack of constraints on the elliptical orbits that the SO(5) symmetry makes itself felt.

There remains the vibrations to be discussed. The variation of Q can be studied in a way analogous to what was done for $E \otimes \varepsilon_g$. The kinetic-energy term in (63) provides a centrifugal contribution to an effective one-dimensional potential in which motion takes place. A synchronous ripple is superposed on the six ellipses as Q oscillates about its mean value Q_e. The ripple is in the plane of each ellipse; but because ω is not, in general, a rational multiple of ω_r, the ripple does not close on itself as the ellipse is fully traversed.

VIII. $T_1 \otimes (\varepsilon_g \oplus \tau_{2g})$

We turn now to a system that has been widely studied because of its close approximation to an electron trapped in an oxygen vacancy in CaO. The sharp absorption line in the ultraviolet corresponding to the transition $s \to p$ possesses a broad absorption band, with several discernible features, on its high-energy side. Among the various experimental investigations, we cite the Faraday-rotation measurements of Kemp et al.[27] and Bessent et al.,[28] the observation of circular dichroism by Merle d'Aubigné and Roussel,[29] and that of linear dichroism by Duran et al.[30] As a result of this work, it has become clear that the two modes ε_g and τ_{2g} describing the local motion of the Ca^{2+} ions possess approximately the same frequency and that their coupling to the p electron is also very similar. A p state transforms like the irreducible representation T_{1u} of O_h, so we have an example of the JT system $T_1 \otimes (\varepsilon_g \oplus \tau_{2g})$ under conditions of equal coupling and equal frequencies— the same conditions, in fact, that have been assumed in the analysis of $\Gamma_8 \otimes (\varepsilon_g \oplus \tau_{2g})$ in Section VII.

A.   Symmetries

The early theoretical work on $T_1 \otimes (\varepsilon_g \oplus \tau_{2g})$ is due to O'Brien,[31,32] Hughes,[33] and Romestain and Merle d'Aubigné.[34] Glauber states have been used to describe the states of the distorted configurations of the octahedron,[35] and a trajectory analysis has been carried out.[36]

The Hamiltonian can be written as

$$H = \sum p_j^2/2\mu + \tfrac{1}{2}\mu\omega^2\left(Q_\theta^2 + Q_\varepsilon^2 + Q_{yz}^2 + Q_{zx}^2 + Q_{xy}^2\right)$$
$$+ V\left(Q_\theta T_\theta + Q_\varepsilon T_\varepsilon + Q_{yz}T_{yz} + Q_{zx}T_{zx} + Q_{xy}T_{xy}\right) \tag{68}$$

where T_{ij} is defined as in Section III.A, and, to match such equations as $\langle p_x | T_{xy} | p_y \rangle = -1$, the companion operators T_θ and T_ε, in analogy to their earlier definitions (25) and (55), now can be assumed to have the forms

$$T_\theta = \left(\tfrac{1}{3}\right)^{1/2}\left[3l_z^2 - l(l+1)\right], \qquad T_\varepsilon = l_x^2 - l_y^2 \qquad (69)$$

in which $l = 1$. The form of the Hamiltonian (68) is very similar to (57) once the conditions of equal frequencies and equal couplings are applied to the latter. We might be tempted to conclude that the symmetry group for (68) is SO(5). However, the electronic operators $T_\theta, T_\varepsilon, \ldots, T_{xy}$ of (68) are different from the corresponding operators of (57). The former act between the T_1 states of a p electron; the latter between the components of the Γ_8 state. The way in which the nature of the electronic state determines the symmetry group has been discussed by Pooler.[37,38] The principal idea can be grasped by applying it to the JT system in hand. In order for H [defined in (68)] to be scalar with respect to the operations of the group SO(5), the five components $Q_\theta, Q_\varepsilon, \ldots, Q_{xy}$ of the putative SO(5) vector must be combined with five similar components of an SO(5) vector. These components belong to the irreducible representation of SO(5) denoted by (10).[10] Now, the electronic operator with components $T_\theta, T_\varepsilon, \ldots, T_{xy}$ is a set of linear combinations of the nine operators $|i\rangle\langle j|$, where $i, j = x$, y, or z. However, neither the three kets $|i\rangle$ nor the three bras $\langle j|$ span a representation of SO(5): there is no three-dimensional irreducible representation of this group, and if we were to consider either the three kets or the three bras as being superpositions of three irreducible representations of dimension unity, the combinations $|i\rangle\langle j|$ would simply be scalars in SO(5) and thus unacceptable for constructing an SO(5) vector. The situation for $\Gamma_8 \otimes (\varepsilon_g \oplus \tau_{2g})$ is different because the four states $|S, M\rangle$ (for $S = \tfrac{3}{2}$ and $M = -\tfrac{3}{2}, -\tfrac{1}{2}, \tfrac{1}{2}$, and $\tfrac{3}{2}$) form the basis for the four-dimensional irreducible representation $\left(\tfrac{1}{2}\tfrac{1}{2}\right)$ of SO(5). The 16 products $|S, M\rangle\langle S, M'|$ can be linearly combined to form an SO(5) vector because

$$\left(\tfrac{1}{2}\tfrac{1}{2}\right) \times \left(\tfrac{1}{2}\tfrac{1}{2}\right) = (11) + (10) + (00)$$

and the required representation (10) appears in the decomposition on the right-hand side. [An additional condition, pointed out by Pooler,[37] that in decompositions of this sort the desired representation occur in the even part of the product can be made to exclude operators W'_{ij} of the type appearing in (42).]

The upshot is that SO(5) is not a symmetry group for the Hamiltonian of $T_1 \otimes (\varepsilon_g \oplus \tau_{2g})$ even under equal-coupling and equal-frequency conditions. However, the three kets do transform like the irreducible representation $\mathcal{D}_1$ of SO(3), that is, like an ordinary vector. The five operators $T_\theta, T_\varepsilon, \ldots, T_{xy}$ belong to the quadrupolar part $\mathcal{D}_2$ of the Kronecker product $\mathcal{D}_1 \times \mathcal{D}_1$, since, under the reduction $SO(3) \rightarrow O$, we have $\mathscr{D}_2 \rightarrow E + T_2$. The normal coordinates can also be embedded in $\mathscr{D}_2$ of $SO(3)$, since their labels, ε_g and τ_{2g}, are of the same kind. Thus the JT coupling term in (68) is an $SO(3)$ scalar, as are the remaining terms and hence the Hamiltonian as a whole.

B.   Coordinates

Having established the $SO(3)$ symmetry of the Hamiltonian, it is sometimes convenient to replace the five components $Q_\theta, Q_\varepsilon, \ldots, Q_{xy}$ by their linear combinations $Q_m^{(2)}$ that transform under rotations like the spherical harmonics Y_{2m}. The components $Q_m^{(2)}$ constitute a second-rank tensor, $\mathbf{Q}^{(2)}$. This notation enables us to bring the apparatus of angular-momentum theory to bear on the problem (see Edmonds[19]). The potential-energy term in H now assumes the compact form

$$\tfrac{1}{2}\mu\omega^2 \mathbf{Q}^{(2)} \cdot \mathbf{Q}^{(2)} + V(\mathbf{Q}^{(2)} \cdot \mathbf{T}^{(2)}) \tag{70}$$

where $\mathbf{T}^{(2)}$ is now a tensor of rank 2 acting in the electronic space with an amplitude defined by

$$(p\|T^{(2)}\|p) = (10)^{1/2} \tag{71}$$

The magnitude of the reduced matrix element of $\mathbf{T}^{(2)}$ given in (71) is consistent with Eqs. (69).

The $SO(3)$ symmetry of $T_1 \otimes (\varepsilon_g \oplus \tau_{2g})$ becomes apparent as soon as the static displacements are sought. If we continue to use the basis $|x\rangle$, $|y\rangle$, and $|z\rangle$, and if we maintain the old forms for the normal coordinates, the secular determinant (13) is untouched except for the addition of the three terms $VQ_\theta/\sqrt{3} - VQ$, $VQ_\varepsilon/\sqrt{3} + VQ$, and $-2VQ$ on the diagonal, and the substitution of V_T for V. However, in the expansion of the determinant, it is difficult to make it self-evident that the cubic term in the normal coordinates is, in fact, a scalar with respect to $SO(3)$. Using the tensor notation, we get

$$(U - E)^3 - (U - E)V^2\mathbf{Q}^{(2)} \cdot \mathbf{Q}^{(2)} - \left(\tfrac{70}{27}\right)^{1/2}V^3\mathbf{Q}^{(2)} \cdot \mathbf{Q}^{(2)} \cdot \mathbf{Q}^{(2)} = 0 \tag{72}$$

where

$$(\mathbf{Q}^{(2)} \cdot \mathbf{Q}^{(2)} \cdot \mathbf{Q}^{(2)}) = \sum_{qrs} \begin{pmatrix} 2 & 2 & 2 \\ q & r & s \end{pmatrix} Q_q^{(2)} Q_r^{(2)} Q_s^{(2)} \tag{73}$$

This expression is proportional to the scalar product of $\mathbf{Q}^{(2)}$ and the tensor $(\mathbf{Q}^{(2)}\mathbf{Q}^{(2)})^{(2)}$ that is formed by coupling two tensors $\mathbf{Q}^{(2)}$ to a resultant rank of 2. As such it is transparently an $SO(3)$ scalar.

The demands $\partial E / \partial Q_m^{(2)} = 0$ lead to an equation of the type

$$\mathbf{Q}^{(2)} = \Omega (\mathbf{Q}^{(2)} \mathbf{Q}^{(2)})^{(2)} \tag{74}$$

Not surprisingly, these equations do not lead to unique values of the normal coordinates describing the displacements. We write

$$\mathbf{Q}^{(2)} = Q \mathbf{C}^{(2)}(\theta, \phi) \tag{75}$$

where the components $C_m^{(2)}$ of the tensor $\mathbf{C}^{(2)}$ are related to the spherical harmonics through the equation

$$C_m^{(2)} = \left(\frac{4\pi}{5} \right)^{1/2} Y_{2m}$$

We can show[39] that

$$(\mathbf{C}^{(2)} \mathbf{C}^{(2)})^{(2)} = \sqrt{5} \begin{pmatrix} 2 & 2 & 2 \\ 0 & 0 & 0 \end{pmatrix} \mathbf{C}^{(2)} = -(2/7)^{1/2} \mathbf{C}^{(2)}$$

and hence Eq. (75) provides a solution to (74). It is straightforward to verify that the value of Q in (75), which we now write as Q_e, is given by $Q_e = -(4/3)^{1/2} V / \mu \omega^2$.

Admirable though they are for describing the static solution [and also the pure rotations corresponding to the SO(3) symmetry], the expressions for the components $Q_m^{(2)}$ given in (75) cannot be used to describe motions away from the potential-energy minimum. A generalization of (75) has been introduced by O'Brien.[32] It runs

$$Q_0^{(2)} = Q_\theta = Q \left[\tfrac{1}{2} (3\cos^2\theta - 1)\cos\alpha + \tfrac{1}{3} 3 \sin^2\theta \sin\alpha \cos\beta \right]$$

$$\sqrt{2}\, Q_{\pm 2}^{(2)} = Q_\epsilon \pm i Q_{xy} = Q \left[\tfrac{1}{2}\sqrt{3}\, \sin^2\theta \cos\alpha + \tfrac{1}{2}(1 + \cos^2\theta)\sin\alpha\cos\beta \right.$$

$$\left. \pm i \sin\alpha \sin\beta \cos\theta \right] e^{\pm 2i\phi} \tag{76}$$

$$\mp\sqrt{2}\, Q_{\pm 1}^{(2)} = Q_{zx} \pm i Q_{yz} = Q \left[\tfrac{1}{2}\sqrt{3} \sin 2\theta \cos\alpha - \tfrac{1}{2}\sin 2\theta \sin\alpha \cos\beta \right.$$

$$\left. \mp i \sin\theta \sin\alpha \sin\beta \right] e^{\pm i\phi}$$

The four angles $(\alpha, \beta, \theta, \phi)$ and the radial coordinate Q permit a complete parametrization of the normal coordinates of the modes ϵ_g and τ_{2g}. The components $Q_m^{(2)}$, as defined in (76), form a tensor of rank 2 with respect to the angular-momentum vector l, whose components are given by

$$l_x \pm il_y = ie^{\pm i\phi}\left(\cot\theta\frac{\partial}{\partial\phi} \mp i\frac{\partial}{\partial\theta} - \frac{2}{\sin\theta}\frac{\partial}{\partial\beta}\right)$$

$$l_z = -i\frac{\partial}{\partial\phi} \tag{77}$$

The scalar product $\mathbf{Q}^{(2)}\cdot\mathbf{Q}^{(2)}$ remains equal to Q^2, but, as O'Brien recognized, the cubic SO(3) invariant, given in (73), is proportional to $\cos 3\alpha$. In detail,

$$(\mathbf{Q}^{(2)}\cdot\mathbf{Q}^{(2)}\cdot\mathbf{Q}^{(2)}) = -\left(\tfrac{2}{35}\right)^{1/2}Q^3\cos 3\alpha \tag{78}$$

To avoid taking triple products of the components $Q_m^{(2)}$ in the derivation of this result, we may use angular-momentum techniques; the details are given in Appendix I. An alternative method makes use of the fact that $\mathbf{Q}^{(2)}\cdot\mathbf{Q}^{(2)}\cdot\mathbf{Q}^{(2)}$ is an SO(3) scalar. We may thus pick the special values $\theta = \phi = \beta = 0$ and evaluate the components of $\mathbf{Q}^{(2)}$ in (76). We find $Q_0^{(2)} = Q\cos\alpha$, $Q_{\pm 1}^{(2)} = 0$, and $Q_{\pm 2}^{(2)} = Q\sqrt{\tfrac{1}{2}}\sin\alpha$, in agreement with Eqs. (A4) and (A5), in Appendix I, for the special values of the angles. From the components $Q_m^{(2)}$, we can find $\mathbf{Q}^{(2)}\cdot\mathbf{Q}^{(2)}\cdot\mathbf{Q}^{(2)}$. By these methods, (72) becomes

$$(U-E)^3 - (U-E)V^2Q^2 + 2(27)^{-1/2}V^3Q^3\cos 3\alpha = 0 \tag{79}$$

C. Trajectories

We begin the analysis of the trajectories by taking the static solutions (75). When the tensors $\mathbf{C}^{(2)}$ are converted to spherical harmonics and the equations on the left-hand sides of (22) and (32) are used to find the displacements, we get

$$\begin{aligned}
x_1 &= -x_4 = Q_e(12)^{-1/2}(3\sin^2\theta\cos^2\phi - 1)\\
y_1 &= -y_4 = Q_e\left(\tfrac{3}{4}\right)^{1/2}\sin^2\theta\sin\phi\cos\phi\\
z_1 &= -z_4 = Q_e\left(\tfrac{3}{4}\right)^{1/2}\sin\theta\cos\theta\cos\phi\\
x_2 &= -x_5 = Q_e\left(\tfrac{3}{4}\right)^{1/2}\sin^2\theta\sin\phi\cos\phi\\
y_2 &= -y_5 = Q_e(12)^{-1/2}(3\sin^2\theta\sin^2\phi - 1)\\
z_2 &= -z_5 = Q_e\left(\tfrac{3}{4}\right)^{1/2}\sin\theta\cos\theta\sin\phi\\
x_3 &= -x_6 = Q_e\left(\tfrac{3}{4}\right)^{1/2}\sin\theta\cos\theta\cos\phi\\
y_3 &= -y_6 = Q_e\left(\tfrac{3}{4}\right)^{1/2}\sin\theta\cos\theta\sin\phi\\
z_3 &= -z_6 = Q_e(12)^{-1/2}(3\cos^2\theta - 1)
\end{aligned} \tag{80}$$

In these equations Q_e has been written for Q to stress the fact that attention is restricted to the static solutions. Very remarkably, we can easily eliminate θ and ϕ from the coordinates (x_i, y_i, z_i) for a given ligand i. The result is that the displaced ligands lie on the surfaces of the spheres given by the equations

$$\left(x_1 - Q_e(48)^{-1/2}\right)^2 + y_1^2 + z_1^2 = \frac{3Q_e^2}{16}$$

$$x_2^2 + \left(y_2 - Q_e(48)^{-1/2}\right)^2 + z_2^2 = \frac{3Q_e^2}{16}$$

$$x_3^2 + y_3^2 + \left(z_3 - Q_e(48)^{-1/2}\right)^2 = \frac{3Q_e^2}{16}$$

$$\left(x_4 + Q_e(48)^{-1/2}\right)^2 + y_4^2 + z_4^2 = \frac{3Q_e^2}{16} \tag{81}$$

$$x_5^2 + \left(y_5 + Q_e(48)^{-1/2}\right)^2 + z_5^2 = \frac{3Q_e^2}{16}$$

$$x_6^2 + y_6^2 + \left(z_6 + Q_e(48)^{-1/2}\right)^2 = \frac{3Q_e^2}{16}$$

We have now to work out the geometry of these equations. Let the undisplaced position $(a, 0, 0)$ of ligand 1 be called E_1. Its displaced position, which we denote by L_1, is $(a + x_1, y_1, z_1)$. The ligand lies on a sphere of radius $Q_3\sqrt{3}/4$, centered at $(a + Q_e(48)^{-1/2}, 0, 0)$, a point denoted by O_1. This sphere intersects the x axis at an interior point T_1 with coordinates $(a - Q_e(48)^{-1/2}, 0, 0)$. From (80) we may readily show that the line L_1T_1 possesses the direction cosines

$$(\sin\theta\cos\phi, \quad \sin\theta\sin\phi, \quad \cos\theta)$$

That is, L_1T_1 is parallel to the vector $\mathbf{C}^{(1)}$ with polar coordinates (θ, ϕ). The same is true for every line L_iT_i defined (for ligand i) in an analogous way to L_1T_1.

From the coordinates given above, the direction cosines of L_1O_1 are found to be

$$(2\sin^2\theta\cos^2\phi - 1, \quad 2\sin^2\theta\sin\phi\cos\phi, \quad \sin 2\theta\cos\phi). \tag{82}$$

Those for L_2O_2 and L_3O_3 are

$$(2\sin^2\theta\sin\phi\cos\phi, \quad 2\sin^2\theta\sin^2\phi - 1, \quad \sin 2\theta\sin\phi) \tag{83}$$

and

$$(\sin 2\theta \cos \phi, \quad \sin 2\theta \sin \phi, \quad \cos 2\theta) \tag{84}$$

respectively. It is at once found that $O_1 L_1$, $O_2 L_2$, and $O_3 L_3$ are mutually perpendicular. Since the displacements of ligands 4, 5, and 6 are simply the inverses of those for ligands 1, 2, and 3, we see that the set of six ligands is tightly constrained. In fact, if we translate $O_1 L_1$, $O_2 L_2,\ldots, O_6 L_6$ so that the centers O_i of the six spheres coincide, the translated ligands lie at the vertices of a regular octahedron. This is illustrated in Fig. 12. As the point (θ, ϕ) moves on a great circle, the lines $L_i T_i$ rotate in planes. Since a plane cuts a sphere in a circle, the trajectories of the ligands are circles (rather than ellipses) of various diameters. The dynamics of the rotation are studied in greater detail in Section IX.B.

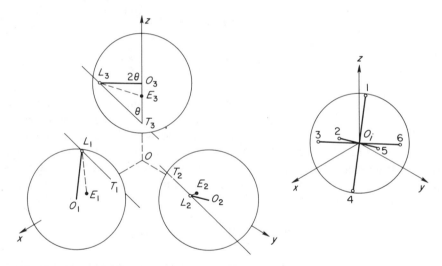

Fig. 12.   Ligand displacements. On the left is shown the static displacements of ligands L_1, L_2, and L_3 for $T_1 \otimes (\varepsilon_g \oplus \tau_{2g})$ from the positions E_1, E_2, and E_3 that they would occupy in the absence of any JT coupling. Under conditions of equal frequencies and equal couplings for the ε and τ_2 modes, the points L_i lie on the surfaces of spheres centered at O_i such that the lines $L_i T_i$, where T_i is the interior intersection of the sphere centered at O_i with a coordinate axis, all possess the same direction cosines $(\sin \theta \cos \phi, \sin \theta \sin \phi, \cos \theta)$. The figure is drawn for the set $(2/\sqrt{14}, -1/\sqrt{14}, 3/\sqrt{14})$. The lines $O_1 L_1$, $O_2 L_2$, and $O_3 L_3$ form an orthogonal set. The distances $O_i O$, where O is the origin of the coordinate system, are much reduced in length for illustrative purposes. The diagram on the right shows the effect of translating these lines so that the points O_i coincide. With the companion ligands L_4, L_5, and L_6, we obtain a rigid regular octahedron. This octahedron rotates in the trough of the potential-energy surface, thereby exhibiting the SO(3) symmetry of the system.

To examine vibrations, we return to (79). The usefulness of the angle α as a parameter at once becomes apparent, since the cubic equation breaks up into three linear factors:

$$E = \tfrac{1}{2}\mu\omega^2 Q^2 + \left(\tfrac{4}{3}\right)^{1/2} VQ\cos\alpha$$
$$E = \tfrac{1}{2}\mu\omega^2 Q^2 + \left(\tfrac{4}{3}\right)^{1/2} VQ\cos\left(\alpha + \tfrac{2}{3}\pi\right) \tag{85}$$
$$E = \tfrac{1}{2}\mu\omega^2 Q^2 + \left(\tfrac{4}{3}\right)^{1/2} VQ\cos\left(\alpha - \tfrac{2}{3}\pi\right)$$

We must add the kinetic-energy term $\sum p_j^2/2\mu$ to the right-hand sides of these equations to get the Hamiltonians corresponding to motion associated with a particular electronic state. Since the static solutions (75) correspond to $\alpha = 0$, we restrict ourselves to small values of α, corresponding to vibrations of small amplitude. The coordinate Q must also be restricted to small oscillations about its equilibrium position Q_e, which, as previously found, is given by $Q_e = -(4/3)^{1/2}V/\mu\omega^2$. We write $Q - Q_e = q$. The effective Hamiltonian for these small oscillations is

$$\sum p_j^2/2\mu - 2V^2/3\mu\omega^2 + \tfrac{1}{2}\mu\omega^2\left(q^2 + \alpha^2 Q_e^2\right) \tag{86}$$

Since $(q^2 + \alpha^2 Q_e^2)^{1/2}$ is just the distance of a point from the bottom of the potential well at $q = \alpha = 0$, the last term in (86) suggests that the motion corresponds to that of an isotropic two-dimensional oscillator. However, the individual components of $\mathbf{Q}^{(2)} - \mathbf{Q}_e^{(2)}$ involve β as well as α and q (and, for that matter, θ and ϕ too). In fact, in the limit of small α, the angle β always appears in either the combination $\alpha\cos\beta$ or $\alpha\sin\beta$. We should regard the last term in (86) as

$$\tfrac{1}{2}\mu\omega^2\left(q^2 + Q_e^2\alpha^2\cos^2\beta + Q_e^2\alpha^2\sin^2\beta\right) \tag{87}$$

corresponding to an isotropic three-dimensional oscillator. The angle β plays the role of an azimuthal angle. In view of the result of the analysis of Section III.B for $T_1 \otimes \tau_{2g}$, we should check that the upper electronic state does not affect the isotropic nature of (87). This is done in Appendix II.

The effect of taking

$$q = q_0\cos\left(\omega t - \sigma_q\right)$$
$$Q_e\alpha\cos\beta = q_X\cos\left(\omega t - \sigma_X\right) \tag{88}$$
$$Q_e\alpha\sin\beta = q_Y\cos\left(\omega t - \sigma_Y\right)$$

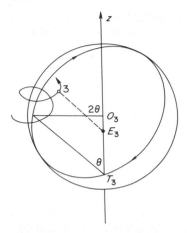

Fig. 13.   The trajectory of ligand 3 for $T_1 \otimes (\varepsilon_g \oplus \tau_{2g})$ under equal-coupling and equal-frequency conditions. An elliptical motion is superposed on the circular motion defined by the intersection of a plane with the sphere centered at O_3. The plane passes through T_3.

—corresponding to motion in the potential well appearing in (86)—is to superpose an elliptical motion on the circular motion of every ligand. This can be seen simply by noting that every component of $\mathbf{Q}^{(2)} - \mathbf{Q}_e^{(2)}$ is a linear combination of the cosine terms of (88) in the limit of small q and α. This is illustrated in Fig. 13. The planes of the ellipses turn as the circular motion takes place, and there does not seem to be any very simple connection between the motions of different ligands.

IX.   ELECTRONIC COUPLING

A.   Doubled Frequencies

In the process of solving the secular equations, expressions are obtained for the electronic eigenfunctions corresponding to particular distortions of the octahedral complex. Equation (31), for example, gives the lower eigenfunction $|\phi_-\rangle$ for the JT system $E \otimes \varepsilon_g$. It is a feature of such eigenfunctions that they involve half-angles ($\frac{1}{2}\phi$ in the present case), whereas the positions of the ligands are defined by whole angles. It thus appears that the electronic eigenfunctions rotate at only half the rate of the distorted ligand structures. On the face of it, this is very odd, because the calculation is carried out for a tight coupling between the two parts of the JT system.

The regular octahedron formed from the displacements of the JT system $T_1 \otimes (\varepsilon_g \oplus \tau_{2g})$, which is drawn in Fig. 12, has its orientation defined by the set of direction cosines given in (82), (83), and (84). It is straightforward to show that these nine quantities are identical to the rotation matrix that takes the coordinates (x, y, z) through the transformation corresponding to the Euler angles $(\phi, 2\theta, \pi - \phi)$. The angle θ occurs doubled; and although ϕ occurs without the factor 2, its appearance in both the first and third Euler angle doubles its effect. In fact, if θ and ϕ are incrementally increased by $\delta\theta$ and $\delta\phi$, the octahedron of Fig. 12 is rotated through an angle $\delta\Omega$, where

$$(\delta\Omega)^2 = 4(\delta\theta)^2 + 4\sin^2\theta(\delta\phi)^2 \tag{89}$$

The ground electronic state, on the other hand, can be written as $\mathbf{C}^{(1)} \cdot |\mathbf{p}\rangle$ for a distortion characterized by θ and ϕ (see Appendix II). In full, it runs

$$\sin\theta\cos\phi|x\rangle + \sin\theta\sin\phi|y\rangle + \cos\theta|z\rangle \tag{90}$$

showing that the axis of the p orbital coincides with the common direction of the $T_i L_i$. Being defined by the polar angles θ and ϕ, the axis rotates through an angle $\delta\Omega'$ (if $\theta \to \theta + \delta\theta$ and $\phi \to \phi + \delta\phi$) that is given by

$$(\delta\Omega')^2 = (\delta\theta)^2 + \sin^2\theta(\delta\phi)^2 \tag{91}$$

Comparison with (89) shows that we again find that the displaced ligands rotate at twice the rate of the electronic state.

The apparent paradox of having two parts of a strongly coupled system rotating at different rates is resolved by noting that the centers of the two rotating parts do not coincide. For $T_1 \otimes (\varepsilon_g \oplus \tau_{2g})$, the p orbital is indeed rigidly locked on to the direction $T_i L_i$ for every ligand L_i as it moves on its circle. From this point of view, the frequency doubling of the ligands arises because the angle marked 2θ in Figs. 12 and 13 is twice that marked θ.

The fact that the three Euler angles describing the orientation of the regular octahedron of displacements $L_i O_i$ depend on just the two angles θ and ϕ means that the rotational motion of the octahedron is not entirely arbitrary. In fact, the expression given in (89) enables us to write down the relation

$$T = 2I(\dot\theta^2 + \sin^2\theta\dot\phi^2) \tag{92}$$

for the kinetic energy of the rotating octahedron, where its moment of inertia is I. Since T has the same form as that of a particle moving freely on the surface of a sphere S, the trajectory of the point (θ, ϕ) is a great circle on S. The polar direction of that point is the same as that of the p orbital and of the lines $T_i L_i$. As each $T_i L_i$ rotates, the point L_i is carried round on the sphere of radius $\sqrt{3}\,Q_e/4$ centered at O_i, and it runs into T_i. This happens for L_3, for example, when $\theta = \frac{1}{2}\pi$. We know that θ must pass through this value because it describes (with ϕ) a great-circle trajectory on S. Thus the circular trajectory of each ligand L_i passes through T_i. This is shown in Fig. 13 for ligand 3.

B. Eigenfunctions

The difference in rotational frequencies for ligands and for the electronic state has repercussions on the acceptable eigenfunctions of the quantized JT system. Our guiding principle is that a given physical configuration should have a single-valued eigenfunction. This rule is easy enough to apply to the ligands: their positions are defined by the displacement coordinates and, if required, their momenta can be obtained by differentiation. The electronic state, however, has only been described in terms of certain kets that form a basis for an irreducible representation of the symmetry group of the Hamiltonian. Any multiplicative factor of the type $e^{i\varphi}$ associated with the electronic state can have no physical significance. For example, the state $|\phi_-\rangle$, given in (31) for the JT system $E \otimes \varepsilon_g$, describes the same physical situation as $-|\phi_-\rangle$.

Having said this, we note that such a state is produced by augmenting ϕ by 2π in the expression for $|\phi_-\rangle$, while the substitution $\phi \to \phi + 2\pi$ brings the ligands through a complete cycle. Evidently the physical situations for ϕ and $\phi + 2\pi$ are identical; but $|\phi_-\rangle$ changes sign. Any ligand eigenfunction $|\psi\rangle$ must change sign too in order to ensure that $|\psi\rangle|\phi_-\rangle$ is single valued. If we choose $|\psi\rangle = e^{im\phi}$, which is diagonal with respect to the term $p_\phi^2/2\mu Q^2$ in (36), the quantum number m is limited to the sequence $m = \pm\frac{1}{2},\ \pm\frac{3}{2}, \pm\frac{5}{2},\ldots$.

Of course, as soon as the product $e^{im\phi}|\phi_-\rangle$ is formed, we no longer have an eigenfunction of $p_\phi^2/2\mu Q^2$, since $|\phi_-\rangle$, like $e^{im\phi}$, is a function of ϕ. However, if we convert the interaction term by writing

$$Q_\theta T_\theta + Q_\varepsilon T_\varepsilon = Q \cos\phi(|\varepsilon\rangle\langle\varepsilon| - |\theta\rangle\langle\theta|) + Q\sin\phi(|\varepsilon\rangle\langle\theta| + |\theta\rangle\langle\varepsilon|)$$

we may verify that it commutes with j_z, defined by

$$j_z = -i\frac{\partial}{\partial\phi} + \frac{1}{2}i(|\theta\rangle\langle\varepsilon| - |\varepsilon\rangle\langle\theta|) \tag{93}$$

as does the entire Hamiltonian H. We may also show that

$$j_z e^{im\phi}|\phi_-\rangle = m e^{im\phi}|\phi_-\rangle \tag{94}$$

so our product function $e^{im\phi}|\phi_-\rangle$, although not an eigenfunction of H, is an eigenfunction of the operator j_z that commutes with H. Thus any corrections to $e^{im\phi}|\phi_-\rangle$ cannot come from other functions $e^{im'\phi}|\phi_-\rangle$ but only from interactions with states associated with the upper electronic branch $|\phi_+\rangle$. These effects can be assumed to be small.

A similar analysis can be carried out for $T_1 \otimes (\varepsilon_g \oplus \tau_{2g})$. We maintain the conditions of equal frequencies and equal couplings. The substitutions $\theta \to$

$\pi - \theta$ and $\phi \to \phi + \pi$ leave the displacements (80) invariant, corresponding to a complete cycle in the ligand motions, while the direction of the p orbital is reversed. That is, $\mathbf{C}^{(1)} \cdot |\mathbf{p}\rangle$ changes sign. To determine an acceptable ligand eigenfunction, we first use (92) to find the canonical momenta p_θ and p_ϕ:

$$
\begin{aligned}
p_\theta &= \frac{\partial T}{\partial \dot{\theta}} = 4I\dot{\theta} \to -\frac{i\hbar\,\partial}{\partial\theta} \\
p_\phi &= \frac{\partial T}{\partial \dot{\phi}} = 4I\sin^2\theta\,\dot{\phi} \to -\frac{i\hbar\,\partial}{\partial\phi}
\end{aligned}
\tag{95}
$$

The rotational part of the quantum mechanical Hamiltonian is

$$
-\frac{\hbar^2}{8I}\left[\frac{1}{\sin\theta}\frac{\partial}{\partial\theta}\left(\sin\theta\frac{\partial}{\partial\theta}\right) + \frac{1}{\sin^2\theta}\frac{\partial^2}{\partial\phi^2} \right]
\tag{96}
$$

The eigenfunctions of (96) are the spherical harmonics Y_{lm} or their differently normalized equivalents, $C_m^{(l)}$. In analogy to $e^{im\phi}|\phi_-\rangle$, we choose

$$
C_m^{(l)}\big(\mathbf{C}^{(1)} \cdot |\mathbf{p}\rangle\big)
\tag{97}
$$

As before, we have not formed an eigenfunction of H; but (97) *is* an eigenfunction of $\mathbf{J}^2$, where

$$
\mathbf{J} = \mathbf{l} + \mathbf{l}_p
\tag{98}
$$

The components of $\mathbf{l}$ are defined in (77), and the second term is the angular momentum of the p electron, analogous to the second term of (93). In terms of the bras and kets that we have been using, the only nonvanishing matrix elements of $\mathbf{l}_p$ are

$$
\langle x|l_{py}|z\rangle = -\langle z|l_{py}|x\rangle = i
\tag{99}
$$

together with all cyclic permutations of x, y, and z.

The condition that (97) be single-valued under the substitutions $\theta \to \pi - \theta$ and $\phi \to \phi + \pi$ leads at once to the condition that l be odd, since inversion takes $\mathbf{C}^{(l)}$ into $(-1)^l\mathbf{C}^{(l)}$. Thus the rotational band based on the ground electronic state, with no vibrational excitations, corresponds to the sequence of quantum numbers $l = 1, 3, 5, \ldots$.

C. Energy Levels for $E \otimes \varepsilon_g$ and Its Equivalents

Our analysis of trajectories and of the associated eigenfunctions has brought us very close to being able to give explicit expressions for the energy levels of JT systems. It is of interest to compare the purely analytical approach with what could be called a descriptive one—that is, one based on a physical knowledge of the trajectories.

We begin by applying the analytical method to $E \otimes \varepsilon_g$. For rotations in the potential-energy trough, the kinetic-energy term $p_\phi^2/2\mu Q^2$ is equivalent to $-(\hbar^2/2\mu Q^2)(\partial^2/\partial \phi^2)$, for which we can easily show that

$$\langle \phi_- | e^{-im\phi} \left(-\frac{\hbar^2}{2\mu Q^2} \right) \left(\frac{\partial^2}{\partial \phi^2} \right) e^{im\phi} | \phi_- \rangle = \left(\frac{\hbar^2}{2\mu Q^2} \right) \left(m^2 + \frac{1}{4} \right) \quad (100)$$

The term $\frac{1}{4}$ in the second parenthesis on the right-hand side of (100) disappears when the Hamiltonian is subjected to a unitary transformation that converts the expression $Q^{-1}(\partial/\partial Q)(Q\partial/\partial Q)$, which is the correct form to use for p_Q^2 when polar coordinates are employed, to $\partial^2/\partial Q^2$. (This is a necessary step to ensure that the vibrational motion is truly one-dimensional in Q.) Thus (100) yields a rotational band of doubly degenerate levels at energies $\hbar^2 m^2/2\mu Q^2$, where $m = \pm\frac{1}{2}, \pm\frac{3}{2}, \pm\frac{5}{2}, \ldots$. The corrections due to interactions with the upper electronic state $|\phi_+\rangle$ have been obtained to high orders in perturbation theory by O'Brien and Pooler.[17]

The expression $\hbar^2 m^2/2\mu Q^2$ is in accord with what we would expect for the rotating dumbbell of Section IV.B, although in that case the theory of a plane rotator leads to the sequence $m = 0, \pm 1, \pm 2, \ldots$ (see, for example, Pauling and Wilson[40]). The different boundary conditions for $E \otimes \varepsilon_g$ have led to a change in the nature of m.

Our knowledge of trajectories finds an immediate application to $\Gamma_8 \otimes \tau_{2g}$ and $\Gamma_8 \otimes (\varepsilon_g \oplus \tau_{2g})$ (for equal couplings and equal frequencies). Both JT systems correspond to trajectories that are constrained to a two-dimensional space embedded in a space of a higher number of dimensions (three for $\Gamma_8 \otimes \tau_{2g}$; five for $\Gamma_8 \otimes (\varepsilon_g \oplus \tau_{2g})$). We would therefore expect the energy levels to follow basically the same pattern. This is borne out by detailed calculations. The only difference lies in a displaced origin of the energy levels (owing to the zero-point motions extending into the unused dimensions of the trajectory space) and a change in m. In its place it is convenient to use the angular-momentum quantum number j for $\Gamma_8 \otimes \tau_{2g}$, since this JT system possesses SO(3) symmetry. It turns out that j is half-integral.[41] The lowest band of rotational levels is obtained by replacing m by $j + \frac{1}{2}$ in the formula for $E \otimes \varepsilon_g$. Because of the unused dimension in trajectory space, the degeneracies of the levels are $2(2j+1)$ for $\Gamma_8 \otimes \tau_{2g}$ rather than the double degeneracies of $E \otimes \varepsilon_g$. In the case of $\Gamma_8 \otimes (\varepsilon_g \oplus \tau_{2g})$, the SO(5) symmetry leads to the use of the irreducible representations $(j\frac{1}{2})$ as labels for the energy levels. The degeneracies are now $(2j+1)(2j+3)(2j+5)/12$.[42] As with $\Gamma_8 \otimes \tau_{2g}$, the quantum number j is half-integral.[26]

An analogous treatment could be presented at this point for $T_1 \otimes (\varepsilon_g \oplus \tau_{2g})$. However, this system is more complicated than $E \otimes \varepsilon_g$, and the discussion seems best deferred until some preliminary analysis is worked out.

X. TENSOR METHODS FOR $T_1 \otimes (\varepsilon_g \oplus \tau_{2g})$

It was established in Section VIII.C that the excitation of vibrations for the system $T_1 \otimes (\varepsilon_g \oplus \tau_{2g})$ produces elliptical motions that are superposed on the circular trajectories of the ligands associated with their motion in the potential-energy trough. To examine these ellipses in detail, we can, of course, use the detailed coordinates of the ligands given in (76). However, many of the relations that we can develop involve sums over the ligands; and, in the act of carrying out these sums, it is found that many of the trigonometric functions combine in a very nice way. It is impossible not to feel that a more direct method must exist. The consistent use of tensors provides such a method. Although it is more abstract than the techniques that we have been using hitherto, the experience gained in the straightforward approach provides an alternative that we can always turn to if the interpretation of the mathematics seems obscure.

A. Displacement Vectors

The origin of much of the complexity of our work lies in Eqs. (22) and (32), where the displacements x_i, y_i, and z_i are related to the normal coordinates $Q_\theta, Q_\varepsilon, Q_{yz}, Q_{zx}, Q_{xy}$. Can these 18 equations be compressed into just one? We know that the five normal coordinates can be thought of (in various combinations) as the components of a tensor $\mathbf{Q}^{(2)}$. We need to be able to use the tensorial character of $\mathbf{Q}^{(2)}$ to construct vectors $\mathbf{r}_i$ whose components are (x_i, y_i, z_i). It is clear that the equilibrium positions of the ligands i must play a role: we therefore introduce the unit vectors $\mathbf{e}_i$ that run out from the origin O of the octahedral complex towards ligand i. They form an orthogonal system: each ligand i, in its undisplaced position, is situated at the point $a\mathbf{e}_i$, in the notation of Fig. 7. For example, $\mathbf{e}_2 = (0,1,0)$, and ligand 2 has coordinates $(0, a, 0)$.

From $\mathbf{Q}^{(2)}$ and $\mathbf{e}_i$ we can form the vector $(\mathbf{e}_i^{(1)} \mathbf{Q}^{(2)})^{(1)}$. Might this be related to $\mathbf{r}_i$? Pick atom 3. We find, for example, that

$$\left(\mathbf{e}_3^{(1)} \mathbf{Q}^{(2)} \right)_1^{(1)} = \left(\mathbf{e}_3^{(1)} \right)_0 Q_1^{(2)} (10, 21 | 11)$$

$$= \left(\tfrac{3}{20} \right)^{1/2} (Q_{zx} + iQ_{yz})$$

To correspond to the conventional phase choices of angular-momentum theory,[19] we would like this to be related to $(\mathbf{r}_3)_1^{(1)}$, that is, to $-(x_3 + iy_3)/\sqrt{2}$.

Since, from (22), $x_3 = \frac{1}{2}Q_{zx}$ and $y_3 = \frac{1}{2}Q_{yz}$, the connection

$$\mathbf{r}_i^{(1)} = -\left(\tfrac{5}{6}\right)^{1/2}\left(\mathbf{e}_i^{(1)}\mathbf{Q}^{(2)}\right)^{(1)} \tag{101}$$

suggests itself. It is straightforward to confirm that this single equation gives all the 18 displacements and is thus the result that we require.

B.   Classical Angular Momentum

The expressions for $\mathbf{Q}^{(2)}$ in terms of the coordinates $(Q, \alpha, \beta, \theta, \phi)$ are set out explicitly in (76) and in tensorial form in Eq. (A4) in Appendix I. To calculate the positions of the ligands in the elliptical orbits relative to the centers of each ellipse, we need to calculate the differences $\mathbf{d}^{(2)}$, where

$$\mathbf{d}^{(2)} = Q\left(\mathbf{D}^{(22)}\left(\phi, \theta, \tfrac{1}{2}\beta\right)\mathbf{X}^{(02)}(\alpha)\right)^{(20)}$$
$$- Q_e\left(\mathbf{D}^{(22)}(\phi, \theta, 0)\mathbf{X}^{(02)}(0)\right)^{(20)} \tag{102}$$

Following the discussion of Section VIII.C, we restrict our attention to small α and for $|Q - Q_e| \ll Q_e$. The three-dimensional oscillator introduced in that section is described by the three coordinates that we now write as (X, Y, Z), where

$$X = Q_e\alpha\cos\beta, \qquad Y = Q_e\alpha\sin\beta, \qquad Z = Q - Q_e = q \tag{103}$$

In order to write $\mathbf{d}^{(2)}$ in terms of X, Y, and Z, we remove the third Euler angle from the first $\mathbf{D}$ function in (102) by writing

$$D_{mn}^{(22)}\left(\phi, \theta, \tfrac{1}{2}\beta\right) = D_{mn}^{(22)}(\phi, \theta, 0)e^{-i\beta n/2}$$

The exponential is now incorporated into the tensor $\mathbf{X}^{(02)}(\alpha)$ by defining a new tensor $\mathbf{\Xi}^{(02)}$ whose components are

$$\Xi_{00}^{(02)} = Z, \qquad \Xi_{0\pm1}^{(02)} = 0, \qquad \Xi_{0\pm2}^{(02)} = (X \pm iY)/\sqrt{2} \tag{104}$$

It may seem strange to see the components X, Y, and Z of a vector appearing as the components of a tensor of rank 2. However, we are at liberty to specify the components of any tensor, just as we picked the components of the vector $\mathbf{e}_2$ in Section X.A as $(0, 1, 0)$.

With these preliminaries, it is not difficult to convert (102) to

$$\mathbf{d}^{(2)} = \left(\mathbf{D}^{(22)}\mathbf{\Xi}^{(02)}\right)^{(20)} \tag{105}$$

with the understanding that the third Euler angle in $\mathbf{D}^{(22)}$ is zero. The actual ligand displacements $\mathbf{s}_i^{(1)}$ relative to their static displacements are given by

$$\mathbf{s}_i^{(1)} = \mathbf{r}_i^{(1)}(Q, \alpha, \beta, \theta, \phi) - \mathbf{r}_i^{(1)}(Q_e, 0, 0, \theta, \phi)$$

$$= -(5/6)^{1/2}(\mathbf{e}_i^{(1)}\mathbf{d}^{(2)})^{(1)} \tag{106}$$

The classical angular momentum of ligand i relative to its static displacement is $(\mathbf{s}_i \times \dot{\mathbf{s}}_i)/\mu$. In tensorial notation, this is $-i\sqrt{2}\,(\mathbf{s}_i^{(1)}\dot{\mathbf{s}}_i^{(1)})^{(1)}/\mu$. Thus we get

$$(\mathbf{s}_i \times \dot{\mathbf{s}}_i)/\mu = -\left(5\sqrt{2}/6\right) i\left(\left(\mathbf{e}_i^{(1)}\mathbf{d}^{(2)}\right)^{(1)}\left(\mathbf{e}_i^{(1)}\dot{\mathbf{d}}^{(2)}\right)^{(1)}\right)^{(1)}/\mu$$

It is to our advantage to bring together the two $\mathbf{e}_i^{(1)}$ vectors on the right-hand side of this equation. The coupled product $(\mathbf{e}_i^{(1)}\mathbf{e}_i^{(1)})^{(t)}$ vanishes for $t = 1$ (since it corresponds to the vector product of two identical vectors). Moreover, $(\mathbf{e}_i^{(1)}\mathbf{e}_i^{(1)})^{(2)}$ also vanishes when the summation over i is performed, since an object of octahedral symmetry cannot possess a quadrupolar moment. Only $(\mathbf{e}_i^{(1)}\mathbf{e}_i^{(1)})^{(0)}$ remains. It evaluates to $-1/\sqrt{3}$. In the course of bringing the $\mathbf{e}$ vectors together, we need the recoupling coefficient[19]

$$((12)1(12)1, 1|(11)0(22)1, 1) = 3\sqrt{3}\begin{Bmatrix} 1 & 2 & 1 \\ 1 & 2 & 1 \\ 0 & 1 & 1 \end{Bmatrix}$$

$$= -3\begin{Bmatrix} 2 & 1 & 1 \\ 1 & 2 & 1 \end{Bmatrix} = \left(\tfrac{3}{20}\right)^{1/2}$$

So

$$\sum_{i=1}^{6} \frac{(\mathbf{s}_i \times \dot{\mathbf{s}}_i)}{\mu} = \frac{\left(\tfrac{5}{2}\right)^{1/2} i(\mathbf{d}^{(2)}\dot{\mathbf{d}}^{(2)})^{(1)}}{\mu} \tag{107}$$

Using (105) and (A3), we get

$$(\mathbf{d}^{(2)}\dot{\mathbf{d}}^{(2)})^{(1)} = \left((\mathbf{D}^{(22)}\mathbf{\Xi}^{(02)})^{(20)}(\mathbf{D}^{(22)}\dot{\mathbf{\Xi}}^{(02)})^{(20)}\right)^{(10)}$$

$$= ((20)2(20)2, 1|(22)1(00)0, 1)$$

$$\times ((22)0(22)0, 0|(22)1(22)1, 0)$$

$$\times \left((\mathbf{D}^{(22)}\mathbf{D}^{(22)})^{(11)}(\mathbf{\Xi}^{(02)}\dot{\mathbf{\Xi}}^{(02)})^{(01)}\right)^{(10)}$$

$$= \left(\mathbf{D}^{(11)}(\mathbf{\Xi}^{(02)}\dot{\mathbf{\Xi}}^{(02)})^{(01)}\right)^{(10)} \tag{108}$$

From the definition of $\Xi^{(02)}$ we can quickly verify that

$$(\Xi^{(02)}\dot{\Xi}^{(02)})^{(01)}_{00} = -\left(\tfrac{2}{5}\right)^{1/2} i(X\dot{Y} - Y\dot{X}) \tag{109}$$

with all other components vanishing. Now, it can be shown that[43]

$$D^{(11)}_{m0} = -\sqrt{3}\,C^{(1)}_m(\theta, \phi)$$

and so, collecting our results, we get

$$\sum_{i=1}^{6} \frac{(\mathbf{s}_i \times \dot{\mathbf{s}}_i)}{\mu} = \frac{\mathbf{C}^{(1)}(X\dot{Y} - Y\dot{X})}{\mu} \tag{110}$$

This extraordinary equation proves that the elliptical motions of the six ligands of $T_1 \otimes (\varepsilon_g \oplus \tau_{2g})$ provide an angular momentum equal to the angular momentum L_Z of the isotropic oscillator with coordinates X, Y, and Z; and this angular momentum is directed along an axis defined by the polar angles (θ, ϕ). This is the same direction as the lines $T_i L_i$ of Fig. 12, and it is also the direction of the orbital axis of the p electron. It goes without saying that we have completely avoided all manipulations with trigonometric functions in this derivation. The price for this advantage is the time spent absorbing and putting into effect the techniques of angular-momentum theory; but because of its application to many fields of chemistry and physics, it is time that no one should begrudge. A derivation of (110) by conventional methods can be found in the literature.[36]

C. Eigenfunctions

The states of the lowest rotational band of $T_1 \otimes (\varepsilon_g \oplus \tau_{2g})$ have already been given in (97). These are appropriate for the case when there are no vibrational excitations: the octahedron of Fig. 12 remains rigid as it rotates. We have now to take into account the elliptical motions of the ligands that are superposed on their circular motions, as described in Section X.B in terms of the coordinates X, Y, and Z. In isolation, the three-dimensional oscillator would be described by kets of the type $|n, L, m\rangle$, where the labels refer to the principal quantum number, that of the angular momentum, and that of its Z component. In view of (105), which rotates the coordinates of the oscillator in order to produce the actual differences $\mathbf{d}^{(2)}$, it is natural to anticipate that the eigenfunctions of the entire JT system would reflect a similar rotation. The angle β presents us with a choice: we can either include it in the representation of the kets $|n, L, m\rangle$ and take $\mathbf{D}$ tensors of the form $\mathbf{D}^{(JJ)}(\phi, \theta, 0)$ or else extract it from the kets by writing

$$|[n, L, m]\rangle \equiv |n, L, m\rangle e^{-im\beta}$$

and take $\mathbf{D}^{(JJ)}(\phi, \theta, \tfrac{1}{2}\beta)$. If we adopt the second course, the generalization of (97) becomes $|J, M, N\rangle$, where

$$|J, M, N\rangle = D_{MN}^{(JJ)}(\phi, \theta, \gamma)|[n, L, m]\rangle(\mathbf{C}^{(1)} \cdot |\mathbf{p}\rangle), \tag{111}$$

in which $\gamma = \tfrac{1}{2}\beta$. For the β dependence to be properly represented by the $\mathbf{D}$ function, we must have $e^{-iN\gamma} = e^{im\beta}$, so

$$-N = 2m = 0, \pm 2, \pm 4, \ldots$$

The ket $|[n, L, m]\rangle$ is a scalar with respect to both l of (77) and λ of (A1), and so the state (111) is an eigenstate of $\mathbf{J}^2$ [where $\mathbf{J}$ is defined in (98)] with eigenvalue $J(J+1)$.

As shown in Appendix III, the kinetic-energy term in the Hamiltonian can be written, after a unitary transformation, as $-\hbar^2 U^{-1} \nabla^2 U/2\mu$. The state (111) is an eigenstate of the first three terms on the right-hand side of Eq. (A14) with eigenvalue $n + \tfrac{3}{2}$. It is straightforward to show that

$$\left(\lambda^2 - \lambda_\zeta^2 - 2\right)|J, M, N\rangle = \left[J(J+1) - N^2\right]|J, M, N\rangle \tag{112}$$

The term -2 is the analog of the term $\tfrac{1}{4}$ in (100); both are associated with the unitary transformation that has the effect of changing the dimensionality of the oscillator part of the Hamiltonian. The fact that both $\tfrac{1}{4}$ and -2 are absorbed in other terms and effectively disappear from view as soon as the energies are calculated suggests that some physical principle is involved. A good candidate is the fact that a rotational band possesses no zero-point energy, so the origin of the energy-level structure defined by the algebraic form $J(J+1)$ must be given by $d\hbar\omega/2$, where d is the dimensionality of the oscillator space. Such a requirement would automatically squeeze out spurious constant terms.

In the limits $\alpha \to 0$ and $Q \to Q_e$, Eq. (112) leads to the contribution

$$E_{JN} = \frac{\hbar^2\left[J(J+1) - N^2\right]}{6\mu Q_e^2} \tag{113}$$

to the energy. The terms $\lambda_\pm^2$ in (A14) drop out when $\alpha \to 0$. Thus the states $|J, M, N\rangle$ are eigenstates of $U^{-1} \nabla^2 U$ in that limit, and the energies E_{JN} of (113) give the rotational structure that is superposed on every vibrational state n. For $n = 0$ and 1, we must have $N = 0$; but for $n \geq 2$, additional bands occur for which $|N| \geq 2$.

D. Energies from Trajectories

The foregoing analysis gives the expression E_{JN} for the energy of a rotational band associated with a particular vibrational state n. Although the image of the circular motion and the superposed elliptical motion of the ligands has been introduced to give some sense to the direction of the mathematics, no immediate use has been made of our knowledge of the trajectories to actually produce a formula for E_{JN}. As a final example of the value of knowing the motions of the ligands, the energies E_{JN} are now derived for the JT system $T_1 \otimes (\varepsilon_g \oplus \tau_{2g})$ by a descriptive approach.

We begin with the circular motion of the ligands composing the rigid octahedron of Fig. 12. If this object were free to rotate without constraint, the energies would be given by $\hbar^2 l(l+1)/2I$ and each level would have a degeneracy of $(2l+1)^2$. The restriction of the Euler angles to $(\pi - \phi, 2\theta, \phi)$ has the effect of limiting the spatial freedom to that of the polar vector (θ, ϕ), and the degeneracy drops to $2l+1$. The change of the formula for the energy levels to $\hbar^2 l(l+1)/8I$ can be seen immediately from (96). To calculate the moment of inertia I of the octahedron, we note first that for an object of cubic or octahedral symmetry the inertia tensor is isotropic. Taking the axis as lying through opposite vertices of the octahedron of Fig. 12, we have

$$I = 4\mu \left(\frac{3Q_e^2}{16} \right) = \frac{3\mu Q_e^2}{4}$$

Thus we arrive at the expression

$$E_l = \frac{\hbar^2 l(l+1)}{6\mu Q_e^2} \tag{114}$$

which agrees with the expression (113) for E_{JN} when we limit the latter to $N = 0$, for which $J = l$. An equivalent form of (114) has been given by O'Brien.[31]

Turning now to the ellipses superposed on the circular motions, we need the result, obtained in Section X.B, that these ellipses, although variously oriented, contribute an angular momentum $\mathbf{L}$ equal to L_Z along the direction of the polar vector (θ, ϕ). Let the angular momentum of the circular motions of the ligands be called l'. Then the total angular momentum $\mathbf{J}'$ is given by $\mathbf{J}' = l' + \mathbf{L}$. It has already been established that the planes of the circular motions are parallel to the polar vector (θ, ϕ), as can be seen from Fig. 13. Thus $l' \cdot \mathbf{L} = 0$, and we have

$$\mathbf{J}'^2 = l'^2 + \mathbf{L}^2 = l'^2 + L_Z^2 \tag{115}$$

The energy $l'^2/2I$ of the rotating octahedron can now be written as $(\mathbf{J}'^2 - L_Z^2)/2I$. Classically, this is as far as we can go. However, if we replace L_Z by $-i\hbar\partial/\partial\beta$, with eigenvalues $\hbar m$, or, equivalently, $\frac{1}{2}\hbar N$, where $N = 0, \pm 2, \pm 4, \ldots$, then the term $-\hbar^2 N^2/8I$ tallies nicely with the corresponding term in E_{JN} as given in (113).

The last step is to show that $\langle J'^2 \rangle = \hbar^2 J(J+1)/4$. The factor $\frac{1}{4}$ arises because the vector $\mathbf{J}$, given in (98), involves the angular-momentum vector l, which is defined in terms of the angles ϕ, θ, and β in (77). The rotation of the octahedron of Fig. 12, on the other hand, is effectively described by the double angles 2θ and 2ϕ, as the kinetic energy (92) shows. Operators like $\partial^2/\partial\phi^2$ in the kinetic energy are more appropriately divided by 4, and this leads to the factor $\frac{1}{4}$.

A more precise derivation of this result uses (A8) in Appendix II in the limit of small α to generalize (92) to

$$T = 2I(\dot{\theta}^2 + \dot{\phi}^2\sin^2\theta) + \tfrac{1}{2}\mu\alpha^2 Q^2(\dot{\beta} + 2\dot{\phi}\cos\theta)^2$$

We have set $\dot{\alpha} = \dot{Q} = 0$ because our interest lies in the incorporation of L_Z (represented solely through $\dot{\beta}$) into T. We find

$$p_\theta = \frac{\partial T}{\partial\dot{\theta}} = 4I\dot{\theta}$$

$$p_\phi = \frac{\partial T}{\partial\dot{\phi}} = 4I\dot{\phi}\sin^2\theta + 2\mu\alpha^2 Q^2(\dot{\beta} + 2\dot{\phi}\cos\theta)\cos\theta$$

$$p_\beta = \frac{\partial T}{\partial\dot{\beta}} = \mu\alpha^2 Q^2(\dot{\beta} + 2\dot{\phi}\cos\theta)$$

Thus the terms in T involving $1/I$ are

$$\frac{\left[p_\theta^2 + \operatorname{cosec}^2\theta\left(p_\phi - 2p_\beta\cos\theta\right)^2\right]}{8I}$$

$$= \frac{\left[p_\theta^2 + p_\phi^2\operatorname{cosec}^2\theta - 4p_\phi p_\beta\cot\theta\operatorname{cosec}\theta + 4p_\beta^2\operatorname{cosec}^2\theta - 4p_\beta^2\right]}{8I}$$

$$(116)$$

The transformation to quantum mechanics through

$$p_\theta \to -\frac{i\hbar\partial}{\partial\theta}, \qquad p_\phi \to -\frac{i\hbar\partial}{\partial\phi}, \qquad p_\beta \to -\frac{i\hbar\partial}{\partial\beta}$$

converts (116) to $(\hbar^2 l^2 - 4L_Z^2)/8I$. From (98) we can write

$$l^2 = \mathbf{J}^2 - 2\mathbf{J}\cdot l_p + l_p^2$$

We can rapidly confirm from (99) that

$$\left(\mathbf{C}^{(1)} \cdot \langle \mathbf{p}|\right) l_p \left(\mathbf{C}^{(1)} \cdot |\mathbf{p}\rangle\right) = 0$$

and so $\langle l^2 \rangle = \langle \mathbf{J}^2 \rangle + 2$. Dropping the constant term $+2$ [an act that can be justified formally by following the analysis leading to (112)], we find, for the eigenstate $|J, M, N\rangle$ of (111), that $\langle l^2 \rangle = J(J+1)$ and $4L_Z^2 = 4\hbar^2 m^2 = \hbar^2 N^2$. We thus obtain the result of (113).

XI. FINAL REMARKS

The use of trajectories has provided vivid images of the dynamics of complicated JT systems. There is no substitute for seeing what is actually going on. It is not difficult to see how the kind of analyses presented here can be extended to other JT systems. It would be of considerable interest to find the trajectories for the ligands comprising the icosahedral systems discussed by Pooler,[38] for example.

An extension in another direction would be to examine the cooperative JT effect. It is not difficult to see, for example, that a cubic solid comprising E states interacting with ε strains (such as are described by Gehring and Gehring[44]) can distort into an infinite number of energetically equivalent rectangular parallelepipeds, whose dimensions x, y, and z are coupled according to equations of the type

$$x - a = f\cos\left(\phi - \tfrac{2}{3}\pi\right)$$
$$y - a = f\cos\left(\phi + \tfrac{2}{3}\pi\right)$$
$$z - a = f\cos\phi$$

which parallel the Eqs. (32) for the isolated $E \otimes \varepsilon_g$ system. The analogs of the trajectories are the pulsating faces of the cube, each one $\tfrac{2}{3}\pi$ out of phase with its neighbors.

Another aspect to the study of actual physical motions is that it directs attention to classical interpretations of quantum mechanics. The interplay between these two approaches is always interesting and often very instructive.

APPENDIX I: CUBIC INVARIANT FOR $T_1 \otimes (\varepsilon_g \oplus \tau_{2g})$

To derive (78), we need to set up the problem so that we can use angular-momentum theory to its greatest advantage. Much of the basic theory that follows has been given elsewhere.[45] We begin by defining an angular-momentum vector analogous to l of (77):

$$\lambda_\xi \pm i\lambda_\eta = ie^{\mp i\beta/2}\left(2\cot\theta\frac{\partial}{\partial\beta} \pm i\frac{\partial}{\partial\theta} - \frac{1}{\sin\theta}\frac{\partial}{\partial\phi}\right)$$

$$\lambda_\zeta = 2i\frac{\partial}{\partial\beta}$$

(A1)

It is easy to show that all components of l commute with all components of λ, so operators can be classified according to their tensorial properties with respect to both of these angular-momentum vectors. The corresponding ranks are given in sequence. Thus the double tensor $\mathbf{T}^{(k\kappa)}$ denotes the $(2\kappa+1)(2k+1)$ components $T_{q\pi}^{(k\kappa)}$ that behave as a tensor of rank k with respect to l and as one of rank κ with respect to λ.

The rotation matrices $\mathscr{D}_{MN}^{J}(\omega)$ that describe how the components of a state $|J\rangle$ or a tensor $\mathbf{T}^{(J)}$ behave under rotation through the Euler angles ω are well known in nuclear theory.[46] They can be regarded as double tensors $\mathbf{D}^{(JJ)}$ if we make the identification

$$D_{M,-N}^{(JJ)} = (-1)^{J-N}(2J+1)^{1/2}\mathscr{D}_{MN}^{J}(\omega)^*$$

(A2)

where the asterisk denotes a complex conjugate, and the Euler angles are interpreted as $(\phi, \theta, \tfrac{1}{2}\beta)$. Properties of the rotation matrices can be cast in tensorial form. For example,

$$(\mathbf{D}^{(AA)}\mathbf{D}^{(BB)})^{(TT')} = \delta(T,T')\left[\frac{(2A+1)(2B+1)}{(2T+1)}\right]^{1/2}\mathbf{D}^{(TT)}$$

(A3)

where the quantity on the left is a coupled product.

With these preliminaries, we can start on the problem in hand by noting that the tensor $\mathbf{Q}^{(2)}$ of Eqs. (76) can be written as

$$Q_m^{(2)} = Q(\mathbf{D}^{(22)}\mathbf{X}^{(02)})_{m0}^{(20)}$$

(A4)

where

$$X_{02}^{(02)} = X_{0-2}^{(02)} = \sqrt{\tfrac{1}{2}}\sin\alpha, \qquad X_{0\pm1}^{(02)} = 0, \qquad X_{00}^{(02)} = \cos\alpha$$

(A5)

To actually prove (A4), we need the components of $\mathbf{D}^{(22)}$ as functions ϕ, θ, and β, as well as the SO(3) Clebsch–Gordan coefficients for coupling $\mathbf{D}^{(22)}$ to $\mathbf{X}^{(02)}$ to the final ranks $(2,0)$. It should be stressed that $\mathbf{X}^{(02)}$ is a quantity whose components are defined in (A5): it is not a tensor in the sense that its commutation relations with λ or l are those of a true double tensor. There is no difficulty here: we can define a vector in ordinary three-dimensional space by specifying three numbers—which would, of course, commute with all components of an angular-momentum vector. The situation for $\mathbf{X}^{(02)}$ is analogous to that.

Having said this, we can now proceed as follows. We need three $\mathbf{Q}^{(2)}$ tensors to form $\mathbf{Q}^{(2)} \cdot \mathbf{Q}^{(2)} \cdot \mathbf{Q}^{(2)}$ which, from its definition (73), can be written as

$$\left(\mathbf{Q}^{(2)} (\mathbf{Q}^{(2)} \mathbf{Q}^{(2)})^{(2)} \right)^{(0)}$$

Now we see

$$(\mathbf{Q}^{(2)} \mathbf{Q}^{(2)})^{(2)} = Q^2 \left((\mathbf{D}^{(22)} \mathbf{X}^{(02)})^{(20)} (\mathbf{D}^{(22)} \mathbf{X}^{(02)})^{(20)} \right)^{(20)}$$

$$= Q^2 ((20)2(20)2,2 | (22)2(00)0,2)$$

$$\times ((22)0(22)0,0 | (22)2(22)2,0)$$

$$\times \left((\mathbf{D}^{(22)} \mathbf{D}^{(22)})^{(22)} \mathbf{Y}^{(02)} \right)^{(20)} \tag{A6}$$

where $\mathbf{Y}^{(02)} = (\mathbf{X}^{(02)} \mathbf{X}^{(02)})^{(02)}$. The two recoupling coefficients in (A6) are converted to 9-j symbols (see Edmonds[19]); they turn out to be 1 and $5^{-1/2}$, respectively. The quantity $\mathbf{Y}^{(02)}$ is easily shown to have components as follows:

$$Y^{(02)}_{0\pm 2} = \sqrt{\tfrac{1}{7}} \sin 2\alpha, \qquad Y^{(02)}_{0\pm 1} = 0, \qquad Y^{(02)}_{00} = -\sqrt{\tfrac{2}{7}} \cos 2\alpha$$

We can now use (A3) to replace $(\mathbf{D}^{(22)} \mathbf{D}^{(22)})^{(22)}$ by $\sqrt{5}\, \mathbf{D}^{(22)}$. At a stroke the enormous tedium of combining high powers of trigonometric functions is totally circumvented.

Proceeding in a similar way, and using $\mathbf{D}^{(00)} = 1$, we get

$$\mathbf{Q}^{(2)} \cdot \mathbf{Q}^{(2)} \cdot \mathbf{Q}^{(2)} = Q^3 \left((\mathbf{D}^{(22)} \mathbf{X}^{(02)})^{(20)} (\mathbf{D}^{(22)} \mathbf{Y}^{(02)})^{(20)} \right)^{(00)}$$

$$= Q^3 (\mathbf{X}^{(02)} \mathbf{Y}^{(02)})^{(00)}$$

$$= Q^3 \sum_m (2m, 2-m|00) X^{(02)}_{0m} Y^{(02)}_{0-m}$$

$$= -\left(\tfrac{2}{35} \right)^{1/2} Q^3 \cos 3\alpha$$

as we require.

APPENDIX II: NULL COUPLING BETWEEN UPPER AND LOWER BRANCHES OF $T_1 \otimes (\varepsilon_g \oplus \tau_{2g})$

From the secular determinant for $T_1 \otimes (\varepsilon_g \oplus \tau_{2g})$, we may confirm that the ground electronic state for this JT system, under conditions of equal frequencies and equal couplings, is $(\mathbf{C}^{(1)} \cdot |\mathbf{p}\rangle)$, a result first obtained (in, however, a slightly different form) by O'Brien.[31] The two orthogonal companions must be contained in the components of $(\mathbf{C}^{(1)} |\mathbf{p}\rangle)^{(1)}$, since

$$(\mathbf{C}^{(1)} \langle \mathbf{p} |)^{(0)} (\mathbf{C}^{(1)} | \mathbf{P} \rangle)^{(1)} \simeq (\mathbf{C}^{(1)} \mathbf{C}^{(1)})^{(1)} \langle \mathbf{p} | \mathbf{p} \rangle = 0$$

because the vector product of two identical vectors is zero. To parallel the analysis of Section III.B, we have to study

$$(\mathbf{C}^{(1)} \langle \mathbf{p} |)^{(0)} (\mathbf{Q}^{(2)} - \mathbf{Q}_e^{(2)}) \cdot \mathbf{T}^{(2)} (|\mathbf{p}\rangle \mathbf{C}^{(1)})^{(1)} \tag{A7}$$

The part involving $\mathbf{Q}_e^{(2)}$ automatically vanishes, since the electronic eigenfunctions are determined for $\mathbf{Q} \equiv \mathbf{Q}_e$. The process of integration entailed in evaluating $\langle \mathbf{p} | \mathbf{T}^{(2)} | \mathbf{p} \rangle$ automatically selects only those parts of the triple product of $\langle \mathbf{p} |$, $\mathbf{T}^{(2)}$, and $|\mathbf{p}\rangle$ that are scalar. Thus (A7) reduces to terms of the type

$$\left(\mathbf{Q}^{(2)} (\mathbf{C}^{(1)} \mathbf{C}^{(1)})^{(t)} \right)^{(1)}$$

Since $(\mathbf{C}^{(1)} \mathbf{C}^{(1)})^{(t)} = 0$ for $t = 1$ or 3, we are left with the term proportional to $(\mathbf{Q}^{(2)} \mathbf{C}^{(2)})^{(1)}$. We know that if (75) were satisfied (as it is for $\mathbf{Q}^{(2)} \equiv \mathbf{Q}_e^{(2)}$), this coupled product would vanish. In the general case, we need (A4) together with the relation[43]

$$C_m^{(2)}(\theta, \phi) = 5^{-1/2} D_{m0}^{(22)}(\phi, \theta, \tfrac{1}{2}\beta)$$

to convert $(\mathbf{Q}^{(2)} \mathbf{C}^{(2)})_q^{(1)}$ to

$$5^{-1/2} Q \left((\mathbf{D}^{(22)} \mathbf{X}^{(02)})^{(20)} \mathbf{D}^{(22)} \right)_{q0}^{(12)}$$

This is proportional to

$$\left((\mathbf{D}^{(22)} \mathbf{D}^{(22)})^{(11)} \mathbf{X}^{(02)} \right)_{q0}^{(12)} \simeq (\mathbf{D}^{(11)} \mathbf{X}^{(02)})_{q0}^{(12)}$$

On expanding this coupled product, we can only get

$$D_{q0}^{(11)} X_{00}^{(02)} (1q, 00 | 1q)(10, 20 | 20)$$

since $X_{0\pm1}^{(02)} = 0$. However, the Clebsch–Gordan coefficient $(10, 20|20)$ is zero, so the entire result vanishes and, in particular, there is no coupling between the upper and the lower electronic states even when $\mathbf{Q}$ departs from its static equilibrium value $\mathbf{Q}_e$. This is in contrast to the system $T_1 \otimes \tau_{2g}$ taken alone.

APPENDIX III: FIVE-DIMENSIONAL LAPLACIAN

To calculate the kinetic energy of the ligands of the JT system $T_1 \otimes (\varepsilon_g \oplus \tau_{2g})$, we need to express the five-dimensional Laplacian ∇^2 in terms of derivatives with respect to the coordinates $(Q, \alpha, \beta, \theta, \phi)$ introduced in (76) and expressed in tensorial form in (A4). The first step is to calculate the line element ds. We may work out each $dQ_m^{(2)}$ from (76), and then form $d\mathbf{Q}^{(2)} \cdot d\mathbf{Q}^{(2)}$. Alternatively, we can simply refer to the result of O'Brien.[32] Putting $\frac{1}{2}\beta = \gamma$, we get

$$ds^2 = dQ^2 + Q^2 d\alpha^2 + 4Q^2 \sin^2\!\alpha \, d\nu_3^2 + 4Q^2 \sin^2\!\left(\alpha + \tfrac{2}{3}\pi\right) d\nu_2^2$$
$$+ 4Q^2 \sin^2\!\left(\alpha - \tfrac{2}{3}\pi\right) d\nu_1^2 \tag{A8}$$

where

$$d\nu_3 = d\gamma + \cos\theta \, d\phi$$
$$d\nu_2 = \cos\gamma \, d\theta + \sin\theta \sin\gamma \, d\phi \tag{A9}$$
$$d\nu_1 = \sin\gamma \, d\theta - \sin\theta \cos\gamma \, d\phi$$

The increments $d\nu_i$ can be recognized as the angles of rotation about the principal axes of an object that is subjected to an infinitesimal rotation characterized by the changes $(d\phi, d\theta, d\gamma)$ in the Euler angles (ϕ, θ, γ) that define the orientation of the principal axes.[47] Thus we can make the identifications

$$\frac{\partial}{\partial \nu_1} = i\lambda_\xi, \qquad \frac{\partial}{\partial \nu_2} = i\lambda_\eta, \qquad \frac{\partial}{\partial \nu_3} = i\lambda_\zeta \tag{A10}$$

where λ is defined in (A1).

The line element ds^2, as expressed in (A8), has the form of one defined for an orthogonal coordinate system $(Q, \alpha, \nu_1, \nu_2, \nu_3)$. The matrix formed from the coefficients g_{ij} in $ds^2 = g_{ij}x^i x^j$ in the usual way (with summations over i and j implied) yields the determinant g given by

$$\sqrt{g} = -8Q^4 \sin\alpha \sin\!\left(\alpha - \tfrac{2}{3}\pi\right)\sin\!\left(\alpha + \tfrac{2}{3}\pi\right)$$
$$= 2Q^4 \sin 3\alpha \tag{A11}$$

We thus obtain

$$
\nabla^2 = g^{-1/2} \frac{\partial}{\partial x^i} \left(g^{1/2} \frac{\partial}{\partial x^j} \right)
$$

$$
= \frac{1}{Q^4} \frac{1}{\partial Q} \left(Q^4 \frac{\partial}{\partial Q} \right) + \frac{1}{Q^2 \sin 3\alpha} \frac{\partial}{\partial \alpha} \left(\sin 3\alpha \frac{\partial}{\partial \alpha} \right)
$$

$$
- \sum_{k=1}^{3} \frac{1}{4Q^2 \sin^2\left(\alpha - \frac{2}{3}k\pi \right)} \lambda_k^2 \tag{A12}
$$

where $(\lambda_1, \lambda_2, \lambda_3) \equiv (\lambda_\xi, \lambda_\eta, \lambda_\zeta)$. If we wish to use the shift operators λ_ζ, λ_+, and λ_-, where $\lambda_\pm = \lambda_\xi \pm i\lambda_\eta$, Eq. (A12) becomes

$$
\nabla^2 = \frac{1}{Q^4} \frac{\partial}{\partial Q} \left(Q^4 \frac{\partial}{\partial Q} \right) + \frac{1}{Q^2 \sin 3\alpha} \frac{\partial}{\partial \alpha} \left(\sin 3\alpha \frac{\partial}{\partial \alpha} \right)
$$

$$
+ \frac{\sqrt{3}}{2} \sin 2\alpha \left(\frac{\sin \alpha}{Q \sin 3\alpha} \right)^2 (\lambda_+^2 + \lambda_-^2) - \frac{1}{4Q^2 \sin^2\alpha} \lambda_\zeta^2
$$

$$
- (\sin^2\alpha + 3\cos^2\alpha) \left(\frac{\sin \alpha}{Q \sin 3\alpha} \right)^2 (\lambda^2 - \lambda_\zeta^2) \tag{A13}
$$

Equation (A12) agrees with the expression used in nuclear theory[48] for the coordinates describing the quadrupolar deformation of a rotating nucleus. Equation (A13) is in agreement with equation (16) of Judd and Vogel[35] provided a correction is made to their term in λ_ζ^2. A partial correction to this term has already been noted.[36]

If we make the transformation $\nabla^2 \to U^{-1} \nabla^2 U$, where

$$
U = \frac{(\sin \alpha / \sin 3\alpha)^{1/2}}{Q}
$$

we get

$$
U^{-1} \nabla^2 U = \frac{1}{Q^2} \frac{\partial}{\partial Q} \left(Q^2 \frac{\partial}{\partial Q} \right) + \frac{1}{Q^2 \sin \alpha} \frac{\partial}{\partial \alpha} \left(\sin \alpha \frac{\partial}{\partial \alpha} \right)
$$

$$
- \frac{1}{Q^2 \sin^2\alpha} \left(\frac{\lambda_\zeta}{2} \right)^2 + \frac{\sqrt{3}}{2} \sin 2\alpha \left(\frac{\sin \alpha}{Q \sin 3\alpha} \right)^2 (\lambda_+^2 + \lambda_-^2)
$$

$$
- (\sin^2\alpha + 3\cos^2\alpha) \left(\frac{\sin \alpha}{Q \sin 3\alpha} \right)^2 (\lambda^2 - \lambda_\zeta^2 - 2) \tag{A14}
$$

The first three terms on the right-hand side of this equation are identical to the usual three-dimensional Laplacian in polar coordinates (Q, α, β). Multiplication by $-\hbar^2/2\mu$ gives the associated kinetic energy. If the first of the three potentials (85) is written as

$$\tfrac{1}{2}\mu\omega^2\left[(Q\cos\alpha - Q_e)^2 + (Q\sin\alpha)^2\right] - \tfrac{1}{2}\mu\omega^2 Q_e^2$$

and added to the kinetic energy, we obtain an isotropic three-dimensional oscillator centered a distance Q_e along the Z axis. The terms in (A14) involving λ and its components give rotational bands superposed on the vibrational structure.

Acknowledgments

Thanks go to colleagues and students who have contributed to my understanding of the JT effect. In particular, Dr. C. DeW. Van Siclen is thanked for discussions on the topics of Section VII. This is probably a good place too to express appreciation for a lecture on $E \otimes \varepsilon$, given during a Bangalore afternoon in the summer of 1971 by Prof. C. J. Ballhausen, which brought something of the elegance and subtlety of the JT effect to my attention.

REFERENCES

1. H. A. Jahn and E. Teller, *Phys. Rev.* **49**, 874 (1936).

2. W. Moffitt and W. Thorson, *Phys. Rev.* **108**, 1251 (1957).

3. L. D. Landau and E. M. Lifshitz, *Mechanics*, Pergamon Press, Oxford, England, and Addison-Wesley, Reading, Mass., 1960, p. 70.

4. A. D. Liehr, *J. Phys. Chem.* **67**, 389 (1963).

5. G. F. Koster, J. O. Dimmock, R. G. Wheeler, and H. Statz, *Properties of the Thirty-Two Point Groups*, MIT Press, Cambridge, Mass., 1963.

6. E. Bright Wilson, J. C. Decius, and P. C. Cross, *Molecular Vibrations*, McGraw-Hill, New York, 1955, p. 21.

7. J. H. Van Vleck, *J. Chem. Phys.* **7**, 72 (1939).

8. U. Öpik and M. H. L. Pryce, *Proc. Roy. Soc.* (London) **A238**, 425 (1957).

9. R. Englman, *The Jahn–Teller Effect in Molecules and Crystals*, Wiley (Interscience), New York, 1972.

10. B. R. Judd, *Can. J. Phys.* **52**, 999 (1974).

11. H. C. Longuet-Higgins, U. Öpik, M. H. L. Pryce, and R. A. Sack, *Proc. Roy. Soc.* (London) **A244**, 1 (1958).

12. H. Barentzen, G. Olbrich, and M. C. M. O'Brien, *J. Phys.* **A14**, 111 (1981).

13. B. R. Judd, *J. Phys.* **C12**, 1685 (1979).

14. H. G. Reik, H. Nusser, and L. A. Amarante Ribeiro, *J. Phys.* **A15**, 3491 (1982).

15. M. C. M. O'Brien, *J. Phys.* **C9**, 2375 (1976).

16. C. J. Ballhausen, *Chem. Phys. Lett.* **93**, 407 (1982).

17. M. C. M. O'Brien and D. R. Pooler, *J. Phys.* **C12**, 311 (1979).

18. C. A. Bates, *Phys. Rep. (Phys. Lett. C)* **35**, 187 (1978).

19. A. R. Edmonds, *Angular Momentum Theory in Quantum Mechanics*, Princeton Univ. Press, Princeton, New Jersey, 1957.

20. S. L. Loney, *The Elements of Coordinate Geometry. Part II: Trilinear Coordinates etc.*, Macmillan, New York, 1946.

21. J. C. Slonczewski, *Phys. Rev.* **131**, 1596 (1963).

22. J. C. Slonczewski and V. L. Moruzzi, *Physics* **3**, 237 (1967).

23. M. S. Child, *Phil. Trans. Roy. Soc. London* **A255**, 31 (1962).

24. M. C. M. O'Brien, private communication.

25. B. R. Judd, *Coll. Int. C.N.R.S.* No. 255 (1976).

26. D. R. Pooler and M. C. M. O'Brien, *J. Phys.* **C10**, 3769 (1977).

27. J. C. Kemp, W. M. Ziniker, J. A. Glaze, and J. C. Cheng, *Phys. Rev.* **171**, 1024 (1968).

28. R. G. Bessent, B. C. Cavenett, and I. C. Hunter, *J. Phys. Chem. Solids* **29**, 1523 (1968).

29. Y. Merle d'Aubigné and A. Roussel, *Phys. Rev.* **B3**, 1421 (1971).

30. J. Duran, Y. Merle d'Aubigné, and R. Romestain, *J. Phys.* **C5**, 2225 (1972).

31. M. C. M. O'Brien, *Phys. Rev.* **187**, 407 (1969).

32. M. C. M. O'Brien, *J. Phys.* **C4**, 2524 (1971).

33. A. E. Hughes, *J. Phys.* **C3**, 627 (1970).

34. R. Romestain and Y. Merle d'Aubigné, *Phys. Rev.* **B4**, 4611 (1971).

35. B. R. Judd and E. E. Vogel, *Phys. Rev.* **B11**, 2427 (1975).

36. B. R. Judd, *J. Chem. Phys.* **68**, 5643 (1978).

37. D. R. Pooler, *J. Phys.* **A11**, 1045 (1978).

38. D. R. Pooler, *J. Phys.* **C13**, 1029 (1980).

39. B. R. Judd, *Operator Techniques in Atomic Spectroscopy*, McGraw-Hill, New York, 1963.

40. L. Pauling and E. Bright Wilson, *Introduction to Quantum Mechanics*, McGraw-Hill, New York, 1935, p. 177.

41. W. Thorson and W. Moffitt, *Phys. Rev.* **168**, 362 (1968).

42. See Ref. 39, p. 131.

43. B. R. Judd, *Angular Momentum Theory for Diatomic Molecules*, Academic Press, New York, 1975, p. 25.

44. G. A. Gehring and K. A. Gehring, *Rep. Prog. Phys.* **38**, 1 (1975), Sec. 4.5.

45. See Ref. 43, Chap. 1.

46. D. M. Brink and G. R. Satchler, *Angular Momentum*, Oxford University Press, New York, 1968.

47. See Ref. 3, Eqs. (35.1).

48. T. M. Corrigan, F. J. Margetan, and S. A. Williams, *Phys. Rev.* **C14**, 2279 (1976).

STRUCTURE, DYNAMICS,
AND DISSIPATION IN HARD-CORE
MOLECULAR LIQUIDS

E. LIPPERT, C. A. CHATZIDIMITRIOU-DREISMANN,
AND K.-H. NAUMANN

I. N. Stranski Institute for Physical and Theoretical Chemistry
Technical University of Berlin
West Berlin, Federal Republic of Germany

Dedicated to Eugene P. Wigner,

Who graduated and taught physics at our university until 1933, when he was compelled to leave Germany, eventually to become Thomas D. Jones Professor of Mathematical Physics at Princeton University;

Who refused to accept any reparations from the Federal Republic of Germany after World War II but used his claims to establish and still pay for a foundation that enables graduates of our university to spend some time in the United States of America or in Israel for scientific research; and

Who especially merits the sincere gratitude of all three authors of this article since he favoured our aims in quantum statistics and sponsored a postdoctoral fellowship for our research, one of us having the honor of serving as the chairman of the Eugene P. Wigner Foundation Committee.

CONTENTS

311

Most physicists and chemists (and all students) prefer to work with streamlined formalisms such as projection operator techniques and linear response theory, which tacitly assume the existence of gross variables endowed with all the properties desired by the authors. ...

The question is: How do macroscopic equations relate to the microscopic ones, and how does one describe the fluctuations?[1]

I. INTRODUCTION: STRUCTURE AND MOTION

Up to now there exists no completely consistent theory of molecular liquids. Starting from well-known results which could nowadays be considered classical and most of which are cited in Refs. 2 and 3, we wish to present some new developments in the statistical theory of molecular liquids based on the most simple model of hard-core bodies; recent experimental and statistical advances justify this purpose.

A. Contact Pair Interaction in Condensed Matter

One of the main aims of equilibrium statistical mechanics is to devise a sound theoretical background for prediction of the structural and thermodynamical properties of pure liquids, of their mixtures, and of solutions of gaseous and solid components in these liquids. An important step toward this goal, which might well be considered a major breakthrough, has been achieved with the realization that the behavior of dense nonassociated fluids is closely related to that of properly chosen hard-particle reference systems.[4-6]

With strong specific attractive interactions [see A. D. Buckingham (Chap. 19) and F. Kohler (Chap. 15) in Ref. 2] like hydrogen bonding,[7,8] electron donor–acceptor complexing,[9] etc. being absent, the liquid structure is dominated by harshly repulsive forces, or, in other words, by the characteristics of molecular shape. For densities more than twice the critical density, the role of slowly varying attractive interactions can be described as creating a background potential that is responsible for the cohesive energy and thus stabilizes the liquid phase at a particular pressure. However, under equilibrium conditions the particles will not feel this background potential, since almost no forces are exerted by it. As a result, the manner in which local molecular arrangements are formed depends mainly on sterical or packing effects.[10,11]

A powerful tool for implementing these ideas into the framework of statistical mechanics is provided by the methods of thermodynamic perturbation theory.[12–17]

The statistical-mechanical description of high-density fluids using the hard-core model has become possible with the decomposition of the pair potential function into a potential branch of repulsive forces and a potential branch of attractive forces.[16]

If a Lennard-Jones-type potential is applied (with u^* and r^* as its minimum coordinates and $r = \sigma$ if $u = 0$), the separation reads

$$u_0(r) = 4u^*\left[\left(\frac{\sigma}{r}\right)^{12} - \left(\frac{\sigma}{r}\right)^6\right] + u^*$$

and

$$u_1 = -u^* \qquad \text{if } r < r^*,$$

but

$$u_1(r) = 4u^*\left[\left(\frac{\sigma}{r}\right)^{12} - \left(\frac{\sigma}{r}\right)^6\right]$$

and

$$u_0 = 0 \qquad \text{if } r > r^*$$

A typical distribution function is shown in Fig. 1.

The total potential energy of a hard-body system is pairwise additive; thus, all its thermodynamic properties can be expressed in terms of two particle distribution functions.[18]

In binary mixtures of components the σ parameter of which differs by a factor of 2, that is, $\sigma_{11}/\sigma_{22} = 2$, the mole fractions vary distinctly in the short-range distance of the particles and exhibit relatively large departures from the macroscopic composition of the mixture.[19]

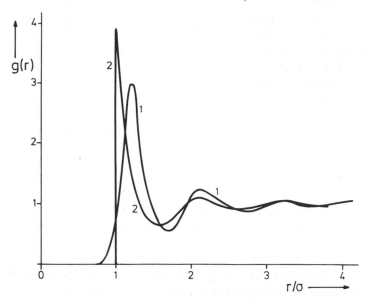

Fig. 1. The total correlation functions for a Lennard-Jones liquid (curve 1) and a hard-sphere fluid (curve 2) at a comparable density.[17] (By permission of *Ber. Bunsenges. Physik. Chem.*)

In thermodynamic perturbation theory of molecular liquids, hard bodies corresponding in shape and size to the effective shape and size of molecules (mainly determined by repulsive forces) are considered as reference particles; nonspherical molecules are most often described by a multicenter interaction site model (ISM) and Kihara noncentral potentials.[20,21]

Recent results of perturbation studies of molecules with anisotropic molecular shape that used hard nonspherical bodies as the reference system seem to justify the assumption that we have already progressed a considerable way from analysis of simple noble gas liquids to that of chemically more interesting models.[22]

A reasonably well selected hard-core reference system (hard spheres, hard convex bodies, etc.), a given density, and some additional thermodynamic perturbation treatment (taking into account the attractive forces) determine, therefore, the structure of a simple molecular liquid.[23]

Also during the last few years some antithetical methods, commonly classified as group contribution concepts, have been proposed and successfully applied in certain areas of thermodynamics and spectroscopy and especially in chemical engineering. They all stem from the idea that site–site attractive intermolecular forces are sufficient to describe the behavior of liquids. But it soon turned out that the basic assumptions in approaches like ASOG,[24] quasi-chemical approximation,[25-27] UNIFAC,[28] and contact pair[29] cannot be justified by rigorous liquid state statistical mechanics. Consequently, every improvement requires additional adjustable parameters. Nevertheless these approaches have proved to be useful for the treatment of special problems such as the calculation of vapor–liquid equilibria, resulting in growing interest – especially among chemical engineers. This has led some investigators to question whether the group contribution concept is really inconsistent with our present-day knowledge of statistical thermodynamics. It is not, as will be shown in Section II: by introducing effective site–site pair potentials that are independent of mutual orientation but are solely functions of temperature, molecular liquids can formally be described as mixtures of chemically functional groups (or larger atoms); and by a simple transformation method it is possible to deal with the different functional groups of the molecules in the liquid as if they were independent from each other. This independent molecular interaction sites (IMIS) model[30] represents a first major step towards bridging the gap between liquid state physical chemistry and semiempirical methods. By comparison of interaction-site pair distribution functions obtained from Monte Carlo studies and with predictions from RISM (reference interaction site model) theory, it has been concluded that the IMIS model represents a close approximation to the behavior of real liquids.[31] By methods similar in spirit, temperature and solvent effects on intramolecular degrees of freedom can be investigated[32] using hard convex body reference systems.[33]

An application of these ideas is the estimation of the static molecular orientation in liquid benzene, furan, and thiophene ($40° < \phi < 60°$) as compared with pyridine, pyrimidine, and pyridazine ($20° < \phi < 40°$), where ϕ is the mean angle between the main symmetry axes as revealed from proton magnetic resonance (PMR) chemical shift measurements at room temperature.[34] In binary mixtures of these compounds with acetonitrile the angle of the main symmetry axes increases with increasing concentration of the aromatic compound, and this increase is proportional to the increase of the molar volume of the mixture. From the equation of state for hard convex molecules and their partition functions a relation to PMR solvent shifts has been developed.[34]

A major advance in our knowledge of the mean (equilibrium) structure of molecular liquids has been achieved by X-ray[35] and coherent neutron[36,37] scattering.

Such types of studies in experimental thermodynamics and the statistical interpretation of their results are useful not only in basic research but also in enabling investigators to predict the quantities needed in chemical engineering. This work is especially important in that it permits storage of data on the thermodynamic properties of liquid mixtures and hence the computerized verification and refinement of such numerical values for use in industrial applications.[38-40]

B.   Orientational Relaxation Times

Experimental sources for information on reorientational dynamics in molecular liquids[41] (in contrast to vibrational relaxation[42]) as described by orientational correlation functions and revealed by various spectroscopic techniques (in the range from near-ultraviolet to very long wavelengths) have been compiled and classified by Versmold,[36] and values of rotational diffusion coefficients calculated from orientational relaxation times obtained with these methods have been tabulated and compared by Kohler.[43]

The methods specified in the aforementioned tables consist of measurements of dielectric loss,[44] far-infrared intensity, infrared line shape,[45] nuclear magnetic resonance (NMR) relaxation time T_1,[46] depolarized Rayleigh and Raman scattering,[47] but without incoherent neutron scattering (which has been dealt with in Refs. 36 and 48) and without fluorescence spectroscopic results (see Section II.D, below). These experimental methods are assumed to be well known by the reader. Detailed discussions may be found in Refs. 49 and 50. Since our aim in this section is to review some advances in classical and quantum statistics in order to lead into our consideration of the theory of molecular liquids, the aforementioned experimental methods will be critically examined only as far as is necessary for our purpose.

Within a rotational diffusion model[51] the Lorentzian band shape in optical spectra is given by

$$S_l(\omega) = \frac{1}{\pi} \frac{(l+1)D_r}{l^2(l+1)^2 D_r^2 + \omega^2}$$

where l refers to the term number in Legendre polynomials P_l and can be readily identified as proportional to the change of angular momentum. The rotational diffusion coefficient D_r is defined by

$$D_r \nabla^2 G(\Omega_0, \Omega; t) = \frac{\partial}{\partial t} G(\Omega_0, \Omega; t)$$

including the probability

$$G = \frac{1}{16\pi^2} \sum_l (2l+1) F_l(t) P_l(\cos\theta)$$

to find the transition moment in the solid angle element $d\Omega_0$ at $t = 0$ and in $d\Omega$ after a time t, respectively.

The rotational diffusion coefficients D_r are further denoted by the subscript $\perp$ for end-over-end rotation of the main symmetry axis (i.e., a long axis like that of H_2C—Cl_2 in methylene chloride or the C_6 axis in benzene, etc.) or by the subscript $\parallel$ for rotation around the main symmetry axis.

The time-dependent function $F_l(t)$ is related to D_r by

$$F_l(t) = \exp\left[- l(l+1) D_r t \right]$$

where $F_1(t)$ is accessible from infrared spectra, $F_2(t)$ is accessible from Raman spectroscopy, etc.[52,53]

For small time intervals

$$F_1(t) = 1 - \frac{kT}{I} t^2 + \cdots$$

The bigger the moment of inertia, I, the wider is the starting parabola of the function $F_1(t)$.

The higher terms of the power series represent the steric hindrance (i.e., where $F_1(t)$ remains far above the correlation function of the free rotor), and this depends on the molar volume of the solvent: the larger the quasi-lattice of the solvent, the freer is the rotation of small solute molecules. For an example of the "microscopic theory of viscosity" and further references, see Ref. 54. The *infrared* (not far-infrared; see Section III) band shape is given by the extinction K as a function of frequency ω and time t. For rigid linear molecules[55]

$$K(\omega) = \left(\frac{8\pi^2}{3\hbar} \right) \tanh\left(\frac{\hbar\omega}{2kT} \right) \int_{-\infty}^{+\infty} \exp(-i\omega t) C_1(t) \, dt$$

$$C_1(t) = 4\pi \left(\frac{N}{V} \right) \left(\frac{\partial \mu}{\partial q} \right)^2 \langle Y_{l,m}(\Omega(0)) Y^*_{l,m}(\Omega(t)) \rangle \langle \mu(0)\mu(t) \rangle$$

with $Y_{l,m}$ as the normalized spherical polynomials and $C_1(t)$ as the symmetrized time correlation function. The corresponding formula for the various types of light scattering are given in Refs. 56 and 57.

Small molecules like acetonitrile show rotational structure in their infrared absorption bands in the neat liquid as well as in solution.[58,59]

By dielectric measurements a macroscopic decay time T_0 is obtained from the real or imaginary part of the dielectric constant (relative permittivity):

$$\varepsilon = \varepsilon' + i\varepsilon''$$

$$\varepsilon' = \varepsilon_\infty + \frac{\varepsilon_0 - \varepsilon_\infty}{1 + \omega^2 T_0^2}$$

$$\varepsilon'' = \frac{\omega T_0(\varepsilon_0 - \varepsilon_\infty)}{1 + \omega^2 T_0^2}$$

The loss factor $\varepsilon''(\omega)$ possesses a maximum at $\omega = 1/T_0$ that is called the critical frequency ω_c.[60]

The macroscopic relaxation time T_0 can be related to an exponential decay of the dipole autocorrelation function

$$\gamma(t) = \frac{\langle \mu_i(0) \cdot \mu_i(t) \rangle}{\langle \mu_i^2(0) \rangle} = \exp\left(\frac{-t}{\tau} \right)$$

for the ith molecular dipole. This relation is given by

$$T_0 = \frac{3\varepsilon_0}{2\varepsilon_0 + \varepsilon_\infty} \tau$$

The value of τ for polar molecules in inert solvents near room temperature usually amounts to 1–50 ps, depending on molecular size and shape.

Rotational diffusion is assumed to be due to intermolecular interactions and collisions in analogy with translational diffusion (Brownian motion). The fluctuations in orientation due to small-step rotational diffusion between an outer magnetic field H_0 and an internal vector of the molecule, for example, a C—H bond of a rigid molecule, could be followed in NMR spectroscopy.[61]

Measurements of T_1 spin lattice relaxation of a carbon-13 nucleus in combination with nuclear Overhauser enhancement allow one to estimate the dipole relaxation time T_{1D} of carbon-13 and eventually the rotational diffusion times τ_i for the ith molecular main symmetry axis.[49,62]

In oxygen-17 enriched water at 25°C a rotational correlation time (1.71 ± 0.07) ps has been determined for the reorientation of the O—H vector by measuring the contribution of the proton–(oxygen-17) dipolar interaction to the proton longitudinal relaxation rate, taking advantage from the fact that due to the short OH-distance the OH-dipolar relaxation contribution is determined almost completely by *intra*molecular interaction.[63]

For scattering experiments[36] it should be pointed out that at least two types of experiment (such as X-ray, coherent and incoherent neutron, and depolarized Rayleigh scattering), or one type of scattering experiment and molecular dynamic calculations,[64] or one set of scattering experiments of only one type but for different concentrations in a binary mixture of molecules of almost equal size and shape (like X-ray with a $SiCl_4/GeCl_4$ mixture[65]) are necessary to obtain a complete set of information even for static orientational correlation in order to separate the molecular center distribution from orientational correlation.

Depolarized Rayleigh scattering techniques have recently been used to determine structure and correlation times and to demonstrate that hydrogen bonding provokes deviations from Stokes–Einstein–Debye plots.[66,67]

If the intermolecular reorientation process is connected with the breakup of the surrounding cage, as should be assumed for large molecules constituting sterically hindered structure-limited liquids, the *jump model* should be applied instead of the rotational diffusion model. After each jump there follows a period of residence in the cage of the neighboring molecules. For this model the correlation functions also decay exponentially with time. In the rotational diffusion model, due to the factor $l(l+1)$ in the exponent, the exponential decay of $F_2(t)$ is three times faster than that of $F_1(t)$. In the jump model the decay of both correlation functions is the same. A jump model is appropriate for liquids consisting of "large" molecules such as aromatic compounds.[68]

For chlorobenzene the ratio τ_1/τ_2 from dielectric and Rayleigh scattering, respectively, amounts to 1.1, but for acetonitrile to 2.9; therefore, in the latter case the application of the rotational diffusion model seems to be justified, but in the former case not.

This result must be taken into account if molecular reorientation is measured in solutions of aromatic compounds by some fluorescence spectroscopic method. In most cases under consideration the solute is a "large" molecule in the sense mentioned before, but it is reorientated in a very dilute solution, that is, in a cage of solvent molecules only. Even if the solvent molecules are relatively small (like acetonitrile, methylene chloride, or even cyclohexane, which all possess a τ_1/τ_2 ratio of about 3), their large cage around a solute does not allow for end-over-end rotation with each jump but only step by step along small angles. The rotation is not limited by collisions between more or less spherical molecules but by the structure of the cage. It seems not to be justified, therefore, to deal with rotational diffusion coefficients in fluorescence spectroscopic investigations of solutions of aromatic compounds but only with exponential fluorescence decay times, τ_F, noting that these are molecular relaxation times in a system far from equilibrium.

C. Direct Methods

Are there any direct experimental methods that permit us to follow orientational relaxation processes? Consider a liquid in its thermodynamic equilibrium state; now we must disturb that equilibrium a little bit—and this has to be done in a volume element of only molecular dimensions! This task can be performed by irradiating the liquid with light for which the liquid is almost transparent but with whose photons it is nevertheless in resonance. After absorption of such a photon by a particular molecule, the equilibrium of the liquid is disturbed only in the volume element around that molecule. Now what methods have we at our disposal to investigate the relaxation process by which the volume element under consideration—the absorbing molecule and its cage of surrounding molecules—returns to the appropriate equilibrium state, that is, to the mean structural and dynamical behavior?

The program outlined above implies that the thermodynamic system can act as a bath, since the amount of dissipated energy is small and well located. There are two ways to perform such direct measurements, namely, by band shape analysis of far-infrared (FIR) absorption (or corresponding Raman scattering) or by luminescence spectroscopy.

Let the complete Hamiltonian H_c of the microscopic system of the volume element under consideration be given by

$$H_c = H_0 + V$$

The term H_0 represents the relevant or diagonal part of the Hamiltonian H_c and should be assumed to be a constant of motion during a certain time interval Δt. The second term, V, is the so-called off-diagonal part of H_c and describes the coupling of the (fictitious) H_0 system with the bath. This splitting of H_c into two parts can be considered as physically meaningful for short time intervals $\Delta t \ll \tau_v$, where τ_v is the relaxation time due to the (strong) coupling V. This short-time limit allows for an adiabatic approximation concerning the degrees of freedom of the bath, which then play the role of external parameters. In the long-time limit, $\Delta t > \tau_v$, however, a time-smoothing and/or coarse-graining procedure can be carried out for the degrees of freedom of the bath, in order to remove the corresponding variables from the Hamiltonian. In any case H_c plays a dual role: it determines the energy levels of the fictitious H_0 system as well as its time evolution. These limiting time ranges are described by the *sudden approximation* and *adiabatic approximation*, respectively.[89-92]

In many problems of the physics of molecular liquids the time interval of observation Δt is of the order of the characteristic time τ_v, and neither the energy spectrum nor the time evolution of the system are readily described.[69,70]

To this category of processes that are connected with the treatment of the Hamiltonian H_c during $\Delta t \approx \tau_v$ belong not only molecular reorientational relaxation itself but also effects like depopulation of excited states, intramolecular energy transfer, and time-dependent spontaneous fluctuations, as well as the energy dissipation connected with these and other processes.[70]

As has already been pointed out in Section I.B, it is well known from dielectric loss and other indirect measurements that the orientational relaxation times of molecules dissolved in organic solvents like methylene chloride or acetonitrile at about room temperature is in the nanosecond-to-picosecond range. This is the order of the lifetime of electronic excited states of aromatic compounds in dilute solutions in such solvents, either due to spontaneous or to induced emission. Time-resolved fluorescence spectroscopic methods of almost every type, including ultrashort laser pulse techniques, are adequate to measure the orientational relaxation times for the solvents used. Some results revealed by such luminescence spectroscopic techniques will, therefore, be reviewed below.

D. Fluorescence and Laser Spectroscopic Studies

Molecular reorientational relaxation times in liquids were directly studied for the first time by Chuang and Eisenthal of the IBM Research Laboratories in San Jose, California.[71] Specifically, they studied rhodamine 6G dissolved in various solvents using picosecond light pulses. An anisotropy in the orientational distribution of the solute molecules, (i.e., a dichroism) was induced by a strongly polarized and very short excitation pulse, and the return of the system to its isotropic distribution was monitored with an attenuated pulse. The molecular rotational motion was found to depend on the structure of the liquid, for instance, on the shape of rigid solvent molecules, on molecular dipole—dipole interactions, and on the existence of hydrogen bonding networks. When dissolved in glycol, the rhodamine 6G molecule does not respond to the full frictional effects of the quasi-polymeric structure of that solvent, which, of course, are indicated by the value of its viscosity. Therefore the Debye–Stokes–Einstein theory of viscosity is not applicable to most of the solvents used by the aforementioned authors. The rotational diffusion time of that relatively large solute molecule was found to vary between 0.1 and 3 ns, in agreement with results obtained with picosecond laser pulses used in a transient grating method.[72]

The rotational diffusion time of some other fluorescing aromatic compounds in organic solvents, that is, of relatively large solute molecules as compared with the size of a solvent monomer, has been measured by the induced dichroism method, mentioned above[73-75] as well as by direct fluorescence depolarization[76-78] or indirectly by phase fluorimetry methods that permit investigation of even the shortest relaxation times,[79,80] and especially by use of synchrotron radiation.[81]

The time scale of excimer formation of neat benzene and toluene has been studied recently by picosecond two-photon absorption and photolysis using the third harmonic of a mode-locked YAG laser. This is an example of the possible use of two-photon excitation of solvents in general.[82]

For further literature on the use of light pulses and related topics, including the optical Kerr effect, see Ref. 83.

The reader should keep in mind that all the methods used in the above-mentioned fluorescence polarization measurements imply the applicability of the model of a solute molecule possessing a certain size and structure, dissolved in an isotropic continuum possessing certain hydrodynamic properties (see, e.g., Ref. 358). From a critical point of view it would be better, therefore, to go the other way around and to determine those rather artificial model parameters:

The degree of polarization $p = (I_\parallel - I_\perp)/(I_\parallel + I_\perp)$ of the fluorescence of a π-electronic system depends on its fluorescence decay time τ_f and on its rotational relaxation time τ_r in that solvent

$$\frac{1}{p} - \frac{1}{3} = \left(\frac{1}{p_0} - \frac{1}{3}\right)\left(1 + 3\frac{\tau_f}{\tau_r}\right)$$

where p_0 is the value of p in the absence of depolarization factors.[359] By combining this Perrin equation with the Debye–Stokes–Einstein equation[360]

$$\tau_r = \frac{\eta v}{kT}$$

we can apply fluorescence polarization measurements to determine the product $(\eta v)_m$ of the effective microscopic viscosity η_m and the hydrodynamically effective molecular volume v_m for an aromatic compound dissolved in an organic solvent as a function of temperature T:

$$(\eta v)_m = kT\tau_f p \frac{3 - p_0}{p_0 - p}$$

In most examples we expect a rather complicated temperature dependence since not only molecular reorientation rates and the value of p but also the rates of fluorescence quenching processes and the value of τ_f depend on temperature.

It may be useful now to explore the possibilities inherent in fluorescence spectroscopic methods for the determination of molecular reorientational relaxation times in some more detail since the application of synchrotron radiation offers real and substantial advances. For the user the synchrotron radiation being beamed on the sample is exactly[362] known in all its physical properties, especially the total spectral photon distribution and the distribution of the two linear polarization components parallel and vertical to the plane of the electron storage ring, as well as its time structure.[84] The broader the spectral band width allowed by the user for the irradiation of the probe, the more the intensity becomes comparable with that of strongest commercial lasers (and in the FIR spectral region it is more intense than any light source commercially available), and these well-known conditions will last for several hours each time bunches of electrons have been injected.

For single-bunch operation with BESSY (*B*erlin *E*lectron *S*torage Ring for *S*ynchrotron Radiation) the light pulses possess a half-width of several picoseconds and a sequence of several nanoseconds. Now let us supply such synchrotron radiation to a sample which shows *dual* luminescence (i.e., a sample in which an adiabatic photoreaction happens after excitation[85]), at first an ultraviolet (UV) light pulse for excitation of a molecule in the solution, and after some time delay an adequate pulse of longer wavelength to supply the excited system with additional activation energy for the adiabatic photoreaction. The shorter the time delay, the higher will be the long-wavelength fluorescence intensity of the phonon-induced adiabatic reaction product as compared with the fluorescence intensity of the absorbing species, and the reaction rate can be measured directly from the fluorescence intensity ratio as a function of the time delay. If molecular reorientation determines the photoreaction rate, the reorientation relaxation time can be measured directly.[86,87]

But even if there is only *one* fluorescence band, its time resolution gives us some insight into the orientational dynamics of the solution if the rates of fluorescence and reorientation compete with each other, since the longer a species remains in its fluorescing state the longer is its time for relaxation and the eventually emitted fluorescence photon contributes to the long-wavelength part of the fluorescence band.[88]

E. Dissipation in the Far-Infrared Spectral Region

The nanosecond-to-picosecond region in the time domain corresponds to the nanometer region of molecular spectroscopy, that is, the domain of rototranslation in liquids. Analyses of the band shape of FIR absorption bands and of inhomogeneous line broadening in corresponding Raman spectra are appropriate means to determine entropy production by dissipation, as will be outlined in Section III.

If a liquid contains small polar molecules, it possesses a long-wavelength IR absorption band which is due to a dipole—dipole coupling between the time-dependent vector of the electric field of external irradiation and the time-dependent vector of the molecular dipole moment.

The time dependency of the molecular dipole moment vector becomes effective in that part of the spectrum in which the orientational relaxation time of interacting molecules matches the reciprocal frequency of the perturbing field.[89-91]

The intermolecular interaction between a polar molecule and the cage in which it is embodied gives rise to fluctuations in the induced rotational motion, and therefore off-diagonal elements of the density matrix do not vanish; hence the so-called adiabatic approximation cannot be justified in this physical context,[92] as has been verified in the case of CH_2Cl_2 by computer simulation[93].

In the present work the interacting forces in a first-order approximation are solely taken as repulsions between hard-core molecules, although rototranslational intermolecular vibrations due to attractive forces can be treated by the methods presented in Sections II and III. Dissipation by inelastic collisions and collision-induced absorption due to molecular polarizability will not be taken into account explicitly.

The temperature dependency of rotational diffusion in liquid CH_2Cl_2 between its melting and boiling point has been calculated by molecular dynamics simulation.[94] As has been pointed out, the FIR absorption spectrum is related only indirectly to dynamical autocorrelation functions, these being strict terms denoting the outcome of fluctuations in the macroscopic polarization. This makes it very difficult to calculate the band shape of an FIR spectrum. In the context of classical statistical mechanics the simulation method is at present one of the most powerful and detailed techniques available, given some necessary checks on the form of the intermolecular pair potential used.

If a certain classical pair potential as delineated in Section I.A is useful in study of a molecular crystal and an imperfect gas, then it may be used via molecular dynamics simulation to produce not only theoretical liquid phase spectra but also a predictive theory of the liquid state.

In Section III, therefore, we shall show explicitly how the aforementioned "anti-Boltzmann" behavior of FIR absorption bands in liquids can be described by linear response theory (LRT). Following some suggestions by van Vliet[95,96] and other authors this procedure leads to a generalization of the LRT and the fluctuation dissipation theorem (FDT), since the so-called golden rule for the first-order approximation of the transition probability as used in Gordon's derivation of FDT is not applicable in the above mentioned context.[97] For golden rule, etc. see Ref. 361.

II.   STRUCTURAL AND THERMODYNAMIC PROPERTIES*

A.   Isotropic and Anisotropic Reference Systems in the Thermodynamic Perturbation Theory of Molecular Fluids

Beginning with the pioneering work of Zwanzig in 1954,[98] thermodynamic perturbation theory has been developed into a powerful tool for predicting the equilibrium properties of simple fluids.[5,12,99] This is particularly true if high-density states are considered and if the division of the pair potential into a repulsive and an attractive branch is performed according to the prescription of Weeks et al.[16] (see also Refs. 100–102). For the treatment of (at least approximately) spherical particles, a properly chosen

*Part of the research reported in this section was completed by K.-H. Naumann while on a postdoctoral fellowship at the Department of Chemical Engineering, Rice University, Houston, Texas. He is indebted to the helpful and stimulating interest of T. W. Leland.

hard-sphere fluid will usually serve as a reference system.[101-104] With accurate pair correlation functions available from the analytical solution of the Percus–Yevick integral equation,[105-108] together with the correction proposed by Verlet and Weis,[101] first-order perturbation calculations (and sometimes even higher order ones[103,104]) are readily feasible today even on the smallest computers.

The straightforward generalization of these ideas to the case of particles interacting via noncentral potentials, such as interaction site[109,110] or Kihara-type[21] molecules, necessitates the consideration of anisotropic reference systems.[111-116] This requirement, however, causes severe problems, as the properties of these reference fluids are not yet generally well known. Despite the considerable amount of computer simulation data already available for hard interaction site molecules, especially for homonuclear and heteronuclear hard dumbbells,[117-122] and to a smaller extent also for hard convex bodies,[123-128] this research is still in its exploratory stage. Nevertheless, the theoretical progress achieved so far[4,22,110,129] justifies the assumption that at least for dense fluids it will be much easier to determine the hard-core reference properties in a first step and to account for the attractive contributions in a second step employing the methods of perturbation theory than to treat the complete system in a more or less direct manner. Once the necessary data on structure and thermodynamics of hard-body fluids can be evaluated readily and with sufficient accuracy, the exploration of systems with realistic interactions will be greatly facilitated. We believe that this should be a strong·motivation for continued efforts toward a better understanding of fluids composed of hard molecules.

A seemingly more convenient way to consider molecular anisotropy within the framework of perturbation theory is the separation of the full potential function into a spherical reference part and a nonspherical perturbation. The expansion of Pople[134] and Gubbins and Gray,[135,136] for example, starts with

$$v_{12}(r_{12}, \Omega_1, \Omega_2, \lambda) = v_{12}^0(r_{12}) + \lambda v_{12}^a(r_{12}, \Omega_1, \Omega_2) \qquad (2.1)$$

where $v_{12}^0(r_{12})$ is obtained by unweighted averaging over the molecular orientations:

$$v_{12}^0(r_{12}) = \langle v_{12}(r_{12}, \Omega_1, \Omega_2) \rangle_{\Omega_1, \Omega_2} \qquad (2.2)$$

Here r_{12} denotes the center-to-center distance, and Ω_1 and Ω_2 are the orientations of molecules 1 and 2, respectively. Setting the coupling parameter λ to zero yields the reference potential, while with $\lambda = 1$ the full interaction is recovered. This method works well, if $v_{12}^a(r_{12}, \Omega_1, \Omega_2)$ consists mainly of idealized polar interaction terms (interactions of point dipoles, point quadrupoles, etc.), particularly in the Padé approximant version suggested by Stell et al.,[104] which partially overcomes the sometimes insufficient convergence behavior of the perturbation series. For cases of strong geometric anisotropy, however, rather poor results are obtained, expecially at high densities.[137]

Despite its merits, the Pople expansion suffers from an inherent weakness, since the anisotropic perturbation cannot be small whenever the particle geometry (as characterized by the harshly repulsive portions of the intermolecular potential) deviates significantly from sphericity. As an example, consider a fluid composed of hard dumbbell molecules. From Eq. (2.2) we obtain a reference system that consists of hard spheres just enveloping the dumbbell bodies. Clearly, with increasing particle anisotropy (i.e., with growing elongation of the dumbbells) this choice will become an increasingly poor one. Thus, it appears to be desirable to introduce as much information about molecular shape as possible into the properties of the reference fluid.

The exponentially averaged effective potential function

$$v_{12}^0(r_{12}) = -kT \ln \left\langle \exp\left[\frac{-v_{12}(r_{12}, \Omega_1, \Omega_2)}{kT} \right] \right\rangle_{\Omega_1, \Omega_2} \qquad (2.3)$$

represents an important step in this direction.[15,22,115,138–145] Equation (2.3) leads to a nonlinear coupling between the reference potential and the perturbation term,

$$v_{12}(r_{12}, \Omega_1, \Omega_2, \lambda) = v_{12}^0(r_{12}) - kT \ln\left[1 + \lambda s_\lambda(r_{12}, \Omega_1, \Omega_2) \right] \qquad (2.4)$$

where

$$s_\lambda(r_{12}, \Omega_1, \Omega_2) = \exp\left\{ \frac{-\left[v_{12}(r_{12}, \Omega_1, \Omega_2) - v_{12}^0(r_{12}) \right]}{kT} \right\} - 1 \qquad (2.5)$$

Equivalently to Eq. (2.4) the Mayer function

$$f_{12}(r_{12}, \Omega_1, \Omega_2) = \exp\left[\frac{-v_{12}(r_{12}, \Omega_1, \Omega_2)}{kT} \right] - 1 \qquad (2.6)$$

may be used as an expansion functional:[139]

$$f_{12}(r_{12}, \Omega_1, \Omega_2, \lambda) = f_{12}^0(r_{12}) + \lambda\left[f_{12}(r_{12}, \Omega_1, \Omega_2) - f_{12}^0(r_{12}) \right] \qquad (2.7)$$

where

$$f_{12}^0(r_{12}) = \langle f_{12}(r_{12}, \Omega_1, \Omega_2) \rangle_{\Omega_1, \Omega_2} \tag{2.8}$$

Therefore, this method is often referred to as the RAM (reference system average Mayer function) theory.[139,144,145]

To first order the molecular pair distribution function is given by

$$g_{12}(r_{12}, \Omega_1, \Omega_2) = g_{12}^0(r_{12})[1 + s_\lambda(r_{12}, \Omega_1, \Omega_2)] + \rho W^{-1} \int g_{123}^0(r_{12}, r_{13}, r_{23})$$
$$\times [s_\lambda(r_{13}, \Omega_1, \Omega_3) + s_\lambda(r_{23}, \Omega_2, \Omega_3)] \, d\mathbf{r}_3 \, d\Omega_3 \tag{2.9}$$

where g_{12}^0 and g_{123}^0 denote the reference pair and triplet correlation functions, respectively, and

$$W = \begin{cases} 4\pi & \text{for linear molecules} \\ 8\pi^2 & \text{for nonlinear molecules} \end{cases} \tag{2.10}$$

The zeroth-order RAM theory is exact in the low-density limit, $\rho \to 0$, in that it predicts the second virial coefficient B_2 correctly. Note that the first-order correction term to Eq. (2.9) vanishes on integrating over Ω_1 and Ω_2. Thus, the zeroth-order orientationally averaged center-to-center correlation function as obtained from the reference pair potential remains correct within a first-order approximation, independent of density.

For liquids composed of hard dumbbell molecules, recent investigations have shown good agreement between $g_{12}^0(r_{12})$ and $\langle g_{12}(r_{12}, \Omega_1, \Omega_2) \rangle_{\Omega_1, \Omega_2}$ as evaluated from computer simulation data.[115,140,141] More important, the reproduction of angular correlations is quite satisfactory in most cases considered so far.[144,145] But despite these encouraging results, some deficiencies still remain.[141,143,144] Obviously, the price to be paid for circumventing the treatment of anisotropic reference systems is a certain limitation of the range of applicability. For particles with distinctly nonspherical shape the RAM predictions are likely to become less reliable, especially at high densities.[141,144] Nevertheless the investigations conducted so far do suggest that the RAM perturbation series converges much faster than the Pople expansion. Thus it appears to be worthwhile to keep the concept of Boltzmann averaging as a valuable guideline when looking for more elaborate and yet numerically tractable reference systems. We shall come back to this point later.

Once $g_{12}(r_{12}, \Omega_1, \Omega_2)$ is known, most of the structural and thermodynamic fluid properties may be evaluated. The configurational part of the internal energy, for example, is given by

$$U_c = 2\pi\rho N \int \langle g_{12}(r_{12}, \Omega_1, \Omega_2) v_{12}(r_{12}, \Omega_1, \Omega_2) \rangle_{\Omega_1, \Omega_2} r_{12}^2 \, dr_{12} \tag{2.11}$$

and the virial equation of state reads

$$\frac{P}{\rho kT} = 1 - \frac{2\pi\rho}{3kT} \int \langle g_{12}(r_{12},\Omega_1,\Omega_2) \frac{\partial}{\partial r_{12}} v_{12}(r_{12},\Omega_1,\Omega_2) \rangle_{\Omega_1,\Omega_2} r_{12}^3 \, dr_{12}$$

(2.12)

Alternatively, the equation of state may be obtained via the compressibility equation

$$\rho kT\kappa = 1 + 4\pi\rho \int \langle g_{12}(r_{12},\Omega_1,\Omega_2) - 1 \rangle_{\Omega_1,\Omega_2} r_{12}^2 \, dr_{12}$$

(2.13)

where

$$\kappa = -\frac{1}{V}\left(\frac{\partial V}{\partial P}\right)_T$$

(2.14)

denotes the isothermal compressibility.

Inspecting Eqs. (2.2) and (2.3) we note that no specific assumptions are made concerning the nature of the intermolecular interactions, except that they are reasonably described by pair potentials. On principal, this generality might be considered a highly expedient feature. For practical applications, however, it can be very rewarding to carefully examine the analytical structure as well as the physical implications of the potential model under consideration. Utilizing special symmetry properties or employing appropriate coordinate transforms, for instance, may not only result in more efficient or more convenient formalisms but will sometimes even lead to alternative physical descriptions of the liquid system of interest.

In order to exemplify this statement, we shall now consider a model fluid, where the orientation-dependent molecular pair potential is represented by a sum over contributions from interacting atoms, groups of atoms, point charge centers, etc. (so-called interaction sites[109,146]):

$$v_{12}(r_{12},\Omega_1,\Omega_2) = \sum_{\alpha,\gamma} v_{\alpha,\gamma}\left(|\mathbf{r}_\alpha^1 - \mathbf{r}_\gamma^2|\right)$$

(2.15)

The pair interaction between two particles with m and n interaction sites, respectively, is then described by a set of $m \cdot n$ spherically symmetric site–site potentials $v_{\alpha\gamma}(r_{\alpha\gamma})$. See Figs. 2 and 3. By virtue of Eq. (2.15) the potential energy acting between two molecules depends only on the relative distances between their interaction sites. Thus, in dealing with interaction site systems it is often convenient to express structural and thermodynamic properties in

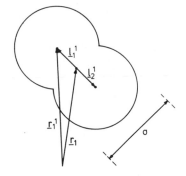

Fig. 2. Example of an interaction site molecule ($\alpha = 1$; $\gamma = 2$). The location of site α of molecule 1 is given by $\mathbf{r}_\alpha^1 = \mathbf{r}_1 + \mathbf{l}_\alpha^1$, where $\mathbf{l}_\alpha^1 = \mathbf{l}_\alpha^1(\Omega_1)$.

terms of site–site distribution functions $g_{\alpha\gamma}(r_{\alpha\gamma})$. These are related to the molecular pair correlation function by[109]

$$g_{\alpha\gamma}(|\mathbf{r}-\mathbf{r}'|) = W^{-2} \int d\mathbf{r}_1 \, d\Omega_1 \int d\mathbf{r}_2 \, d\Omega_2 \, g_{12}(r_{12},\Omega_1,\Omega_2) \, \delta(\mathbf{r}_\alpha^1 - \mathbf{r}) \, \delta(\mathbf{r}_\gamma^2 - \mathbf{r}')$$

(2.16)

The configurational energy, for example, can now be written as

$$U_c = 2\pi\rho N \sum_{\alpha\gamma} \int g_{\alpha\gamma}(r_{\alpha\gamma}) v_{\alpha\gamma}(r_{\alpha\gamma}) r_{\alpha\gamma}^2 \, dr_{\alpha\gamma}$$

(2.17)

which should be compared to Eq. (2.11). The most general route to the equation of state proceeds from the compressibility theorem for interaction site molecules:[146]

$$\rho kT\kappa = 1 + 4\pi\rho \int \left[g_{\alpha\gamma}(r_{\alpha\gamma}) - 1 \right] r_{\alpha\gamma}^2 \, dr_{\alpha\gamma}.$$

(2.18)

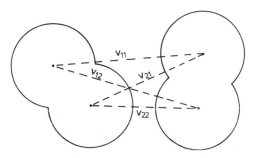

Fig. 3. Site–site interactions between two interaction site molecules ($\alpha = 1$; $\gamma = 2$).

Note that the integral on the right-hand side of Eq. (2.18) does not depend on the particular choice of α and γ.

The results of X-ray or neutron diffraction experiments on molecular liquids are usually discussed in terms of the structure factor $S(k)$. This factor can be expressed as a linear combination of the Fourier transforms of inter- and intramolecular atom–atom or group–group correlation functions:[4]

$$S(k) = \left[\sum_{\eta} a_{\eta} \right]^{-2} \sum_{\alpha\gamma} a_{\alpha}(k) a_{\gamma}(k) \left[\hat{\omega}_{\alpha\gamma}(k) + \rho \hat{h}_{\alpha\gamma}(k) \right] \qquad (2.19)$$

where

$$\hat{h}_{\alpha\gamma}(k) = \int h_{\alpha\gamma}(r) \exp(i\mathbf{k}\cdot\mathbf{r})\, d\mathbf{r} \qquad (2.20)$$

and

$$h_{\alpha\gamma}(r) = g_{\alpha\gamma}(r) - 1 \qquad (2.21)$$

$\omega_{\alpha\gamma}$ denotes the intramolecular correlation function of the sites α and γ. For rigid molecules it possesses the form

$$\omega_{\alpha\gamma}(r) = \frac{\delta\left(r - |\mathbf{r}_{\alpha}^1 - \mathbf{r}_{\gamma}^2|\right)}{4\pi|\mathbf{r}_{\alpha}^1 - \mathbf{r}_{\gamma}^2|^2} = \frac{\delta(r - L_{\alpha\gamma})}{4\pi L_{\alpha\gamma}^2} \qquad (2.22)$$

and its Fourier transform is given by

$$\hat{\omega}_{\alpha\gamma}(k) = \frac{\sin(kL_{\alpha\gamma})}{kL_{\alpha\gamma}} \qquad (2.23)$$

The values of the weighting coefficients $a_{\eta}(k)$ depend upon the type of experiment being performed. If X-ray scattering is considered, a_{η} denotes the atomic or group form factor f_{η}, while in the case of neutron scattering it stands for the coherent scattering length b_{η} of the ηth nucleus in a molecule.

Equation (2.19) establishes a mathematical connection of calculated $h_{\alpha\gamma}(r_{\alpha\gamma})$ functions to the experimentally accessible $S(k)$ and therefore allows the interpretation of diffraction data on the basis of model calculations.[10,11,65,147-155] In cases where there are enough independent (and sufficiently accurate) diffraction experiments available (e.g., by combination of X-ray and neutron techniques and/or by isotropic substitution), it is even possible to determine the $h_{\alpha\gamma}(r_{\alpha\gamma})$ directly from experimental data.[148,154,156,157]

The site–site pair distribution functions $g_{\alpha\gamma}(r_{\alpha\gamma})$ are spherically symmetric functions and thus do not contain explicit informations about molecular orientational correlations. These informations, however, can be recovered by applying the site superposition approximation (SSA) as proposed by Hsu et al.[11] and by Quirke and Tildesley.[158] Similar in spirit to the well-known Kirkwood superposition approximation,[159] the SSA suggests that the probability of finding two molecules in a particular configuration, $(r_{12}, \Omega_1, \Omega_2)$, is given by a normalized product of site–site distribution functions evaluated at the respective $r_{\alpha\gamma}$ distances. This method seems to give fair approximations to full or partial orientation-dependent molecular distribution functions.[11,158,160] In the case of liquid acetonitrile,[150] for example, it confirms the ideas about local structure revealed from X-ray and neutron diffraction data[161,162] and from IR investigations.[58]

Equations (2.17) and especially (2.18) clearly indicate that the implications of utilizing the comparatively simple projections $g_{\alpha\gamma}(r_{\alpha\gamma})$ instead of the rather complex full distribution functions $g_{12}(r_{12}, \Omega_1, \Omega_2)$ go well beyond purely formal considerations. Whenever one chooses to describe the structural properties of a fluid in terms of $g_{12}(r_{12}, \Omega_1, \Omega_2)$, one usually thinks of it as an assembly of molecular entities. Interaction site distribution functions such as $g_{\alpha\gamma}(r_{\alpha\gamma})$ and $\omega_{\alpha\gamma}(r_{\alpha\gamma})$, however, suggest an alternative physical picture, which is intimately related to the idea of regarding "molecular liquids" as mixtures of atoms or atomic groups.

The first theoretical approach to $g_{\alpha\gamma}(r_{\alpha\gamma})$ that emerged from this background is the RISM (reference interaction site model) theory of Chandler and Anderson.[163–165] The central relation in the RISM formalism is the definition of the direct site–site pair correlation function $c_{\alpha\gamma}(r_{\alpha\gamma})$:

$$\hat{\mathbf{h}}(k) = \hat{\omega}(k)\hat{\mathbf{c}}(k)\hat{\omega}(k) + \rho\hat{\omega}(k)\hat{\mathbf{c}}(k)\hat{\mathbf{h}}(k) \tag{2.24}$$

where $\hat{\mathbf{h}}(k)$, $\hat{\mathbf{c}}(k)$, and $\hat{\omega}(k)$ denote matrices with the elements $\hat{h}_{\alpha\gamma}(k)$, $\hat{c}_{\alpha\gamma}(k)$, and $\hat{\omega}_{\alpha\gamma}(k)$, respectively, and

$$\hat{c}_{\alpha\gamma}(k) = \int c_{\alpha\gamma}(r)\exp(i\mathbf{k}\cdot\mathbf{r})\,d\mathbf{r} \tag{2.25}$$

The physical meaning of the generalized Ornstein–Zernike equation, Eq. (2.24), is most easily explained in the low density limit, $\rho \to 0$, where it assumes the form

$$\hat{\mathbf{h}}(k) = \hat{\omega}(k)\hat{\mathbf{c}}(k)\hat{\omega}(k) \tag{2.26}$$

Now consider a system consisting of homonuclear diatomic molecules. We then have (in explicit notation)

$$\begin{vmatrix} \hat{h}_{11} & \hat{h}_{12} \\ \hat{h}_{21} & \hat{h}_{22} \end{vmatrix} = \begin{vmatrix} \hat{\omega}_{11} & \hat{\omega}_{12} \\ \hat{\omega}_{21} & \hat{\omega}_{22} \end{vmatrix} \begin{vmatrix} \hat{c}_{11} & \hat{c}_{12} \\ \hat{c}_{21} & \hat{c}_{22} \end{vmatrix} \begin{vmatrix} \hat{\omega}_{11} & \hat{\omega}_{12} \\ \hat{\omega}_{21} & \hat{\omega}_{22} \end{vmatrix} \tag{2.27}$$

For reasons of symmetry, however,

$$\hat{h}_{11}(k) = \hat{h}_{12}(k) = \hat{h}_{21}(k) = \hat{h}_{22}(k) \tag{2.28}$$

and Eq. (2.26) finally becomes

$$\hat{h}_{11} = \hat{c}_{11} + \hat{\omega}_{12}\hat{c}_{21} + \hat{c}_{12}\hat{\omega}_{21} + \hat{\omega}_{12}\hat{c}_{22}\hat{\omega}_{21} \tag{2.29}$$

This result may be casted into the graphical representation shown in Eq. (2.30):

$$\tag{2.30}$$

Here $\hat{c}$-functions are denoted by straight lines, $\hat{\omega}$-functions by wiggly ones, and the label 1^i stands for site 1 of molecule i ($i = 1, 2$).

Equation (2.30) shows that the RISM theory not only accounts for direct interaction contributions to $\hat{h}_{\alpha\gamma}(k)$ as characterized by the direct site–site correlation function $\hat{c}_{\alpha\gamma}(k)$; additional intermolecular correlations between two particular sites are produced, for example, if one site interacts with a third that is attached to the second site by a chemical bond [characterized by $\hat{\omega}_{\alpha\gamma}(k)$]. Such situations are described by the second and the third graph in Eq. (2.30).

Generally, Eq. (2.24) defines correlation chains (consisting of intermolecular and intramolecular terms) between the interaction sites of different particles. Of course, for densities $\rho > 0$, most of these chains will range over more than two molecules. (For a detailed discussion of interaction site cluster expansions, see Ref. 109.)

From Eq. (2.24) alone $h_{\alpha\gamma}(r_{\alpha\gamma})$ and $c_{\alpha\gamma}(r_{\alpha\gamma})$ are not yet uniquely determined. This is achieved, however, by assuming the Percus–Yevick-like closure relation

$$c_{\alpha\gamma}(r_{\alpha\gamma}) = y_{\alpha\gamma}(r_{\alpha\gamma}) f_{\alpha\gamma}(r_{\alpha\gamma}) \tag{2.31}$$

where

$$y_{\alpha\gamma}(r_{\alpha\gamma}) = g_{\alpha\gamma}(r_{\alpha\gamma}) \exp\left[\frac{v_{\alpha\gamma}(r_{\alpha\gamma})}{kT}\right] \tag{2.32}$$

and

$$f_{\alpha\gamma}(r_{\alpha\gamma}) = \exp\left[\frac{-v_{\alpha\gamma}(r_{\alpha\gamma})}{kT}\right] - 1 \tag{2.33}$$

Equations (2.24), (2.31), (2.32), and (2.33) represent a closed set of relations, which allows the evaluation of the unknown correlation functions $h_{\alpha\gamma}(r_{\alpha\gamma})$, $c_{\alpha\gamma}(r_{\alpha\gamma})$, and $y_{\alpha\gamma}(r_{\alpha\gamma})$ for arbitrary site–site potentials $v_{\alpha\gamma}(r_{\alpha\gamma})$ by iterative numerical procedures.

Although the RISM theory by itself is a nonperturbative approach, it may be used in principle to evaluate the structural properties of anisotropic hard core reference fluids in terms of site–site distribution functions.[116,163,166–168] In fact, the name *reference interaction site model* originated from this perspective.[163] For hard interaction site molecules

$$v_{\alpha\gamma}(r_{\alpha\gamma}) = \begin{cases} \infty, & r_{\alpha\gamma} < \sigma_{\alpha\gamma} \\ 0, & r_{\alpha\gamma} \geq \sigma_{\alpha\gamma} \end{cases} \tag{2.34}$$

where $\sigma_{\alpha\gamma} = (\sigma_\alpha + \sigma_\gamma)/2$ denotes the interaction distance between sites α and γ (see Fig. 2). The closure relations to Eq. (2.24) are then given by

$$h_{\alpha\gamma}(r_{\alpha\gamma}) = -1, \qquad r_{\alpha\gamma} < \sigma_{\alpha\gamma} \tag{2.35}$$

$$c_{\alpha\gamma}(r_{\alpha\gamma}) = 0, \qquad r_{\alpha\gamma} \geq \sigma_{\alpha\gamma} \tag{2.36}$$

Equation (2.35) is, of course, a rigorous statement. Equation (2.36), which follows directly from Eq. (2.31), however, represents an approximation similar in spirit to the Percus–Yevick equation for hard spheres.

Unfortunately, the approximate nature of Eq. (2.36) produces certain shortcomings in the prediction of nearest-neighbor correlations, which are particularly severe at low and moderate densities.[169,170] The RISM theory is nevertheless useful at high densities, where most of structural features of molecular fluids are reproduced in a qualitatively reliable manner.[171] There are some noteworthy exceptions, however, such as the complete failure to predict nonzero values for the orientational correlation parameters

$$G_l = 4\pi\rho \int \langle g_{12}(r_{12}, \Omega_1, \Omega_2) P_l(\cos\theta_{12}) \rangle_{\Omega_1, \Omega_2} r_{12}^2 \, dr_{12}, \qquad l = 1, 2 \tag{2.37}$$

in the case of fluids consisting of linear molecules.[172] In Eq. (2.37) $P_l(\cos\theta_{12})$ denotes the lth-order Legendre polynomial, and θ_{12} represents the angle between the symmetry axes of particles 1 and 2; G_1 determines the value of the dielectric constant for rigid polar molecule fluids, while G_2 is closely related to the integrated intensity of depolarized light scattering experiments.[4]

In order to improve the accuracy of the RISM formalism, Cummings et al.[173] proposed the addition of a short-range Yukawa-type correction to $c_{\alpha\gamma}(r_{\alpha\gamma})$:

$$c_{\alpha\gamma}(r_{\alpha\gamma}) = \frac{A_{\alpha\gamma}\exp\left[-z_{\alpha\gamma}(r_{\alpha\gamma}-\sigma_{\alpha\gamma})\right]}{r_{\alpha\gamma}}, \qquad r_{\alpha\gamma} \geq \sigma_{\alpha\gamma} \qquad (2.38)$$

In much the same spirit as Verlet and Weis corrected the Percus–Yevick pair distribution function for hard spheres,[101] the adjustable parameters $A_{\alpha\gamma}$ and $z_{\alpha\gamma}$ could be chosen in a way that would optimize the agreement of RISM site–site correlation functions with the respective computer simulation results. This semiempirical approach should work for relatively simple molecular geometries such as homonuclear hard dumbbells, provided the particle density is not too high. With increasing density and increasing complexity of molecular structure, however, expression (2.38) is likely to become inadequate.

Of course, there are alternatives to the method of Cummings et al. Instead of searching for possible improvements to the closure relation Eq. (2.36), one might also consider sacrificing Eq. (2.24) itself. In fact, by utilizing diagrammatic techniques, Chandler et al. arrived at a new class of site–site direct correlation functions, which are related to each other through four coupled Ornstein–Zernike-like equations.[174] This approach overcomes many of the problems associated with the RISM theory.[174,175] Unfortunately, the price to be paid for this achievement is a considerable increase in numerical complexity.

Historically, the RISM theory represents the first successful approach to the site–site distribution functions of hard interaction site model fluids, which together with its relative simplicity accounts for its unbroken popularity. It should be emphasized, however, that the exploration of interaction site correlations in molecular fluids is by no means restricted to integral equation theories (such as RISM or related approaches) alone. As a matter of fact, the group contribution idea to regard molecular liquids as mixtures of atoms or atomic groups can be exploited in a much more direct manner by employing the methods of thermodynamic perturbation theory. We (Naumann and Lippert,[30] Naumann[130]) and Nezbeda and Smith[131,132] have pointed out that Boltzmann-averaged effective pair potential functions (the underlying idea of the RAM theory for molecular distribution functions) are especially useful at the site–site level. By taking advantage of this fact one eventually arrives at a new type of reference systems for interaction site model fluids, where molecular liquids are treated as mixtures of formally independent molecular interaction sites (the IMIS model.)[30,31,130] A detailed discussion of the IMIS concept will be presented in Section II.B, below.

Another important potential model frequently used to describe the intermolecular interactions between nonspherical particles is the core potential originally proposed by Kihara.[21,176] The general representation reads

$$v_{12}(r_{12}, \Omega_1, \Omega_2) = v_{12}(l_{12}) \tag{2.39}$$

where $l_{12} = l_{12}(r_{12}, \Omega_1, \Omega_2)$ denotes the shortest distance between the impenetrable molecular cores for a given configuration $(r_{12}, \Omega_1, \Omega_2)$. In most cases of interest the shape of this core region is sufficiently determined by the geometric characteristics of the atomic skeleton,[21] and the functional dependence of v_{12} on l_{12} is usually assumed to be of the Lennard-Jones type.[17,95] Particles that interact via Kihara core potentials are often called "convex" molecules (Fig. 4).[113,114,177] Here again, as in the case of interaction site molecules, we find ourselves confronted with the situation that $g_{12}(r_{12}, \Omega_1, \Omega_2)$ does not necessarily represent the most convenient route to the structural and especially to the thermodynamic properties of the system under consideration. Since $v_{12}(l_{12})$ does not exhibit an explicit dependence on molecular orientations, the knowledge of the one-dimensional orientationally averaged core distribution function $g_{12}^{av}(l_{12})$ is completely sufficient for the evaluation of configurational energy and isothermal compressibility. The resulting expressions are

$$U_c = \tfrac{1}{2} N\rho \int g_{12}^{av}(l_{12}) v_{12}(l_{12}) S_{c+l+c} \, dl_{12} \tag{2.40}$$

$$\rho k T\kappa = 1 + \rho \int \left[g_{12}^{av}(l_{12}) - 1 \right] S_{c+l+c} \, dl_{12} \tag{2.41}$$

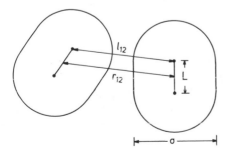

Fig. 4. Geometry of two interacting Kihara-type molecules.

The generalized surface function S_{c+l+c} substitutes for the well-known expression $4\pi r^2$ that appears in problems with spherical symmetry. It is related to the molecular core parameters by

$$S_{c+l+c} = 2S_c + 8\pi R_c^2 + 16\pi R_c l_{12} + 4\pi l_{12}^2 \qquad (2.43)$$

with S_c and R_c representing the core surface area and the mean curvature radius, respectively. For central potentials with point cores

$$S_c = R_c = 0 \qquad (2.44)$$

Thus, in this special case

$$S_{c+l+c} = 4\pi l_{12}^2 \qquad (2.45)$$

results. QED

The natural starting point for the treatment of Kihara-type molecules within the framework of perturbation theory are reference fluids or mixtures composed of hard convex bodies (HCB).[113,114,178-181] This line of research converges with the recent derivation of accurate equations of state by Boublik[18,182] and Naumann et al.,[178,183] which provide a complete description of the thermodynamics of HCB systems. For most practical applications, there is probably little to choose between these two equations, since both are in very good agreement with all available computer simulation results. From a theoretical point of view, however, they clearly imply somewhat different assumptions concerning the interrelations between molecular shape and thermodynamic properties. Additional information on this subtle connection is obviously desirable but hard to come by. We shall reconsider this problem in Section II.D.

With the knowledge of accurate HCB equations of state, the thermodynamic perturbation theory of Kihara core molecules may be used as a powerful and versatile tool in many research areas of chemical physics and engineering. It may be applied, for instance, to study the coupling of conformational equilibria of flexible particles to the structure of the molecular surroundings[32] (see Section II.E, below) or to extract informations on nearest-neighbour correlations in dense liquid mixtures from solvent-induced shifts on PMR signals.[34] As a further example, a generalized conformal solution theory for mixtures of convex molecules[178] will be presented in Section II.C.

B. The Independent Molecular Interaction Sites (IMIS) Model

1. Site–Site Distribution Functions from Effective Site-Centered Pair Potentials: The IMIS Approach

In order to illustrate the basic ideas of the IMIS approach, we shall first consider the behavior of the site–site pair correlation function $g_{\alpha\gamma}(r_{\alpha\gamma})$ in the limiting case of vanishing particle density, that is, $\rho \to 0$. Since

$$g_{12}(r_{12}, \Omega_1, \Omega_2; \rho \to 0) = \exp\left\{\frac{-v_{12}(r_{12}, \Omega_1, \Omega_2)}{kT}\right\} \qquad (2.46)$$

one finds from Eqs. (2.15) and (2.16)

$$g_{\alpha\gamma}(r_{\alpha\gamma}^{12}; \rho \to 0) = W^{-2} \int dr_1 \, d\Omega_1 \int dr_2 \, d\Omega_2 \exp\left\{\frac{-\sum_{\eta\nu} v_{\eta\nu}(r_{\eta\nu}^{12})}{kT}\right\}$$
$$\times \delta(r_\alpha^1) \, \delta(r_\gamma^2 - r_{\alpha\gamma}^{12}) \qquad (2.47)$$

As an example we have evaluated this expression for a simple four-center Lennard-Jones model of carbon tetrachloride. In Fig. 5 the calculated distribution function $g_{ClCl}(r_{ClCl}; \rho \to 0)$ is displayed together with Narten's experimental result on $g_{ClCl}(r_{ClCl})$ for the pure liquid.[156] Note that corresponding maxima and minima are located at almost exactly the same positions. The striking similarity between both curves clearly indicates that the presence of other particles in a dense molecular fluid has rather little effect on the structure of the potential field which acts on adjacent molecules.

On partial or complete overlap of interaction sites belonging to different molecules strong repulsive forces occur. Owing to the exponential weighting implied by Eq. (2.47), such pair configurations yield only negligible contributions to $g_{\alpha\gamma}(r_{\alpha\gamma}; \rho \to 0)$. Specific interactions of nonspherical particles, on the other hand, such as the gearing of Cl—C—Cl groups in CCl_4[156,184] are taken into account correctly. The same mechanisms of configurational selection, however, which essentially determine the form of site–site distribution functions in molecular gases, are also responsible for the appearance of packing effects and translation–rotation coupling in dense liquids. Independent of whether two gas particles approach each other under the influence of attractive forces or whether molecules are forced into contact by the high density in a liquid system, the discrimination between energetically favorable and unfavorable arrangements is mainly determined by geometric aspects.

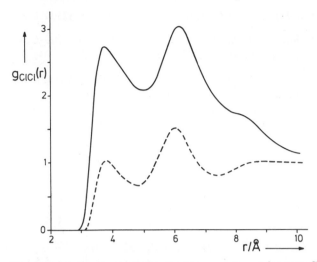

Fig. 5. The atom–atom pair correlation function $g_{ClCl}(r_{ClCl})$ for pure CCl_4. At 293 K: — $\rho = 0$ (calculated); --- $\rho = 0.00624$ Å^{-3} (experimental data from Ref. 156). The calculated zero density correlation function has been obtained from Eq. (2.47) assuming a tetrahedral four-center Lennard–Jones model with the parameters $l_{ClCl} = 1.77$ Å, $\varepsilon/K = 178$ K, and $\sigma_{ClCl} = 3.40$ Å.[30] (By permission of *Ber. Bunsenges. Physik. Chem.*)

This principle characterizes the leading idea of the IMIS concept. Additional evidence for its validity has been presented in a slightly different context,[185,186] and we expect that it holds for most, if not all, nonassociated molecular fluids that are reasonably described by interaction site pair potential models.

The accurate evaluation of expression (2.47) is substantially simplified by the recent development of powerful numerical integration techniques.[187,188] It is therefore tempting to search for a formal framework that makes use of the aforementioned properties of low-density site–site pair correlation functions in the course of approximately predicting equilibrium structure and thermodynamics of molecular liquids and compressed gases.

To proceed in this direction, we shall start with a definition of the temperature- and density-dependent average force potential:

$$\bar{v}_{\alpha\gamma}(r_{\alpha\gamma}) = -kT \ln g_{\alpha\gamma}(r_{\alpha\gamma}) \qquad (2.48)$$

which bears a certain resemblance to the mean force potential in simple liquids.[189] For the limiting case $\rho \to 0$ we find from Eq. (2.47) that

$$\bar{v}_{\alpha\gamma}\left(r^{12}_{\alpha\gamma}; \rho \to 0\right) = -2kT \ln W - kT \ln \int d\mathbf{r}_1 \, d\Omega_1 \int d\mathbf{r}_2 \, d\Omega_2$$

$$\cdot \exp\left\{ -\sum_{\eta\nu} \frac{v_{\eta\nu}\left(r^{12}_{\eta\nu}\right)}{kT} \right\} \delta\left(\mathbf{r}^1_\alpha\right) \delta\left(\mathbf{r}^2_\gamma - \mathbf{r}^{12}_{\alpha\gamma}\right) \quad (2.49)$$

Note that for simple liquids the mean force potential becomes the pair potential as $\rho \to 0$:

$$\bar{v}_{12}\left(r_{12}; \rho \to 0\right) = v_{12}\left(r_{12}\right) \quad (2.50)$$

On comparing Eqs. (2.49) and (2.50) it becomes obvious that $\bar{v}_{\alpha\gamma}(r_{\alpha\gamma}; \rho \to 0)$ has the meaning of an effective site–site pair potential. Equation (2.49) may thus be regarded as the implementation of Boltzmann-averaged potential functions at the site–site level.

The characterization of intermolecular interactions by a set of effective site–site potentials should be advantageous from several points of view. Each $\bar{v}_{\alpha\gamma}(r_{\alpha\gamma}; \rho \to 0)$ contains specifically coded information about *all* site–site interactions between a pair of molecules. Furthermore, each $\bar{v}_{\alpha\gamma}(r_{\alpha\gamma}; \rho \to 0)$ incorporates *all* essential information about the intramolecular correlations (chemical bonds) of the molecules. We can thus avoid the introduction of special intramolecular correlation functions, thereby considerably simplifying the statistical mechanical formalism. These aspects are once again elucidated in Fig. 6, where $\bar{v}_{ClCl}(r_{ClCl}; \rho \to 0)$ for the already mentioned model of CCl_4 is plotted together with the pure Lennard-Jones function $v_{ClCl}(r_{ClCl})$. A comparison of both curves gives us some insight into the amount of information stored in the effective pair potential.

The definition of the effective site–site pair potential $\bar{v}_{\alpha\gamma}(r_{\alpha\gamma}; \rho \to 0)$ now enables us to derive a simple formalism in which molecular fluids are viewed as mixtures of formally independent molecular interaction sites (the IMIS model). Equation (2.49) is the central relation of this approach. The inclusion of intramolecular correlations into the effective pair potentials allows us to calculate site–site pair distribution functions by the same well-known methods[190-193] that are used for atomic liquids and gases. The IMIS–Ornstein–Zernike equation for an interaction site system then reads

$$h_{\alpha\gamma}\left(r_{\alpha\gamma}\right) = c_{\alpha\gamma}\left(r_{\alpha\gamma}\right) + \rho \sum_\eta \int c_{\alpha\eta}\left(r_{\alpha\eta}\right) h_{\eta\gamma}\left(r_{\eta\gamma}\right) d\mathbf{r}_\eta \quad (2.51)$$

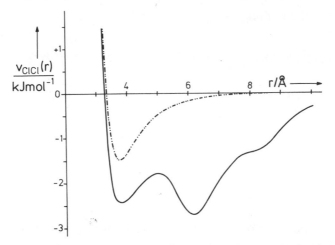

Fig. 6.   Pair potential between two chlorine sites in CCl_4: — effective site–site potential $\bar{v}_{ClCl}(r_{ClCl}; \rho \to 0)$; $-\cdot\cdot-$ Lennard–Jones potential $v_{ClCl}(r_{ClCl})$. (By permission of *Ber. Bunsenges. Physik. Chem.*)

or in Fourier representation

$$\hat{h}_{\alpha\gamma}(k) = \hat{c}_{\alpha\gamma}(k) + \rho \sum_{\eta} \hat{c}_{\alpha\eta}(k)\hat{h}_{\eta\gamma}(k) \qquad (2.52)$$

Equivalent to Eq. (2.52) is the matrix representation

$$\hat{\mathbf{h}}(k) = \hat{\mathbf{c}}(k)[\mathbf{1} - \rho\hat{\mathbf{c}}(k)]^{-1} \qquad (2.53)$$

where $\mathbf{1}$ denotes the identity matrix.

From Eq. (2.51) alone $h_{\alpha\gamma}(r_{\alpha\gamma})$ and $c_{\alpha\gamma}(r_{\alpha\gamma})$ are not yet completely determined. Similar as for simple liquids, the Percus–Yevick equation

$$c_{\alpha\gamma}(r_{\alpha\gamma}) = \bar{y}_{\alpha\gamma}(r_{\alpha\gamma})\bar{f}_{\alpha\gamma}(r_{\alpha\gamma}) \qquad (2.54)$$

(for short-range potentials) or the Hypernetted-Chain equation

$$c_{\alpha\gamma}(r_{\alpha\gamma}) = \bar{y}_{\alpha\gamma}(r_{\alpha\gamma})\left[\bar{f}_{\alpha\gamma}(r_{\alpha\gamma}) + 1\right] - \ln \bar{y}_{\alpha\gamma}(r_{\alpha\gamma}) - 1 \qquad (2.55)$$

(for long-range interactions) provides useful closure relations to Eq.

(2.51).[40,42] The functions

$$\bar{y}_{\alpha\gamma}(r_{\alpha\gamma}) = g_{\alpha\gamma}(r_{\alpha\gamma}) \exp\left\{ \frac{\bar{v}_{\alpha\gamma}(r_{\alpha\gamma}; \rho \to 0)}{kT} \right\} \qquad (2.56)$$

and

$$\bar{f}_{\alpha\gamma}(r_{\alpha\gamma}) = \exp\left\{ \frac{-\bar{v}_{\alpha\gamma}(r_{\alpha\gamma}; \rho \to 0)}{kT} \right\} - 1$$

$$= g_{\alpha\gamma}(r_{\alpha\gamma}; \rho \to 0) - 1 \qquad (2.57)$$

have been introduced here for numerical convenience. For hard interaction site systems the "corrected" Hypernetted-Chain equation appears to be a worthwhile alternative.[132] This special closure relation actually focuses on the difference between $g_{\alpha\gamma}(r_{\alpha\gamma})$ and an accurately known hard-sphere reference distribution function.

We now have a complete formalism for the calculation of site–site pair correlation functions $g_{\alpha\gamma}(r_{\alpha\gamma})$ at our disposal that provides a theoretical access to the structural and thermodynamic properties of molecular fluids. In order to avoid possible confusion, it should be noted, however, that the direct correlation function $c_{\alpha\gamma}$ as used in the IMIS model is not identical with $c_{\alpha\gamma}$ as obtained from the RISM theory. For low-density systems a comparison of Eqs. (2.24) and (2.53) suggests that

$$\hat{c}_{\alpha\gamma}(k)_{\text{IMIS}} \triangleq \left[\hat{\omega}(k)\hat{c}(k)_{\text{RISM}}\hat{\omega}(k) \right]_{\alpha\gamma} \qquad (2.58)$$

but due to the different theoretical foundations of both methods Eq. (2.58) does not necessarily imply numerical equivalence. At high densities Eq. (2.58) is clearly invalid, except for almost spherical particles.

For the special case of homonuclear molecules the IMIS approach is formally related to the RAM theory. To show this, we will consider the site-centered orientation-dependent molecular pair correlation function $g_{12}(\mathbf{r}_{\alpha\alpha}^{12}, \Omega_1, \Omega_2)$. Note that the vector $\mathbf{r}_{\alpha\alpha}^{12}$ connects the α sites of molecules 1 and 2 and not their centers of gravity. By integrating over Ω_1 and Ω_2 one obtains the site–site distribution function $g_{\alpha\alpha}(r_{\alpha\alpha})$:

$$g_{\alpha\alpha}(r_{\alpha\alpha}) = W^{-2} \int d\Omega_1 \int d\Omega_2 \, g_{12}(r_{\alpha\alpha}^{12}, \Omega_1, \Omega_2) \qquad (2.59)$$

To derive a perturbation theory similar in spirit to the RAM theory, we may separate the Mayer function

$$f_{12}(r_{\alpha\alpha}^{12}, \Omega_1, \Omega_2) = \exp\left\{ \frac{-v_{12}(r_{\alpha\alpha}^{12}, \Omega_1, \Omega_2)}{kT} \right\} - 1 \qquad (2.60)$$

into a spherically symmetric reference part and a nonspherical perturbation:

$$f_{12}(r_{\alpha\alpha}^{12}, \Omega_1, \Omega_2, \lambda) = f_{12}^0(r_{\alpha\alpha}^{12}) + \lambda\{f_{12}(r_{\alpha\alpha}^{12}, \Omega_1, \Omega_2) - f_{12}^0(r_{\alpha\alpha}^{12})\} \quad (2.61)$$

With the particular choice

$$f_{12}^0(r_{\alpha\alpha}^{12}) = \bar{f}_{\alpha\alpha}(r_{\alpha\alpha}^{12}) \tag{2.62}$$

the reference system becomes identical to the IMIS fluid.

Expanding the pair distribution function $g_{12}(r_{\alpha\alpha}^{12}, \Omega_1, \Omega_2)$ about $g_{\alpha\alpha}(r_{\alpha\alpha})_{\text{IMIS}}$ and setting λ to unity, we find (cf. Eq. (2.9))

$$g_{12}(r_{\alpha\alpha}^{12}, \Omega_1, \Omega_2) = g_{\alpha\alpha}(r_{\alpha\alpha}^{12})_{\text{IMIS}}[1 + s'_\lambda(r_{\alpha\alpha}^{12}, \Omega_1, \Omega_2)] + \rho W^{-1}$$
$$\times \int g_{\alpha\alpha\alpha}(r_{\alpha\alpha}^{12}, r_{\alpha\alpha}^{13}, r_{\alpha\alpha}^{23})_{\text{IMIS}}\langle s'_\lambda(r_{\alpha\alpha}^{13}, \Omega_1, \Omega_3) + s'_\lambda(r_{\alpha\alpha}^{23}, \Omega_2, \Omega_3)\rangle_{\Omega_3} d\mathbf{r}_\alpha^3$$

$$(2.63)$$

where

$$s'_\lambda(r_{\alpha\alpha}^{12}, \Omega_1, \Omega_2) = \exp\left\{\frac{-[v_{12}(r_{\alpha\alpha}^{12}, \Omega_1, \Omega_2) - \bar{v}_{\alpha\alpha}(r_{\alpha\alpha}^{12}; \rho \to 0)]}{kT}\right\} - 1$$

$$(2.64)$$

and $g_{\alpha\alpha\alpha}(r_{\alpha\alpha}^{12}, r_{\alpha\alpha}^{13}, r_{\alpha\alpha}^{23})_{\text{IMIS}}$ denotes the IMIS reference site–site–site triplet distribution function.

Applying Eq. (2.63) reestablishes a molecular point of view. As long as we are only interested in the spherically symmetric site–site distribution function $g_{\alpha\alpha}(r_{\alpha\alpha})$, we may average expression (2.63) over Ω_1 and Ω_2. It is easy to verify that the first-order correction term will vanish under these circumstances. A straightforward calculation reveals that the same is also true for the general case of heteronuclear interaction site systems.

The formal equivalence between the IMIS model and the RAM approach breaks down, however, if molecules with different types of interaction sites are considered. Please note, that within the framework of the IMIS formalism *all* effective site–site pair potentials $\bar{v}_{\alpha\gamma}(r_{\alpha\gamma}; \rho \to 0)$ are used to establish the reference pair correlation function $g_{\alpha\gamma}(r_{\alpha\gamma})_{\text{IMIS}}$ for a given pair of interaction sites. In fact, it is even possible to introduce even more information about molecular shape into the reference system through the definition of "auxiliary" sites (in analogy to the treatment of Hsu et al.[11]). The same technique could be used, of course, in order to evaluate pair correlation functions for molecular sites that are not interaction sites. The RAM theory, on the other hand, is always based on only one effective pair potential.

2. The IMIS Reference Fluid for Homonuclear Hard Dumbbells

Homonuclear hard-dumbbell particles denote the next complicated class of interaction site molecules beyond hard spheres, to which they degenerate in the case of vanishing bond length $L \rightarrow 0$. In analogy to the hard sphere case, the effective site–site potential $\bar{v}_{ss}(r_{ss}; \rho \rightarrow 0)$ becomes infinitely repulsive as r_{ss} decreases below the diameter σ of the fused spherical groups. However, owing to the configurational selection process defined by Eq. (2.49), an additional softly repulsive tail emerges in the region $\sigma \leq r_{ss} \leq 2L + \sigma$. These facts are illustrated in Fig. 7, where the (temperature-independent) reduced potential function $\bar{v}_{ss}(r_{ss}; \rho \rightarrow 0)/kT$ is shown for the system $L^* = L/\sigma = 0.3$. The appearance of a cusp at $r_{ss} = L + \sigma$ is closely related to the well-known cusps in the interaction site distribution functions of hard-dumbbell fluids. The significance of this phenomenon has been demonstrated by geometric arguments,[120] as well as by a rigorous proof based on the interaction site cluster expansion of Ladanyi and Chandler.[109] Therefore, a reliable theory of molecular liquids is expected to reproduce these cusps correctly.

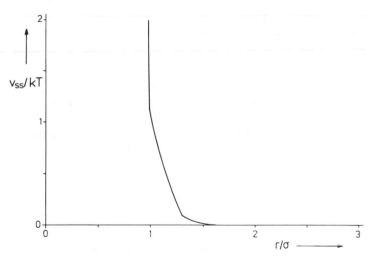

Fig. 7. The reduced effective site–site reference pair potential $\bar{v}_{ss}(r_{ss}; \rho \rightarrow 0)/kT$ for hard dumbbells with reduced bond length $L^* = 0.3$. (By permission of *Ber. Bunsenges. Physik. Chem.*)

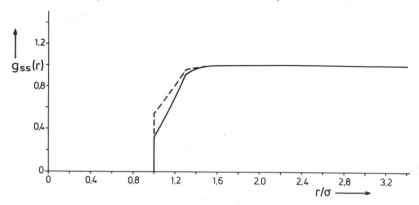

Fig. 8. Zero density site–site distribution function $g_{ss}(r_{ss};\ \rho^* \to 0)$ for hard dumbbells ($L^* = 0.3$): — IMIS model; --- RISM theory. (By permission of *Ber. Bunsenges. Physik. Chem.*)

By virtue of its theoretical foundations the IMIS model yields *exact* zero density site–site distribution functions. Thus, it is perhaps quite instructive to compare the results from the IMIS model and the predictions of the RISM theory in the low-density limit.

From Fig. 8, which illustrates the special case $L^* = 0.3$, it turns out that there are remarkable discrepancies between both methods for values of r_{ss} at or close to contact. The RISM theory obviously underestimates the strong orientational correlations that appear at small interparticle separations, thereby accounting for a lot of physically impossible configurations. As a result, the predicted contact value $g_{ss}(\sigma)$ is about 40% too high. Clearly, the RISM theory cannot be recommended for the treatment of low-density problems. The IMIS model, on the other hand, is perfectly suited for that purpose.

At moderate and high densities the fluid structure of homonuclear hard dumbbells of different elongations has been subject to fairly extensive computer simulation studies.[116,119,120,194] Hard-dumbbell fluids thus represent particularly suitable test objects for a comprehensive investigation of the performance of the IMIS approach. The most direct and unambiguous route to the structure of the IMIS reference fluid is provided by the Monte Carlo technique.[195,196] Simulations covering a range of reduced bond lengths 0.2 $\leq L^* \leq 0.6$ and of reduced densities 0.2 $\leq \rho^* \leq 0.6559$ are available from the work of Nezbeda and Smith[*131,133] and of Naumann et al.,[197] as shown in the accompanying tabulation.

---

*Actually, the approach considered by Nezbeda and Smith is a generalized RAM theory. For homonuclear molecules, however, the IMIS model and the RAM theory are equivalent to each other, as has been shown in the preceding subsection.

IMIS Reference Fluids Considered in Section II.B.2

L/σ	$\rho\sigma^3$	Reference
0.2	0.463	197
0.2	0.6019	197
0.2	0.6559	197
0.4	0.4975	133,197
0.4	0.574	197
0.5	0.2	131
0.5	0.4	131
0.5	0.5	131
0.6	0.3	197
0.6	0.5022	133,197

In Figs. 9–18 the reference site–site distribution functions as predicted from the IMIS model are compared with the respective computer simulation

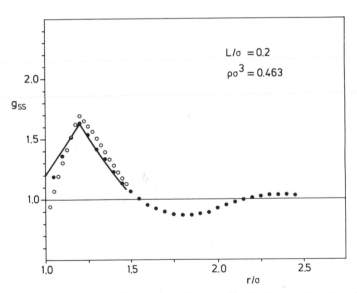

Fig. 9. Site–site distribution function g_{ss} for hard dumbbells as a function of reduced site separation distance r/σ. Open circles are computer simulation results of Streett and Tildesley,[116,120] solid circles are the simulation results for the IMIS reference fluid, and the full line is the result of the RISM theory.

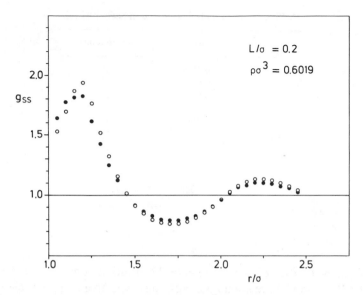

Fig. 10.   See the caption to Fig. 9.

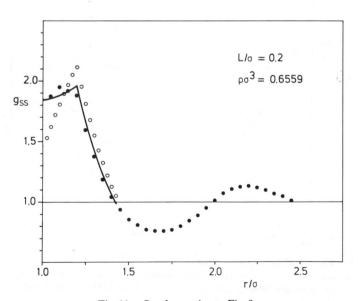

Fig. 11.   See the caption to Fig. 9.

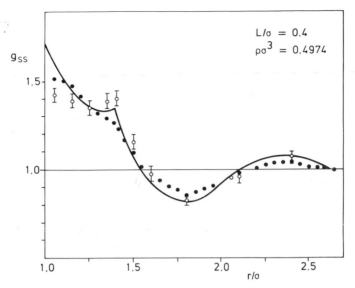

Fig. 12. See the caption to Fig. 9. (By permission of *Mol. Phys.*)

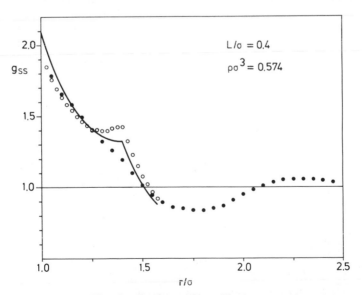

Fig. 13. See the caption to Fig. 9.

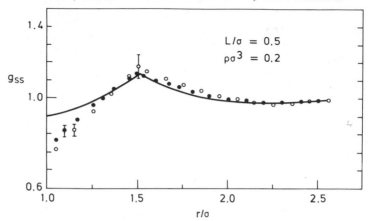

Fig. 14.   See the caption to Fig. 9—, except that here open circles are computer simulation results of Morriss.[194] (By permission of *Chem. Phys. Lett.*)

data for the complete pair potential and with results obtained from the RISM theory. The IMIS results are generally quite satisfactory and overall much better than those of the RISM theory. As expected, the IMIS model works especially well at low and moderate densities, where the RISM approach performs rather poorly.

From a qualitative point of view, the IMIS model succeeds in reproducing all the characteristical features of homonuclear hard-dumbbell site–site

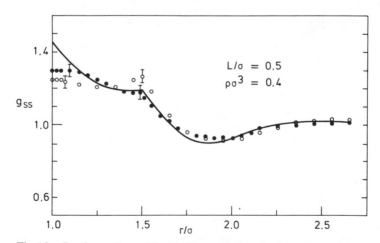

Fig. 15.   See the caption to Fig. 14. (By permission of *Chem. Phys. Lett.*)

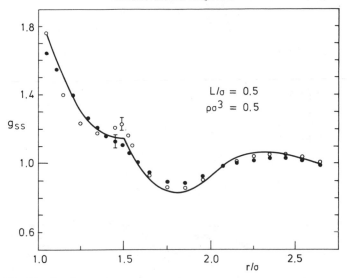

Fig. 16. See the caption to Fig. 14. (By permission of *Chem. Phys. Lett.*)

correlation functions, including the typical slope discontinuities at $r_{ss} = L + \sigma$. This result substantiates the correctness of our basic assumption, namely, that the structure of the effective force field between molecular sites is only weakly dependent on density. Except for thermodynamic states close to the liquid–solid phase transition, this density dependence appears to be almost negligible. Boltzmann-averaged effective site–site pair potentials are

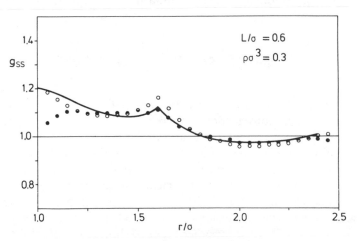

Fig. 17. See the caption to Fig. 9.

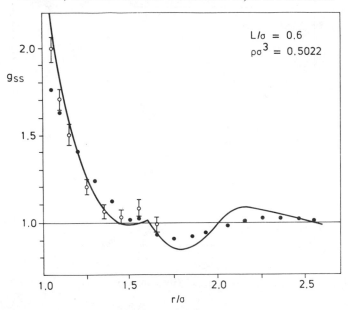

Fig. 18.   See the caption to Fig. 9. (By permission of *Mol. Phys.*)

thus expected to provide a qualitatively accurate model for the actual effective potential acting between the sites of different molecules.

3.   Multicenter Lennard-Jones Molecules

A set of appropriately chosen Lennard-Jones-type site–site potential functions

$$v_{\alpha\gamma}^{LJ}(r_{\alpha\gamma}) = 4\varepsilon_{\alpha\gamma}\left\{\left(\frac{\sigma_{\alpha\gamma}}{r_{\alpha\gamma}}\right)^{12} - \left(\frac{\sigma_{\alpha\gamma}}{r_{\alpha\gamma}}\right)^{6}\right\} \qquad (2.65)$$

where $\varepsilon_{\alpha\gamma}$ and $\sigma_{\alpha\gamma}$ are energy and distance parameters, respectively, provides an analytically convenient and thereby fairly realistic description of the interactions between nonpolar polyatomic molecules.[152,153,186,198,199] It appears to be worthwhile, therefore, to investigate this popular model within the framework of the IMIS formalism. From the basic considerations already presented in Section II.B.1, it is clear that this may be done in a completely straightforward manner without the detour via hard interaction site reference systems. The conclusions to be drawn from this investigation are expected to be equally valid for other types of short-range attractive site–site potentials.

In their molecular dynamics study of homonuclear Lennard-Jones diatomics, Singer et al. considered a model of liquid chlorine near its triple point[186] ($L/\sigma = 0.63$; $\rho\sigma^3 = 0.539$; $kT/\varepsilon = 1.00$). In Fig. 19 we show the site–site distribution $g_{ClCl}(r_{ClCl})$ as obtained from the IMIS model in the Percus–Yevick approximation [see Eq. (2.54)] for this particular system together with the correlation function reported by Singer et al. The agreement between both curves is quite satisfactory except in the vicinity of the second peak, which appears to be partially suppressed by the IMIS model. A similar observation has been made by Nezbeda and Smith, who considered the IMIS reference fluid for a liquid of Lennard-Jones diatomics characterized by $L/\sigma = 0.5471$, $\rho\sigma^3 = 0.5$, and $kT/\varepsilon = 0.59$, utilizing the Monte Carlo technique.[133] Their result should be compared with the distribution function obtained by Streett and Tildesley from a molecular dynamics investigation of the complete system[185] (see Fig. 20). Again there is good agreement except for the region around $r/\sigma = 1.55$, where the shoulder in g_{ss} is slightly smoothed by the IMIS model. Comparing the results displayed in Figs. 19 and 20, we should also note that the Percus–Yevick equation represents an accurate closure relation to the IMIS–Ornstein–Zernike equation (2.51).

A rather demanding test for perturbation theoretical approaches starting from spherically symmetric reference potential functions is provided by the six-center Lennard-Jones model of benzene as proposed by Evans and

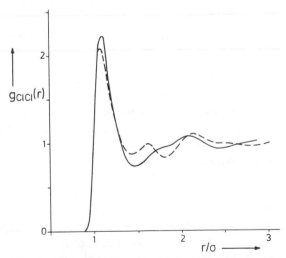

Fig. 19.   Site–site distribution function $g_{ClCl}(r)$ for a Lennard–Jones diatomic model of liquid chlorine: $L/\sigma = 0.63$; $\rho\sigma^3 = 0.539$; $kT/\varepsilon = 1.00$; --- computer simulation results of Singer et al.; — IMIS model in the Percus–Yevick approximation.

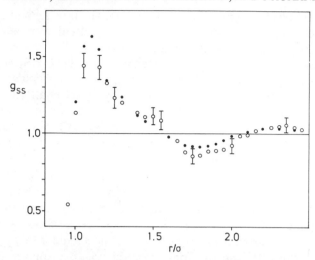

Fig. 20.   Site–site distribution function g_{ss} for homonuclear Lennard–Jones diatomic: $L/\sigma = 0.5471$; $\rho\sigma^3 = 0.5$; $kT/\varepsilon = 0.59$. Open circles are simulation results of Streett and Tildesley,[185] and full circles are the simulation results for the IMIS reference fluid.[133] (By permission of *Mol. Phys.*)

Watts.[198] It consists of six interaction sites placed on the vertices of a regular hexagon of radius $R_b = 1.756$ Å. The Lennard-Jones parameters for this model are $\varepsilon/k = 77$ K and $\sigma = 3.5$ Å. It is perhaps instructive to consider the effective site–site potential $\bar{v}_{ss}(r_{ss}; \rho \to 0)$, which characterizes the IMIS reference fluid for this particular case. Note that the potential function displayed in Fig. 21 incorporates a weighted sum over 36 intermolecular site–site interactions. The thermodynamic state chosen by Evans and Watts in their Monte Carlo study of liquid benzene was $\rho = 0.00633$ Å^{-3}, $T = 328$ K. The computer-simulated site–site distribution function for this model fluid is shown in Fig. 22 together with the one predicted from the IMIS model in the Percus–Yevick approximation. Considering the relative complexity of the intermolecular interactions involved, the agreement is indeed very satisfactory.

The foregoing examples clearly indicate that the IMIS approach opens a reliable direct route to the structural properties of interaction site model fluids with short-range attractive site–site potentials even at high densities. From a theoretical point of view, the good overall agreement between pseudoexperimental and IMIS reference distribution functions provides a stringent proof for the general validity of the underlying ideas of the IMIS concept.

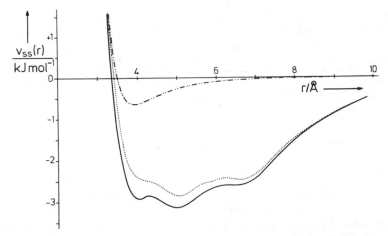

Fig. 21. Pair potential between two sites for six-center Lennard–Jones model of benzene: — effective site–site potential $\bar{v}_{ss}(r_{ss}; \ \rho \to 0)$, $T = 298$ K; $\cdots\cdot$ effective site–site potential $\bar{v}_{ss}(r_{ss}; \ \rho \to 0)$, $T = 328$ K; $-\cdot\cdot-$ Lennard–Jones potential $v_{ss}(r_{ss})$.

We would like to emphasize, however, that for practical applications in connection with multicenter Lennard-Jones or related types of interaction site model fluids the straightforward evaluation of the site–site pair correlation functions $g_{\alpha\gamma}(r_{\alpha\gamma})$ does not necessarily represent the most convenient method. As is well known, the accurate numerical solution of the Percus–Yevick equation for high-density states of attractively interacting

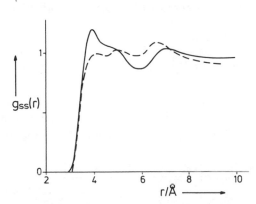

Fig. 22. Site–site distribution function $g_{ss}(r)$ for six-center Lennard–Jones model of benzene: $T = 328$ K; $\rho = 0.00633$ Å^{-3}; --- computer simulation results of Evans and Watts; — IMIS model in the Percus–Yevick approximation.

particles often proves to be a nontrivial task. For purely repulsive systems such as those considered in the preceding subsection, on the other hand, the Percus–Yevick equation or preferably the corrected hypernetted-chain equation are readily solved.[31,130,132] Provided the particle density is high, it will thus be much easier to determine the structural properties of an appropriately chosen hard-core reference system and to account for the attractive contributions by means of perturbation theory. At low and moderate densities, however, the hard-body expansion usually performs rather poorly. For the treatment of such systems, the direct application of the IMIS formalism as described above is clearly preferable.

In Sections II.A and II.B the RISM theory and the IMIS model have been presented as alternative routes toward analyzing and predicting the equilibrium interaction site pair correlations in molecular fluids. We would like to emphasize, however, that these approaches are not antithetical by nature. In fact, it has been shown recently that both methods may be combined, resulting in a highly efficient though semiempirical formalism for calculating site–site pair distribution functions.

In the low density limit $\rho \to 0$ the RISM total site–site pair correlation functions are given by

$$\hat{h}_{\alpha\gamma}^{\mathrm{RISM}}(k) = [\hat{\omega}(k)\hat{c}(k)\hat{\omega}(k)]_{\alpha\gamma} \qquad (2.26)$$

while from the IMIS model we have

$$h_{\alpha\gamma}^{\mathrm{IMIS}}(r) = \bar{f}_{\alpha\gamma}(r) = \exp\left\{\frac{-\bar{v}_{\alpha\gamma}(r; \rho \to 0)}{kT}\right\} - 1$$

If the RISM theory were accurate at low densities, $h_{\alpha\gamma}^{\mathrm{RISM}}$ and $h_{\alpha\gamma}^{\mathrm{IMIS}}$ should be equivalent:

$$[\hat{\omega}(k)\hat{c}(k)\hat{\omega}(k)]_{\alpha\gamma} = \int \tilde{f}_{\alpha\gamma}(r)\exp(i\mathbf{k}\cdot\mathbf{r})\,d\mathbf{r} \qquad (2.58a)$$

We may, however, force the RISM result to become exact as $\rho \to 0$ by solving Eq. (2.58a) for $\hat{c}_{\alpha\gamma}(k)$ and eventually for $c_{\alpha\gamma}(r)$. Finally, from the Percus–Yevick-like closure [Eq. (2.31) and Eq. (2.32)] together with

$$\tilde{\bar{f}}_{\alpha\gamma}(r) = \exp\left\{\frac{-\bar{\bar{v}}_{\alpha\gamma}(r; \rho \to 0)}{kT}\right\} - 1 \qquad (2.33a)$$

a new type of effective site–site pair potential functions $\bar{\bar{v}}_{\alpha\gamma}(r; \rho \to 0)$ may be evaluated. Substituting the "true" site–site potentials $v_{\alpha\gamma}(r)$ by

$\bar{\bar{v}}_{\alpha\gamma}(r; \rho \to 0)$ within the framework of the RISM formalism considerably improves the accuracy of the predicted correlation functions $g_{\alpha\gamma}(r)$ even at highest densities and for large particle anisotropies. Treating highly compressed hard-core systems, $\bar{\bar{v}}_{\alpha\gamma}(r; \rho \to 0)$ is profitably combined with the MSA-like closure

$$h_{\alpha\gamma}(r) = -1; \qquad r < \sigma_{\alpha\gamma} \tag{2.35}$$

$$c_{\alpha\gamma}(r) = \frac{-\bar{\bar{v}}_{\alpha\gamma}(r; \rho \to 0)}{kT}; \qquad r \geq \sigma_{\alpha\gamma} \tag{2.36a}$$

C. The Hard Convex Body Expansion Conformal Solution Theory (HCBE–CST)

In the preceding section we have shown by means of the independent molecular interaction sites model that it can be profitable to expand the properties of a pure fluid with respect to those of an appropriately chosen reference *mixture*. A second example of this type will be presented in connection with the treatment of conformational equilibria of flexible molecules in Section II.E, below.

In contrast to these recent developments, the reverse procedure, that is, the expansion of the properties of a mixture with respect to those of a pure reference fluid, has been known and also applied for a long time.[12,200,201] Theoretical approaches emerging from this particular background are usually classified as "one-fluid conformal solution theories" (one-fluid CST). The so called van der Waals one-fluid conformal solution theory (vdW1–CST), for example, belongs to the standard tools of thermodynamic calculations within chemical engineering applications.[202–204]

In order to apply CST to mixtures of *spherical* molecules, we must only require that all pair interactions within the system of interest are of the common form

$$v_{ik}(r_{ik}) = \varepsilon_{ik} f\left(\frac{r_{ik}}{\sigma_{ik}}\right) \tag{2.66}$$

where ε_{ik} and σ_{ik} have their usual meanings as energy and distance parameters, respectively, and $f(r_{ik}/\sigma_{ik})$ represents an universal function of the reduced particle separation r_{ik}/σ_{ik}. For example, Eq. (2.66) is trivially satisfied for a mixture of Lennard-Jones-type molecules. However, in contrast to other perturbation theoretical approaches such as the well-known Barker–Henderson theory,[14,205] it is not necessary to specify the precise functional dependence of $f(r_{ik}/\sigma_{ik})$ on r_{ik}.

Within the framework of the vdW1–CST, the pair distribution function for a particular pair $i-k$ in a mixture of composition $(x_1, x_2, \ldots, x_n)$ is given by

$$g_{ik}(r_{ik}, \rho, T, \varepsilon_{11}, \varepsilon_{12}, \ldots, \sigma_{11}, \sigma_{12}, \ldots, x_1, x_2, \ldots)$$

$$= g_x\left(\frac{r_{ik}}{\sigma_{ik}}, \frac{\varepsilon_x}{kT}, \rho\sigma_x^3\right) + \delta g_{ik} + \cdots \qquad (2.67)$$

To zeroth order, all pair distribution functions are thus assumed to be equal on a reduced length scale. Individual distribution functions would be regenerated by evaluating the first-order corrections δg_{ik}, but these terms as well as all higher-order ones are usually neglected. The pseudoparameters ε_x and σ_x are given by

$$\varepsilon_x = \frac{\sum\limits_{ik} x_i x_k \varepsilon_{ik} \sigma_{ik}^3}{\sigma_x^3} \qquad (2.68)$$

$$\sigma_x^3 = \sum_{ik} x_i x_k \sigma_{ik}^3 \qquad (2.69)$$

Thermodynamic excess properties of fluid mixtures are readily evaluated from the vdW1–CST with very little numerical effort. If it is compared to the Barker–Henderson theory, however, the vdW1–CST is generally slightly less accurate.

A certain drawback of "traditional" conformal solution approaches such as vdW1 results from the inadequate representation of the structural features of fluid mixtures by means of the truncated Eq. (2.67) or related expressions. Note that Eq. (2.67) does not even hold in the low-density limit. The Barker–Henderson theory, on the other hand, explicitly accounts for the differences between macroscopic and microscopic (i.e., local) composition by starting from an appropriately chosen hard-sphere reference mixture. Thus CST and Barker–Henderson theory might appear to represent basically antithetical approaches. Fortunately this is not the case.

In an attempt to improve the conformal solution concept without sacrificing its numerical simplicity, Mansoori and Leland devised an elegant scheme that bridges the gap between hard-sphere expansion perturbation theory and CST, thereby combining the advantages of both methods.[206] Similar in spirit to the work of Barker and Henderson, the aforementioned scheme started from a perturbation expansion with respect to a reference mixture of hard spheres. However, instead of evaluating all pair distribution functions to zeroth or even to first order, Mansoori and Leland introduced a special assumption concerning the form of the $g_{ik}(r_{ik})$, the so-called mean density approximation:

$$g_{ik}(r_{ik}, \rho, T, \varepsilon_{11}, \varepsilon_{12}, \ldots, \sigma_{11}, \sigma_{12}, \ldots, x_1, x_2, \ldots)$$

$$= g_x\left(\frac{r_{ik}}{\sigma_{ik}}, \frac{\varepsilon_{ik}}{kT}, \rho\sigma_x^3\right) \tag{2.70}$$

Equation (2.70) is exact in the-low density limit and generally much more accurate than Eq. (2.67). If compared to the "exact" computer simulation results for the system argon/krypton, the mean deviation caused by Eq. (2.70) only amounts to about 5%.[207] From a formal point of view, Eq. (2.70) represents the implementation of the conformal solution concept within the framework of the hard-sphere expansion perturbation theory. The resulting approach has thus been named the hard-sphere expansion conformal solution theory (HSE–CST) by Mansoori and Leland.

To second order in $1/kT$, the Helmholtz free energy A_M as predicted from the HSE–CST for a mixture of spherical particles is given by

$$A_M = A_M^{HS} + A_x - A_x^{HS} \tag{2.71}$$

where A_M^{HS} and A_x^{HS} denote the free energy of the hard-sphere reference mixture and the repulsion contribution to the free energy A_x of the pure conformal reference fluid, respectively. Both quantities are readily evaluated from accurate analytical expressions derived by Mansoori et al.[208] and by Carnahan and Starling.[209] Note that only the contribution of attractive forces to the free energy, $A_x - A_x^{HS}$, is actually computed by means of CST. Furthermore, the mean density approximation Eq. (2.70) will also be used only for that particular purpose. The conformal reference fluid is now characterized by the pseudoparameters

$$\varepsilon_x = \frac{\displaystyle\sum_{ik} x_i x_k \varepsilon_{ik}^2 \sigma_{ik}^3}{\displaystyle\sum_{ik} x_i x_k \varepsilon_{ik} \sigma_{ik}^3} \tag{2.72}$$

$$\sigma_x^3 = \frac{\left(\displaystyle\sum_{ik} x_i x_k \varepsilon_{ik} \sigma_{ik}^3\right)^2}{\displaystyle\sum_{ik} x_i x_k \varepsilon_{ik}^2 \sigma_{ik}^3} \tag{2.72}$$

Calculations on Lennard-Jones mixtures suggest that the simple HSE–CST is at least equivalent to the Barker–Henderson theory as long as the interaction parameters ε_{ik} and σ_{ik} are of comparable magnitude for different pairs of molecules.[210]

So far we have only been concerned with mixtures of spherical particles. In general, however, intermolecular interactions are not adequately represented by central potentials. A much more realistic description is provided, for example, by Kihara-type potential models,

$$v_{ik}(r_{ik}, \Omega_i, \Omega_k) = v_{ik}(l_{ik}) \tag{2.39}$$

which have already been briefly considered in Section II.A. Recall that $l_{ik} = l_{ik}(r_{ik}, \Omega_i, \Omega_k)$ denotes the shortest distance between the impenetrable molecular cores for a given configuration $(r_{ik}, \Omega_i, \Omega_k)$. For the sake of generality we will make only one confining assumption concerning the functional dependence of v_{ik} on l_{ik}:

$$v_{ik}(l_{ik}) = \varepsilon_{ik} f'\left(\frac{l_{ik}}{\sigma_{ik}}\right) \tag{2.73}$$

where $f'(l_{ik}/\sigma_{ik})$ denotes a universal function of the reduced distance l_{ik}/σ_{ik} [see Eq. (2.66)].

Following the basic ideas of thermodynamic perturbation theory, we may separate the pair potential (2.73) into a repulsive and an attractive part:

$$v_{ik}(l_{ik}) = v_{ik,0}(l_{ik}) + v_{ik,1}(l_{ik}) \tag{2.74}$$

The repulsive terms $v_{ik,0}$ now define a reference mixture consisting of hard convex bodies. The perturbation expansion for the Helmholtz free energy thus reads

$$A_M = A_M^{\text{HCB}} + A_1 + A_2 + \cdots \tag{2.75}$$

The free energy A_M^{HCB} of the reference system is readily obtained from an accurate analytical equation of state for mixtures of hard convex molecules. A detailed discussion of this topic will be postponed, however, to Section II.D.

The first- and second-order terms in the expansion (2.75) are of the form

$$A_1 = \tfrac{1}{2} N\rho \sum_{ik} x_i x_k \int g_{ik,0}^{\text{av}}(l_{ik}) v_{ik,1}(l_{ik}) S_{i+l_{ik}+k} \, dl_{ik} \tag{2.76}$$

$$A_2 = \frac{N\rho}{4kT} \sum_{ik} x_i x_k \int g_{ik,1}^{\text{av}}(l_{ik}) v_{ik,1}(l_{ik}) S_{i+l_{ik}+k} \, dl_{ik} \tag{2.77}$$

For pairs of different Kihara-type molecules the generalized surface function $S_{i+l_{ik}+k}$ is given by

$$S_{i+l_{ik}+k} = S_i + S_k + 8\pi R_i R_k + 8\pi(R_i + R_k)l_{ik} + 4\pi l_{ik}^2 \qquad (2.78)$$

In the special case $i = k$, Eq. (2.43) of Section II.A is recovered.

In contrast to the situation encountered for hard spheres or hard interaction site molecules, accurate data on the orientationally averaged hard convex body core distribution functions $g_{ik,0}^{av}$ are usually not available except from computer simulation studies. Merely the contact value $g_{ik,0}^{av}(d_{ik})$ may be evaluated analytically.[182] For this reason, a generalized Barker–Henderson theory for mixtures of convex molecules is presently only feasible by means of semiempirical expressions for $g_{ik,0}^{av}$ such as proposed by Boublik and Lu[113] (see also Ref. 114).

In view of this problem, an additional advantage of the conformal solution approach becomes particularly important. In order to apply CST it is entirely sufficient to know how to scale the pair distribution function g_x of a pure reference fluid so as to adjust it to the respective distribution functions g_{ik} of the mixture under consideration. The precise form of the g_{ik} need not be specified, however. Furthermore, CST will release us from the chore of explicitly evaluating the extremely complicated first-order correction term $g_{ik,1}^{av}(l_{ik})$, which depends on two-, three-, and four-particle correlations in the hard convex body reference system. From the viewpoint of CST it is solely important here to remark that, by virtue of Eq. (2.73), $g_{ik,1}^{av}(l_{ik})$ may be written as

$$g_{ik,1}^{av}(l_{ik}) = \frac{\varepsilon_{ik}}{kT}\psi_{ik}(l_{ik}) \qquad (2.79)$$

The most straightforward way toward derivation of a conformal solution theory for Kihara-type molecules similar in spirit to the work of Mansoori and Leland is to assume the following generalized mean density approximation for $g_{ik}^{av}(l_{ik})$:[211]

$$g_{ik}^{av}(l_{ik}, \rho, T, \varepsilon_{11}, \varepsilon_{12}, \ldots, \sigma_{11}, \sigma_{12}, \ldots, x_1, x_2, \ldots)\frac{S_{i+l_{ik}+k}}{S_{i+d_{ik}+k}}$$

$$= g_x^{av}\left(\frac{l_{ik}}{\sigma_{ik}}, \frac{\varepsilon_{ik}}{kT}, \rho_x^*\right)\frac{S_{x+l_x+x}}{S_{x+d_x+x}} \qquad (2.80)$$

The reduced density ρ_x^* is defined as the product of the pure reference fluid particle density and the volume of the reference hard convex bodies, and d_{ik}

and d_x are interaction distances for pairs of hard convex bodies measured along the shortest distance between the molecular cores, respectively. The reduced surface functions $S_{i+l_{ik}+k}/S_{i+d_{ik}+k}$ and S_{x+l_x+x}/S_{x+d_x+x} have been introduced to account for the fact that the different molecular species in the mixture and in the pure conformal reference fluid will generally exhibit different shapes. Note that for the special case of spherical particles the original version of the mean density approximation is recovered [see Eq. (2.70)].

The next step consists of the hard convex body perturbation expansion for the pure reference fluid:

$$A_x = A_x^{HCB} + A_{x,1} + A_{x,2} + \cdots \tag{2.81}$$

The expressions for $A_{x,1}$ and $A_{x,2}$ are similar to those given in Eqs. (2.76) and (2.77), respectively. Introducing the reduced quantities

$$q_{ik} = \frac{l_{ik}}{d_{ik}} \tag{2.82}$$

and

$$S_{ik}(q_{ik}) = \frac{S_{i+l_{ik}+k}}{S_{i+d_{ik}+k}} \tag{2.83}$$

for notational convenience, we eventually arrive at the desired perturbation expansion with respect to a one-component reference system consisting of Kihara-type molecules. To second order in $1/kT$, it is given by

$$
\begin{aligned}
A_M = {} & A_M^{HCB} + A_1 + A_2 + A_x - A_x^{HCB} - A_{x,1} - A_{x,2} \\
= {} & A_M^{HCB} + A_x - A_x^{HCB} \\
& + \tfrac{1}{2} N\rho \left\{ \sum_{ik} x_i x_k \varepsilon_{ik} d_{ik} S_{i+d_{ik}+k} - \varepsilon_x d_x S_{x+d_x+x} \right\} \\
& \cdot \int g_{x,0}^{av}(q_x) f_1'(q_x) S_x(q_x) \, dq_x \\
& + \frac{N\rho}{4kT} \left\{ \sum_{ik} x_i x_k \varepsilon_{ik}^2 d_{ik} S_{i+d_{ik}+k} - \varepsilon_x^2 d_x S_{x+d_x+x} \right\} \\
& \cdot \int \psi_x(q_x) f_1'(q_x) S_x(q_x) \, dq_x
\end{aligned}
\tag{2.84}
$$

It is easy to verify that the first- and second-order perturbation terms are caused to vanish by choosing the pseudoparameters ε_x and d_x as

$$\varepsilon_x = \frac{\sum\limits_{ik} x_i x_k \varepsilon_{ik}^2 d_{ik} S_{i+d_{ik}+k}}{\sum\limits_{ik} x_i x_k \varepsilon_{ik} d_{ik} S_{i+d_{ik}+k}} \tag{2.85}$$

$$d_x = \frac{\left(\sum\limits_{ik} x_i x_k \varepsilon_{ik} d_{ik} S_{i+d_{ik}+k}\right)^2}{\left(\sum\limits_{ik} x_i x_k \varepsilon_{ik}^2 d_{ik} S_{i+d_{ik}+k}\right) S_{x+d_x+x}} \tag{2.86}$$

Though implicit in d_x, Eq. (2.86) is easily solved by means of usual root-finding procedures. The conformal reference system is thus uniquely characterized by Eqs. (2.85) and (2.86), and we have the final result

$$A_M = A_M^{\text{HCB}} + A_x - A_x^{\text{HCB}} \tag{2.87}$$

The hard convex body expansion conformal solution theory (HCBE–CST) as established by the set of equations (2.85)–(2.87) resembles the HSE–CST in that it primarily aims to evaluate the excess property $A_x - A_x^{\text{HCB}}$, which represents the contribution of the attractive forces to the Helmholtz free energy. The formal similarity between Eqs. (2.87) and (2.71) is therefore not accidental.

The major advantage of the HCBE–CST lies in the pronounced shape dependence of the new pseudoparameters ε_x and d_x defined by Eqs. (2.85) and (2.86), respectively, since there is clear experimental evidence that the mixing properties of molecular fluids are generally rather sensitive with respect to the geometric characteristics of the interacting particles.[200] Moreover, this shape dependence is accounted for in a quite obvious manner without introducing empirically adjustable parameters. The HCBE–CST thus provides a sound basis for the computation of the thermodynamic functions of fluid mixtures, which, because of its numerical simplicity, should be particularly useful in connection with chemical engineering applications.

D. An Equation of State for Hard Convex Molecules

Theoretical investigations of the thermodynamics of hard convex body fluids are interesting from several points of view. First, this information is essential in connection with perturbation theories for Kihara-type molecular liquids such as the HCBE–CST presented in the preceding section. Secondly, and probably equally as important, hard convex body systems may serve as expedient vehicles in the course of studying the complex interrelations between molecular geometry and thermodynamic properties. Last but not least, it should be noted that the detour via appropriately chosen hard convex body reference fluids often represents the only means of estimating the thermodynamic properties of assemblies of nonconvex particles such as hard interaction site molecules.

It was realized quite early from the implications of scaled particle theory[212,213] that besides the density ρ and, in the case of mixtures, the composition $(x_1, x_2, \ldots)$ there are essentially three molecular parameters which determine the hard convex body (HCB) equation of state.[214-216] For a given particle species i these are the volume $\tilde{V}_i$, the surface area $\tilde{S}_i$, and the mean curvature radius $\tilde{R}_i$, respectively. These molecular quantities are related to the respective core parameters V_i, S_i, and R_i by

$$\tilde{R}_i = R_i + \frac{d_i}{2} \tag{2.88}$$

$$\tilde{S}_i = S_i + 4\pi R_i d_i + \pi d_i^2 \tag{2.89}$$

$$\tilde{V}_i = V_i + \frac{1}{2} S_i d_i + \frac{\pi}{6} d_i^3 \tag{2.90}$$

where d_i denotes twice the shortest distance between the core and the enveloping molecular surface.

Among the first fairly successful approaches toward a reliable analytical HCB equation of state, the semiempirical expression proposed by Boublik in 1975[33] remains noteworthy. It reads

$$\frac{p}{\rho kT} = \frac{1 + y(3\alpha - 2) + y^2(3\gamma - 3\alpha + 1) - \gamma y^3}{(1 - y)^3} \tag{2.91}$$

The parameters in Eq. (2.91) are given by

$$y = \rho \langle \tilde{V} \rangle \tag{2.92}$$

$$\alpha = \frac{\langle \tilde{R} \rangle \langle \tilde{S} \rangle}{3 \langle \tilde{V} \rangle} \tag{2.93}$$

$$\gamma = \frac{\langle \tilde{R}^2 \rangle \langle \tilde{S} \rangle^2}{9 \langle \tilde{V} \rangle^2} \tag{2.94}$$

where $\langle \ \rangle$ denotes a composition-weighted average over the different molecular species within the mixture. Note that for pure fluids $\gamma = \alpha^2$. For the special case of one-component hard-sphere systems, α becomes equal to unity and Eq. (2.91) reduces to the well-known Carnahan–Starling expression[209]

$$\frac{p}{\rho kT} = \frac{1 + y + y^2 - y^3}{(1-y)^3} \tag{2.95}$$

In the light of recent computer simulation studies,[123-128,217-219] Eq. (2.91) can no longer be regarded as a satisfactory HCB equation of state. It remains nevertheless useful, however, since it appears to represent a fairly accurate semiempirical equation of state for hard interaction site molecules that are reasonably approximated by an enveloping convex body.[18,220-223] If we utilize Eq. (2.91) for this particular purpose, $\tilde{V}$ and $\tilde{S}$ should reflect the true concave geometry of the interaction site particles while $\tilde{R}$ should be evaluated for the closest convex envelope.

A further detail deserving attention concerns the striking similarity between Eqs. (2.91) and (2.95). From a formal point of view Eq. (2.91) may thus be regarded as a generalization of the Carnahan–Starling equation, with the "eccentricity parameter" α accounting for molecular anisotropy. This identification is indeed important, since it suggests a promising route toward an improved description of the thermodynamics of hard convex molecules. In much the same spirit as Carnahan and Starling succeeded in approximately resumming the virial expansion for hard spheres, we might imagine the known results to fit the HCB virial expansion[21,224-227] to the generalized Carnahan–Starling type expression:[182,211]

$$\frac{p}{\rho kT} = (1-y)^{-3}\{1 + c_1 y + c_2 y^2 + c_3 y^3 + c_4 y^4 + \cdots\} \tag{2.96}$$

Equation (2.91) suggests that the knowledge of a limited number of coefficients c_n should be sufficient in order to produce an accurate approximation to the HCB equation of state.

Expanding Eq. (2.96) yields

$$\frac{p}{\rho kT} = 1 + y(c_1 + 3) + y^2(c_2 + 3c_1 + 6) + y^3(c_3 + 3c_2 + 6c_1 + 10)$$
$$+ y^4(c_4 + 3c_3 + 6c_2 + 10c_1 + 15) + \cdots \tag{2.97}$$

which should be compared to the virial series

$$\frac{p}{\rho kT} = 1 + B_2^* y + B_3^* y^2 + B_4^* y^3 + B_5^* y^4 + \cdots \tag{2.98}$$

In terms of the c_n coefficients, the reduced virial coefficients $B_m^* =$

$B_m / \langle \tilde{V} \rangle^{m-1}$ are thus given by

$$B_2^* = c_1 + 3 \tag{2.99}$$

$$B_3^* = c_2 + 3c_1 + 6 \tag{2.100}$$

$$B_4^* = c_3 + 3c_2 + 6c_1 + 10 \tag{2.101}$$

$$B_5^* = c_4 + 3c_3 + 6c_2 + 10c_1 + 15 \tag{2.102}$$

Equations (2.99)–(2.101) clearly indicate that there are severe limitations as to the choice of appropriate analytical expressions for the c_n coefficients. The second virial coefficient B_2, for instance, is rigorously known to be of the form[21]

$$
\begin{aligned}
B_2 &= \sum_{ij} x_i x_j B_{ij} \\
&= \tfrac{1}{2} \sum_{ij} x_i x_j \left(\tilde{V}_i + \tilde{V}_j + \tilde{R}_i \tilde{S}_j + \tilde{R}_j \tilde{S}_i \right)
\end{aligned} \tag{2.103}
$$

In terms of the parameter α we may rewrite Eq. (2.103) as

$$B_2^* = 1 + 3\alpha \tag{2.104}$$

which implies

$$c_1 = 3\alpha - 2 \tag{2.105}$$

With the exception of hard-sphere systems, no exact analytical expression is known for the third virial coefficient B_3. However, Kihara and Miyoshi obtained the following limiting result:[224]

$$
\begin{aligned}
B_3 &= \sum_{ijk} x_i x_j x_k B_{ijk} \\
&= \tfrac{1}{3} \sum_{ijk} x_i x_j x_k \left\{ \tilde{V}_i \tilde{V}_j + \tilde{V}_i \tilde{V}_k + \tilde{V}_j \tilde{V}_k + \tilde{V}_i \left(\tilde{S}_j \tilde{R}_k + \tilde{S}_k \tilde{R}_j \right) \right. \\
&\quad \left. + \tilde{V}_j \left(\tilde{S}_i \tilde{R}_k + \tilde{S}_k \tilde{R}_i \right) + \tilde{V}_k \left(\tilde{S}_i \tilde{R}_j + \tilde{S}_j \tilde{R}_i \right) + G_{ijk} \right\}
\end{aligned} \tag{2.106}
$$

where

$$\frac{1}{4\pi} \tilde{S}_i \tilde{S}_j \tilde{S}_k \le G_{ijk} \le \left(4\pi \tilde{R}_i \tilde{R}_j \tilde{R}_k \right)^2 \tag{2.107}$$

Note the coalescence of upper and lower bounds to G_{ijk} for the case of hard spheres. With increasing molecular anisotropy, however, the discrepancy between both expressions becomes considerable.

Additional information concerning the dependence of G_{ijk} on molecular shape can be extracted from Monte Carlo results on B_3 for hard spherocylinders of different length-to-breadth ratios L'. Assuming

$$G_{ijk} = \frac{1}{4\pi} \tilde{S}_i \tilde{S}_j \tilde{S}_k \qquad (2.108)$$

results in a slight underestimation of the third virial coefficient for $L' > 1$, while

$$G_{ijk} = \tfrac{1}{3}\left(\tilde{R}_i^2 \tilde{S}_j \tilde{S}_k + \tilde{R}_j^2 \tilde{S}_i \tilde{S}_k + \tilde{R}_k^2 \tilde{S}_i \tilde{S}_j \right) \qquad (2.109)$$

[as predicted by Eq. (2.91) and recently reconsidered by Boublik[182]] tends to overestimate B_3 for large particle anisotropies. Good overall agreement with the pseudoexperimental data might thus be expected from the linear combination of Eqs. (2.108) and (2.109), which, together with Eq. (2.106), leads to the following expression for B_3^*:

$$B_3^* = 1 + 6\alpha + \tfrac{3}{2}\gamma(1 + \tau) \qquad (2.110)$$

Table I reveals that this is indeed the case.

TABLE I
Reduced Third Virial Coefficients B_3^* for Hard Spherocylinders
of Length-to-Breadth Ratio L' Obtained from
Eqs. (2.110) and (2.120) and by Monte Carlo
(MC) Simulation[a]

	B_3^*		
L'	Eq. (2.110)	Eq. (2.120)	MC
1.4	10.56	10.61	10.64 ± 0.05
1.8	11.65	11.81	11.84 ± 0.06
2.0	12.28	12.54	12.54 ± 0.07
			12.34 ± 0.03
2.5	14.01	14.51	14.30 ± 0.07
3.0	15.91	16.75	16.27 ± 0.08
			16.20 ± 0.03
4.0	20.04	21.83	20.48 ± 0.10
			20.43 ± 0.04

[a] Data from Ref. 227.

Note that the shape parameter

$$\tau = \frac{\langle \tilde{S} \rangle}{4\pi \langle \tilde{R}^2 \rangle} \tag{2.111}$$

does not appear in Eq. (2.91). We will later show that the introduction of this parameter bears certain important consequences with respect to the uniqueness of HCB equations of state.

In the absence of any theoretically substantiated analytical relation for HCB virial coefficients higher than B_3, computer simulations represent the only reliable source of information about B_4 and B_5. A careful examination of the available Monte Carlo results for hard spherocylinders suggests that except for the extreme case $L' = 4$ the dependence of the fourth virial coefficient on the molecular elongation is accurately reproduced by the empirical expression

$$B_4^* = 1 + 9\alpha + 2\gamma(1 + 3\tau) \tag{2.112}$$

and that a fair approximation to the fifth virial coefficient is provided by

$$B_5^* = 1 + 12\alpha + \tfrac{1}{2}\gamma(41\tau - 11) \tag{2.113}$$

Values obtained from Eqs. (2.112) and (2.113) are always positive. For one-component hard-sphere systems we find $B_4^* = 18.00$ and $B_5^* = 28.00$, in com-

TABLE II

Reduced Fourth Virial Coefficients B_4^* for Hard Spherocylinders
of Length-to-Breadth Ratio L' Obtained from
Eqs. (2.112) and (2.121) and by Monte Carlo
(MC) Simulation[a]

	B_4^*		
L'	Eq. (2.112)	Eq. (2.121)	MC
1.4	19.09	19.01	19.26 ± 0.30
1.8	21.16	20.97	21.50 ± 0.30
2.0	22.36	22.12	22.34 ± 0.55
			22.50 ± 0.25
2.5	25.62	25.28	26.06 ± 0.65
3.0	29.13	28.75	29.15 ± 0.73
			28.00 ± 0.30
4.0	36.67	36.37	31.90 ± 0.32

[a]Data from Ref. 227.

TABLE III
Reduced Fifth Virial Coefficient B_5^* for Hard Spherocylinders
of Length-to-Breadth Ratio L' Obtained from
Eqs. (2.113) and (2.122) and by Monte Carlo
(MC) Simulation[a]

		B_5^*	
L'	Eq. (2.113)	Eq. (2.122)	MC
2.0	33.72	33.4	31.9 ± 1.3
3.0	41.22	40.5	36.8 ± 1.5
4.0	48.01	50.1	39.7 ± 1.6

[a]Data from Ref. 227.

plete analogy with the Carnahan–Starling equation. A detailed comparison between pseudoexperimental and calculated data for B_4^* and B_5^* is provided by Tables II and III, respectively.

Being armed with the knowledge of analytical expressions for all virial coefficients up to B_5^*, we are now able to evaluate the remaining c_n-coefficients from Eqs. (2.100)–(2.102). The resulting relations are

$$c_2 = \tfrac{3}{2}\gamma(\tau + 1) - 3\alpha + 1 \tag{2.114}$$

$$c_3 = \tfrac{1}{2}\gamma(3\tau - 5) \tag{2.115}$$

$$c_4 = 7\gamma(\tau - 1) \tag{2.116}$$

Assuming

$$c_n = 0, \qquad n \geq 5 \tag{2.117}$$

the HCB equation of state finally becomes

$$\frac{p}{\rho kT} = \frac{1 + y(3\alpha - 2) + y^2\left[\tfrac{3}{2}\gamma(\tau + 1) - 3\alpha + 1\right] + \tfrac{1}{2}y^3\gamma(3\tau - 5) + 7y^4(\tau - 1)}{(1 - y)^3} \tag{2.118}$$

Compressibility factors $p/\rho kT$, as evaluated from Eq. (2.118) for several hard spherocylinder model fluids, are presented in Table IV together with the related Monte Carlo results. There is total agreement between theoretically predicted and pseudoexperimental data within the estimated numerical uncertainty of the latter.

TABLE IV

Compressibility Factors $p/\rho kT$ for Hard Spherocylinders of Length-to-Breadth Ratio L' Obtained from Eqs. (2.118) and (2.119) and by Monte Carlo (MC) Simulation[a]

		$p/\rho kT$		
L'	y	Eq. (2.118)	Eq. (2.119)	MC
2.0	0.20	2.66	2.67	2.65 ± 0.02
	0.2454	3.38	3.39	3.37 ± 0.04
	0.30	4.54	4.56	4.48 ± 0.07
	0.3351	5.53	5.55	5.53 ± 0.14
	0.3879	7.50	7.52	7.57 ± 0.26
	0.40	8.06	8.08	8.20 ± 0.10
	0.4460	10.67	10.71	10.74 ± 0.24
	0.5096	16.13	16.22	16.80 ± 0.90
3.0	0.20	3.06	3.09	3.07 ± 0.03
	0.2676	4.48	4.55	4.53 ± 0.18
	0.30	5.40	5.48	5.40 ± 0.10
	0.3058	5.58	5.67	5.52 ± 0.23
	0.35	7.20	7.34	7.17 ± 0.11
	0.3927	9.25	9.46	8.99 ± 0.44
	0.40	9.66	9.89	9.60 ± 0.10
	0.45	13.04	13.44	13.00 ± 0.16
	0.50	17.75	18.50	18.00 ± 0.40

[a] Data from Ref. 227.

It should be emphasized that in the course of deriving expression (2.118) no attempt has been made to adjust the coefficients c_n with respect to the properties of multicomponent systems. The computer simulation results obtained for three binary mixtures of hard spheres and spherocylinders of length-to-breadth ratio $L' = 2$ at different densities thus provide a severe test of the reliability of Eq. (2.118). Table V reveals, however, that Eq. (2.118) nevertheless performs highly satisfactorily.

So far we have mainly focused on geometric quantities such as molecular volume, surface area, and mean curvature radius in order to establish the element of molecular shape within the theoretical description of HCB fluids. Especially among chemical engineers it is common practice, however, to account for the effects of particle anisotropy in terms of dimensionless eccentricity parameters or shape factors. Adopting this view, we find that in the case of pure fluids Eq. (2.118) essentially depends on three parameters. These are the reduced density (or packing fraction) y and the shape factors α and τ, respectively. Equation (2.91), on the other hand, represents a typi-

TABLE V

Compressibility Factors $p/\rho kT$ for Mixtures of Hard Spheres (Mole Fraction x_1) and Hard Spherocylinders of Length-to-Breadth Ratio $L' = 2$ Obtained from Eqs. (2.118) and (2.119) and by Monte Carlo (MC) Simulation[a]

		$p/\rho kT$		
x_1	y	Eq. (2.118)	Eq. (2.119)	MC
0.20	0.3277	5.11	5.13	5.17 ± 0.10
	0.4361	9.63	9.68	9.89 ± 0.20
	0.4983	14.33	14.43	14.34 ± 0.40
0.50	0.3128	4.48	4.49	4.52 ± 0.08
	0.4163	8.06	8.11	8.07 ± 0.15
	0.4757	11.65	11.74	11.59 ± 0.23
0.7143	0.2979	4.01	4.02	4.03 ± 0.07
	0.3965	6.93	6.96	7.02 ± 0.12
	0.4530	9.75	9.81	9.70 ± 0.21

[a] Data from Ref. 182.

cal two-parameter equation of state. In fact, if we set τ equal to unity, Eq. (2.91) may be formally recovered from Eq. (2.118).

When we compare the analytical form of Eqs. (2.91) and (2.118), it becomes obvious that the enhanced accuracy of Eq. (2.118) is mainly due to the introduction of the second shape factor τ. This does not imply a priori, however, that reliable equations of state for hard convex molecules must necessarily be of the three-parameter type. For example, Boublik has recently proposed a numerically accurate expression[182] that assumes a two-parameter form if applied to pure fluids. In our notation, it reads

$$\frac{p}{\rho kT} = \frac{1 + y(3\alpha - 2) + y^2(3\gamma - 3\alpha + 1) + y^3(5\alpha - 6\gamma)}{(1 - y)^3} \quad (2.119)$$

Equation (2.119) implies Eq. (2.104) for B_2^* and

$$B_3^* = 1 + 6\alpha + 3\gamma \quad (2.120)$$

$$B_4^* = 1 + 14\alpha + 3\gamma \quad (2.121)$$

$$B_5^* = 1 + 27\alpha \quad (2.122)$$

for the next three coefficients of the virial series, respectively.

The results predicted from Eqs. (2.119)–(2.122) for hard spherocylinders and for mixtures of hard spherocylinders with hard spheres have been included in Tables I to V for the sake of comparison. It turns out that Eq. (2.119) is only slightly inferior to Eq. (2.118) for these particular systems. In view of our incomplete knowledge concerning the interrelations between particle geometry and thermodynamic properties of molecular fluids, we might therefore be tempted to ask whether it is actually reasonable to devise an HCB equation of state depending on two-shape parameters. The following example will clarify this point.

Consider a spherocylinder of length-to-breadth ratio $L' = 3$ and a cube of same volume. Evaluating the shape parameter α for these particular hard convex bodies, we find $\alpha = 1.5$ in both cases. In consequence, two-parameter equations of state such as Eqs. (2.91) and (2.119) will thus predict exactly identical thermodynamic functions for pure fluids consisting of hard spherocylinders of $L' = 3$ and of hard cubes, respectively. Although there are no computer simulation data available so far for unconstrained hard cube systems, it appears extremely unlikely that spherocylinders and cubes might exhibit the same equation of state. Rather, experience suggests a higher pressure for the edged cubes if compared to the smoothly shaped spherocylinders at a given packing fraction y. By virtue of the additional shape factor τ, Eq. (2.118) succeeds in predicting this different behavior, as can be seen from Table VI, thus indicating the superiority of the three-parameter approach.

Judging from the excellent agreement of the most recently developed HCB equations of state with all available computer simulation results, it appears that we may safely rely on them in most cases of practical interest. In view of its high accuracy and by virtue of its great range of applicability, Eq. (2.118) should provide a valuable source of information concerning the ther-

TABLE VI
Compressibility Factors $p/\rho kT$ Predicted
by Eq. (2.118) for Hard Spherocylinders
of Length-to-Breadth Ratio
$L' = 3$ and for Hard Cubes

	$p/\rho kT$	
y	Spherocylinders	Cubes
0.1	1.743	1.748
0.2	3.057	3.093
0.3	5.397	5.547
0.4	9.658	10.19
0.5	17.75	19.53

modynamic properties of hard convex body reference systems. Furthermore, Eq. (2.118) should also provide a sound basis for the development of reliable equations of state for real fluids and fluid mixtures within the framework of molecular thermodynamics, and these equations could eventually meet the high accuracy requirements of chemical engineering applications.

E. A Theoretical Approach to the Conformational Equilibrium of Small-Chain Molecules in Condensed Phases

It has long been realized that when flexible molecules are dissolved in a dense fluid medium the population of their conformational states may be altered significantly.[228] Most of the traditional attempts to explain this phenomenon have been based on macroscopic dielectric theories and sometimes also on the assumption of specific intermolecular interactions.[229,230] Provided that the electric moments of a molecule vary with its conformational structure, a dielectric medium will tend to stabilize the more polar species relative to the less polar ones. Recent investigations have confirmed, however, that additional mechanisms are operative which may impose similar or even larger solvent shifts on conformational equilibria.[231-239,241] For example, there now exists a considerable amount of experimental evidence that the apparent energy barrier, $\Delta E_{t,g}$, between *trans* and *gauche* states in short alkane chains is definitely lowered in condensed phases relative to the ideal gas state,[231-234] despite the fact that n-alkanes are not appreciably polar. This effect seems to be most pronounced for small molecules such as butane.[231,232]

Additional support for these experimental results comes from the molecular dynamics simulations of Ryckaert and Bellemans[235,236] and Rebertus et al.[237] Investigating the properties of a simple four-center Lennard-Jones model of butane, they observed in each case that for a given temperature the *gauche* population is larger in the condensed phase than in the ideal gas state.

A promising route toward a quantitative understanding of solvent-induced changes of conformational equilibria is offered by thermodynamic perturbation theory. As has been pointed out in the preceding sections, the nearest-neighbor correlations in dense liquid systems are largely dominated by size and shape of the interacting particles. Therefore, if molecules may vary their geometric characteristics because of internal degrees of freedom, the interrelations between liquid structure and conformational distribution must be taken into account. The treatment of a suitably chosen hard-core reference system can serve as a convenient starting point for that purpose.

As an example, let us consider the Ryckaert–Bellemans model for butane: the molecules of the reference fluid consist of four identical fused hard spheres (representing the CH_3— and CH_2— groups, respectively) with one internal rotational degree of freedom characterized by the Scott–Scheraga potential.[240]

By means of classical statistical mechanics, Chandler and Pratt have derived a rigorous theory relating the solvent effect on the conformational structure of chain molecules to the cavity distribution function for the spherical groups.[238] In the case of butane, however, application of this theory requires the knowledge of a four-point distribution function, which is hardly achieved with any reasonable effort. In order to circumvent this problem, Pratt et al.[239] considered a simple two-cavity model, thereby approximating the CH_3-CH_2- units as spherical groups. This is equivalent to viewing the molecule as a dumbbell-like particle with variable bond length. Calculating the two-cavity distribution function from the RISM theory, they obtained the concentrations of *trans* and *gauche* conformers in pure butane at 274 K, in fair agreement with the respective computer simulation result. For butane dissolved in CCl_4, however, the solvent effect is considerably overestimated by this model.[237]

We shall now show that the methods and ideas discussed in the preceding sections suggest an alternative approach to the Pratt–Chandler theory, essentially based on thermodynamic arguments. By taking advantage of Eq. (2.91) as a reliable equation of state for hard interaction site molecules, the difficulties associated with the evaluation of higher-order cavity distribution functions will be avoided. Furthermore, the numerical efforts in estimating solvent effects on conformational equilibria will be substantially reduced.

Assuming a model of butane with rigid C—C bonds and fixed angles between adjacent bonds, the only intramolecular degree of freedom is associated with the dihedral angle ϕ, which measures the relative rotation of the terminal C atoms around the central bond. The conformational structure of the molecule is thus completely described by specifying the normalized distribution function $p(\phi)$. In the ideal gas state, that is, in the absence of intermolecular interactions, $p(\phi)$ is given by

$$p(\phi) = \frac{\exp[-u(\phi)/kT]}{\int_{-\pi}^{\pi} d\phi \exp[-u(\phi)/kT]} \tag{2.123}$$

where $u(\phi)$ denotes the torsional potential. To remain consistent with the available computer simulation studies and with the work of Pratt et al., the semiempirical function

$$\frac{u(\phi)}{k} = (1116 + 1462\cos\phi - 1578\cos^2\phi - 368\cos^3\phi$$

$$+ 3156\cos^4\phi - 3788\cos^5\phi)\ K \tag{2.124}$$

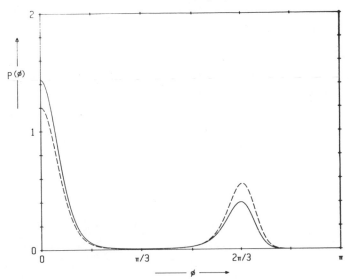

Fig. 23. Conformational distribution function for pure butane: $T = 274$ K; --- $\rho = 0.006053$ Å^{-3}; — ideal gas state.

as proposed by Scott and Scheraga will be adopted here. The resulting ideal gas state distribution function is shown for two different temperatures in Figs. 23 and 24. The pronounced peak around $\phi = 0$ [corresponding to the absolute minimum in $u(\phi)$] represents the population of the *trans* state. The mole fraction of trans conformers is thus given by

$$x_t = \int_{-\pi/3}^{\pi/3} d\phi\, p(\phi) = 2 \int_0^{\pi/3} d\phi\, p(\phi) \qquad (2.125)$$

By integrating over the peaks around $\phi = \pm 2\pi/3$, the fraction of gauche conformers is obtained:

$$x_g = \int_{-\pi}^{-\pi/3} d\phi\, p(\phi) + \int_{\pi/3}^{\pi} d\phi\, p(\phi) = 2 \int_{\pi/3}^{\pi} d\phi\, p(\phi) \qquad (2.126)$$

As soon as the butane molecules are exposed to intermolecular interactions, however, $p(\phi)$ is no longer determined by $u(\phi)$ alone. This fact can be appreciated by defining the potential of mean torsion

$$w(\phi) = u(\phi) + q(\phi) \qquad (2.127)$$

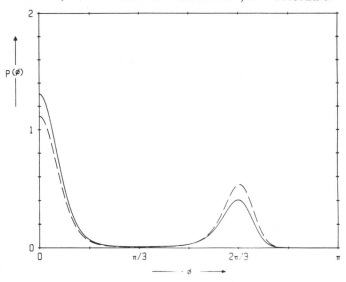

Fig. 24.   Conformational distribution function for butane dissolved in CCl_4: $T = 300$ K; --- $\rho = 0.006303$ Å^{-3}; — ideal gas state.

where $q(\phi)$ accounts for the effects of neighboring particles. In the limiting case of vanishing particle density, $\rho \to 0$, the function $q(\phi)$ will tend to zero as well. The general expression for $p(\phi)$ is therefore

$$p(\phi) = \frac{\exp[-w(\phi)/kT]}{\int_{-\pi}^{\pi} d\phi \exp[-w(\phi)/kT]} \qquad (2.128)$$

So far, Eqs. (2.127) and (2.128) merely represent a redefinition of our problem. The physical nature of $q(\phi)$ has yet to be examined.

With respect to the following considerations it will prove worthwhile to take a slightly different point of view. It has already been demonstrated by means of the IMIS model that it can be rewarding to represent a system of molecular entities by a mixture of pseudoparticles that deliberately emphasize certain special aspects of the description of the real particles (such as the effective potential field experienced by a particular interaction site).

In the present context we will utilize a method that bears a certain resemblance to the underlying ideas of the IMIS approach. Within the framework of classical equilibrium statistical mechanics a system of *flexible* molecules can be equivalently treated as a mixture of formally *rigid* conformers with variable composition, since the configurational partition function will be the same in both cases.

Adopting this alternative picture, the infinitesimally small internal rotation step of a particle from a state ϕ to $\phi + d\phi$ may formally be regarded as a "chemical" reaction (i.e., a rearrangement) of a rigid particle with intramolecular energy $u(\phi)$ to a different particle with intramolecular energy $u(\phi + d\phi)$. From macroscopic thermodynamics it is well known, however, that the equilibrium concentration of a certain species in a reacting system is proportional to the Boltzmann factor of its chemical potential. This implies that

$$p(\phi) \propto \exp\left[\frac{-\mu(\phi)}{kT}\right] \tag{2.129}$$

Expression (2.129) may be rearranged into the equivalent form

$$p(\phi) \propto \exp\left\{\frac{-[u(\phi) + \mu^*(\phi)]}{kT}\right\} \tag{2.130}$$

where $\mu^*(\phi)$ denotes the residual chemical potential of a ϕ conformer (i.e., the chemical potential relative to the ideal gas of the same temperature and density[200]). When we compare Eqs. (2.127), (2.128), and (2.130), we find that

$$q(\phi) = \mu^*(\phi) \tag{2.131}$$

The problem of predicting the medium effect on the conformational equilibrium of short-chain molecules has thus been reduced to the evaluation of the residual chemical potentials of the conformers in the liquid. A powerful tool for actually performing these calculations is provided by perturbation theory.

The thermodynamic properties of the hard-core reference fluid for liquids containing flexible molecules like butane are readily evaluated from Eq. (2.91) by employing the generalized shape functions

$$\langle \tilde{V} \rangle = \sum_i x_i^s \tilde{V}_i + x_B \int_{-\pi}^{\pi} d\phi \, p(\phi) \tilde{V}(\phi) \tag{2.132}$$

$$\langle \tilde{S} \rangle = \sum_i x_i^s \tilde{S}_i + x_B \int_{-\pi}^{\pi} d\phi \, p(\phi) \tilde{S}(\phi) \tag{2.133}$$

$$\langle \tilde{R} \rangle = \sum_i x_i^s \tilde{R}_i + x_B \int_{-\pi}^{\pi} d\phi \, p(\phi) \tilde{R}(\phi) \tag{2.134}$$

Here x_i^s are the mole fractions of the solvent components, while x_B denotes the mole fraction of butane; $\tilde{V}(\phi)$, $\tilde{S}(\phi)$ and $\tilde{R}(\phi)$ represent the molecular volume, the surface area, and the mean curvature radius of a ϕ conformer, respectively.

From Eq. (2.91), the expression for the residual chemical potential of conformers with dihedral angle ϕ follows by means of straightforward thermodynamics. The result is

$$\frac{\mu^*(\phi)}{kT} = \{\gamma[t(\phi)+2s(\phi)-2v(\phi)]-1\}\ln(1-y)$$

$$+ \frac{y\{3\alpha[r(\phi)+s(\phi)-v(\phi)]-(\gamma-1)v(\phi)\}}{(1-y)}$$

$$+ \frac{y\{\gamma[t(\phi)+2s(\phi)-3v(\phi)]+3\alpha v(\phi)\}}{(1-y)^2}$$

$$+ \frac{2\gamma y v(\phi)}{(1-y)^3} \tag{2.135}$$

where

$$r(\phi) = \frac{\tilde{R}(\phi)}{\langle \tilde{R} \rangle} \tag{2.136}$$

$$s(\phi) = \frac{\tilde{S}(\phi)}{\langle \tilde{S} \rangle} \tag{2.137}$$

$$v(\phi) = \frac{\tilde{V}(\phi)}{\langle \tilde{V} \rangle} \tag{2.138}$$

$$t(\phi) = \frac{\tilde{R}(\phi)^2}{\langle \tilde{R}^2 \rangle} \tag{2.139}$$

Considering the structure of Eq. (2.135), the reader will note that in order to calculate $\mu^*(\phi)/kT$ the distribution function $p(\phi)$ must be known beforehand. The only exceptions are infinitely dilute solutions of butane, where x_B approaches zero and $\langle \tilde{R} \rangle$, $\langle \tilde{R}^2 \rangle$, $\langle \tilde{S} \rangle$, and $\langle \tilde{V} \rangle$ reflect only solvent properties. In general, however, $p(\phi)$ has to be determined by iteration, starting from the ideal gas state distribution Eq. (2.123) as a suitable initial guess.

At this point a brief remark concerning the contribution of attractive intermolecular forces to $\mu^*(\phi)$ is in order. These contributions are readily evaluated by means of the HCBE–CST presented above in Section II.C, using a pure Lennard-Jones reference fluid. The results thus obtained clearly indicate, however, that in the case of the Ryckaert–Bellemans model of butane the attractive contributions to $\mu^*(\phi)$ are almost the same for all conformers. In the course of normalizing $p(\phi)$ they largely cancel out and may thus be neglected to a good approximation. This does not imply, however, that the influence of attractive interactions on conformational equilibria is negligible in general.

In order to actually calculate $p(\phi)$ for dense systems such as pure liquid butane or butane dissolved in CCl_4, some additional parameters must now be specified. In accordance with the work of Ryckaert and Bellemans, a constant length of 1.53 Å is assigned to all C—C bonds and the angle between adjacent bonds is assumed to be 109°28'. Ryckaert and Bellemans used two slightly different parameter sets in their computer simulations of liquid butane to describe the intermolecular interactions of the Lennard-Jones centers associated with the carbon sites. These were

$$\frac{\varepsilon}{k} = 72 \text{ K}, \qquad \sigma = 3.923 \text{ Å}$$

for the thermodynamic states

$$T = 199.9 \text{ K}, \qquad \rho = 0.006946 \text{ Å}^{-3}$$

and

$$T = 291.5 \text{ K}, \qquad \rho = 0.006042 \text{ Å}^{-3}$$

but

$$\frac{\varepsilon}{k} = 84 \text{ K}, \qquad \sigma = 3.92 \text{ Å}$$

in the case

$$T = 274.0 \text{ K}, \qquad \rho = 0.006053 \text{ Å}^{-3}$$

The latter parameters were also adopted by Rebertus et al. in their molecular dynamics study of butane in CCl_4 (thermodynamic state for this simulation: $T = 300.0$ K; $\rho = 0.006303$ Å^{-3}; $x_B \rightarrow 0$).

The effective hard-sphere diameters of the four overlapping chain segments can be related to the Lennard-Jones parameters by a simple formula given by Verlet and Weis:[101]

$$d_{\text{eff}} = \sigma(0.3837 + 1.068\varepsilon/kT)(0.4293 + \varepsilon/kT)^{-1} \qquad (2.140)$$

A more sophisticated treatment would also account for the slight density dependence of d_{eff}. Since this correction is rather small compared to the temperature effect, it will be neglected here. Finally, the geometry of the CCl_4 molecules needs to be specified. Consistent with the work of Pratt et al., a spherical model with effective diameter $d_{eff} = 5.27$ Å is assumed.

The most straightforward approach to the parameters $\tilde{R}(\phi)$, $\tilde{S}(\phi)$, and $\tilde{V}(\phi)$ would be to calculate them from the closest convex envelope around the spherical groups of the hard interaction site reference model of butane. A thorough inspection of the particle geometries reveals, however, that this procedure results in a slight overestimation of the molecular volumes $\tilde{V}(\phi)$ in the range $\pi/4 \leq \phi \leq 3\pi/4$ relative to the *trans* conformer. For the *gauche* state ($\phi = 2\pi/3$), this effect amounts to about 3%. Following Boublik and Nezbeda,[220] this problem can be avoided by evaluating $\tilde{V}(\phi)$ and $\tilde{S}(\phi)$ exactly for the fused hard-sphere model. However, to keep the numerics as simple as possible, a slightly different approach will be used here. The relative error in $\tilde{V}(\phi)$ can be corrected to a good approximation by employing the modified volume function

$$\tilde{V}'(\phi) = \tilde{V}(\phi)\left[1 - \frac{0.05|\phi|}{\pi}\right], \qquad -\pi \leq \phi \leq \pi \qquad (2.141)$$

Although the volume of the *cis* conformer ($\phi = \pi$) is now slightly underestimated, this will be of no influence on the final results, since for $\phi \to \pm\pi$ the distribution function $p(\phi)$ is almost completely determined by the very large value of $u(\phi)$.

With the ideal gas state distribution function as an initial guess, the iterative solution process for $p(\phi)$ is found to be rapidly convergent. Only three

TABLE VII
Values of the *trans*–*gauche* Equilibrium Constant $K = x_g/x_t$ Obtained from
Molecular Dynamics Simulations (MD) and from Various Models

System	Temperature (K)	Ideal gas	MD	HCB	Pratt et al.
			K		
Butane	199.9	0.30	—	0.73	—
Butane	274.0	0.48	—	0.77	0.67[c]
Butane	291.5	0.52	0.85[a]	0.80	—
Butane/CCl_4	300.0	0.54	0.80[b]	0.80	1.00[b]

[a] Ref. 236.   [b] Ref. 237.   [c] Ref. 239.

TABLE VIII
Medium Shift of the Reduced Intramolecular Energy $E_{intra}/N\varepsilon$ Obtained
from Molecular Dynamics Simulations (MD) and from
the HCB Approach

| System | Temperature (K) | $E_{intra}/N\varepsilon$ | | |
		Ideal gas	MD	HCB
Butane	199.9	2.71	3.80[a]	3.61
Butane	274.0	3.37	3.77[b]	3.76
Butane	291.5	4.21	5.14[a]	4.61

[a] Ref. 236.　　[b] Ref. 235.

to five cycles are required to reduce the relative deviations of two consecutive results below 10^{-5}. The results thus obtained are given in Figs. 23 and 24 and in Tables VII and VIII.

For pure butane at 291.5 K and butane dissolved in CCl_4 the predicted ratio x_g/x_t is in very good agreement with the corresponding computer simulation data. Unfortunately, this ratio has not been reported for the other molecular dynamics studies of butane.

The solvent shift of the intramolecular energy is predicted accurately for butane at 274 K but somewhat underestimated for $T = 199.9$ K and $T = 291.5$ K. The latter discrepancy is indeed surprising, since the ratio of conformer concentrations is well reproduced for this thermodynamic state. One possible explanation for this phenomenon is that the simulated system did not reach complete equilibrium during the respective molecular dynamics run.[236] It should also be noted, however, that the intramolecular energy is probably more sensitive to the form of the distribution function $p(\phi)$ than the ratio x_g/x_t.

Overall, the approach presented in this section appears to be promising. If applied to molecules that are reasonably described by convex bodies, it is much simpler and at least as accurate as the Chandler–Pratt theory. With the development of accurate equations of state for distinctly nonconvex particles, the extension to such systems would be straightforward. The availability of conceptually simple and numerically convenient methods for the treatment of solvent-induced shifts on conformational equilibria of flexible molecules is not only of theoretical interest, however. Such methods could also serve as useful tools for the interpretation of experimental results in many areas of spectroscopy, thermodynamics, and chemical engineering.

III. A QUANTUM STATISTICAL DESCRIPTION OF MOLECULAR DYNAMICAL PROCESSES*

A. Molecular Motions and Far-Infrared Absorption: Some Experimental Results

1. Dynamics of Molecular Motions and Irradiation Energy Dissipation in Condensed Matter

The basic theoretical problem in the field of dynamics of molecular motion is to describe the "response" of a physical system in thermal equilibrium to a "weak" external field. (The terms in quotation marks will be defined precisely later in this section.) The connection between (a) the band shapes (or line profiles) that are measured spectroscopically and (b) the "motions" of microscopic molecular systems that interact with the applied external field is of basic importance for the above spectroscopic studies. This connection is usually given by what has been called the *fluctuation–dissipation theorem* (FDT); see, for example, Refs. 242–268 and references cited therein.

The FDT (or, specifically, the *different forms* of it) can be derived with the aid of different formalisms, for instance, *time-dependent perturbation theory* or *linear response theory* (LRT); see Sections III.B, III.C, and III.E. Both these formal derivations are well known among spectroscopists (theoreticians, as well as experimentalists). Nevertheless, the derivation with LRT is of particular importance because LRT itself is nowadays the dominant theory for the treatment of the many-body dynamical problem (near the thermal equilibrium).

Kubo's original papers[242–244] are considered as the beginning of LRT. In these papers, he developed a general quantum statistical formalism for expressing transport coefficients in terms of correlation functions of dynamical quantities of the system under consideration. The Kubo results generalized earlier work of Kirkwood,[264] Green,[263] and other authors. The system response to a weak external field is treated very generally by solving the exact Von Neumann–Liouville equation,[269–272] and it is found to be expressible in terms of a time correlation function of the spontaneous fluctuations of a system quantity (dynamical variable), the system being in thermal equilibrium. The Kubo relations can be shown to be equivalent with the FDT, which, in its general form, is due to Callen and Welton,[265] and Greene and

*Part of the work described in this section is part of the postdoctoral research program "Quantum Statistics of Microscopic Irreversibility in the Theory of Condensed Matter" of C. A. Chatzidimitriou-Dreismann. He is indebted to W. Schröer, K. M. van Vliet, and R. E. Wilde for helpful discussions.

Callen;[266,267] Kubo's findings are also closely related to earlier results of Einstein[273] and Nyquist.[274]

In this section we shall discuss the quantum statistical aspects of intermolecular motions in liquids (with respect to time-dependent perturbation theory and LRT) as well as their connection with spectroscopic experiments in the FIR region. We shall review recent developments in theoretical and experimental research on "irradiation energy dissipation in disordered condensed matter as revealed by quantum statistics," the initial reports of which have appeared in Refs. 89–92, 97, 275, and 276.

The starting point for these considerations is the numerical estimation of the parameters characterizing the physical state of molecular liquids (at normal experimental conditions). This is necessary because we intend to discuss critically the applicability of the standard FDT in the case under consideration. Some FIR spectroscopic experimental results are reported below (in Section III.A.3) that seem to be in contradiction with the Boltzmann distribution law. In the subsequent discussion we shall show that the existing physical conditions imply that *quantum* statistical mechanics is needed for a proper description of the dynamics of the aforementioned molecular systems. The connection with the time–energy uncertainty relation will become obvious. Recently two *classical* statistical treatments of the problem under consideration were presented by Evans et al.[50,277] The interested reader should consult these two excellent reviews on the subject.

In Sections III.B and III.C the derivation of the "standard" FDT will be reported. A careful examination of the *necessary* conditions for the validity of the FDT will be carried out in Section III.D. As a result, it will become obvious that the physical conditions that obtain in the experimental context under consideration are in strict contradiction to the presuppositions necessary for the derivation of FDT. Additionally, some critical remarks of van Kampen,[278,280] ter Haar,[279] and van Vliet[95,96] concerning LRT will be reported and taken into account.

In Sections III.E and III.F the derivation of the *generalized* FDT[89] (within standard LRT formalism) will be presented and discussed. As a result, the aforementioned "anti-Boltzmann" experimental data (see Section III.A.3) are revealed to be physically clear, because (with respect to the generalized FDT) no ad hoc assumptions are necessary for their interpretation. The corresponding discussion will clarify the following: the quantum molecular systems that interact with the external electromagnetic field and "absorb" irradiation energy are *open* systems in (strong) interaction with their environment. This is the very reason for the weakness of the standard form of FDT with respect to the physical context under consideration.[97]

Another recent paper of theoretical importance is that of van Vliet,[95] who stated that in the original formalism of LRT, as given by Kubo, no dissipation is manifest. This will be confirmed in Section III.E.1 with respect to the physical problem of interest.

In our view van Kampen's criticism[278] of LRT is of fundamental importance, at least in the physical context under consideration (discussed next, in Section III.A.2).

2. Characteristic Parameters for Rotational Relaxation and FIR Absorption in Liquids

In the FIR spectral domain (typically $\tilde{\nu}$ ranges from 10 to 300 cm^{-1}), dipolar molecular liquids show a characteristic broad absorption band. Usually one rather prefers to consider dilute solutions of dipolar molecules in nonpolar solvents. This makes it possible to connect the observed FIR-band shape with the time-correlation function of the *molecular* dipole moment **m**. As mentioned above, this connection is established by the FDT. This theorem will be discussed and extended in following sections of this chapter.

For reasons of clarity as well as for illustration, therefore, the magnitude of physical parameters characterizing the FIR absorption process must be taken into account explicitly. In the following considerations, small dipolar molecules (like CH_2Cl_2 or CH_3CN) are regarded as being dissolved in nonpolar solvents (like CS_2 or CCl_4). Some real experimental results are presented in Subsection III.A.3.

The reorientational molecular motion (of the dipolar molecules) and the corresponding absorption of FIR irradiation energy are governed by some parameters in the time domain:

First, it is very well known that at room temperature the rotational relaxation time τ_m of the dissolved molecules with permanent dipole moment **m** is of the order of 1 ps:

$$\tau_m \sim 10^{-12} \, s \qquad (3.A1)$$

(e.g., see Refs. 277 and 281–286).

The reason for this molecular reorientational motion is given by the *thermal motion* of the molecules. Each dipolar molecule is coupled with the molecules in its environment. Thus, the thermal fluctuations in the environment disturb the motion of each dipolar molecule under consideration. Let τ_v be the characteristic time of these thermal fluctuations. (The subscript v indicates the "potential" between molecules.) The energy of each degree of freedom of the dipolar molecule fluctuates by the amount $k_B T/2$, where k_B is the Boltzmann constant and T is the absolute temperature. This rough estimation and the standard energy–time uncertainty relation[287] yield[97,275]

$$\tau_v \sim \frac{2\hbar}{k_B T} \qquad (3.A2)$$

Here we are considering standard experimental conditions. Thus, the quantity on the right-hand side of this relation is also of the order of 1 ps. Therefore it follows that

$$\tau_m \sim \tau_v \qquad (3.A3)$$

This relation has a plain physical meaning, namely, that the orientational relaxation process under consideration is due to the thermal fluctuations of the system consisting of the dipolar molecule and its environment. Thus, the three foregoing relations are consistent with each other.

Now, the period T_{max} of the FIR electromagnetic wave at the maximum of typical FIR absorption bands is also of the order of 1 ps:

$$T_{max} \equiv \frac{2\pi}{\omega_{max}} \sim 10^{-12} \text{ s} \qquad (3.A4)$$

(see examples in the next subsection). (Remember that 66.6 cm^{-1} correspond with 0.5×10^{-12} s.) Another parameter is given by the time τ_t that the molecular system under consideration takes to effectuate a transition.[95] This time is of the order of magnitude

$$\tau_t \sim \frac{2\pi}{\omega} \sim \frac{2\pi}{\omega_{max}} \qquad (3.A5)$$

where ω is the cyclic frequency of the absorbed wave. Here τ_t can be regarded as the time "needed by the molecular system to absorb a photon $\hbar\omega$ and make a transition."

From the aforementioned numerical estimations it follows immediately that

$$\frac{2\pi}{\omega_{max}} \sim \tau_m \qquad (3.A6)$$

This relation has a clear physical meaning, too. If one formally considers the field with ω_{max} and the reorientational molecular motion to correspond to two different fictive oscillators, relation (3.A6) says: these two oscillators couple best with each other if they have equal frequencies. This is, of course, the usual "resonance condition" of mechanics.

As a result,[91,97] we have

$$\frac{2\hbar}{k_B T} \sim \tau_v \sim \tau_m \sim \tau_t \sim \frac{2\pi}{\omega} \sim 10^{-12} \text{ s} \qquad (3.A7)$$

(at least in cases concerning FIR absorption in liquids at normal experimental conditions).

Of particular importance for the treatment of the generalized FDT (see Section III.E) is the result $\tau_m \sim \tau_t$. In physical terms, this means that the coupling V of the "absorbing" dipolar molecule with its environment is *strong*. As a rule a strong interaction implies that the "relaxation times" and the "transition times" can be of the same order of magnitude.[95] From $\tau_m \sim \tau_t$ it also follows that one cannot distinguish between the two time ranges:

$$\text{(a)} \quad \tau_t \ll t \ll \tau_m \quad \text{and} \quad \text{(b)} \quad t \gg \tau_m \qquad (3.A8)$$

These relations are usually referred to as the "small time range" and the "large time range," respectively. As a consequence, neither the *van Hove limit*[288] nor the *van Vliet limit*[95] can be applied successfully in the case under consideration.

Another immediate consequence from relations (3.A7) is that the usual *adiabatic* FDT (e.g., see Refs. 52 and 289–291 and Sections III.B and III.C) cannot be of physical relevance and/or reliability *within the physical context under consideration.*[97]

Another noteworthy point is that a proper treatment of the molecular reorientational relaxation must make use of *quantum* theory. As a matter of fact, the relations (3.A7) are strongly linked to the range of applicability of the *energy–time uncertainty relation*; this has already been pointed out earlier. To put it more simply: in the picosecond time scale [corresponding to relations (3.A7)] the concept of molecular trajectory becomes meaningless.

3.   An "Anomalous" Temperature Dependence of the FIR Absorption Band of Dipolar Molecules in Liquids: Some Experimental Results

An anomalous "red shift" of the FIR absorption band of dilute solutions of acetonitrile (CH_3CN) in *n*-heptane was first observed by Kroon and van der Elsken.[292] In this case the FIR band shifts continuously to *lower* frequencies with *increasing* temperature (see Figs. 25a, b). As these authors pointed out: "This now is contrary to what one could expect for a pure rotation band of a gas and is also in disagreement with existing theories [of R. G. Gordon[52]] for rotation bands in liquids" (cf. Ref. 292, p. 287).

As a matter of fact, this effect also seems to contradict the validity of the Boltzmann distribution law.[91,275] That is, with *increasing* temperature, it seems that the *lower* (quasi-)rotational energy levels of the absorbing molecular systems become more populated at the expense of the population of the higher energy levels. Clearly, this last remark is a contradiction to standard statistical mechanics, and this is due to the unwarranted "applicability" of

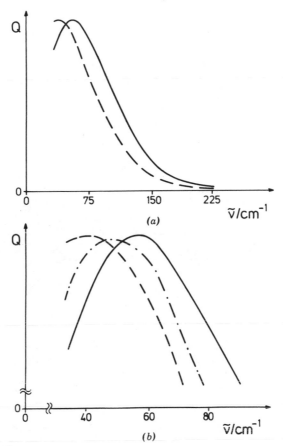

Fig. 25. (*a*) FIR absorption spectra of CH_3CN in *n*-heptane at various temperatures: concentration, 0.19 mol/liter; --- +60°C; — −20°C. (Absorption coefficient Q in arbitrary units; see the text.) (*b*) Detail of Fig. 25*a*: --- +60°C; −·−· +20°C; — −20°C.

the usual version of the FDT in the physical context under consideration (see Section III.E.2, below).

The aforementioned "anti-Boltzmann" effect of acetonitrile has also been observed in other nonpolar solvents.[275] Some results are shown in Figs. 26, 27, and 28. In these diagrams, the ordinate gives the absorption coefficient (or the rate of absorbed irradiation energy) in arbitrary units. Additionally, the absorption contours at various temperatures are "normalized" to have the peak nearly at the same level; the reason for this is to highlight the temperature effect mentioned above. For more experimental details, as well as the absolute values of the absorption coefficients, see Ref. 276.

The aforementioned temperature effect of CH_3CN has also been reported and/or confirmed in various papers (e.g., see Refs. 293–295). The same

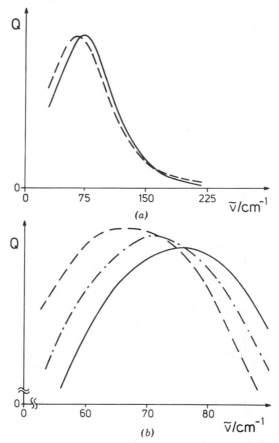

Fig. 26. (*a*) FIR absorption spectra of CH_3CN in CS_2: concentration, 0.1 mol/liter; ---
+20°C; — -27°C. (*b*) Detail of Fig. 26*a*: --- +20°C; -·-· 0°C; — -27°C.

anti-Boltzmann temperature effect has been demonstrated in the FIR absorption bands of $CHCl_3$, C_6H_5Cl, and C_6H_5F,[296] and also CH_2Cl_2.[297]

The interpretation of the temperature effect under consideration is reviewed in Section III.E.2, where the original references are also cited. It turns out that this effect is by no means an "anomalous" one and that it is a direct consequence of the basic physical conditions (3.A7) characterizing the FIR absorption process under consideration.

B. Perturbational Approach: Gordon's Treatment

A simple derivation of the FDT was proposed by Gordon (see Ref. 52 and references cited therein) and is now well known in many spectroscopic disciplines. As mentioned in the previous section, this theorem connects an (observable) band shape in a FIR experiment with the (microscopic) re-

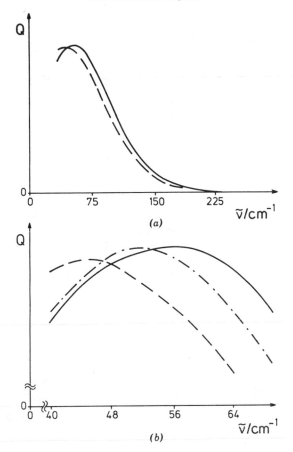

Fig. 27. (a) FIR absorption spectra of CH_3CN in cyclohexane: concentration, 0.19 mol/liter; --- $+60°C$; — $+20°C$. (b) Detail of Fig. 27a: --- $+60°C$; —·—· $+40°C$; — $+20°C$.

orientational relaxation time of an absorbing molecule. The following brief outline of Gordon's derivation is based upon Refs. 52 and 97.

Let

$$\mathbf{E}(t) = \tfrac{1}{2}E_0\varepsilon(e^{i\omega t} + e^{-i\omega t}), \qquad |\varepsilon| = 1 \qquad (3.B1)$$

be a monochromatic field that interacts with the electric dipole moment

$$\mathbf{M} = \sum_{i=1}^{N} \mathbf{m}_i \qquad (3.B2)$$

of a system consisting of N polar molecules (see Section III.A, above).

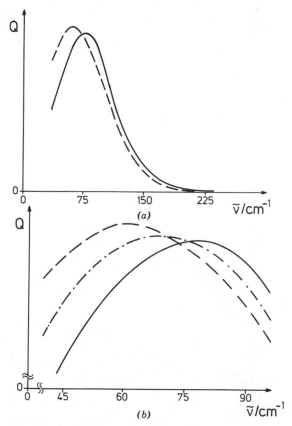

Fig. 28. (*a*) FIR absorption spectra of CH_3CN in CCl_4: concentration, 0.19 mol/liter; --- $+60°C$; — $-15°C$. (*b*) Detail of Fig. 28*a*: --- $+60°C$; $-\cdot-\cdot$ $+20°C$; — $-15°C$.

Ansatz (3.B2) is a good approximation within the physical context of interest, because of the fact that collision-induced components of the dipole moment are very small compared to the permanent dipole moment of a molecule (like CH_3CN).

The interaction Hamiltonian describing the interaction between the field and the molecules can be written as

$$\mathbf{H}_I = -\mathbf{M} \cdot \mathbf{E}(t) \tag{3.B3}$$

(see also Section III.C; remember that the wavelength is large compared with molecular dimensions).

The starting point in Gordon's treatment is the so-called golden rule.:

$$P_{f \leftarrow i}(\omega) = \frac{\pi E_0^2}{2\hbar^2} |\langle f|\boldsymbol{\varepsilon}\cdot\mathbf{M}|i\rangle|^2 \{\delta(\omega_{fi} - \omega) + \delta(\omega_{fi} + \omega)\} \quad (3.B4)$$

This formula is nothing but the *first-order approximation* to the transition probability per second from state i (for initial) to state f (for final) within standard time-dependent perturbation theory. The corresponding Bohr frequency is denoted by

$$\omega_{fi} = \frac{E_f - E_i}{\hbar}, \quad (3.B5)$$

where the E's represent the energies of two quantum mechanical states of the *unperturbed* Hamiltonian $\mathbf{H}_0$ of the *macro*scopic system (with dipole moment $\mathbf{M}$). This system is considered to be isolated.

At this stage the concept of the *probability* ρ_i of finding the system in the state i is introduced. Then it is argued that one measures the rate of *irradiation energy loss* $-\dot{E}_{\text{rad}}$, rather than the transition rate $P_{f \leftarrow i}(\omega)$. Thus, it follows from Eq. (3.B4)

$$-\dot{E}_{\text{rad}} = \sum_{f,i} \hbar\omega_{fi} P_{f \leftarrow i}(\omega)\rho_i$$

$$= \frac{\pi E_0^2}{2\hbar} \sum_{f,i} \omega_{fi}\rho_i |\langle f|\boldsymbol{\varepsilon}\cdot\mathbf{M}|i\rangle|^2 \{\delta(\omega_{fi} - \omega) + \delta(\omega_{fi} + \omega)\} \quad (3.B6)$$

Since the summations i and f go over all quantum states, by interchanging these indices in the summation over the second δ function one obtains

$$-\dot{E}_{\text{rad}} = \frac{\pi E_0^2}{2\hbar} \sum_{f,i} \omega_{fi}(\rho_i - \rho_f) |\langle f|\boldsymbol{\varepsilon}\cdot\mathbf{M}|i\rangle|^2 \delta(\omega_{fi} - \omega) \quad (3.B7)$$

The equilibrium *canonical* ensemble of thermostatics is now introduced into the formalism

$$\rho_f = \rho_i \exp\left(\frac{-\hbar\omega_{fi}}{k_B T}\right) \quad (3.B8)$$

where k_B is the Boltzmann constant and T is the absolute temperature. Thus, with $\beta = 1/k_B T$, it follows

$$-\dot{E}_{\text{rad}} = \frac{\pi E_0^2}{2\hbar} (1 - e^{-\beta\hbar\omega}) \omega \sum_{f,i} \rho_i |\langle f|\boldsymbol{\varepsilon}\cdot\mathbf{M}|i\rangle|^2 \delta(\omega_{fi} - \omega) \quad (3.B9)$$

An "absorption coefficient" can be defined by

$$\alpha(\omega) = \frac{-\dot{E}_{rad}}{E_{rad}}$$

$$= \frac{\varepsilon''(\omega) \cdot \omega}{nc} \tag{3.B10}$$

(e.g., see Ref. 285). Here ε'' is the imaginary part of the dielectric constant, n is the index of refraction, and c is the vacuum velocity of light. The incident energy flux E_{rad} is given by the magnitude of the Poynting vector:

$$E_{rad} = \frac{c}{8\pi} n E_0^2 \tag{3.B11}$$

Thus

$$\alpha(\omega) = \frac{4\pi^2}{\hbar cn} \omega (1 - e^{-\beta\hbar\omega}) \sum_{f,i} \rho_i |\langle f | \boldsymbol{\varepsilon} \cdot \mathbf{M} | i \rangle|^2 \delta(\omega_{fi} - \omega) \tag{3.B12}$$

It is also convenient[52] to define an "absorption band shape" $I(\omega)$ by

$$I(\omega) \equiv \frac{3\hbar cn}{4\pi^2 \omega (1 - \exp[-\beta\hbar\omega])} \alpha(\omega)$$

$$= 3 \sum_{f,i} \rho_i |\langle f | \boldsymbol{\varepsilon} \cdot \mathbf{M} | i \rangle|^2 \delta(\omega_{fi} - \omega) \tag{3.B13}$$

Now, $I(\omega)$ can be transformed by the following mathematical trick. One has

$$\delta(\omega) = \frac{1}{2\pi} \int_{-\infty}^{\infty} dt \exp(i\omega t) \tag{3.B14}$$

for the δ function. The Heisenberg operator of the dipole moment is

$$\mathbf{M}(t) \equiv \exp\left(\frac{i\mathbf{H}_0 t}{\hbar}\right) \mathbf{M} \exp\left(\frac{-i\mathbf{H}_0 t}{\hbar}\right) \tag{3.B15}$$

Remember that $\mathbf{H}_0$ represents an isolated system. As E_i and E_f are eigenvalues of $\mathbf{H}_0$, it holds

$$\exp\left(-\frac{iE_i t}{\hbar}\right) | i \rangle = \exp\left(-\frac{i\mathbf{H}_0 t}{\hbar}\right) | i \rangle \tag{3.B16}$$

From Eq. (B13) some trivial rearrangements yield:

$$I(\omega) = \frac{3}{2\pi} \int_{-\infty}^{\infty} dt\, e^{-i\omega t} \sum_i \rho_i \langle i|\boldsymbol{\varepsilon}\cdot\mathbf{M}(0)\boldsymbol{\varepsilon}\cdot\mathbf{M}(t)|i\rangle \qquad (3.\text{B17})$$

The summation over the eigenstates of $\mathbf{H}_0$ yields the ensemble average of the appearing operator:

$$I(\omega) = \frac{3}{2\pi} \int_{-\infty}^{\infty} dt\, e^{-i\omega t} \langle \boldsymbol{\varepsilon}\cdot\mathbf{M}(0)\boldsymbol{\varepsilon}\cdot\mathbf{M}(t)\rangle \qquad (3.\text{B18})$$

Finally, note that for an *isotropically* absorbing system the same result is obtained for any direction of the polarization vector $\boldsymbol{\varepsilon}$. Therefore, in this case

$$I(\omega) = \frac{1}{2\pi} \int_{-\infty}^{\infty} dt\, e^{-i\omega t} \langle \mathbf{M}(0)\cdot\mathbf{M}(t)\rangle \qquad (3.\text{B19})$$

If we consider solutions of dipolar molecules in nonpolar solvents that are sufficiently dilute, the cross terms $\langle \mathbf{m}_i(0)\cdot\mathbf{m}_j(t)\rangle$ between different molecules, $i \neq j$, may be neglected. Therefore we have the result

$$I(\omega) = \frac{N}{2\pi} \int_{-\infty}^{\infty} dt\, e^{-i\omega t} \langle \mathbf{m}(0)\cdot\mathbf{m}(t)\rangle \qquad (3.\text{B20})$$

This result represents the FDT in Gordon's treatment.

The physical relevance of the above derivation with respect to the physical conditions under consideration, as stated in Section III.A, will be critically examined in Section III.D (cf. also Ref. 97).

C. Linear Response Treatment in the Adiabatic Approximation

1. Elements of Kubo's Linear Response Theory

This section is devoted to the presentation of those elements of *Kubo's original presentation of linear response theory* (LRT)[242,275] that are necessary for a critical discussion of this matter (see also Section III.D).

Let $\mathbf{H}_0$ be the Hamiltonian of an *isolated* system. Kubo called the dynamical motion of the system determined by $\mathbf{H}_0$ the "natural motion." It is supposed that an external force $F(t)$ is applied to the system, the effect of which is represented by the perturbation Hamiltonian

$$\mathbf{H}_I = -\mathbf{A}\, F(t) \qquad (3.\text{C1})$$

The operator $\mathbf{A}$ represents a system quantity that effectuates the coupling of the physical system under consideration with the field $F(t)$, which is represented by a (complex) time function (c-number).

The field $F(t)$ is assumed to be turned on at time $t = -\infty$. The initial density operator $\rho_0(t)$ describes an *unperturbed* ensemble of systems with Hamiltonian H_0; that is, it obeys the Von Neumann–Liouville equation of motion

$$i\frac{\partial}{\partial t}\rho_0(t) = \frac{1}{\hbar}[H_0, \rho_0(t)]$$

$$\equiv L_0 \rho_0(t) \tag{3.C2}$$

The second equation defines the Liouvillean L_0 of the natural motion. The time evolution of the *complete* density operator $\rho_c(t)$ is given by

$$i\frac{\partial}{\partial t}\rho_c(t) = \frac{1}{\hbar}[H_0 - A\,F(t), \rho_c(t)]$$

$$\equiv L_c \rho_c(t) \tag{3.C3}$$

where the *complete* Hamiltonian

$$H_c = H_0 + H_I \equiv H_0 - A\,F(t) \tag{3.C4}$$

is taken into account.

In the following we can take for granted that $\rho_0(t)$ is well known and expresses the complete density operator $\rho_c(t)$ by the Volterra integral equation:

$$\rho_c(t) \equiv \rho_0(t) + \Delta\rho(t)$$

$$= \rho_0(t) + \frac{i}{\hbar}\int_{-\infty}^{t} ds\,\exp\left\{-\frac{i}{\hbar}H_0(t-s)\right\}[A\,F(s), \rho_c(s)]$$

$$\times \exp\left\{+\frac{i}{\hbar}H_0(t-s)\right\} \tag{3.C5}$$

This equation is the exact formal equivalent to Eq. (3.C3). Additionally, the initial condition

$$F(-\infty) = 0 \tag{3.C6}$$

is taken into account. (At time $t = -\infty$ the perturbation $F(t)$ is inserted "adiabatically.")

Equation (3.C5) can be approximated in arbitrary order by the usual successive approximation procedure.

In the following discussion we consider the case of a "weak" perturbation $F(t)$. (The physical meaning of this limitation will be considered in Section III.D.) Kubo argues that the *linear approximation* $\rho_L(t)$ for $\rho_c(t)$ is then given by

$$\rho_L(t) = \rho_0(t) + \frac{i}{\hbar} \int_{-\infty}^{t} ds \exp\left\{-\frac{i}{\hbar} H_0(t-s)\right\} [A F(s), \rho_0(s)]$$

$$\times \exp\left\{+\frac{i}{\hbar} H_0(t-s)\right\}$$

$$\equiv \rho_0(t) + \Delta\rho_{\mathrm{lin}}(t) \tag{3.C7}$$

This expression follows from Eq. (3.C5) by replacing the complete density operator occurring on the right-hand side with the unperturbed operator $\rho_0(s)$; that is,

$$\textit{Linear approximation:} \quad \int_{-\infty}^{t} \cdots \rho_c(s) \cdots \rightarrow \int_{-\infty}^{t} \cdots \rho_0(s) \cdots$$

$$\tag{3.C8}$$

At this stage of the formal derivation it is not necessary to assume that the unperturbed ensemble is stationary.

Let **B** be an operator (in the Schrödinger representation) representing a dynamical quantity. The ensemble overage of this quantity on time t, $\langle B(t) \rangle$, is then given by

$$\langle B(t) \rangle \equiv \mathrm{Tr}\,\rho(t)\mathbf{B} \tag{3.C9}$$

In most cases we are interested in the "disturbance" of this average corresponding with the external field $F(t)$. Thus we should rather consider the expression

$$\langle \Delta B(t) \rangle = \mathrm{Tr}\,\rho(t)\mathbf{B} - \mathrm{Tr}\,\rho_0(t)\mathbf{B}$$

$$\equiv \mathrm{Tr}\,\Delta\rho(t)\mathbf{B} \tag{3.C10}$$

In the linear approximation it holds that $\rho(t) \equiv \rho_L(t)$; see Eq. (3.C7). It follows immediately

$$\langle \Delta B(t) \rangle_L = \frac{i}{\hbar} \mathrm{Tr}\left(\int_{-\infty}^{t} ds\, \mathbf{B} \exp\left\{-\frac{i}{\hbar} H_0(t-s)\right\} [A F(s), \rho_0(s)] \right.$$

$$\left. \times \exp\left\{+\frac{i}{\hbar} H_0(t-s)\right\} \right) \tag{3.C11}$$

We can use the fact that the trace is invariant under cyclic permutation, $\mathrm{Tr}(\mathbf{ABC}\ldots) = \mathrm{Tr}(\mathbf{BC}\ldots\mathbf{A})$, etc. Thus it holds that

$$\langle \Delta B(t) \rangle_L = \frac{i}{\hbar} \mathrm{Tr}\left(\int_{-\infty}^{t} ds\, \mathbf{B}(t-s)[\mathbf{A}, \rho_0(s)] F(s) \right) \qquad (3.\mathrm{C}12)$$

where the Heisenberg operator (with respect to $\mathbf{H}_0$)

$$\mathbf{B}(t) \equiv \exp\left(\frac{i}{\hbar} \mathbf{H}_0 t \right) \mathbf{B} \exp\left(-\frac{i}{\hbar} \mathbf{H}_0 t \right) \qquad (3.\mathrm{C}13)$$

has been introduced. The term $F(s)$ is a c-number and, therefore, has been taken out of the commutator on the right-hand side of Eq. (3.C12). The corresponding *exact* result for the ensemble average (3.C10) is

$$\langle \Delta B(t) \rangle_c = \frac{i}{\hbar} \mathrm{Tr}\left(\int_{-\infty}^{t} ds\, \mathbf{B}(t-s)[\mathbf{A}, \rho_c(s)] F(s) \right) \qquad (3.\mathrm{C}14)$$

The transition from the exact expression (3.C14) to the linear approximation (3.C12) amounts to replacing the exact $\rho_c(s)$ by the unperturbed $\rho_0(s)$ on the right-hand side of the last equation.

It should be realized that both results (3.C12) and (3.C14) do not correspond to usual formulas neither in the Schrödinger nor in the Heisenberg representation; the reason is that here a dynamical time dependence occurs in the operator $\mathbf{B}$ as well as in the density operator ρ_0 (or ρ_c).

Equation (3.C12) is a formal result of basic importance in standard LRT. From a *formal* point of view, it can be stated in a slightly different way as follows:[242] One confines the above derivation to ensembles that are in *thermal equilibrium* before the perturbation $F(t)$ is turned on. As is well known from standard statistical mechanics, such ensembles are usually described by a *static* (or stationary) density operator ρ_{eq}, for $\partial \rho_{eq}/\partial t = 0$, which implies that ρ_{eq} commutes with the corresponding Hamiltonian.

Thus, in standard LRT we can make use of the specification

$$\rho_0(t) \to \rho_{eq} \qquad (3.\mathrm{C}15)$$

$$[\rho_{eq}, \mathbf{H}_0] = 0 \qquad (3.\mathrm{C}16)$$

for the unperturbed ensemble. For the linear response $\langle \Delta B(t) \rangle_L$, it then follows that

$$\langle \Delta B(t) \rangle_L = \frac{i}{\hbar} \mathrm{Tr}\left(\int_{-\infty}^{t} ds\, \mathbf{B}(t-s)[\mathbf{A}, \rho_{eq}] F(s) \right)$$

$$\equiv \frac{i}{i\hbar} \mathrm{Tr}\left(\int_{-\infty}^{t} ds\, [\mathbf{A}, \mathbf{B}(t-s)] \rho_{eq} F(s) \right) \qquad (3.\mathrm{C}17)$$

Note that the relation $\mathrm{Tr}(\mathbf{A}[\mathbf{B}, \mathbf{C}]) = \mathrm{Tr}([\mathbf{A}, \mathbf{B}]\mathbf{C})$ is valid.

The *aftereffect function*[242] $\phi_{BA}(t)$ within LRT is defined by

$$\phi_{BA}(t) \equiv \frac{1}{i\hbar} \text{Tr}[\mathbf{A}, \mathbf{B}(t)] \rho_{eq}$$

$$= \frac{1}{i\hbar} \text{Tr}[\rho_{eq}, \mathbf{A}] \mathbf{B}(t) \qquad (3.\text{C}18)$$

and thus

$$\langle \Delta B(t) \rangle_L = \int_{-\infty}^{t} ds' \, \phi_{BA}(t - s') F(s')$$

$$= \int_{0}^{\infty} ds \, \phi_{BA}(s) F(t - s) \qquad (3.\text{C}19)$$

Here the variable transformation $s = t - s'$ has been used. Please note that the "commutation"

$$\text{Tr} \int ds \ldots = \int ds \, \text{Tr} \ldots \qquad (3.\text{C}20)$$

has been used by the derivation of Eq. (3.C19).

At this stage the *canonical* density operator for the unperturbed ensemble

$$\rho_{eq} = \frac{\exp(-\beta \mathbf{H}_0)}{\text{Tr}\{\exp(-\beta \mathbf{H}_0)\}} \qquad (3.\text{C}21)$$

with $\beta = 1/k_B T$, where k_B is the Boltzmann constant, will be introduced into the formalism explicitly. Primarily this makes possible the conversion of the commutator expressions derived above into time-correlation expressions that are more convenient for our present purposes (see Sections III.A and III.D). Now Kubo's identity[242]

$$\frac{d}{d\lambda} (e^{\lambda \mathbf{H}_0} \mathbf{A} e^{-\lambda \mathbf{H}_0}) = e^{\lambda \mathbf{H}_0} [\mathbf{H}_0, \mathbf{A}] e^{-\lambda \mathbf{H}_0} \qquad (3.\text{C}22)$$

is used; it is easily verified by formal differentiation on the left-hand side by the (complex) parameter λ. From this by integration from 0 to β it follows

$$e^{\beta \mathbf{H}_0} \mathbf{A} e^{-\beta \mathbf{H}_0} - \mathbf{A} = \int_{0}^{\beta} d\lambda \, e^{\lambda \mathbf{H}_0} [\mathbf{H}_0, \mathbf{A}] e^{-\lambda \mathbf{H}_0} \qquad (3.\text{C}23)$$

A remultiplication by $\exp(-\beta \mathbf{H}_0)$ yields

$$[\mathbf{A}, e^{-\beta \mathbf{H}_0}] = \int_{0}^{\beta} d\lambda \, e^{-\beta \mathbf{H}_0} (e^{\lambda \mathbf{H}_0} [\mathbf{H}_0, \mathbf{A}] e^{-\lambda \mathbf{H}_0}) \qquad (3.\text{C}24)$$

The expression in the parenthesis can be modified in the following way: one defines an "imaginary time" by

$$t \equiv -i\hbar\lambda \tag{3.C25}$$

Then the Heisenberg equation of motion

$$\frac{d}{dt}\mathbf{A}(t) = \frac{1}{i\hbar}[\mathbf{A}(t), \mathbf{H}_0] \tag{3.C26}$$

is considered. With the aid of Eq. (3.C13), the Heisenberg operator $\mathbf{A}(t)$ is defined,

$$\mathbf{A}(t) = \exp\left(\frac{1}{\hbar}\mathbf{H}_0 t\right)\mathbf{A}\ \exp\left(-\frac{1}{\hbar}\mathbf{H}_0 t\right) \tag{3.C13a}$$

From Eq. (3.C26) it follows that

$$\dot{\mathbf{A}}(t) \equiv \frac{d}{dt}\mathbf{A}(t) = \frac{1}{i\hbar}\exp\left(\frac{i}{\hbar}\mathbf{H}_0 t\right)[\mathbf{A}, \mathbf{H}_0]\exp\left(-\frac{i}{\hbar}\mathbf{H}_0 t\right) \tag{3.C27}$$

The right-hand side of this equation and the expression in the parenthesis of Eq. (3.C24) are revealed to be similar when the formal definition (3.C25) is taken into account. Equation (3.C24), therefore, can be written in the form

$$[\mathbf{A}, \exp(-\beta\mathbf{H}_0)] = -i\hbar\int_0^\beta d\lambda \exp(-\beta\mathbf{H}_0)\dot{\mathbf{A}}(-i\hbar\lambda) \tag{3.C28}$$

With this important result we obtain for the aftereffect function

$$\phi_{BA}(t) = \int_0^\beta d\lambda \operatorname{Tr}\exp(-\beta\mathbf{H}_0)\dot{\mathbf{A}}(-i\hbar\lambda)\mathbf{B}(t)/\operatorname{Tr}\exp(-\beta\mathbf{H}_0)$$

$$\equiv \int_0^\beta d\lambda \langle \dot{\mathbf{A}}(-i\hbar\lambda)\mathbf{B}(t)\rangle \tag{3.C29}$$

Here the *ensemble average* $\langle \cdots \rangle$ is defined. This formula is a starting one for the following applications. Comparison of Eq. (3.C29) with Eq. (3.C18) yields the formal result

$$\int_0^\beta d\lambda \langle \dot{\mathbf{A}}(-i\hbar\lambda)\mathbf{B}(t)\rangle = \frac{1}{i\hbar}\langle[\mathbf{A}(0), \mathbf{B}(t)]\rangle \tag{3.C30}$$

where $\mathbf{A}(0) \equiv \mathbf{A}$ [see Eq. (3.C13)]. The expression on the left-hand side is sometimes called the *Kubo transform* of the time-correlation function $\langle \mathbf{A}(0)\mathbf{B}(t)\rangle$.

2. The Adiabatic Fluctuation – Dissipation Theorem

The connection between the shape of a spectroscopic band (or line) and microdynamic behavior of molecules absorbing electromagnetic energy is given by the FDT. The mathematical derivation of the FDT in the *adiabatic* approximation within LRT is presented in this subsection. For critical remarks concerning the physical relevance and reliability of the FDT with respect to the physical context under consideration, see Sections III.D and III.F, below.

The complete Hamiltonian [see Eq. (3.C4)] of a *macro*scopic absorbing specimen is now

$$\mathbf{H}_c = \mathbf{H}_0 + \mathbf{H}_I = \mathbf{H}_0 - \mathbf{M}\,E(t) \tag{3.C31}$$

where $\mathbf{M}$ represents the electric dipole moment operator of a system consisting of N molecules

$$\mathbf{M} \equiv \sum_{i=1}^{N} \mathbf{m}_i \tag{3.C32}$$

and $E(t)$ is the electric component of the acting external field. The interaction Hamiltonian $\mathbf{H}_I = -\mathbf{M}E(t)$ is the nonrelativistic dipole approximation for the interaction between electromagnetic field and electric charges.[298,299]

The response of the system under consideration to the external perturbation $E(t)$ is accompanied by the absorption of irradiation energy. This follows because under the influence of the external perturbation the system changes state. In the case under consideration, the absorbed irradiation energy will be dissipated within a time of the order of the reorientational relaxation time τ_m of the molecular dipole moment $\mathbf{m}$ (see Section III.A). The irradiation energy dissipated per second Q_{ad} is thus equal to the time rate of change of the system's energy. Thus,

$$Q_{ad} = \mathrm{Tr}\left(\rho \cdot \frac{d\mathbf{H}_c}{dt} \right) \equiv \left\langle \frac{\partial \mathbf{H}_c}{\partial t} \right\rangle \tag{3.C33}$$

for the stationary absorption process under consideration. But an important theorem of Hamiltonian mechanics states that

$$\frac{d\mathbf{H}_c}{dt} = \frac{\partial \mathbf{H}_c}{\partial t} \tag{3.C34}$$

(e.g., see Ref. 300). Therefore one obtains with Eq. (3.C31) the result[301]

$$Q_{ad} = \mathrm{Tr}\left(\rho \cdot \frac{\partial \mathbf{H}_I}{\partial t} \right) = -(\mathrm{Tr}\,\rho\,\mathbf{M}) \cdot \frac{\partial E}{\partial t}$$
$$\equiv -\langle \mathbf{M} \rangle \dot{E}(t) \tag{3.C35}$$

In actual calculations a monochromatic and polarized field

$$E(t) \equiv \tfrac{1}{2} \{ E \exp(i\omega t) + E^* \exp(-i\omega t) \} \qquad (3.C36)$$

is used. The reason is that the superposition principle, with respect to the response quantity Q, is valid within LRT. It appears that then the right-hand side of Eq. (3.C35) contains some oscillating terms with cyclic frequency ω having no physical meaning (see below). Therefore one would rather use the expression[301]

$$\overline{Q_{\mathrm{ad}}(\omega)}^T \equiv -\frac{1}{T} \int_0^T dt \, \langle \mathbf{M} \rangle \dot{E}(t)$$

$$\equiv \overline{\langle \partial \mathbf{H}_I / \partial t \rangle}^T \qquad (3.C37)$$

where the artificial oscillating terms are canceled. Here it holds that $T = 2\pi/\omega$, T being the period of $E(t)$.

In the following discussion $Q_{\mathrm{ad}}(\omega)$ is calculated with the aid of formulas derived earlier in Section III.C.1 (see also Refs. 89 and 275). The aftereffect function

$$\phi_{MM}(t) = \int_0^\beta d\lambda \, \langle \dot{\mathbf{M}}(-i\hbar\lambda) \mathbf{M}(t) \rangle \qquad (3.C38)$$

is introduced [see Eq. (3.C29)]. With the aid of result (3.C19), we obtain from Eq. (3.C35)

$$Q_{\mathrm{ad}}(\omega) = -\int_0^\infty ds \, \phi_{MM}(s) \dot{E}(t) E(t-s) \qquad (3.C39)$$

For the factor $\dot{E}(t)E(t-s)$, we obtain from Eq. (3.C36)

$$\dot{E}(t)E(t-s) = \tfrac{1}{2} \mathrm{Re}\{ i\omega EE \exp(2i\omega t - i\omega s) \}$$

$$+ \tfrac{1}{2} \mathrm{Re}\{ -i\omega EE^* \exp(-i\omega s) \} \qquad (3.C40)$$

where $\mathrm{Re}\{\cdots\}$ indicates the real part of the expression in the párenthesis. The first term on the right-hand side is an oscillating one, as mentioned above. Thus it follows that

$$\overline{Q_{\mathrm{ad}}(\omega)}^T = -\int_0^\infty ds \, \phi_{MM}(s) \mathrm{Re}\{ -\tfrac{1}{2} i\omega EE^* \exp(-i\omega s) \} \quad (3.C41)$$

The Fourier decomposition

$$\phi_{MM}(s) = \int_{-\infty}^{+\infty} d\Omega \, \hat{\phi}_{MM}(\Omega) \exp(i\Omega s) \qquad (3.\text{C}42)$$

is now introduced. Thus

$$\overline{Q_{ad}(\omega)}^T = + \int_0^\infty ds \int_{-\infty}^{+\infty} d\Omega \, \hat{\phi}_{MM}(\Omega) \exp(i\Omega s) \text{Re} \left\{ \frac{i\omega}{2} EE^* \exp(-i\omega s) \right\}$$

$$(3.\text{C}43)$$

The first integral can be evaluated with the aid of the distribution relation

$$\delta_+(\Omega) \equiv \int_0^\infty ds \exp(i\Omega s)$$

$$= \pi\delta(\Omega) + i\mathscr{P}\left(\frac{1}{\Omega}\right) \qquad (3.\text{C}44)$$

Here δ denotes Dirac's delta function and $\mathscr{P}$ the principal value. As usual, it will be taken for granted that $\phi_{MM}(\Omega)$ is *analytic* in the *lower* complex Ω plane. This makes it possible to evaluate the integral over Ω in Eq. (3.C43) by making use of the integration path indicated in Fig. 29. With Eq. (3.C44) it follows from Eq. (3.C43) that

$$\overline{Q_{ad}(\omega)}^T = \int_{-\infty}^{+\infty} d\Omega \, \hat{\phi}_{MM}(\Omega) \left\{ \frac{i\omega}{4} EE^* \delta_+(\Omega - \omega) - \frac{i\omega}{4} EE^* \delta_+(\Omega + \omega) \right\}$$

$$(3.\text{C}45)$$

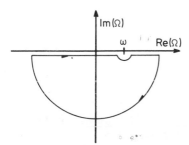

Fig. 29.   Integration path for the calculation of the integrals in Eqs. (3.C45) and (3.E16).

From the analytic property mentioned above, we have

$$\lim_{|\Omega| \to \infty} \hat{\phi}_{MM}(\Omega) = 0 \qquad \text{for} \quad \text{Im}\{\Omega\} \leq 0 \qquad (3.C46)$$

where $\text{Im}\{\cdots\}$ indicates the imaginary part of the term in the parenthesis. This condition allows us to replace the real integral over Ω in Eq. (3.C.45) by a complex integration over the contour indicated in Fig. 29. Thus, from Cauchy's theorem we obtain the intermediate result

$$\mathscr{P} \int_{-\infty}^{+\infty} d\Omega \frac{\hat{\phi}_{MM}(\Omega)}{(\Omega - \omega)} = -i\pi \hat{\phi}_{MM}(\omega) \qquad (3.C47)$$

From Eq. (3.C.45), therefore, it follows with (3.C.44) and (3.C.47) that

$$\overline{Q_{ad}(\omega)}^T = \int_{-\infty}^{+\infty} d\Omega \, \hat{\phi}_{MM}(\Omega) \frac{i\omega}{4} EE^* [\pi\delta(\Omega - \omega) - \pi\delta(\Omega + \omega)]$$

$$+ \frac{i\omega}{4} EE^* [\pi\hat{\phi}_{MM}(\omega) - \pi\hat{\phi}_{MM}(-\omega)]$$

$$= \tfrac{1}{2}\pi i\omega EE^* [\hat{\phi}_{MM}(\omega) - \hat{\phi}_{MM}(-\omega)] \qquad (3.C48)$$

The function $\phi_{MM}(t)$ can be expressed with the aid of the symmetrized quantum mechanical time-correlation function

$$C_{MM}(t) \equiv \tfrac{1}{2}\text{Tr}\,\rho_{eq}[\mathbf{M}(0)\mathbf{M}(t) + \mathbf{M}(t)\mathbf{M}(0)] \qquad (3.C49)$$

For this purpose it will be proved that the following *theorem* holds:

$$\hat{\phi}_{MM}(\omega) = -\frac{2i}{\hbar} \tanh\left(\frac{\beta\hbar\omega}{2}\right) \hat{C}_{MM}(\omega) \qquad (3.C50)$$

(The proof is given in Section III.C.3.) The last term on the right-hand side is the Fourier transform of $C_{MM}(t)$. With the aid of this theorem, the final result[302] is obtained:

$$\overline{Q_{ad}(\omega)}^T = +\frac{2\pi}{\hbar}|E|^2 \cdot \omega \cdot \tanh\left(\frac{\beta\hbar\omega}{2}\right) \cdot \hat{C}_{MM}(\omega) \qquad (3.C51)$$

which constitutes the adiabatic FDT.

This is the quantum statistical result that follows immediately within LRT from the *adiabatic* assumption (3.C33); see Sections III.D and III.E, below. It will be shown that the adequate expression for the rate of irradiation energy dissipated because of molecular motion contains an additional term that depends on the quantal time-correlation function

$$C_{\dot{M}\dot{M}}(t) \equiv \tfrac{1}{2}\mathrm{Tr}\,\rho_{eq}[\dot{M}(0)\dot{M}(t)+\dot{M}(t)\dot{M}(0)] \qquad (3.C52)$$

of the operator $\dot{M} \equiv dM/dt$ [see Eq. (3.E23), below].

3.   Some Theorems

Since we have tried to make Section III largely self-contained, some theorems of LRT are proved that can be found in various parts of Refs. 95, 242, 243, and 275.

In this subsection the abbreviation

$$\langle AB \cdots \rangle \equiv \mathrm{Tr}\,\rho_{eq} AB \cdots \qquad (3.C53)$$

is used, where ρ_{eq} always means the canonical density operator (3.C21). The Fourier transformation to be used is given by

$$\chi(t) = \int_{-\infty}^{\infty} d\Omega\, \hat{\chi}(\Omega)\exp(i\Omega t)$$

$$\hat{\chi}(\Omega) = \frac{1}{2\pi} \int_{-\infty}^{\infty} dt\, \chi(t)\exp(-i\Omega t) \qquad (3.C54)$$

Let A and B be two dynamical operators. It is assumed that $\langle A \rangle = 0 = \langle B \rangle$, which means that the steady state parts of these quantities vanish. Additionally, the *mixing property*

$$\lim_{t \to \infty} \langle A(0)B(t) \rangle = \langle A(0) \rangle \langle B(t) \rangle$$

$$= \langle A \rangle \langle B \rangle$$

$$= 0 \qquad (3.C55)$$

is assumed.

At first it holds with $Q \equiv \mathrm{Tr}\,\rho_{eq}$

$$\chi(t) \equiv \langle A(-i\hbar\beta)B(t) \rangle = \mathrm{Tr}\,\rho_{eq} A(-i\hbar\beta)B(t)$$

$$= \mathrm{Tr}(Q^{-1}e^{-\beta H_0})(e^{\beta H_0}A(0)e^{-\beta H_0})B(t)$$

$$= \mathrm{Tr}\,Q^{-1}A(0)e^{-\beta H_0}B(t)$$

$$= \mathrm{Tr}\,\rho_{eq} B(t)A(0) \equiv \langle B(t)A(0) \rangle, \qquad (3.C56)$$

see Eq. (3.C13) with $t \equiv -i\hbar\beta$. Because of the stationarity condition

$$\langle \mathbf{A}(-i\hbar\beta)\mathbf{B}(t) \rangle = \langle \mathbf{A}(0)\mathbf{B}(t + i\hbar\beta) \rangle \qquad (3.\text{C}57)$$

Additionally, one requires the validity of the so-called *Kubo–Martin–Schwinger condition*:[242,303] $\hat{\chi}(\Omega)$ is analytic at least in the open strip $0 < \text{Im}\{\Omega\} < \hbar\beta$ and continuous in the closed strip $0 \le \text{Im}\{\Omega\} \le \hbar\beta$. Then it holds that

$$\hat{\chi}(\Omega) \equiv \frac{1}{2\pi} \int_{-\infty}^{\infty} dt \exp[-\Omega(t + i\hbar\beta - i\hbar\beta)]\langle \mathbf{A}(0)\mathbf{B}(t + i\hbar\beta) \rangle \quad (3.\text{C}58)$$

which follows from Eqs. (3.C54) and (3.C57). Let now be $t' \equiv t + i\hbar\beta$; then it follows that

$$\hat{\chi}(\Omega) \equiv \frac{1}{2\pi} \int_{-\infty}^{\infty} dt\, e^{-\hbar\Omega\beta} e^{-i\Omega t'}\langle \mathbf{A}(0)\mathbf{B}(t') \rangle$$

$$= \frac{1}{2\pi} e^{-\hbar\Omega\beta} \int_{-\infty + i\hbar\beta}^{\infty + i\hbar\beta} dt'\, e^{-i\Omega t'}\langle \mathbf{A}(0)\mathbf{B}(t') \rangle$$

$$= \frac{1}{2\pi} e^{-\hbar\Omega\beta} \int_{-\infty}^{\infty} dt\, e^{-i\Omega t}\langle \mathbf{A}(0)\mathbf{B}(t) \rangle \qquad (3.\text{C}59)$$

The last equation follows from Cauchy's theorem and the Kubo–Martin–Schwinger condition. With result (3.C56) and the definition of the *one-sided* time-correlation function

$$S_{BA}(t) \equiv \langle \mathbf{A}(0)\mathbf{B}(t) \rangle \qquad (3.\text{C}60)$$

the result

$$\hat{S}_{BA}(\Omega) = e^{+\hbar\Omega\beta} \int_{-\infty}^{\infty} dt\, e^{-i\Omega t}\langle \mathbf{B}(t)\mathbf{A}(0) \rangle \qquad (3.\text{C}61)$$

has been proved.

Let the following Kubo form be defined by

$$K_{BA}(t) \equiv \int_{0}^{\beta} d\lambda \langle \mathbf{A}(-i\hbar\lambda)\mathbf{B}(t) \rangle \qquad (3.\text{C}62)$$

By the above procedure we obtain

$$K_{BA}(\Omega) \equiv \frac{1}{2\pi} \int_{-\infty}^{\infty} dt\, e^{-i\Omega t} K(t)$$

$$= \frac{1}{2\pi} \int_{-\infty}^{\infty} dt \int_{0}^{\beta} d\lambda \exp[-i\Omega(t + i\hbar\lambda - i\hbar\lambda)] \langle A(0)B(t + i\hbar\lambda)\rangle$$

$$= \frac{1}{2\pi} \int_{0}^{\beta} d\lambda \int_{-\infty + i\hbar\lambda}^{\infty + i\hbar\lambda} dt'\, \exp[-i\Omega t'] \langle A(0)B(t')\rangle \exp[-\lambda\hbar\Omega]$$

$$= \frac{1}{2\pi} \int_{-\infty}^{\infty} dt\, e^{-i\Omega t} \int_{0}^{\beta} d\lambda \langle A(0)B(t)\rangle \exp[-\lambda\hbar\Omega]$$

$$= \frac{1 - \exp[-\beta\hbar\Omega]}{\hbar\Omega} \hat{S}_{BA}(\Omega) \tag{3.C63}$$

Equations (3.C63) and (3.C61) can be combined, and the important theorem

$$\hat{K}_{BA}(\Omega) = \frac{2}{\hbar\Omega} \tanh\left(\frac{\beta\hbar\Omega}{2}\right) \hat{C}_{BA}(\Omega) \tag{3.C64}$$

is immediately proved.

Another theorem follows from Eq. (3.C30). By definition (3.C62), it holds that

$$i\hbar\hat{K}_{B\dot{A}}(\Omega) = \frac{1}{2\pi} \int_{-\infty}^{\infty} dt\, e^{-i\Omega t} \{\langle A(0)B(t)\rangle - \langle B(t)A(0)\rangle\} \tag{3.C65}$$

and with Eq. (3.C61)

$$\frac{\hbar}{i} \hat{K}_{B\dot{A}}(\Omega) = \{e^{-\beta\hbar\Omega} - 1\} \hat{S}_{BA}(\Omega) \tag{3.C66}$$

Also with Eq. (3.C61) it follows immediately that

$$\hat{C}_{BA}(\Omega) \equiv \frac{1}{2\pi} \int_{-\infty}^{\infty} dt\, e^{-i\Omega t} \frac{1}{2} \{\langle A(0)B(t)\rangle + \langle B(t)A(0)\rangle\}$$

$$= \tfrac{1}{2} \{e^{-\beta\hbar\Omega} + 1\} \hat{S}_{BA}(\Omega) \tag{3.C67}$$

Combination of the last two equations yields

$$\hat{K}_{B\dot{A}}(\Omega) = -\frac{2i}{\hbar} \tanh\left(\frac{\beta\hbar\omega}{2}\right) \hat{C}_{BA}(\Omega) \tag{3.C68}$$

Theorem (3.C50) mentioned above [in the derivation of the adiabatic FDT, (3.C51)] is a special case of this equation. This can be seen with $A \equiv B \equiv M$ and definition (3.C38) of the aftereffect function

$$\phi_{MM}(t) \equiv K_{M\dot{M}}(t) \tag{3.C69}$$

[cf. Eq. (3.C62)]. Additionally, we must take into account that the autocorrelation function $C_{AA}(t)$ is an even function of the time:[289,290]

$$C_{AA}(t) = C_{AA}(-t) \tag{3.C70}$$

Thus it follows that

$$\hat{C}_{AA}(\Omega) = \hat{C}_{AA}(-\Omega) \tag{3.C71}$$

Last but not at least, it should be pointed out that in the calculations of this section it has *never* been presupposed that $(\mathrm{Tr}\rho_{eq}\ldots)$ *commutes* with $(d\ldots/dt)$. This caveat will be of importance in later considerations.

D.   On the Applicability of the Adiabatic Fluctuation–Dissipation Theorem

1.   On the Perturbational Treatment

In Section III.B, the derivation of the adiabatic FDT within first-order time-dependent perturbation theory was reported. Recently it has been pointed out that this derivation *cannot* be justified within quantum statistical theory.[97] This statement is supported by the following argument.

The starting point in Gordon's derivation of FDT is the "golden rule," Eq. (3.B4). This formula plainly represents the first-order approximation for the transition probabilities (between states of the considered unperturbed Hamiltonian) within standard perturbation theory.[299] Now it must be emphasized that the necessary conditions for the golden rule to be valid are in *contradiction* to the physical conditions of the FIR absorption process under consideration [see relations (3.A7)]. As a matter of fact, the golden rule holds for times t that are sufficiently large compared with $2\pi/\omega$, ω being the cyclic frequency of the disturbance $E(t)$ [Eq. (3.B1)]:

$$t \gg \frac{2\pi}{\omega} \tag{3.D1}$$

(For a detailed treatment see, e.g., Ref. 299.) The reason for this is that only in the limiting case (3.D1) one can obtain the δ functions occurring in the golden rule. These δ functions represent the conservation of energy in cases where the unperturbed system emits or absorbs the quantity of energy $\hbar\omega \equiv \hbar\omega_{fi}$. We should note that the existence of these δ functions is necessary for the introduction of the Heisenberg operator $\mathbf{M}(t)$ into the formulism and, therefore, the derivation of Eq. (3.B17), that is, the adiabatic FDT.

One realizes that condition (3.D1) is in strict contradiction with relations (3.A7), which describe the characteristic time scale of the process under consideration. As a matter of fact, one is interested in the evolution of the system during a time interval of the order $\Delta t \sim 2\pi/\omega$.

Moreover, Gordon's derivation (presented in Section III.B) shows an additional serious weakness, which consists in the following: The rate of irradiation energy loss $-\dot{E}_{rad}$ [see Eq. (3.B6)] is given with the aid of "logical" or "intuitive" remarks that seem to be "physically clear." As a whole, Eq. (3.B6) is strongly related to the Pauli master equation.[304] This remark permits us to apply van Kampen's criticism[305-307] of that equation in the case under consideration as well.[97] First, it is obvious that the probabilities ρ_i, appearing in Eq. (3.B6), correspond with the diagonal elements of the density operator being proper for the quantum statistical treatment of the problem. It follows that all nondiagonal elements of the density operator are discarded during the *whole time interval* Δt that is relevant for the process under consideration.

Thus the crucial step of discarding the quantum mechanical coherencies and introducing "dissipation" is achieved by an ad hoc assumption. Moreover, to discard the nondiagonal elements of ρ during the whole Δt "is an unwarranted mutilation of quantum mechanics" (see Ref. 305).

2. On the Linear Response Treatment

In Section III.C.2 we presented the standard derivation of the adiabatic FDT within linear response theory (LRT). That derivation yielded the result (3.C51), which obviously is analogous to Eq. (3.B19) in Gordon's treatment.

An "unessential" difference between these two formulas concerns the ω-dependent factor, which connects the absorption coefficient (or rate of absorbed irradiation energy) $Q(\omega)$ with the Fourier transform of the time autocorrelation function of the dipole moment operator $M(t)$. This difference is due to the fact that in Gordon's result the one-sided correlation function $\langle \mathbf{M}(0)\mathbf{M}(t) \rangle$ appears, whereas the standard LRT result makes use of the *symmetrized* quantal correlation function $C_{MM}(t)$.* It can be shown that the perturbational treatment of the FDT can be worked out in such a way that $C_{MM}(t)$ appears and that, therefore, the two treatments mentioned above yield identical results for the FDT.[308]

---

*In this context the symmetrized correlation function, rather than the one-sided correlation function, is of physical significance.

In the previous subsection it was pointed out that the perturbational treatment of FDT is subject to certain serious weaknesses. Thus one would expect that the LRT treatment might have analogous serious drawbacks. That this is indeed true can be shown in the following way. That derivation started with Eqs. (3.C33) and (3.C35):

$$Q_{\text{ad}} = \left\langle \frac{\partial \mathbf{H}_C}{\partial t} \right\rangle = \left\langle \frac{\partial \mathbf{H}_I}{\partial t} \right\rangle \tag{3.D2}$$

This formula is usually believed to be "exact," in the sense that it is proved with the aid of standard statistical mechanics. [As an example, see Ref. 289, p. 74. This "proof" was presented in Section III.C.2 by means of Eqs. (3.C31)–(3.C35).]

But this is by no means correct. The above formula is physically meaningful *only if* the external disturbance $E(t)$ varies very slowly in time (see, e.g., Ref. 309, §§ 11 and 125; Ref. 310, § 34). This necessary condition physically means that the period $2\pi/\omega$ of the (monochromatic) disturbance $E(t)$ should be much greater than the characteristic relaxation time of the system under consideration,

$$\frac{2\pi}{\omega} \gg \tau_m \tag{3.D3}$$

In this case one can take for granted that the system under consideration, being initially in thermal equilibrium, persists in (quasi-)equilibrium during the disturbance. Thus one can make use of the ergodic hypothesis and, therefore, replace the actual time average (of quantity $\partial \mathbf{H}_c/\partial t$) with an ensemble average, in agreement with statistical mechanics.

But condition (3.D3) is strictly in *contradiction* to the physical conditions concerning the process under consideration[275] [see relations (3.A7)]. For the FIR absorption experiment one needs an electromagnetic field with cyclic frequency ω satisfying the relation $2\pi/\omega \approx \tau_m$, and by no means (3.D3).

An external mechanical disturbance $\mathbf{H}_I$ that meet requirement (3.D3) is called an *adiabatic* (or quasi-static adiabatic[311]) *process*. Therefore, the FDT derived in Section III.C was called the adiabatic FDT (see also Ref. 91).

3.   On the Kubo Linearization of the Density Operator

In Section III.C.1, the Kubo linearization procedure for the complete density operator $\rho_c(t)$

$$\rho_c(t) \rightarrow \rho_{\text{eq}} \tag{3.D4}$$

has been reported. Here ρ_{eq} is the canonical density operator with respect to the unperturbed Hamiltonian H_0. Recently, this linearization has been very critically discussed by van Kampen[278,280] and van Vliet[95] (see also Refs. 92, 275, and 279).

Some of those critical considerations are related to the physical context of present interest as well. First, we note that the Kubo relations (Section III.C) have a special appeal since they deal directly with the *microscopic* quantum mechanical motion of a process. Moreover, the derivation of the Kubo formulas for the system response is by many believed to be *exact*, except for the linearization in the applied external field.

As is well known (see Section III.C.1), this "linearization" consists in the replacement (3.D4). As van Kampen has pointed out, linearity of the macroscopic response $\langle B(t) \rangle$ does *not* mean that the equation of motion for $\rho_c(t)$ should also be solved to first order in the external field $F(t)$.

The equation of motion for $\rho_c(t)$, however, is equivalent to the Schrödinger equation for the whole many-body system and contains therefore all the details of the microscopic motion of all individual particles. It is clear that this microscopic motion is much more affected by a given field $F(t)$ than is the macroscopic response $\langle B(t) \rangle$; in other words, linearity of the microscopic motion is entirely different from macroscopic linearity. Thus, the linearization procedure (3.D4) that is so crucial to the Kubo theory seems to be physically unjustified.[278,280]

Recently, van Vliet has suggested that the linearization procedure (3.D4) becomes reasonable if one can greatly reduce the time interval Δt over which this linearization must work. If the system of interest shows a small relaxation time (τ_m, for the case under consideration), then the above linearization is possibly good for Δt of the order of the relaxation time.[95] Thus, the physical context under consideration (Section III.A.2) can be "approximately" treated with LRT—at least as regards the Kubo linearization of the density operator.

Another serious shortcoming of the general LRT formalism consists in the *absence of dissipation*. As van Vliet states: "Linear response theory speaks of dissipation and associated transport coefficients, but nowhere is the dynamics commensurate with dissipation introduced."[95] With respect to the physical context of interest, we know that the molecular reorientational relaxation is caused by the (strong) coupling V between molecular system and its environment. Furthermore, the irradiation energy absorbed and dissipated by the molecular system is a process that accompanies the aforementioned relaxation. It must be pointed out that dissipation (relaxation, and other irreversible processes too) are *not caused* by the mechanical disturbance $H_I = -AF(t)$ (see Ref. 95, p. 1353).

In this context it becomes clear that the coupling V must be taken into account explicitly in the LRT formalism. This has been achieved recently, for the limiting case of weak coupling V.[95,96,312] In the physical context of interest, however, the coupling $\mathbf{V}$ is strong, and thus the linearization (3.D4) as well as the adiabatic FDT show an essential weakness. It is clear that the fluctuating environment of the molecular system under consideration implies that the proper system Hamiltonian $\mathbf{H}_0$ is not a constant of the motion during Δt [see relations (3.A7)]. Of course, this is due to the strong coupling $\mathbf{V}$ and the thermal motion of the molecules. Therefore, one expects that the proper density operator "fluctuates" also. This means that there is an additive time-dependent contribution $\delta\rho(t)$ to the canonical density operator ρ_{eq} (see Sections III.E.1 and III.F.3).

This physical argument is of importance in many other cases where the interaction of molecular systems in condensed matter with external fields are also considered. The above objection against the canonical form ρ_{eq} of the density operator has sometimes been answered thus: one can define a greater relevant system that contains all interacting parts; that is, one considers the *whole* spectroscopic specimen to be the relevant system with Hamiltonian $\mathbf{H}_0$. (Now the coupling with the environment is negligible, at least during Δt $\sim 10^{-12}$ s.) But this trick also does not work successfully, for the following two reasons:

1. As Landau and Lifschitz have pointed out,[309] macroscopic bodies cannot be characterized properly by a static (or stationary) density operator. This is due to the fact that the eigenvalues of the Hamiltonian $\mathbf{H}_0$ are distributed in a quasi-continuous manner. Thus, the standard energy–time uncertainty relation "guarantees" that there are correlations between eigenstates of similar energy. This means that the density operator cannot be assumed to be diagonal in the $\mathbf{H}_0$ representation, as ρ_{eq} does.

2. Now let take for granted that (3.D4) holds (in some approximative manner). Then the adiabatic FDT, Eq. (3.C51), contains the autocorrelation function of the macroscopic dipole moment $\mathbf{M}$. But for the interpretation of the experimental data in terms of microscopic (i.e., molecular) motions, one must extract the molecular autocorrelation function $C_{mm}(t)$ from $C_{MM}(t)$. It is clear that this problem is equivalent to the definition of some proper density operator for the molecular systems constituting the macroscopic spectroscopic probe.[92,97]

From the considerations of this section it follows that the adiabatic FDT must be amended before the treatment of the FIR spectroscopic band shapes in terms of molecular motions can be carried out in a physically meaningful way.

E.   Generalized Linear Response Treatment of Rotational Relaxation

1.   Derivation of the Generalized FDT

In the last section it was pointed out that in the physical context under consideration the necessary conditions for the validity of the adiabatic FDT,

Eq. (3.C51), were violated. Thus a generalization of this theorem had to be found. This was achieved within LRT recently.[89]

We start with the argument that it is not reasonable to regard the absorbing molecules as *isolated* systems. Then we formulate the interaction Hamiltonian

$$\mathbf{H}_I = -\mathbf{m} E(t) \tag{3.E1}$$

Here, $\mathbf{m}$ represents the dipole moment operator of one molecular quantum system and $E(t)$ is the acting external field [Eq. (3.C36)].

As discussed extensively in the last section, we must take into account the fact that Kubo's linearization procedure

$$\rho_c(t) \rightarrow \rho_0(t) \rightarrow \rho_{eq} \tag{3.E2}$$

(see Section III.C.1) is not an accurate description of the molecular relaxation process under consideration. Thus, the linear response treatment must show that the density operator of the problem is not a static quantity, at least in the microscopic time scale; see relations (3.A7). For this reason, the following *ansatz* for the linearization ρ_L of the density operator $\rho_c(t)$ may be useful (see also Section III.F.3):

$$\rho_c(t) \rightarrow \rho_0(t) \rightarrow \rho_{eq} + \delta\rho(t) \equiv \rho_L \tag{3.E3}$$

where $\delta\rho(t)$ represents a fluctuating term representing the fact that the molecular systems under consideration are open systems. (In other words, the environment of a dipolar molecule "fluctuates" in the picosecond time scale; see Section III.A.2.) The fluctuating contribution $\delta\rho(t)$ *must not be confused with* the response term $\Delta\rho(t)$ in Eqs. (3.C5) and (3.C7).

In the following discussion, $\delta\rho(t)$ will be considered to be very small during a time interval Δt of *microscopic* order of magnitude (e.g., some picoseconds). But it will not be assumed that its time derivative $d\delta\rho(t)/dt$ is small, because this assumption would contradict the physical situation described in Section III.A.2. Of course, this statement also has physical significance if one is exclusively interested in the time evolution of the system during the time Δt of the order τ_m or τ_v.

These remarks show that the commutation of $(d\ldots/dt)$ with $\rho_L \equiv \rho_{eq} + \delta\rho(t)$ is problematic. As a precaution, it is supposed that

$$\frac{d}{dt}\mathrm{Tr}(\rho_L \cdot \mathbf{A}) \neq \mathrm{Tr}\left(\rho_L \cdot \frac{d}{dt}\mathbf{A}\right) \tag{3.E4}$$

Recall that the theorems derived in Section III.C.3 do not make use of this commutation.

We can now proceed to find the proper response quantity for the description of the FIR absorption process. As relations (3.A7) are valid, the following illustrative description of the process under consideration can be presented:

During a time interval Δt of the order of 1 ps the molecular system (a) absorbs an FIR photon and (b) relaxes back to equilibrium; thus (c) the absorbed amount of irradiation energy has been transported from the molecular system to the environment (thermal bath), which plainly means that (d) absorbed irradiation energy will be *dissipated during* Δt.

Points (c) and (d), above, make use explicitly of van Vliet's remark concerning the importance of the coupling $\mathbf{V}$ between the molecular system and the environment. Because of this coupling, the molecular system undergoes a fluctuating motion that is determined by $\mathbf{V}$ and *not* by the Hamiltonian

$$\mathbf{H}_c = \mathbf{H}_0 - \mathbf{m}E(t) \tag{3.E5}$$

with $\mathbf{H}_0$ representing the isolated molecular system. During Δt, the amount of the field $E(t)$ as well as the orientation of the molecular dipole moment $\mathbf{m}$ will be changed. The energy transport between field and molecular system can therefore be formally represented by

$$\dot{\mathbf{H}}_I \equiv \frac{d}{dt}\mathbf{H}_I = -\dot{\mathbf{m}}E(t) - \mathbf{m}\dot{E}(t) \tag{3.E6}$$

This formula leads to a generalization of the adiabatic *ansatz* (3.C35), which followed immediately from the term $-\mathbf{m}\dot{E}(t)$. In that treatment the term $-\dot{\mathbf{m}}E(t)$ did not appear because the system was assumed to be isolated. It must be pointed out that the dynamical behavior of the "reorientation" operator $\dot{\mathbf{m}}$ is not determined through the Hamiltonian (3.E5) exclusively, but rather by the Hamiltonian[89-93,275]

$$\mathbf{H}_{\text{open}} = \mathbf{H}_0 + \mathbf{V} - \mathbf{m}E(t) \tag{3.E7}$$

in fully agreement with van Vliet's ideas.[95]

Until now we have referred to $\dot{\mathbf{m}}$ as the "reorientation" of the molecular dipole moment resulting from the many-body coupling $\mathbf{V}$. This remark must be completed by the following: As is well known,[313] owing to the interaction of molecular systems with their environment, there is also an additional *collision-induced* dipole moment $\delta\mathbf{m}$. Usually this contribution to the total dipole moment is of small magnitude compared with the permanent dipole moment of molecular systems (for examples, see Ref. 281). But it does not imply that also the "time derivative" $\delta\dot{\mathbf{m}}$ must be small. [The arguments are the same as in the case of the fluctuating contribution $\delta\rho(t)$ to the canonical density operator ρ_{eq}.] This term also contributes to the "nonadiabatic" part $-\dot{\mathbf{m}}E(t)$ of $\dot{\mathbf{H}}_I$ in Eq. (3.E6).

In the following, the calculation of the response term $\langle -\dot{\mathbf{m}}E(t)\rangle$ is presented.[89,275] The *irradiation energy dissipated per second* Q is given by the ensemble average of the time derivative $\dot{\mathbf{H}}_I$ of the interaction Hamiltonian. Thus it holds that

$$Q = \mathrm{Tr}\,\rho\dot{\mathbf{H}}_I = -\langle\dot{\mathbf{m}}\rangle E(t) - \langle\mathbf{m}\rangle\dot{E}(t) \qquad (3.E8)$$

or, more accurately [see Eq. (3.C37)],

$$\overline{Q(\omega)}^T = -\frac{1}{T}\int_0^T dt\,[\langle\dot{\mathbf{m}}\rangle E(t) + \langle\mathbf{m}\rangle\dot{E}(t)] \qquad (3.E9)$$

This is, of course, a straightforward generalization of the adiabatic *ansatz* (3.C33) or (3.C37). Within the linear response treatment we have

$$\rho = \rho_L \approx \rho_{\text{eq}}$$

but then the possibility of noncommutation (3.E4) must be carefully taken into account.

The second term in the last equation can be formally treated like the quantity $-\langle\mathbf{M}\rangle\dot{E}(t)$ in Section III.C. For the calculation of the first term, the aftereffect function

$$\phi_{\dot{m}m}(t) = \int_0^\beta d\lambda\,\langle\dot{\mathbf{m}}(-i\hbar\lambda)\dot{\mathbf{m}}(t)\rangle \qquad (3.E10)$$

is defined. With the aid of (3.C19) we immediately obtain for the "nonadiabatic" contribution $\Delta Q(\omega)$ to the rate of dissipated irradiation energy

$$\Delta Q(\omega) = -\int_0^\infty ds\,\phi_{\dot{m}m}(s)E(t)E(t-s) \qquad (3.E11)$$

Verification of the equation is straightforward:

$$E(t)E(t-s) = \tfrac{1}{2}\mathrm{Re}\{EE\exp(2i\omega t - i\omega s)\}$$
$$+ \tfrac{1}{2}\mathrm{Re}\{EE^*\exp(-i\omega s)\} \qquad (3.E12)$$

which follows from definition (3.C36). The first term contains an oscillating factor and thus does not contribute to (3.E11) or (3.E9). Therefore it follows that

$$\overline{\Delta Q(\omega)}^T = -\int_0^\infty ds\, \phi_{\dot{m}m}(s)\mathrm{Re}\{\tfrac{1}{2}EE^*\exp(-i\omega s)\} \qquad (3.E13)$$

The Fourier decomposition of the aftereffect function $\phi_{\dot{m}m}$ is now introduced,

$$\phi_{\dot{m}m}(s) = \int_{-\infty}^\infty d\Omega\, \hat{\phi}_{\dot{m}m}(\Omega)\exp(i\Omega s) \qquad (3.E14)$$

Thus one has

$$\overline{\Delta Q(\omega)}^T = -\int_0^\infty ds \int_{-\infty}^\infty d\Omega\, \hat{\phi}_{\dot{m}m}(\Omega)\exp(i\Omega s)\mathrm{Re}\{\tfrac{1}{2}EE^*\exp(-i\omega s)\}$$
$$(3.E15)$$

The integral over s can be evaluated with the aid of the distribution relation (3.C44).

Also it is assumed here that $\phi_{\dot{m}m}(\Omega)$ is analytic in the lower complex Ω plane. This makes it possible to calculate the integral over ω with the same procedure presented in Section III.C.2:

First, one has

$$\overline{\Delta Q^*(\omega)}^T = \int_{-\infty}^\infty d\Omega\, \hat{\phi}_{\dot{m}m}(\Omega)\left\{ \frac{EE^*}{4}\delta_+(\Omega-\omega) + \frac{EE^*}{4}\delta_+(\Omega+\omega)\right\} \qquad (3.E16)$$

From the analytic property of $\hat{\phi}_{\dot{m}m}(\Omega)$ mentioned above it follows that

$$\lim_{|\Omega|\to\infty} \hat{\phi}_{\dot{m}m}(\Omega) = 0 \qquad \text{for}\quad \mathrm{Im}\{\Omega\}\le 0 \qquad (3.E17)$$

[see (3.C46)]. With the aid of integration over the contour shown in Fig. 29 we obtain an analogous way as in Section III.C.2 the intermediate result

$$\mathscr{P}\int_{-\infty}^\infty d\Omega\, \hat{\phi}_{\dot{m}m}\frac{(\Omega)}{(\Omega-\omega)} = -i\pi\hat{\phi}_{\dot{m}m}(\omega) \qquad (3.E18)$$

and, therefore,

$$\overline{\Delta Q(\omega)}^T = -\tfrac{1}{2}\pi EE^*\left[\hat{\phi}_{\dot{m}m}(\omega) + \hat{\phi}_{\dot{m}m}(-\omega)\right] \qquad (3.E19)$$

The aftereffect function $\phi_{\dot{m}m}(t)$ can be expressed through the symmetrized quantum mechanical time-correlation function $C_{\dot{m}\dot{m}}(t)$, Eq. (3.C52). With definition (3.C62) and (3.E10) we have

$$\hat{\phi}_{\dot{m}m}(\Omega) \equiv \hat{K}_{\dot{m}\dot{m}}(\Omega) \tag{3.E20}$$

From theorem (3.C64) we obtain

$$\hat{\phi}_{\dot{m}m}(\Omega) = \frac{2}{\hbar\Omega}\tanh\left(\frac{\beta\hbar\Omega}{2}\right)\hat{C}_{\dot{m}\dot{m}}(\Omega) \tag{3.E21}$$

With the aid of the symmetric property (3.C71), we obtain (with $T = 2\pi/\omega$)

$$\overline{\Delta Q(\omega)}^{-2\pi/\omega} = -\frac{2\pi}{\hbar}|E|^2 \cdot \omega^{-1} \cdot \tanh\left(\frac{\beta\hbar\omega}{2}\right)\hat{C}_{\dot{m}\dot{m}}(\omega) \tag{3.E22}$$

This term completes the adiabatic approximation (C.51). Thus, from (3.E9) there follows the *generalized fluctuation–dissipation theorem*:[89,275]

$$\overline{Q(\omega)}^{-2\pi/\omega} = \frac{2\pi}{\hbar}|E|^2 \cdot \tanh\left(\frac{\beta\hbar\omega}{2}\right) \cdot \left[\omega\hat{C}_{mm}(\omega) - \omega^{-1}\hat{C}_{\dot{m}\dot{m}}(\omega)\right] \tag{3.E23}$$

Let us once again note that the derivation of the generalized FDT given above does take into account the possibility (3.E4), namely, that it is unreasonable to assume the formal commutation

$$\mathrm{Tr}\rho\frac{d\mathbf{A}}{dt} \equiv \frac{d}{dt}\mathrm{Tr}\rho\mathbf{A}$$

Remember that this formal commutation contradicts the validity of the physical conditions underlying the experimental context under consideration, as presented in Section III.A.2.

The generalized FDT reveals an expression for the *microscopic* description of *dissipation* of irradiation energy by quantum systems of molecular dimensions. For the microscopic entropy production $P_{\mathrm{micro}}(\omega)$, we can define[90]

$$P_{\mathrm{micro}}(\omega) = \frac{1}{T}\overline{Q(\omega)}^{-2\pi/\omega} \tag{3.E24}$$

with T being the absolute temperature (of the ensemble).

It has been known since 1909, when Carathéodory defined the term "integrating factor" for the quantity $1/T$, that the term "temperature" is not a primary quantity in thermodynamics but rather a property of an ensemble under certain equilibrium conditions.[314]

Some important physical consequences as well as the weakness of the adiabatic FDT (in the physical context under consideration) will be discussed in Section III.F.

2. Interpretation of the "Anti-Boltzmann" Temperature Dependence of FIR Absorption Bands

In Section III.A.3, some experimental results were presented that seem to contradict the well-known Boltzmann distribution law. Namely, with *increasing* temperature it seems that the *lower* (quasi-)rotational energy levels of the molecular systems become more and more populated at the expense of the population of the higher energy levels. Of course, this conclusion is correct *only if* one assumes the validity of the "usual" form of the FDT, that is, the adiabatic one. The proper treatment of molecular *open* systems yields the generalized FDT (3.E23) that explains the aforementioned experiments without ad hoc assumptions or artificial suppositions. Indeed, according to (E.23) it is the matter of the *difference*

$$\omega \hat{C}_{mm}(\omega) - \omega^{-1}\hat{C}_{\dot{m}\dot{m}}(\omega)$$

which may in fact cause the red shift of the absorption band with increasing temperature.

This can happen because both correlation functions appearing in this expression show a natural "blue shift"; that is, they spread out with increasing temperature. The observed red shift of the FIR absorption band plainly confirms the physical significance of the new term [with $\hat{C}_{\dot{m}\dot{m}}(\omega)$] in the generalized FDT. In other words, this observed anti-Boltzmann temperature effect is *anomalous only with respect to the adiabatic FDT*, Eq. (3.C51).

For completeness, a few additional comments are needed concerning alternative interpretations of the aforementioned experimental result.

Sometimes, Kubo's theory of stochastic modulation[315,316] is referred to in order to explain "narrowing effects" on spectral bands. One could conjecture that the temperature dependence of the band shapes under consideration can be a special case of such effects, in the limiting case of the so-called *fast modulation*. A simple consideration of the important relations (3.A7), however, shows that the necessary conditions for fast modulation are not to be satisfied in this physical context.[89]

Another attempt to understand the red shift of the FIR bands by increasing temperature is offered by the following: A rigorous *coupling between rotation and translation* is suggested.[292] This coupling would make it possible that the observed FIR absorption bands are due to rotations *and* translational motions of the polar molecules. The above temperature effect is contrary to that anticipated for a purely rotational band (with respect to the *adiabatic* FDT). Thus, this effect seems to emphasize the translational character of the absorption bands under consideration.

This interpretation of the effect under consideration has recently been repeated and even emphasized (see Ref. 277, pp. 422–425).

It is interesting to consider the quantum mechanical procedure that would allow one to take into account that rotation–translation interaction: one would have to introduce an additional contribution to the complete molecular Hamiltonian representing this interaction. This means that the coupling operator V in Eq. (3.E7) would contain a component appropriate to the rotation–translation coupling. It must be pointed out that the Hamiltonian H_I [Eq. (3.E1)], which represents the interaction between polar molecules and the acting external field, is not affected by the rotation–translation coupling under consideration (at least, in a first approximation). Therefore, the generalized FDT, Eq. (3.E23), also takes into account the possibility that the aforementioned rotation–translation interaction does really exist. The corresponding interaction Hamiltonian would then entail a special relation between the correlation functions $C_{mm}(t)$ and $C_{\dot{m}\dot{m}}(t)$ appearing in the generalized FDT.[317]

F. Irradiation Energy Dissipation in the Far-Infrared

1. On Vanishing Dissipation Within the Standard LRT

Recently, van Vliet and colleagues (see Van Vliet[95,96] and Charbonneau et al.[312]) presented a "revisited linear response theory" that does not show the weakness of the standard version of LRT, as discussed in Section III.D. Unfortunately, the given physical conditions (see Section III.A.2) do not permit to make use of this improved LRT in the case under consideration. Nevertheless, the generalized FDT makes contact with basic concepts of the theory.

As van Vliet states,[95] "Linear response theory speaks of dissipation and associated transport coefficients, but nowhere is the dynamics commensurate with dissipation introduced." And additionally, "In the formalism [of LRT] as it stands no dissipation is manifest."

With the aid of the generalized FDT it will be shown that the rate, $Q(\omega)$, of irradiation energy dissipated in the molecular process under consideration vanishes identically in the standard version of LRT. For this purpose we consider a very dilute gas consisting of dipolar molecules. Now the coupling V is null and the density operator can be chosen to be equal to the canonical operator ρ_{eq} exactly. Furthermore, it holds that

$$\dot{m}(t) = \frac{1}{i\hbar}[m(t), H_0] \tag{3.F1}$$

Now the calculation is straightforward:

$$\begin{aligned}
\frac{d^2}{dt^2}\text{Tr}\rho_{eq}m(0)m(t) &= \left(\frac{1}{i\hbar}\right)^2 \text{Tr}\rho_{eq}m(0)\big[[m(t), H_0], H_0\big] \\
&= \left(\frac{1}{i\hbar}\right)^2 \text{Tr}[m(t), H_0]\big[H_0, \rho_{eq}m(0)\big] \\
&= \frac{1}{i\hbar}\text{Tr}\dot{m}(t)\big\{H_0\rho_{eq}m - \rho_{eq}m(0)H_0\big\} \\
&= \frac{1}{i\hbar}\text{Tr}\dot{m}(t)\rho_{eq}[H_0, m(0)] \\
&= \text{Tr}\dot{m}(t)\rho_{eq}(-\dot{m}(0)) \\
&= -\text{Tr}\rho_{eq}\dot{m}(0)\dot{m}(t)
\end{aligned} \tag{3.F2}$$

The second equation holds because $\text{Tr}A[B, C] = \text{Tr}B[C, A]$; the fourth equation follows from $[\rho_{eq}, H_0] = 0$. For the symmetrized time-correlation function $C_{mm}(t)$, therefore, we have

$$\frac{d^2}{dt^2}C_{mm}(t) = -C_{\dot{m}\dot{m}}(t) \tag{3.F3}$$

Here it holds that $\partial\rho_{eq}/\partial t \equiv 0$, and thus, by Fourier transformation, it follows that

$$\omega^2\hat{C}_{mm}(\omega) = +\hat{C}_{\dot{m}\dot{m}}(\omega) \tag{3.F4}$$

In other words, the generalized FDT, (3.E23), yields[90,91,275]

$$\overline{Q(\omega)}^{2\pi/\omega} \equiv 0 \tag{3.F5}$$

Thus, within standard LRT there is no dissipation of irradiation energy. At first sight this result is unexpected and seems to contradict the validity of the generalized FDT. But an examination revealing the *physical* conditions for (3.F5) to be valid shows that this result is in fully agreement with basic physical concepts. Namely, Eq. (3.F1) states that the systems under consideration are *isolated* molecules. Such systems can *absorb* FIR irradiation energy and make rotational transitions; but they *cannot dissipate* this energy amount owing to the trivial fact that there is no contact with a thermal bath. Note that (a) "excitation of well defined quantum states by *absorption* of irradiation energy" and (b) "*dissipation* of absorbed irradiation energy" are *not identical* physical processes, at least conceptually. Dissipation (b) is a typical *irreversible* process and cannot be described meaningfully by the one-body process treated above. As van Vliet states, dissipation is due to the many-body interaction that must be embodied in the formalism of LRT.

Of course, process (a) belongs to the domain of the adiabatic FDT. The above result (3.F5) also plainly shows the weakness of this FDT with respect to the dissipative process (b), because one now has $\overline{Q_{ad}(\omega)} > 0$. It follows that Eq. (3.F51) possibly describes "absorption" processes but not dissipative processes, for example, the process that has been described explicitly in Section III.A.2.

These remarks, and especially Eq. (3.F5), confirm van Vliet's statement concerning vanishing dissipation in the standard LRT formalism.

2. Non-Markovian Coupling Operator and Dissipation

In the following discussion we consider a molecular system during time intervals Δt of a microscopic order. Under the experimental conditions concerning liquids (at standard temperature and pressure) this roughly means

$$\Delta t \sim 10^{-12} \text{ s} \qquad (3.F6)$$

[see Section III.A.2]. Because the molecular system is strongly coupled with its environment [see relations (3.A7)], it is impossible to determine a system Hamiltonian $\mathbf{H}_0$ that is proper for the dynamical description of the system during the *whole* time interval Δt. The formal reason for this is that, during Δt, the projectors $\mathbf{P}_j(t)$ as well as the energy eigenvalues $E_j(t)$ defining the Hamiltonian, for instance

$$\mathbf{H}_0 = \sum E_j(t)\mathbf{P}_j(t) \qquad (3.F7)$$

are not constants of motion. Additionally, it is crucial that now neither the *sudden* nor the *adiabatic* approximation[299] for the evolution operator of the molecular system can be applied.

In the case under consideration, therefore, some coarse graining and/or time-smoothing procedures must be carried out on the bath degrees of freedom, which appear in the system–bath coupling Hamiltonian $\mathbf{V}$. Because this coupling is strong, these "approximations" are not easy to work out with sufficient accuracy.

The foregoing remarks show that it is reasonable to make the *ansatz*

$$\frac{d}{dt}\mathbf{m}(t) = -\int_0^t \phi(t-s)\mathbf{m}(s)\,ds + \mathbf{f}(t) \qquad (3.F8)$$

for the dynamics of the operator **m** that couples the molecular system with the external field. This equation sometimes is called the *generalized Langevin equation*. Here occurs the memory kernel ϕ that is due to the many-body interaction between system and thermal bath.

Here it should be stressed that Eq. (3.F8) must not be confused with Mori's equation.[318] Thus, the memory function ϕ and the "nonsecular" force $\mathbf{f}(t)$ are not related through the *second fluctuation dissipation theorem*.[248,250] This is because the Mori equation of motion (being exact) is derived with respect to *isolated* systems. In the physical context under consideration, the molecular systems are definitely not isolated systems. Thus, the so-called Zwanzig–Mori[318-324] projection methods cannot formally prove the validity of (3.F8).

Usually Eq. (3.F8) is called *non-Markovian* because it involves properties of **m** not only at a single time t.

In the following it will be assumed that the density operator ρ_L appearing in the time-correlation functions $C_{mm}(t)$ and $C_{\dot{m}\dot{m}}(t)$ is *exactly* a *static* quantity. In other words, we will assume the commutation (**A**: arbitrary operator, $\rho_L = \rho$)

$$\frac{d}{dt}\operatorname{Tr}\rho\mathbf{A} = \operatorname{Tr}\rho\left(\frac{d}{dt}\mathbf{A}\right) \tag{3.F9}$$

(The physical meaning of this relation was discussed critically in Section III.E.1.)

From Eqs. (3.F8) and (3.F9) a *contradiction* can be shown, especially in the *short-time range* (3.F6); this result, of course, plainly reconfirms that assumption (3.F9) is physically meaningless within the physical context under consideration.

First, a formal "multiplication" of Eq. (3.F8) with $\operatorname{Tr}\rho\mathbf{m}(t_0)$ yields

$$\frac{d}{dt}\operatorname{Tr}\rho\mathbf{m}(t_0)\mathbf{m}(t) = -\int_0^t \phi(t-s)\left[\operatorname{Tr}\rho\mathbf{m}(t_0)\mathbf{m}(s)\right]\,ds + \operatorname{Tr}\rho\mathbf{m}(t_0)\mathbf{f}(t)$$

$$\tag{3.F10}$$

The equation obtained from Eq. (3.F10) by neglecting the last term on the right-hand side is not to be confused with the well-known equation describing the time evolution of the correlation function $C_{mm}(t)$; see Ref. 325. Now let us consider the short-time limit of Eq. (3.F10), which here means

$$0 < t - t_0 \leq \tau_\phi \tag{3.F11}$$

where τ_ϕ is the decay time of the memory function ϕ. Additionally it is assumed that

$$t \gg \tau_\phi \tag{3.F12}$$

Since stationarity holds, the left-hand side of Eq. (3.F10) reads (with $t - t_0 = \varepsilon$)

$$\frac{d}{dt} \mathrm{Tr} \rho \mathbf{m}(t_0) \mathbf{m}(t) = \frac{d}{d\varepsilon} \mathrm{Tr} \rho \mathbf{m}(0) \mathbf{m}(\varepsilon) \tag{3.F13}$$

This quantity is of the order of ε, and it vanishes if $\varepsilon \to 0$. The first term on the right-hand side of Eq. (3.F10)

$$-\int_0^t \phi(t - s) [\mathrm{Tr} \rho \mathbf{m}(t - \varepsilon) \mathbf{m}(s)] \, ds \tag{3.F14}$$

remains much greater than ε, and in the limit $\varepsilon \to +0$ it tends to a constant value independent of t if the limit (3.F12) is taken into consideration.

The second term on the right-hand side of Eq. (3.F10) is usually assumed to vanish identically for $t_0 = 0$, but for other values, $t_0 \neq 0$, it is not a constant but rather a complicated time function, which will not be further specified here. Thus, a *contradiction* is derived, because the left-hand side is of the order of ε for each t obeying (3.F12), whereas the right-hand side generally depends on the special value of the parameter t, Q.E.D.

The physical meaning of this formal result gains in clearness if we consider the *Markovian* limit of vanishing memory, which here means the limit

$$\phi(t) \to a(t) \cdot \delta(t), \qquad a(t) \neq 0.$$

From Eq. (3.F10) it follows that

$$\frac{d}{dt} \mathrm{Tr} \rho \mathbf{m}(t_0) \mathbf{m}(t) = -\int_0^t a(t - s) \cdot \delta(t - s) [\mathrm{Tr} \rho \mathbf{m}(t_0) \mathbf{m}(s)] \, ds$$
$$+ \mathrm{Tr} \rho \mathbf{m}(t_0) \mathbf{f}(t) \tag{3.F15}$$

For $t_0 \to t - \varepsilon$ and

$$|t - t_0| \leq \tau_m \tag{3.F16}$$

there is a *contradiction* again: the left-hand side has the order of magnitude of ε, whereas the right-hand side will be equal to

$$a(0) \cdot [\mathrm{Tr} \rho \mathbf{m}(t - \varepsilon) \mathbf{m}(t)] + \mathrm{Tr}[\rho \mathbf{m}(t - \varepsilon) \mathbf{f}(t)]$$

Additionally, it is instructive to consider the "long-time" limit $|t - t_0| \gg \tau_m$, τ_ϕ; $t \to \infty$. From Eqs. (3.F10) or (3.F15) it follows that

$$0 = \text{Tr}\,\rho\,\mathbf{m}(t_0)\mathbf{f}(t) \tag{3.F17}$$

This result is *physically meaningful*. It says that for "long times" the non-secular force $\mathbf{f}(t)$ can be considered as "smoothed out" or "fully randomized." In this case one can neglect the nonsecular force.

These formal results show the following: Within the physical context under consideration one *cannot* assume the validity of the commutation (3.F9). Furthermore, it follows immediately that the mathematical proof (3.F2) of Eq. (3.F4)

$$\omega^2 \hat{C}_{mm}(\omega) = \hat{C}_{\dot{m}\dot{m}}(\omega)$$

is now irrelevant because it makes use of the commutation (3.F9).

In other words, in the short-time range considered above, the generalized FDT yields a rate of irradiation energy dissipation that does not vanish identically. From the second law of thermodynamics one obtains the relation

$$\omega\hat{C}_{mm}(\omega) \geq \omega^{-1}\hat{C}_{\dot{m}\dot{m}}(\omega) \tag{3.F18}$$

This is a novel relation between the time autocorrelation functions of the molecular operators $\mathbf{m}$ and $\dot{\mathbf{m}}$.[89]

3. Nondiagonal Density Operator and Dissipation

In previous sections it was pointed out that for the "linear approximation" ρ_L of the complete density operator ρ_c the *ansatz* (3.E3)

$$\rho_L(t) \equiv \rho_{eq} + \delta\rho(t)$$

must be considered (at least in the physical context presented in Section III.A.2). Here the fluctuating term $\delta\rho(t)$ must not be confused with the "response" $\Delta\rho(t)$ within standard LRT, as presented in Section III.C. Remember that ρ_{eq} represents the *canonical* density operator. This operator is diagonal in the representation of the unperturbed Hamiltonian $\mathbf{H}_0$, and is thus a static quantity (with respect to standard LRT).

For the derivation of the generalized FDT (see Section III.E.), a different viewpoint (compared with standard LRT) has been taken, as follows: we take into account the fact that the coupling operator $\mathbf{V}$ between *molecular* systems and their environment is not small; but if we are interested in the time evolution of the molecular system during a *small* time interval Δt (that is, $\Delta t \sim 10^{-12}$ s), then we can choose a Hamiltonian $\mathbf{H}_0$ that describes the molecular system as being quasi-isolated. In other words, for the molecular system Hamiltonian we have the *ansatz*

$$\mathbf{H}_0 + \lambda\mathbf{V} \tag{3.F19}$$

with *small* parameter λ, during Δt.

This *ansatz* permits us to take into account van Vliet's criticism (see Section III.D.3). With respect to a small Δt, therefore, the *ansatz* (3.E3) for the density operator in the "linear approximation" becomes physically meaningful: the fluctuating term $\delta\rho(t)$ will now represent a small contribution to the proper density operator $\rho_L(t)$; the main contribution ρ_{eq} is assumed to be diagonal in the $\mathbf{H}_0$ representation (see Ref. 95, p. 1353).

Clearly, ρ_L cannot be taken for granted as being exactly diagonal (in the $\mathbf{H}_0$ representation) at all times (in Δt). This is the case because then one would have $[\rho_L, \mathbf{H}_0] = 0$; from the Von Neumann–Liouville equation it then would follow that ρ_L is determined by $\lambda\mathbf{V}$ alone, which is clearly wrong.[326]

Of course, the above physical arguments concerning the justification of *ansatz* (3.E3) can be formulated in more formal terms, too.[275] Let $\mathbf{P}$ be the superprojector extracting the diagonal part ρ_d of the density operator $\rho_L(t)$:

$$\mathbf{P}\rho_L(t) \equiv \rho_d(t)$$
$$(\mathbf{1} - \mathbf{P})\rho_L(t) \equiv \rho_{nd}(t) \tag{3.F20}$$

With the aid of the Liouvillean $\mathbf{L}$,

$$\mathbf{L}\cdots \equiv \frac{1}{\hbar}[\mathbf{H}_0 + \lambda\mathbf{V}, \ldots] \tag{3.F21}$$

we straightforwardly obtain[96] from the Von Neumann–Liouville equation

$$\frac{\partial}{\partial t}\rho_L(t) + \frac{1}{i\hbar}[\rho_L(t), \mathbf{H}_0 + \lambda\mathbf{V}] = \frac{1}{i\hbar}[\rho_L(t), \mathbf{m}]E(t) \tag{3.F22}$$

the coupled pair

$$\frac{\partial}{\partial t}\rho_d + i\mathbf{P}\mathbf{L}\rho_{nd} = \frac{1}{i\hbar}E(t)\mathbf{P}[\rho_d + \rho_{nd}, \mathbf{m}] \tag{3.F23}$$

and

$$\frac{\partial}{\partial t}\rho_{nd} + i\mathbf{L}\rho_d + i(\mathbf{1} - \mathbf{P})\mathbf{L}\rho_{nd} = \frac{1}{i\hbar}E(t)(\mathbf{1} - \mathbf{P})[\rho_d + \rho_{nd}, \mathbf{m}] \tag{3.F24}$$

[Here the abbreviations $\rho_d \equiv \rho_d(t)$ and $\rho_{nd} \equiv \rho_{nd}(t)$ have been used.] Thus it is clear that the time-independent diagonal matrix elements of ρ_L constitute the operator ρ_{eq}, whereas the nondiagonal matrix elements of ρ_L (as well as the time dependent part of ρ_d) constitute the "fluctuating" operator $\delta\rho(t)$.

Hence, it becomes obvious that the "proof" (3.F2) cannot be carried out in the case under consideration. The reason is that the time derivative

$$\frac{d}{dt}\,\delta\rho(t)$$

also must not be small in cases where $\delta\rho(t)$ itself can be taken for granted to be small. More precisely, from the smallness of the ensemble average $\mathrm{Tr}\,\delta\rho(t)A$ it *does not* follow that

$$\mathrm{Tr}\,A\!\left(\frac{d}{dt}\,\delta\rho(t)\right) \ll \mathrm{Tr}\,\dot{A}\rho_{eq}$$

The term on the left-hand side is of importance if the "relaxation time" corresponding to A is of the same order of magnitude as the characteristic fluctuation time of $\delta\rho(t)$ and/or λV. Thus we conclude that in the case under consideration [see (3.A7)] it holds that

$$\frac{d}{dt}\,\mathrm{Tr}\,\rho_L(t)m \approx \mathrm{Tr}\,\rho_{eq}\!\left(\frac{d}{dt}\,m\right) + \mathrm{Tr}\,m\!\left(\frac{d}{dt}\,\delta\rho(t)\right) \qquad (3.\mathrm{F}25)$$

Thus, the noncommutation (E.4), being crucial for the derivation of the generalized FDT, has been justified. Therefore, the novel inequality (3.F18) must be valid.

It is clear that the time dependence of $\delta\rho(t)$ cannot be determined through the Von Neumann–Liouville equation with the Liouvillean

$$L_0 \cdots = \frac{1}{\hbar}[H_0, \ldots]$$

As concerns the time dependence of the proper density operator, *ansatz* (3.E3), the following point should be noted. In standard LRT calculations, there is a change from the Schrödinger to the Heisenberg representation that permits us to derive the closed formulas for the system response [e.g., (3.C18), (3.C19), and (3.C29)]. But it is not this change of representation that permits the linear response density operator to become a time-independent quantity. It is rather the ad hoc specification (3.C15)

$$\rho_0(t) \to \rho_{eq}$$

for the unperturbed density operator (ρ_{eq} being the canonical density operator) which entirely removes its time dependency; see Section III.C.1.

It is also worthwhile to note that the fluctuating term $\delta\rho(t)$ can be of importance in other spectroscopic fields. This term is due to the coupling λV, which implies specific coherences (or synergetic effects) accompanying the process under consideration. In the case of, say, an emission process like luminescence, the Hamiltonian λV effectuates specific fluctuations in the emitted photon current. These fluctuations, having a dynamical origin, are of nonstochastic character. This has been shown by Chatzidimitriou-Dreismann, with the aid of the Prigogine theory of star-unitary transformations; additionally, this author has indicated how they can be measured by a special wavelength-selective photon-counting experiment.[327]

G. Outlook

The foregoing discussion clearly showed the many-body character of the problem under consideration, that is, molecular motions and dynamics in liquids (and, more generally, in disordered condensed matter). The physical parameters and/or conditions discussed in Section III.A.2 indicate that the aforementioned problem is directly linked with the problem of explaining irreversibility on a mechanical basis. If the coupling V between molecular quantum systems and their environment is strong, the standard derivation of FDT cannot be physically justified. Moreover, the applicability of standard LRT formalism in the context under consideration appears to be doubtful. In this regard, however, it should be mentioned that van Vliet's work discussed above as well as some recent work by Rhodes[328,329] explicitly show how the coupling V can be embodied in the LRT formalism.

In future papers it is intended to study the infrared and Raman vibrational lines of molecules in liquids with the aid of generalized FDT. It is worthwhile to emphasize that vibrational modes of even small molecules (like CH_3I) usually decay very fast, that is, in the picosecond time range.[332-334] In this case, the well-known treatment of vibrational bands (i.e., in terms of the Gordon theory) must be amended as well.

It is just beginning to be appreciated that the term "density operator" must be considered as the *primary* concept (with respect to the term "state vector") in statistical mechanics.[333-338] In view of this remark it should be clear that the canonical form ρ_{eq} (as used within standard LRT) corresponds to some mutilation of the density operator ρ in the sense that ρ_{eq} does not take into account the quantum mechanical coherencies between the projectors corresponding to ρ_{eq}. As discussed earlier, the existence of a strong coupling V implies that the above mutilation of ρ is a "nonsatisfactory approximation," at least for the description of the physical process under consideration. It is assumed that a more appealing as well as "exact" treatment of such problems as FIR absorption, molecular motion, and excited state depopulation can be achieved by means of Prigogine's theory of star-unitary transformations[338-344] (see also Refs. 69 and 70).

IV. CONCLUDING REMARKS: FREEZING AND MELTING

Theoretical approaches to the microscopic description of freezing can be developed only as our understanding progresses concerning the structure and dynamics of molecular liquids.

The kinetics of freezing differs from the kinetics of melting at least in systems of nonspherical molecules. While melting could be assumed to be a stepwise process, the freezing of molecular liquids should be assumed to involve cooperative phenomena, perhaps proceeding via metastable solid states, and the nucleation might even be phonon induced in order to transform a certain subassembly of the molecular liquid into the crystalline orientational symmetry.

In contrast to common theories of the nucleation of rare gases and of inorganic ionic materials, in the present context one should ask for the factors that favor supercooling, for the nature of the phase transition from liquids to glasses, and for the mechanisms by which liquid crystals and other mesophases are formed. It is anticipated that, by applying some time-dependent perturbation treatment to hard-core molecular liquid models, some of the underlying problems may well be solved in the near future.

Some first attempts at computer simulation methods have been reviewed in Ref.345. Nucleation in a three-dimensional, amorphous Lennard-Jones system was first observed accidentally by molecular dynamics.[346] Later on, freezing was analyzed in the framework of the Landau theory, reasoning that, close to the melting line, fluctuations in a BCC-like fluid are uniquely favored on the basis of general symmetry arguments; the Landau theory suggests that for weak first-order transitions BCC should be the stable solid phase at coexistence and, even when other solid phases are actually more stable, nucleation will occur preferentially into a metastable BCC phase.[347]

In order to demonstrate that the kinetics of freezing are not the reverse of the kinetics of melting, in the following discussion attention will be drawn to the behavior of acetonitrile. It will be shown that any assessment of theoretical molecular mechanisms of melting and freezing must include the ability to describe hysteresis effects. This behavior of acetonitrile, of course, is by no means a special one. Comparable effects have been observed for arsine, AsH_3, and quantitatively explained by a memory function approach and by a phonon contribution to the dipolar autocorrelation function.[348]

The pair correlation function for acetonitrile at room temperature (Figs. 30 and 31)[162] allows us to describe the structure of the liquid by antiparallel

oriented molecular chains, at least for the first shell around a certain molecule (Fig. 30), but for outer parts of the cage around that molecule the deviations from the chain structure of the liquid become significant.

There are at least two solid state phases known for acetonitrile, the
α-phase and the β-phase.[349] An additional γ-phase has been postulated by
Jakobsen and Mikawe[350] on the basis of pressure-dependent infrared spectroscopic observations at low temperature. The α-phase is orthorombic with
the lattice parameters: $a = 11.1$ Å, $b = 8.12$ Å, and $c = 5.09$ Å; it has eight
molecules per unit cell, which are all orientated axially and antiparallel, as
has been observed at $-192°$C by Pace and Noe.[351] The intermolecular distances in two directions are similar to those in the liquid state, but in the
perpendicular axial direction the distance is significantly shorter, so that
specific interactions should be discussed. The corresponding antiparallel intermolecular vibration increases its wave number from 87 cm^{-1} in the liquid
state to 116 cm^{-1} in the β-phase and at least 119 cm^{-1} in the α-phase,

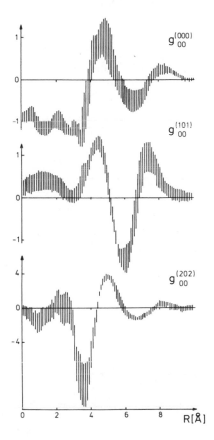

Fig.. 30. Expansion coefficients of the
molecular pair correlation function for liquid
acetonitrile at 20°C.[162] (By permission of *Mol.
Phys.*)

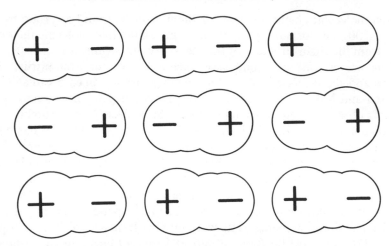

Fig. 31. Acetonitrile: axial and antiparallel orientation of the molecular dipoles—libration perpendicular to the axis.[58] (By permission of John Wiley & Sons Ltd.)

whereas the wave number of the axial intermolecular vibration remains almost constant with phase transitions. No X-ray crystal structure determination of the β-phase has yet been achieved. Casabella and Bray[352] have observed the solidification by nuclear quadrupole resonance measurements. The process of solidification was also studied by Putnam et al.[353] via calorimetric measurements. They found a decrease of molar enthalpy for the β-phase (Fig. 32).

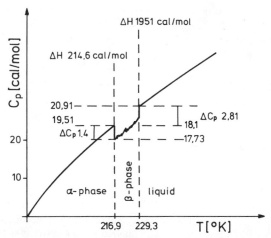

Fig. 32. The molar enthalpy of acetonitrile versus temperature.[58] (By permission of John Wiley & Sons Ltd.)

The temperature dependence of the extinction in certain absorption regions of acetonitrile along the phase transitions in question, as shown in Figs. 33 and 34, possesses some properties that should be investigated further and analyzed much more carefully. The curves seem to depend on the nature and on the class of symmetry of the corresponding vibrational modes, but those effects will not be discussed here in detail. Rather, we wish to draw the attention of the reader to the fact that the curves for freezing and melting in all cases are definitely not the same but show a hysteresis effect.[58]

The gradients and the points of inflection of the intensity of almost all normal and combination modes versus temperature suggest that there are presolid states and metastable processes. As has been reported in Ref. 58, by cooling very slowly one can observe a presolid state with a lifetime of less than half a second. It is characterized by a rapid increase of the intensities of the normal modes, coupled with a rapid decrease of the intensities of the combination modes. A definite melting point is not found when reheating, but rather a transition region with a shape similar to a titration curve. The

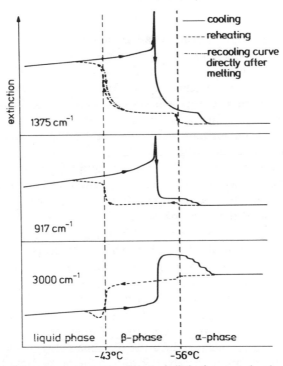

Fig. 33. Extinction versus temperature of acetonitrile in the spectral regions of acetonitrile phase transitions: ν_3 symmetric CH_3 deformation, ν_4 C—C stretch, and ν_5 CH_3 asymmetric stretch, for cooling and reheating.[58] (By permission of John Wiley & Sons Ltd.)

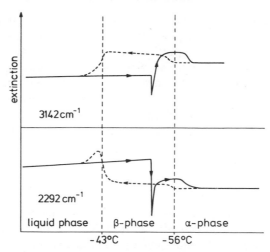

Fig. 34.   Extinction versus temperature of acetonitrile in the special regions of acetonitrile phase transition: $\nu_2 + \nu_4$ (C—N + C—C), $\nu_3 + \nu_4$ (CH$_3$ def + C—C), for cooling and re-heating.[58] (By permission of John Wiley & Sons Ltd.)

beginning of the phase transition can already be observed about 8° (celsius) before melting. The actual melting region is contained within 2°. Directly after melting, the liquid structure is different from that of the solid just before freezing at the same temperature. After some minutes or after reheating, the old situation returns. On cooling directly after melting, the shape of the solidification is similar to the melting curve and there is no sign of a presolid state. After melting there might still be some antiparallel charge-transfer complexing, and these "complexes" influence the process of solidification, which might be described via an equilibrium of aggregates. In the stable liquid state there is no sign of these antiparallel complexes; they seem to be hindered by the free rotation around the axis, which appears to be a typical attribute of the liquid phase.

The presolid state can be observed just before freezing and does not depend on the temperature of freezing. This state seems to be necessary for freezing when cooling from the stable state of the liquid. The sudden increase of the maximum intensities of the normal modes accompanied by a sudden decrease of the maximum intensities of the combination modes is a sign of steep and harmonic potentials. It seems to be due to cooperative phenomena originating from a coupling of molecular oscillations. This cooperative oscillation might reduce the distance between the molecules and destroy the free rotation, thereby permitting the properties of the solid state. The observed sensitivity of the solidification should then be a question of the sensitivity of the cooperative effect to perturbations.

Collective modes might be the basis of the cooperative phenomena. The idea of collective modes in polar liquids has been proposed by Lobo et al.[354] From the reflection of obliquely incident linearly polarized FIR radiation, Ascarelli[355] detected a longitudinal and a transverse wave in liquid nitromethane and compared them with computer calculations. Acetonitrile should have collective modes, too, because of its higher dipole moment. A distinct coupling between molecular oscillators can be deduced from the local asymmetry in the structure of liquid acetonitrile, as revealed by X-ray analyses; Kratochwill et al. have calculated an asymmetry in the perpendicular intermolecular distance, although the perpendicular potentials of the molecules should be symmetric.[161]

Even in 1976 the dynamic behavior of coupled quartic oscillators was investigated, instead of a Hamiltonian of linearly coupled anharmonic oscillators; for single-well potentials and weak coupling, the displacement spectral function shows a single peak due to phonon vibrations.[356]

Most recently Gareth J. Evans compared the frequencies of the collective modes in the β solid phase of acetonitrile with the composite structure of the 100 cm^{-1} absorption band of the liquid phase (which eventually is removed, and a broad profile is obtained when acetonitrile is dissolved in a nonpolar solute), and his results strongly support our above-mentioned proposals,[58] namely (1) the suggestion of a strong local lattice structure, (2) the prediction of the existence of collective modes, and (3) the postulation of the existence of strong cooperative phenomena originating from a coupling of the molecular oscillations in the pure liquid, and Evans concluded: "We cannot over-emphasize the extreme importance of these observations for the elucidation of the liquid state of matter in general if they are substantiated".[357]

Acknowledgments

Support of this work by the Bundesminister für Forschung und Technologie, Förderkennzeichen 05286 LI, by the Deutsche Forschungsgemeinschaft, and by the Fonds der Chemischen Industrie is gratefully acknowledged.

REFERENCES

1. N. G. van Kampen, in *Perspectives in Statistical Physics*, H. J. Raveché, ed., North-Holland Publs., Amsterdam, 1981.

2. A. D. Buckingham, E. Lippert, and S. Bratos, eds., *Organic Liquids: Structure, Dynamics and Chemical Properties*, Wiley, Chichester, 1978.

3. F. Kohler, *The Liquid State*, Verlag Chemie, Weinheim, W. Germany, 1972.

4. D. Chandler, *Ann. Rev. Phys. Chem.* **29**, 441 (1978).

5. H. C. Andersen, D. Chandler, and J. D. Weeks, *Advan. Chem. Phys.* **34**, 105 (1976).

6. J. P. Hansen and J. R. McDonald, *Theory of Simple Liquids*, Academic Press, New York, 1976, p. 141ff.

7. E. Lippert, in *The Hydrogen Bond*, P. Schuster, G. Zundel, and C. Sandorfy, eds., North-Holland Publs., Amsterdam, 1976, Chap. I.

8. I. R. McDonald and M. L. Klein, *Discuss. Faraday Soc.* **66**, 48 (1978).

9. R. S. Mulliken and W. B. Person, *Molecular Complexes*, Wiley, New York, 1969.

10. L. J. Lowden and D. Chandler, *J. Chem. Phys.* **61**, 5228 (1974).

11. C. S. Hsu, D. Chandler, and L. J. Lowden, *Chem. Phys.* **14**, 213 (1976).

12. W. R. Smith, in *Statistical Mechanics*, Vol. 1, Specialist Periodical Reports, The Chemical Society, London, 1973, Chap. 2.

13. D. Henderson and J. A. Barker, in *Physical Chemistry*, Vol. 8, Pt. A, H. Eyring, D. Henderson, and W. Jost, eds., Academic Press, New York, 1971, Chap. 6.

14. D. Henderson and P. J. Leonard, in *Physical Chemistry*, Vol. 8, Pt. B, H. Eyring, D. Henderson, and W. Jost, eds., Academic Press, New York, 1971, Chap. 7.

15. W. R. Smith, *Can. J. Phys.* **52**, 2022 (1974).

16. J. D. Weeks, D. Chandler, and H. C. Andersen, *J. Chem. Phys.* **54**, 5237 (1971).

17. J. S. Rowlinson, *Ber. Bunsenges. Phys. Chem.* **85**, 970 (1981).

18. T. Boublik, *Ber. Bunsenges. Physik. Chem.* **85**, 1038 (1981).

19. C. Hoheisel, *Ber. Bunsenges. Physik. Chem.* **85**, 1054 (1981).

20. J. O. Hirschfelder, C. F. Curtiss, and R. B. Bird, *Molecular Theory of Gases and Liquids*, Wiley, New York, 1954.

21. T. Kihara, *Advan. Chem. Phys.* **5**, 147 (1963).

22. F. Kohler, *Ber. Bunsenges. Physik. Chem.* **85**, 937 (1981).

23. T. Kihara, *Intermolecular Forces*, Wiley, Chichester, 1978, Chap. 7.

24. E. L. Derr and C. H. Deal, *Ind. Chem. Eng. Symp. Ser. 32* (*Inst. Chem. Engrs., London*) **3**, 40 (1969).

25. J. A. Barker, *J. Chem. Phys.* **20**, 1526 (1952).

26. J. A. Barker, *J. Chem. Phys.* **21**, 1399 (1953).

27. J. A. Barker and F. Smith, *J. Chem. Phys.* **22**, 375 (1954).

28. A. Fredenslund, J. Gmehling, and P. Rasmussen, *Vapor – Liquid Equilibria Using UNIFAC*, Elsevier, New York, 1977.

29. W. Schröer, W. D. Domke, K.-H. Naumann, and E. Lippert, Chap. 18 in Ref 2.

30. K.-H. Naumann, E. Lippert, *Ber. Bunsenges. Physik. Chem.* **85**, 650 (1981).

31. K.-H. Naumann, *Ber. Bunsenges. Physik. Chem.* **86**, 519 (1982).

32. K.-H. Naumann, *J. Mol. Struct.* **84**, 293 (1982).

33. T. Boublik, *J. Chem. Phys.* **63**, 4084 (1975).

34. B. Boyadjiev, Thesis, D. 83, Technical University of Berlin, 1982.

35. H. W. Zimmermann, Chap. 1 in Ref. 2.

36. H. Versmold, *Ber. Bunsenges. Physik. Chem.* **85**, 979 (1981).

37. M. D. Zeidler, *Stud. Phys. Theor. Chem.* (Budapest) **13**, 271 (1981).

38. Y. P. Chen, P. Angelo, K.-H. Naumann, and T. W. Leland, in *Proceedings of the Eighth Symposium on Thermophysical Properties*, Vol. 1, J. D. Sengers, ed., American Society of Mechanical Engineers, New York, 1982, p. 24ff.

39. J. M. Prausnitz, *Ber. Bunsenges. Physik. Chem.* **81**, 900 (1977).

40. H. Knapp and S. I. Sandler, eds., *Proceedings of the 2nd International Conference on Phase Equilibria and Fluid Properties in the Chemical Industry*, DECHEMA, Frankfurt am Main, 1980.

41. S. Bratos, Chap. 20 in Ref. 2.

42. Th. Dorfmüller, W. Mersch, D. Samios, and G. Fytas, Chap. 6, and A. Lauberau and W. Kaiser, Chap. 7 in Ref. 2.

43. See, e.g., Table 11.1, p. 215 of Ref. 3.

44. M. Davies, Chap. 10; E. U. Franck, Chap. 11; A. Gerschel, Ch. 12; and J. Huvwic, Chap. 13 in Ref. 2.

45. M. Perrot and J. Lascombe, Chap. 3, and J. Vincent-Gaisse and J. Joussen-Jacob, Chap. 4 in Ref. 2.

46. R. M. Lynden-Bell, Chap. 8, and H. W. Spiess, Chap. 9 in Ref. 2.

47. H. Versmold, Chap. 5 in Ref. 2.

48. F. Volino and A. J. Dianoux, Chap. 2 in Ref. 2.

49. W. A. Steele, *Advan. Chem. Phys.* **34**, 1 (1976).

50. M. Evans, G. Evans, and R. Davies, *Advan. Chem. Phys.* **44**, 255 (1980).

51. C. F. J. Böttcher and P. Bordewijk, *Theory of Electric Polarization*, 2nd ed., Vol. II, Elsevier, New York, 1978, p. 202ff.

52. R. G. Gordon, *Advan. Magnetic Resonance* **3**, 1 (1968).

53. R. M. Lynden-Bell and J. R. Mcdonald, *Mol. Phys.* **43**, 1429 (1981).

54. W. Rettig, *J. Phys. Chem.* **86**, 1970 (1982).

55. J. D. Poll, in *Intermolecular Spectroscopy and Dynamical Properties of Dense Systems*, J. van Kranendonk, ed., North-Holland Publs., Amsterdam, 1980, p. 45.

56. B. J. Berne and R. Pecora, *Dynamic Light Scattering*, Wiley, New York, 1976.

57. D. Frenkel and J. P. Mctague, *J. Chem. Phys.* **72**, 2801 (1980).

58. H. Michel and E. Lippert, Chap. 17 in Ref. 2.

59. J. Yarwood, *Proceedings of the Daresbury Laboratory Study Weekend*, Jan. 1979.

60. K. S. Cole and R. H. Cole, *J. Chem. Phys.* **9**, 341 (1941).

61. N. Bloembergen, E. M. Purcell, and R. D. Pound, *Phys. Rev.* **73**, 679 (1946).

62. M. Sass, Dissertation, D. 83, Technical University of Berlin, 1979.

63. D. Lankhorst, J. Schriever, and J. C. Leyte *Ber. Bunsenges. Physik. Chem.* **86**, 215 (1982).

64. L. A. Woodcock, Chap. 14 in Ref. 2.

65. H.-R. Baier and J.-U. Weidner, *Ber. Bunsenges. Physik. Chem.* **85**, 1044 (1981).

66. G. Fytas and Th. Dorfmüller, *Ber. Bunsenges. Physik. Chem.* **85**, 1064 (1981).

67. R. Vilhjalmsson, *Ber. Bunsenges. Physik. Chem.* **85**, 1071 (1981).

68. F. J. Bartoli and T. A. Litovitz, *J. Chem. Phys.* **56**, 404 (1972).

69. I. Prigogine and C. George, *Int. J. Quantum Chem.* **12**, Suppl. 1, 177 (1977); and see references therein.

70. C. A. Chatzidimitriou-Dreismann and E. Lippert, *Int. J. Quantum Chem. (Quantum Chem. Symp.)* **16**, 195 (1982).

71. T. J. Chuang and K. B. Eisenthal, *Chem. Phys. Lett.* **11**, 386 (1971).

72. D. W. Phillion, D. J. Kuizenga, and A. E. Siegmann, *Appl. Phys. Lett.* **27**, 85 (1975).

73. A. von Jena and H. E. Lessing, *Chem. Phys. Lett.* **78**, 187 (1981); **36**, 517 (1975); *Chem. Phys.* **40**, 245 (1979), etc.

74. D. Huppert, D. C. Douglas, and P. M. Rentzipis, *J. Chem. Phys.* **72**, 2841 (1980).

75. D. P. Millar, R. Shah, and A. H. Zewail, *Chem. Phys. Lett.* **66**, 435 (1979).

76. G. Porter, P. J. Sadkowski, and C. J. Tredwell, *Chem. Phys. Lett.* **49**, 416 (1977).

77. G. R. Fleming, G. W. Robinson, et al., *Chem. Phys. Lett.* **49**, 1 (1977).

78. S. A. Rice and G. A. Kenney-Wallace, *Chem. Phys.* **47**, 161 (1980).

79. R. W. Wijnaendts van Resandt and L. de Maeyer, *Chem. Phys. Lett.* **78**, 219 (1981).

80. U. K. A. Klein, H.-P. Haar, *J. Mol. Struct.* **61**, 11 (1980); *Chem. Phys. Lett.* **63**, 40 (1979); **58**, 531 (1978); H.-P. Haar, U. K. A Klein, F. W. Hafner, and M. Hauser, *Chem. Phys. Lett.* **49**, 563 (1977).

81. I. A. Munro, J. Pecht, and L. Styler, *Proc. Nat. Acad. Sci. U.S.* **76**, 56 (1979); I. A. Munro, and A. P. Sabersky, in *Synchrotron Radiation Research*, H. Winick and S. Domach, Eds, Plenum Press, New York, 1982, ch.9.

82. H. Masuhara, H. Miyasaka, N. Ikeda, and N. Mataga, *Chem. Phys. Lett.* **82**, 59 (1981).

83. K. B. Eisenthal, in *Ultrashort Light Pulses*, S. L. Shapiro, ed., Springer-Verlag, New York, 1977.

84. See, e.g., J. R. Stevenson, H. Ellis, and R. Bartlett, *Appl. Opt.* **12**, 2884 (1973) for tantalus I at Madison, Wisconsin.

85. E. Lippert, in *Organic Molecular Photophysics*, J. Birks, ed., Vol. 2, Wiley, New York, 1975, Chap. 1.

86. See, e.g., W. Rettig and E. Lippert, *J. Mol. Struct.* **61**, 17 (1980).

87. Th. Kindt and E. Lippert, in *Excited States in Organic Chemistry and Biochemistry*, B. Pullmann and H. Goldblum, eds., Reidel, Dordrecht, Netherlands, 1977, p. 221.

88. See, e.g., H. H. Klingenberg, E. Lippert, and W. Rapp, *Chem. Phys. Lett.* **18**, 417 (1973); B. Gronau, E. Lippert, and W. Rapp, *Ber. Bunsenges. Physik. Chem.* **76**, 432 (1972).

89. C. A. Chatzidimitriou-Dreismann and E. Lippert, *Ber. Bunsenges. Physik. Chem.* **84**, 775 (1980).

90. C. A. Chatzidimitriou-Dreismann and E. Lippert, *Croat. Chem. Acta* **55**, 23 (1982).

91. E. Lippert and C. A. Chatzidimitriou-Dreismann, *Int. J. Quantum Chem. (Quantum Chem. Symp.)* **16**, 183 (1982).

92. C. A. Chatzidimitriou-Dreismann and E. Lippert, *Ber. Bunsenges. Physik. Chem.* **85**, 1078 (1981).

93. Myron W. Evans, private communication.

94. J. K. Vij, C. J. Reid, G. J. Evans, M. Ferrario, and M. W. Evans, *Advan. Mol. Relaxation* **22**, 79 (1982).

95. K. M. van Vliet, *J. Math. Phys.* **19**, 1345 (1978).

96. K. M. van Vliet, *J. Math. Phys.* **20**, 2573 (1979).

97. C. A. Chatzidimitriou-Dreismann, *J. Mol. Struct.* **84**, 213 (1982).

98. R. W. Zwanzig, *J. Chem. Phys.* **22**, 1420 (1954).

99. J. A. Barker and D. Henderson, *Rev. Mod. Phys.* **48**, 587 (1976).

100. H. C. Andersen, J. D. Weeks, and D. Chandler, *Phys. Rev.* **A4**, 1597 (1971).

101. L. Verlet and J. J. Weis, *Phys. Rev.* **A5**, 939 (1972).

102. L. Verlet and J. J. Weis, *Mol. Phys.* **24**, 1013 (1972).

103. G. Stell, J. C. Rasaiah, and N. Narang, *Mol. Phys.* **23**, 393 (1972).

104. G. Stell, J. C. Rasaiah, and N. Narang, *Mol. Phys.* **27**, 1393 (1974).

105. M. S. Wertheim, *Phys. Rev. Lett.* **10**, 321 (1963).
106. M. S. Wertheim, *J. Math. Phys.* **5**, 643 (1964).
107. E. Thiele, *J. Chem. Phys.* **39**, 474 (1963).
108. W. R. Smith and D. Henderson, *Mol. Phys.* **19**, 411 (1970).
109. B. M. Ladanyi and D. Chandler, *J. Chem. Phys.* **62**, 4308 (1975).
110. D. Chandler, in *The Liquid State of Matter*, E. W. Montroll and J. L. Lebowitz, eds., North-Holland Publs., Amsterdam, 1982.
111. K. C. Mo and K. E. Gubbins, *Chem. Phys. Lett.* **27**, 144 (1974).
112. K. C. Mo and K. E. Gubbins, *J. Chem. Phys.* **63**, 1490 (1975).
113. T. Boublik and B. C. Y. Lu, *J. Phys. Chem.* **82**, 2801 (1978).
114. T. Boublik, *Fluid Phase Equilibria* **3**, 85 (1979).
115. F. Kohler, N. Quirke, and J. W. Perram, *J. Chem. Phys.* **71**, 4128 (1979).
116. D. J. Tildesley, *Mol. Phys.* **41**, 341 (1980).
117. B. C. Freasier, *Chem. Phys. Lett.* **35**, 280 (1975).
118. W. B. Streett and D. J. Tildesley, *Proc. Roy. Soc.* (London) **A348**, 485 (1976).
119. D. Jolly, B. C. Freasier, and R. J. Bearman, *Chem. Phys. Lett.* **46**, 75 (1977).
120. W. B. Streett and D. J. Tildesley, *J. Chem. Phys.* **68**, 1275 (1978).
121. D. J. Tildesley and W. B. Streett, *Mol. Phys.* **41**, 85 (1980).
122. P. Cummings, I. Nezbeda, W. R. Smith, and G. Morriss, *Mol. Phys.* **43**, 1471 (1981).
123. J. Viellard-Baron, *Mol. Phys.* **28**, 809 (1974).
124. T. Boublik, I. Nezbeda, and O. Trnka, *Czech. J. Phys.* **B26**, 1081 (1976).
125. D. W. Rebertus and K. M. Sando, *J. Chem. Phys.* **67**, 2585 (1977).
126. P. A. Manson and M. Rigby, *Chem. Phys. Lett.* **58**, 122 (1978).
127. I. Nezbeda and T. Boublik, *Czech. J. Phys.* **B28**, 353 (1978).
128. I. Nezbeda, *Czech. J. Phys.* **B30**, 601 (1980).
129. W. B. Streett and K. E. Gubbins, *Ann. Rev. Phys. Chem.* **28**, 373 (1977).
130. K.-H. Naumann, Thesis, D. 83, Technical University of Berlin, 1979.
131. I. Nezbeda and W. R. Smith, *Chem. Phys. Lett.* **81**, 79 (1981).
132. I. Nezbeda and W. R. Smith, *Chem. Phys. Lett.* **82**, 96 (1981).
133. I. Nezbeda and W. R. Smith, *Mol. Phys.* **45**, 681 (1982).
134. J. A. Pople, *Proc. Roy. Soc.* (London) **A221**, 498 (1954).
135. K. E. Gubbins and C. G. Gray, *Mol. Phys.* **23**, 187 (1972).
136. M. S. Ananth, K. E. Gubbins, and C. G. Gray, *Mol. Phys.* **28**, 1005 (1974).
137. C. G. Gray, K. E. Gubbins, and C. H. Twu, *J. Chem. Phys.* **69**, 182 (1978).
138. J. W. Perram and L. R. White, *Mol. Phys.* **27**, 527 (1974).
139. W. R. Smith, I. Nezbeda, T. W. Melnyk, and D. W. Fitts, *Discuss. Faraday Soc.* **66**, 130 (1978).
140. W. R. Smith and I. Nezbeda, *Chem. Phys. Lett.* **64**, 146 (1979).
141. F. Kohler, W. Marius, N. Quirke, J. W. Perram, K. Hoheisel, and H. Breitenfelder-Manske, *Mol. Phys.* **38**, 2057 (1979).
142. T. W. Melnyk and W. R. Smith, *Mol. Phys.* **40**, 317 (1980).
143. N. Quirke, J. W. Perram, and G. Jacucci, *Mol. Phys.* **39**, 1311 (1980).
144. W. R. Smith and I. Nezbeda, *Mol. Phys.* **44**, 347 (1981).

145. T. W. Melnyk, W. R. Smith, and I. Nezbeda, *Mol. Phys.* **46**, 629 (1982).

146. P. T. Cummings and G. Stell, *Mol. Phys.* **46**, 383 (1982).

147. S. I. Sandler and A. H. Narten, *Mol. Phys.* **32**, 1543 (1976).

148. A. H. Narten, *J. Chem. Phys.* **67**, 2102 (1977).

149. A. H. Narten, S. I. Sandler, and T. Rensi, *Discuss. Faraday Soc.* **66**, 39 (1978).

150. C. S. Hsu and D. Chandler, *Mol. Phys.* **36**, 215 (1978).

151. C. S. Hsu and D. Chandler, *Mol. Phys.* **37**, 299 (1979).

152. O. Steinhauser and M. Neumann, *Mol. Phys.* **37**, 1921 (1979).

153. O. Steinhauser and M. Neumann, *Mol. Phys.* **40**, 115 (1980).

154. A. H. Narten, E. Johnson, and A. Habenschuss, *J. Chem. Phys.* **73**, 1248 (1980).

155. A. Habenschuss, E. Johnson, and A. H. Narten, *J. Chem. Phys.* **74**, 5234 (1981).

156. A. H. Narten, *J. Chem. Phys.* **65**, 573 (1976).

157. A. H. Narten, W. E. Thiessen, and L. Blum, *Science* **217**, 1033 (1982).

158. N. Quirke and D. J. Tildesley, *Mol. Phys.* **45**, 811 (1982).

159. J. G. Kirkwood, *Phys. Rev.* **44**, 31 (1933).

160. B. M. Ladanyi, T. Keyes, W. B. Streett, and D. J. Tildesley, *Mol. Phys.* **39**, 645 (1980).

161. A. Kratochwill, J. U. Weidner, and H. Zimmermann, *Ber. Bunsenges. Physik. Chem.* **77**, 408 (1973).

162. H. Bertagnolli and M. D. Zeidler, *Mol. Phys.* **35**, 177 (1978).

163. D. Chandler and H. C. Andersen, *J. Chem. Phys.* **57**, 1930 (1972).

164. D. Chandler, *J. Chem. Phys.* **59**, 2742 (1973).

165. D. Chandler, *Mol. Phys.* **31**, 1213 (1976).

166. M. Lombardero, J. F. L. Abascal, and S. Lago, *Mol. Phys.* **42**, 999 (1981).

167. M. Lombardero and J. F. L. Abascal, *Chem. Phys. Lett.* **85**, 117 (1982).

168. J. F. L. Abascal and M. Lombardero, *Trans. Faraday Soc. II* **78**, 965 (1982).

169. L. J. Lowden and D. Chandler, *J. Chem. Phys.* **59**, 6587 (1973).

170. G. P. Morriss and E. R. Smith, *J. Statist. Phys.* **24**, 607 (1981).

171. D. Chandler, C. S. Hsu, and W. B. Streett, *J. Chem. Phys.* **66**, 5231 (1977).

172. D. E. Sullivan and C. G. Gray, *Mol. Phys.* **42**, 443 (1981).

173. P. T. Cummings, G. P. Morriss, and C. C. Wright, *Mol. Phys.* **43**, 1299 (1981).

174. D. Chandler, R. Silbey, and B. M. Ladanyi, *Mol. Phys.* **46**, 1335 (1982).

175. D. Chandler, C. G. Joslin, and J. M. Deutch, *Mol. Phys.* **47**, 871 (1982).

176. T. Kihara, *Rev. Mod. Phys.* **25**, 831 (1953).

177. T. Kihara, *Chem. Phys. Lett.* **92**, 175 (1982).

178. K.-H. Naumann, Y. P. Chen, and T. W. Leland, *Ber. Bunsenges. Physik. Chem.* **85**, 1029 (1981).

179. T. Boublik, *Mol. Phys.* **32**, 1737 (1976).

180. S. Lago and T. Boublik, *Coll. Czech. Chem. Commun.* **45**, 3051 (1980).

181. A. Kreglewski and S. S. Chen, *Ber. Bunsenges. Physik. Chem.* **81**, 1048 (1977).

182. T. Boublik, *Mol. Phys.* **42**, 209 (1981).

183. K.-H. Naumann and T. W. Leland, to be published.

184. O. Steinhauser and H. Bertagnolli, *Ber. Bunsenges. Physik. Chem.* **85**, 45 (1981).

185. W. B. Streett and D. J. Tildesley, *Proc. Roy. Soc.* (London) **A355**, 239 (1977).

186. K. Singer, A. Taylor, and J. V. L. Singer, *Mol. Phys.* **33**, 1757 (1977).

187. C. B. Haselgrove, *Math. Computation* **15**, 323 (1961).

188. H. Conroy, *J. Chem. Phys.* **52**, 5531 (1970).

189. C. A. Croxton, *Liquid State Physics*, Cambridge Univ. Press, New York, 1974.

190. R. O. Watts, in *Statistical Mechanics*, Vol. 1, Specialist Periodical Reports, The Chemical Society, London, 1973, Chap. 1.

191. W. R. Smith and D. Henderson, *J. Chem. Phys.* **69**, 319 (1978).

192. P. J. Rossky and W. D. T. Dale, *J. Chem. Phys.* **73**, 2457 (1980).

193. W. G. Madden and S. A. Rice, *J. Chem. Phys.* **72**, 4208 (1980).

194. G. P. Morriss, Ph.D. Thesis, University of Melbourne, 1980.

195. W. W. Wood, in *Physics of Simple Liquids*, H. N. V. Temperley, J. S. Rowlinson, and G. S. Rushbrooke, eds., North-Holland, Publs., Amsterdam, 1968, Chap. 5.

196. J. P. Valleau and S. G. Whittington, in *Statistical Mechanics*, Pt. A, B. J. Berne, ed., Modern Theoretical Chemistry, Plenum Press, New York, 1977, Chap. 4.

197. K.-H. Naumann, R. Bansal, and W. Bruns, *Ber. Bunsenges. Physik. Chem.* **87**, 664 (1983).

198. D. J. Evans and R. O. Watts, *Mol. Phys.* **32**, 93 (1976).

199. J. G. Powles and K. E. Gubbins, *Chem. Phys. Lett.* **38**, 405 (1976).

200. J. S. Rowlinson and F. L. Swinton, *Liquids and Liquid Mixtures*, Butterworth, Woburn, Mass., 1982.

201. I. R. McDonald, in *Statistical Mechanics*, Vol. 1, Specialist Periodical Reports, The Chemical Society, London, 1973, Chap. 3.

202. T. W. Leland, J. S. Rowlinson, and G. A. Sather, *Trans. Faraday Soc.* **64**, 1447 (1968).

203. T. W. Leland, P. S. Chappelear, and B. W. Gamson, *AIChE (Amer. Inst. Chem. Engrs.) J.* **8**, 482 (1962).

204. R. C. Reid and T. W. Leland, *AIChE (Amer. Inst. Chem. Engrs.) J.* **11**, 228 (1965).

205. E. W. Grundke, D. Henderson, J. A. Barker, and P. J. Leonard, *Mol. Phys.* **25**, 883 (1973).

206. G. A. Mansoori and T. W. Leland, *Trans. Faraday Soc. II* **68**, 320 (1972).

207. J. B. Gonsalves and T. W. Leland, *AIChE (Amer. Inst. Chem. Engrs.) J.* **24**, 279 (1978).

208. G. A. Mansoori, N. F. Carnahan, K. E. Starling, and T. W. Leland, *J. Chem. Phys.* **54**, 1523 (1971).

209. N. F. Carnahan and K. E. Starling, *J. Chem. Phys.* **51**, 635 (1969).

210. J. I. C. Chang, F. S. S. Hwu, and T. W. Leland, in *Equations of State in Engineering and Research* (Advances in Chemistry Ser. No. 182), K. C. Chao and R. L. Robinson, eds., American Chemical Society, Washington, D.C., 1979, Chap. 4.

211. K.-H. Naumann and T. W. Leland, to be published.

212. H. L. Frisch, *Advan. Chem. Phys.* **6**, 229 (1964).

213. H. Reiss, *Advan. Chem. Phys.* **9**, 1 (1965).

214. R. M. Gibbons, *Mol. Phys.* **17**, 81 (1969).

215. R. M. Gibbons, *Mol. Phys.* **18**, 809 (1970).

216. T. Boublik, *Mol. Phys.* **27**, 1415 (1974).

217. T. Boublik and I. Nezbeda, *Czech. J. Phys.* **B30**, 121 (1980).

218. P. A. Manson and M. Rigby, *Mol. Phys.* **39**, 977 (1980).

219. I. Nezbeda and T. Boublik, *Czech. J. Phys.* **B30**, 953 (1980).

220. T. Boublik and I. Nezbeda, *Chem. Phys. Lett.* **46**, 315 (1977).

221. I. Nezbeda and T. Boublik, *Czech. J. Phys.* **B27**, 1071 (1977).

222. E. Enciso, J. Gil, and S. Lago, *Chem. Phys. Lett.* **88**, 333 (1982).

223. W. B. Streett and D. J. Tildesley, *Discuss. Faraday Soc.* **66**, 27 (1978).

224. T. Kihara and K. Miyoshi, *J. Statist. Phys.* **13**, 1337 (1975).

225. I. Nezbeda, *Chem. Phys. Lett.* **41**, 55 (1976).

226. M. Rigby, *J. Chem. Phys.* **53**, 1021 (1970).

227. I. Nezbeda, J. Pavliček, and S. Labik, *Coll. Czech. Chem. Commun.* **44**, 3555 (1979).

228. W. J. Orville-Thomas, ed., *Internal Rotation in Molecules*, Wiley, New York, 1974.

229. R. J. Abraham and E. Bretschneider, in Ref. 228.

230. P. A. Park, R. J. D. Pethrick, and B. H. Thomas, in Ref. 228.

231. L. Colombo and G. Zerbi, *J. Chem. Phys.* **73**, 2013 (1980).

232. S. Kint, J. R. Scherer, and R. G. Synder, *J. Chem. Phys.* **73**, 2599 (1980).

233. A. L. Verma, W. F. Murphy, and H. J. Bernstein, *J. Chem. Phys.* **60**, 1540 (1974).

234. J. R. Durig and D. A. C. Compton, *J. Phys. Chem.* **83**, 265 (1979).

235. J. P. Ryckaert and A. Bellemans, *Chem. Phys. Lett.* **30**, 123 (1975).

236. J. P. Ryckaert and A. Bellemans, *Discuss. Faraday Soc.* **66**, 95 (1978).

237. D. W. Rebertus, B. J. Berne, and D. Chandler, *J. Chem. Phys.* **70**, 3395 (1979).

238. D. Chandler and L. R. Pratt, *J. Chem. Phys.* **65**, 2925 (1976).

239. L. R. Pratt, C. S. Hsu, and D. Chandler, *J. Chem. Phys.* **68**, 4202 (1978).

240. R. A. Scott and H. A. Scheraga, *J. Chem. Phys.* **44**, 3054 (1966).

241. W. L. Jorgensen, *J. Chem. Phys.* **77**, 5757 (1982).

242. R. Kubo, *J. Phys. Soc. Japan* **12**, 570 (1957).

243. R. Kubo, in *Lectures in Theoretical Physics*, W. E. Britten and L. G. Durham, eds., Vol. I, Wiley (Interscience), New York, 1959.

244. R. Kubo, *Can. J. Phys.* **34**, 1274 (1956).

245. R. Kubo and K. Tomita, *J. Phys. Soc. Japan* **9**, 888 (1954).

246. H. Takahasi, *J. Phys. Soc. Japan* **7**, 439 (1952).

247. R. Kubo, M. Yokota, and S. Nakajima, *J. Phys. Soc. Japan* **12**, 1203 (1957).

248. R. Kubo, *Rep. Progr. Phys.* **29**, 255 (1966).

249. R. Kubo, *Int. J. Quantum Chem. (Quantum Chem. Symp.)* **16**, 25 (1982).

250. R. Kubo, in *Tokyo Summer Lectures in Theoretical Physics*, Pt. 1, R. Kubo, ed., Benjamin, New York, 1966.

251. R. Kubo, in *Statistical Mechanics of Equilibrium and Non-equilibrium*, J. Meixner, ed., North-Holland Publs., Amsterdam, 1965.

252. M. S. Green, *J. Chem. Phys.* **19**, 1036 (1951).

253. H. Mori, *Phys. Rev.* **112**, 1829 (1958).

254. H. Mori, *Progr. Theor. Phys.* (Kyoto) **28**, 763 (1962).

255. J. A. McLennan, *Phys. Fluids* **3**, 493 (1960).

256. L. P. Kadanoff and P. C. Martin, *Ann. Phys.* **24**, 419 (1963).

257. J. M. Luttinger, *Phys. Rev.* **A135**, 1505 (1964).

258. P. Resibois, *J. Chem. Phys.* **41**, 2979 (1964).

259. K. M. Case, *Trans. Theor. Statist. Phys.* **2**, 129 (1972).

260. R. Zwanzig, *Ann. Rev. Phys. Chem.* **16**, 67 (1965).

261. G. V. Chester and A. Thellung, *Proc. Phys. Soc.* **73**, 745 (1960).

262. S. Fujida and R. Abe, *J. Math. Phys.* **3**, 350 (1962).

263. M. S. Green, *J. Chem. Phys.* **20**, 1281 (1952); **22**, 398 (1954).

264. J. G. Kirkwood, *J. Chem. Phys.* **14**, 180 (1946).

265. H. B. Callen and T. A. Welton, *Phys. Rev.* **83**, 34 (1951).

266. H. B. Callen and R. F. Greene, *Phys. Rev.* **86**, 702 (1952).

267. R. F. Greene and H. B. Callen, *Phys. Rev.* **88**, 1387 (1952).

268. W. Bernard and H. B. Callen, (a) *Phys. Rev.* **118**, 1466 (1960); (b) *Rev. Mod. Phys.* **31**, 1017 (1959).

269. J. Von Neumann, *Mathematical Foundations of Quantum Mechanics*, Princeton Univ. Press, Princeton, N.J., 1955.

270. U. Fano, *Rev. Mod. Phys.* **29**, 74 (1957).

271. D. ter Haar, *Rep. Progr. Phys.* **24**, 304 (1961).

272. I. Prigogine, Non-equilibrium Statistical Mechanics, Wiley, New York, 1962.

273. A. Einstein, *Ann. Phys.* **33**, 1275 (1910).

274. H. Nyquist, *Phys. Rev.* **29**, 614 (1927); **32**, 110 (1928).

275. C. A. Chatzidimitriou-Dreismann, "Quantenstatistische Grundlagen Infrarot-absorptionsspektroskopischer Untersuchungen über Bewegungsvorgänge in organischen Flüssigkeiten," research report, Technical University of Berlin, 1982.

276. R. Goslich, C. A. Chatzidimitriou-Dreismann, and E. Lippert, *Ber. Bunsenges. Physik. Chem.* **87**, 396 (1983).

277. M. Evans, G. J. Evans, W. T. Coffey, and P. Grigolini, *Molecular Dynamics and the Theory of Broad Band Spectroscopy*, Wiley, New York, 1982.

278. N. G. van Kampen, *Phys. Norvegica* **5**, 279 (1971).

279. D. ter Haar, *Lectures on Selected Topics in Statistical Mechanics*, Pergamon Press, Elmsford, N.Y., 1977.

280. N. G. van Kampen, "A Discussion on Linear Response Theory," preprint.

281. K. D. Möller and W. G. Rothschild, *Far-Infrared Spectroscopy*, Wiley, New York, 1971.

282. J. H. R. Clarke, in *Advances in Infrared and Raman Spectroscopy*, Vol. 4, Heyden, London, 1978.

283. J. Dupuy and A. J. Dianoux, *Microscopic Structure and Dynamics of Liquids*, Plenum Press, New York, 1978.

284. S. Bratos and R. M. Pick, *Vibrational Spectroscopy of Molecular Liquids and Solids*, Plenum Press, New York, 1980.

285. A. A. McQuarrie, *Statistical Mechanics*, Harper & Row, New York, 1976.

286. J. Yarwood and R. Arndt, in *Molecular Association*, R. Forster, ed., Vol. 2, Academic Press, New York, 1979.

287. A. Messiah, *Quantum Mechanics*, Vol. I, North-Holland Publs., Amsterdam, 1967.

288. L. van Hove, *Physica* **21**, 517 (1955); **23**, 441 (1957).

289. B. J. Berne and G. D. Harp, *Advan. Chem. Phys.* **17**, 63 (1970).

290. B. J. Berne, in *Physical Chemistry: An Advanced Treatise*, D. Henderson, ed., Vol. 8, Pt. B, Academic Press, New York, 1971.

291. B. J. Berne and D. Forster, *Ann. Rev. Phys. Chem.* **22**, 563 (1971).

292. S. G. Kroon and J. van der Elsken, *Chem. Phys. Lett.* **1**, 285 (1967).

293. J. Yarwood, P. L. James, G. Döge, and R. Arndt, *Faraday Discuss. Chem. Soc.* **66**, 252 (1978).

294. E. Knözinger, D. Leutloff, and R. Wittenbeck, *J. Mol. Struct.* **60**, 115 (1980).

295. J. R. Birch, M. N. Afsar, J. Yarwood, and P. L. James, *Infrared Phys.* **21**, 9 (1981).

296. A. Gerschel, I. Darmon, and C. Brot, *Mol. Phys.* **23**, 317 (1972).

297. See. Ref. 277, p. 424.

298. See, e.g., Ref. 268(b), p. 90.

299. A. Messiah, *Quantum Mechanics*, Vol. II, North-Holland Publs., Amsterdam, 1970.

300. H. Goldstein, *Classical Mechanics*, Addison-Wesley, Reading, Mass., 1953.

301. See, e.g., Ref. 289, p. 74.

302. See, e.g., Refs. 52 and 289–291.

303. P. C. Martin and J. Schwinger, *Phys. Rev.* **115**, 1342 (1959).

304. W. Pauli, in *Festschrift zum 60. Geburtstage A. Sommerfelds*, Hirzel, Leipzig, 1928.

305. N. G. van Kampen, *Physica* **20**, 603 (1954).

306. N. G. van Kampen, *Fortschr. Physik* **4**, 405 (1956).

307. N. G. van Kampen, in *Fundamental Problems in Statistical Mechanics*, E. G. D. Cohen, ed., North-Holland Publs., Amsterdam, 1962.

308. C. A. Chatzidimitriou-Dreismann, unpublished work.

309. L. D. Landau and E. M. Lifschitz, *Statistische Physik*, Akademie, Berlin, 1966.

310. R. Becker, *Theorie der Wärme*, Springer-Verlag, New York, 1966.

311. R. Kubo, H. Ichimura, T. Usui, and N. Hashitsume, *Statistical Mechanics*, North-Holland Publs., Amsterdam, 1967, p. 14.

312. M. Charbonneau, K. M. van Vliet, and P. Vasilopoulos, *J. Math. Phys.* **23**, 318 (1982).

313. G. Birnbau, B. Guillot, and S. Bratos, *Advan. Chem. Phys.* **51**, 49 (1982); see also references cited therein.

314. C. Carathéodory, *Math. Ann.* **67**, 355 (1909).

315. R. Kubo, *J. Math. Phys.* **4**, 174 (1963).

316. R. Kubo, in *Fluctuations, Relaxation and Resonance in Magnetic Systems*, D. ter Haar, ed., Plenum Press, New York, 1962.

317. C. A. Chatzidimitriou-Dreismann, to be published.

318. H. Mori, *Progr. Theor. Phys.* (Kyoto) **33**, 423 (1965); **34**, 399 (1965).

319. R. Zwanzig, *J. Chem. Phys.* **33** , 1338 (1960).

320. R. J. Swenson, *J. Math. Phys.* **3**, 1017 (1962).

321. S. Nakajima, *Progr. Theor. Phys.* (Kyoto) **20**, 948 (1958).

322. R. Résibois, *Physica* **27**, 541 (1961); **29**, 721 (1963).

323. I. Prigogine and P. Résibois, *Physica* **27**, 629 (1961).

324. R. Zwanzig, *Physica* **30**, 1109 (1964).

325. R. Zwanzig, in *Lectures in Theoretical Physics*, W. E. Britten, B. W. Downs, and J. Downs, eds., Vol. 3, Wiley (Interscience), New York, 1961.

326. See Ref. 95, p. 1357.

327. C. A. Chatzidimitriou-Dreismann, *Int. J. Quantum Chem.* **23**, 1505 (1983); *J. Chem. Phys.* **80**, 561 (1984).

328. W. Rhodes, *J. Chem. Phys.* **75**, 2588 (1981); **86**, 2657 (1982).

329. W. Rhodes, "The Case for Linear Response Theory," preprint.

330. A. Laubereau and W. Kaiser, *Rev. Mod. Phys.* **50**, 607 (1978); see also references cited therein.

331. A. Fendt, S. F. Fischer, and W. Kaiser, *Chem. Phys.* **57**, 55 (1981); *Chem. Phys. Lett.* **82**, 350 (1981).

332. K. Spanner, A. Laubereau, and W. Kaiser, *Chem. Phys. Lett.* **44**, 88 (1976).

333. H. D. Zeh, *Found. Phys.* **1**, 69 (1970).

334. E. P. Wigner, "Review of the Quantum Mechanical Measurement Problem," preprint (1981).

335. W. Band and J. L. Park, *Found. Phys.* **8**, 677 (1978).

336. P. O. Löwdin, *Int. J. Quantum Chem.* **12**, Suppl. 1, 197 (1977); **21**, 275 (1982).

337. P. O. Löwdin, *Int. J. Quantum Chem.* (*Quantum Chem. Symp.*) **16**, 485 (1982).

338. I. Prigogine, C. George, F. Henin, and L. Rosenfeld, *Chem. Scr.* (Sweden) **4**, 5 (1973).

339. C. George and I. Prigogine, *Int. J. Quantum Chem.* (*Quantum Chem. Symp.*) **8**, 335 (1974).

340. I. Prigogine, F. Mayné, C. George, and M. de Haan, *Proc. Nat. Acad. Sci. U.S.* **74**, 4152 (1977).

341. Cl. George, F. Henin, F. Mayné, and I. Prigogine, *Hadronic J.* **1**, 520 (1978).

342. B. Misra, I. Prigogine, and M. Courbage, *Physica* **A98**, 1 (1979).

343. C. George and I. Prigogine, *Physica* **A99**, 369 (1979).

344. I. Prigogine (Nobel Lecture), *Angew. Chem.* **90**, 704 (1978).

345. D. Frenkel and J. P. McTague, *Ann. Rev. Phys. Chem.* **31**, 491 (1980).

346. A. Rahman, M. J. Mandell and J. P. McTague, *J. Chem. Phys.* **64**, 1564 (1976).

347. S. Alexander and J. P. McTague, *Phys. Rev. Lett.* **41**, 702 (1978).

348. R. E. Wilde and S. S. Cohen, *J. Chem. Phys.* **67**, 4279 (1977); **68**, 1138, 1237 (1978); *Chem. Phys. Lett.* **53**, 156 (1978).

349. T. E. Bell and J. Jones, *J. Chem. Phys.* **53**, 3315 (1970).

350. R. J. Jakobsen and Y. Mikawe, *Appl. Opt.* **9**, 17 (1979).

351. E. L. Pace and L. J. Noe, *J. Chem. Phys.* **49**, 5317 (1968).

352. P. A. Casabella and P. J. Bray, *J. Chem. Phys.* **29**, 1182 (1958).

353. W. E. Putnam, D. M. McEachern, and J. E. Kulpatrick, *J. Chem. Phys* **42**, 749 (1965).

354. R. Lobo, J. E. Robinson, and S. Rodriguez, *J. Chem. Phys.* **59**, 5992 (1973).

355. G. Ascarelli, *Chem. Phys. Lett.* **39**, 23 (1976).

356. H. Beck, *J. Phys. C: Solid State Phys.* **9**, 33 (1976).

357. G. J. Evans, *Chem. Phys. Lett.* **99**, 173 (July 1983).

358. J. L. Dote and D. Kivelson, *J. Phys. Chem.* **87**, 3889 (1983).

359. F. Perrin, *Ann. Phys.* **12**, 169 (1929).

360. P. Debye, *Polare Molekeln*, S. Hirzel Verlag, Leipzig, 1929.

361. W. G. Rothschild, *Dynamics of Molecular Liquids*, Wiley, New York, 1984.

362. J. Schwinger, *Phys. Rev.* **75**, 1912–25 (1949).

AUTHOR INDEX

Numbers in parentheses are reference numbers and indicate that the author's work is referred to although his name is not mentioned in the text. Numbers in *italics* show the pages on which the complete references are listed.

SUBJECT INDEX